DATE DUE

			PRINTED IN U.S.A.

ENCYCLOPEDIA OF

Global Warming
and Climate Change

ENCYCLOPEDIA OF

Global Warming and Climate Change

S. George Philander

GENERAL EDITOR

VOLUME 3

SAGE

Los Angeles • London • New Delhi • Singapore

A SAGE Reference Publication

For information:

SAGE Publications, Inc.
2455 Teller Road
Thousand Oaks, California 91320
E-mail: order@sagepub.com

SAGE Publications Ltd.
1 Oliver's Yard
55 City Road
London EC1Y 1SP
United Kingdom

SAGE Publications India Pvt. Ltd.
B 1/ I 1 Mohan Cooperative Industrial Area
Mathura Road, New Delhi 110 044
India

SAGE Publications Asia-Pacific Pte. Ltd.
33 Pekin Street #02-01
Far East Square
Singapore 048763

Library of Congress Cataloging-in-Publication Data

Encyclopedia of global warming and climate change / S. George Philander.
    p. cm.
Includes bibliographical references and index.
ISBN 978-1-4129-5878-3 (cloth)
  1. Global warming–Encyclopedias. 2. Climatic changes–Encyclopedias.
  I. Philander, S. George.
QC981.8.G56E47 2008
363.738'7403--dc22   2008006238

This book is printed on acid-free and recycled paper.
08 09 10 11 12 10 9 8 7 6 5 4 3 2 1

GOLSON BOOKS, LTD.		SAGE REFERENCE	
President and Editor	J. Geoffrey Golson	Vice President and Publisher	Rolf A. Janke
Creative Director	Mary Jo Scibetta	Project Editor	Tracy Buyan
Managing Editor	Susan Moskowitz	Cover Production	Janet Foulger
Copyeditor	Mary Le Rouge	Marketing Manager	Amberlyn Erzinger
Layout Editors	Kenneth W. Heller	Editorial Assistant	Michele Thompson
	Stephanie Larson	Reference Systems Manager	Leticia Gutierrez
	Oona Hyla Patrick		
Proofreaders	Deborah Green		
	Summer G. Ventis		
	Barbara Paris		
Indexer	J S Editorial	Photo credits are on page I-79.	

ENCYCLOPEDIA OF

Global Warming and Climate Change

CONTENTS

List of Articles

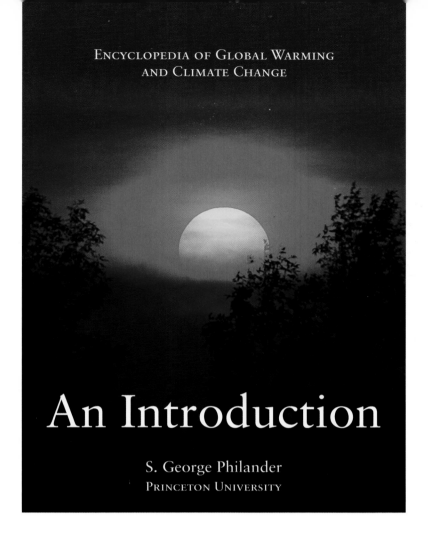

ENCYCLOPEDIA OF GLOBAL WARMING
AND CLIMATE CHANGE

An Introduction

S. George Philander
PRINCETON UNIVERSITY

IN ITS 2007 report, the Intergovernmental Panel on Climate Change (IPCC), a large, international panel of scientists, all experts on the Earth's climate, concluded that human activities, specifically those that cause an increase in the atmospheric concentration of carbon dioxide, have started affecting the Earth's climate. The panel further predicted that far more significant climate changes are imminent. This report, and Al Gore's documentary *An Inconvenient Truth* are persuading a rapidly increasing number of people that human activities can lead to possibly disastrous global climate changes.

Those nonscientists are passionate about being wise and responsible stewards of the Earth, but at present they are handicapped because they take the words of the scientists on faith, and accept the reality of the threat of global warming without grasping the scientific reasons. This is most unfortunate, because our response to the threat of global warming is far more likely to be effective if it were motivated, not merely by the alarms scientists sound, but also by knowledge of how this very complex planet maintains the conditions that suit us so well. We need an awareness of how extremely fortunate we are to be the Earth's inhabitants at this moment in its long and eventful history, and an understanding of how our current activities are putting us at risk. The purpose of this encyclopedia is to help the reader learn about the intricate processes that make ours the only planet known to be habitable. This encyclopedia covers, in addition to the science of global warming, its social and political aspects that are of central importance to the ethical dilemmas that global warming poses: (1) How do we find a balance between regulations and freedom? (2) How do we find a balance between our responsibilities to future generations, and our obligations to the poor suffering today?

The first dilemma, which generates strong emotions, has caused an unfortunate polarization of a complex, multifaceted issue. The extremists who find regulations abhorrent assert that there is no evidence of global warming. (They are sometimes referred to as deniers or skeptics.) Their opponents, the believers, claim that global warming is underway, and is already causing environmental disasters. For believers, the second dilemma assumes global warming is already contributing to the suffering of the poor and therefore is an urgent priority for everyone. They refuse to accept that, for the many people who are so poor that they have nothing to lose, global warming is not an urgent issue. Dilemmas 1 and 2 call for compromises and hence for an objective assessment of the scientific results.

> *Earth's temperatures fluctuate in a relatively narrow range; the Earth, unlike its neighbors Venus and Mars, is neither too warm nor too cold.*

The IPCC reports, which provide such an assessment, are explicit about uncertainties in the available results and hence favor neither the deniers nor the believers. The magnitudes of the uncertainties vary, depending on the time and region under consideration, and depending on whether we focus on temperature, the height of the ocean surface, rainfall or some other parameter.

The following is a very brief synopsis—a bird's eye view—of the discussion of these topics in the numerous entries of this encyclopedia. This information hopefully

TINY FROM AFAR: In our solar system, Earth, third planet from the sun at left, is dwarfed by giants Jupiter and Saturn. The order of the planets starts with Mercury, which is closest to the Sun, then Venus, Earth, Mars, Jupiter, Saturn, Uranus, Neptune, and controversial Pluto.

provides a basis for the development of an effective response to the threat of global warming.

Let us assume that we are aliens from another galaxy, in search of a habitable planet. On entering this particular solar system, our attention is at first drawn to the large, spectacular planets Jupiter and Saturn which are adorned with splendid rings and many moons. Earth, tiny by comparison, is a faint, blue dot from afar. Closer inspection shows that two of the Earth's main features are chaotically swirling white clouds, and vast oceans that cover nearly 70 percent of the surface. Both are vitally important to the Earth's most impressive feature of all: a great diversity of life forms that require water in liquid form. The abundance of liquid water means that, on the Earth, temperatures fluctuate in a relatively narrow range; the Earth, unlike its neighbors Venus and Mars, is neither too warm nor too cold.

The Earth's main source of energy is the sun, but this planet would be far too cold for most of its inhabitants were it not for its atmosphere, the thin veil of transparent gases that covers the globe. (If the Earth were an apple, its atmosphere would have the thickness of the peel.) This veil, by means of an intricate interplay between photons of light and molecules of air, serves as a shield that provides protection from dangerous ultraviolet rays in sunlight. The atmosphere serves as a parasol that reflects sunlight, thus keeping the planet cool; and as a blanket that traps heat from the Earth's surface, thus keeping us warm. The blanket is the greenhouse effect, which depends not on the two gases nitrogen and oxygen that are most abundant, but on trace gases that account for only a tiny part of the atmosphere, .035 percent in the case of carbon dioxide.

The most important greenhouse gas is water vapor, which is capable of engaging in escalating tit-for-tats (or positive feedbacks in engineering terms.) If atmospheric

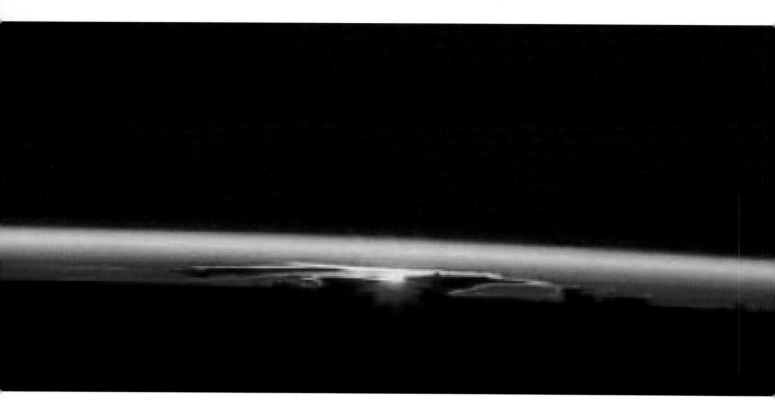

SATELLITE VIEW: A photograph from space of a setting sun shows how thin the atmosphere is. If the Earth were an apple, its atmosphere would have the thickness of the peel.

temperatures were to increase by a modest amount, then evaporation from the oceans will increase, thus increasing the concentration of water vapor in the atmosphere. The result is an enhanced greenhouse effect that increases temperatures further, causing more evaporation, even higher temperatures, and so on. The consequence could be a runaway greenhouse effect—this is thought to be the reason why Venus has no water today. The Earth was spared this fate because it is further from the sun than Venus, and is sufficiently cool for the air to become saturated with water vapor, in which case clouds form. Clouds present the following question: Is their net effect cooling, because of the sunlight they reflect, or warming because of their greenhouse effect? The answer depends on the type of cloud. Occasional glances at the sky reveal that there are many, many types. Uncertainties about future global warming stem mainly from uncertainties concerning the types of clouds that are likely in a warmer world. Simulating these fantastical, ephemeral objects is the biggest challenge for scientists trying to reproduce climate in computer models.

> *The westerly jet streams are so intense that some bands of latitude are known as the Roaring Forties and the Screaming Fifties.*

If the atmosphere were static, we would be confined to a band of mid-latitudes, because the tropics would be too hot, the polar regions too cold. Fortunately, the atmosphere has winds that redistribute heat and also moisture, cooling off the lower latitudes, while warming up higher latitudes. The circulation that effects this redistribution includes surface winds that are easterly (westward) in the tropics, where they converge onto the regions of maximum surface temperature at the equator. There the air rises into tall cumulus towers that provide plentiful rain. Aloft, the air flows poleward, cools, and sinks over the subtropical deserts. Some of the air continues further poleward to join the westerly jet streams that are so intense that some bands of latitude are known as the Roaring Forties and the Screaming Fifties. This atmospheric circulation, despite its chaotic aspects that we refer to as weather, creates distinctive climatic zones—jungles and deserts, prairies and savannahs—that permit enormous biodiversity.

In the tropics, the atmospheric circulation, and hence the pattern of climatic zones, are strongly dependent on patterns of sea surface temperature that influence how much moisture the winds take (evaporate) from the ocean, and then deposit in rain-bearing clouds. The most surprising feature in the sea surface temperature patterns is the presence of very cold surface waters right at the equator in the eastern Pacific Ocean. (When he visited the Galapagos Islands, Charles Darwin commented on the curiously cold water at the equator where sunlight is most intense.) To explain this we need to explore the oceans, the thin film of water that covers much of the globe.

The average depth of the ocean, 3.1 miles or 5 kilometers, is negligible in comparison with the radius of the Earth, which is more than 3,700 miles or 6,000 kilometers. Both the atmosphere and ocean are very thin films of fluid, one air, the other water. Measurements made on expeditions from Antarctica to Alaska show that the ocean

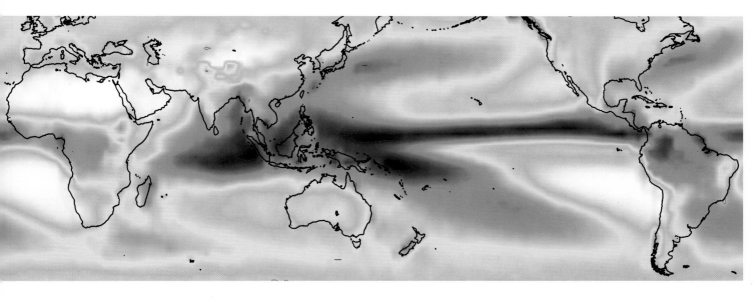

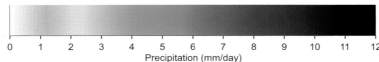

0 1 2 3 4 5 6 7 8 9 10 11 12
Precipitation (mm/day)

PRECIPITATION MAP: There is a strong relationship between amount of precipitation and ocean temperature. Charles Darwin remarked on the surprisingly cold waters off the Galapagos Islands.

EARTH LIGHTS FROM SPACE: This map by NASA shows a composite image of lights on Earth, but both the landforms and lights appear brighter than would be visible to an unaided observer in space. Researchers were able to produce this map of lights showing urban surface activity.

BRIGHT LIGHTS, BIG CITY: What becomes remarkably clear in this image is the energy usage in the United States, western Europe, and Japan, as compared to Africa and the rest of the world. The major national and regional contributors to greenhouse gas emissions are evident.

is composed of a very shallow layer of warm water that floats on a much colder, deep layer. So shallow is the warm layer that, at the equator near the dateline where the surface waters are warmest, the average temperature of a vertical column of water is barely above freezing. An important consequence is that the winds blowing in the right direction can easily expose cold water to the surface by driving oceanic currents in the right direction. The westward trade winds do this along the equator. They drive the warm surface water westward, causing cold water to appear near the Galapagos Islands. Winds parallel to the western coasts of Africa and the Americas, north and south, similarly drive currents that bring cold water to the surface.

Some of the oceanic currents are very slow and deep, others are swift and shallow and include the Gulf Stream and Kuroshio—narrow, rivers of warm water that flow poleward. These currents redistribute heat and chemicals, thus determining patterns of sea surface temperature and oceanic climatic zones that are evident in satellite photographs of the distribution of chlorophyll at the surface of the Earth. Chlorophyll is produced by phytoplankton, literally plants that wander. Those plants, and other life forms that depend on them, are most abundant near the ocean surface, because they need light that penetrates only tens of feet or meters below the ocean surface. When that living matter dies, it sinks and decomposes so that the cold, deep ocean is rich in nutrients.

It follows that ideal conditions for biological productivity—an abundance of light and nutrients—exists where the deep water rises to the surface. These are known as the oceanic upwelling zones, where surface waters are cold, such as off the western coasts of the Americas and Africa. The absence of a layer of warm surface waters around Antarctica makes the Southern Ocean another highly productive zone. Note that the subtropical ocean basins are in effect oceanic deserts with very few plants, because there is practically no exchange between the warm surface waters and the cold water at depth.

The plants on land and at sea, by means of photosynthesis, capture carbon dioxide from the atmosphere during their growing season, and return it when they die and

JULY AND JANUARY: True color composite satellite maps of the Earth's surface in July (above) and January 2004 (at right) from NASA illustrate the significance of seasonal snowfall.

decay. This continual flow of carbon between the ocean, atmosphere, and biosphere (the assemblage of all life on Earth) causes variations in the atmospheric concentration of carbon dioxide. Many people think of the composition of the atmosphere as fixed, in the way that water in a glass is composed of two parts hydrogen and one part oxygen. In reality the atmospheric composition changes continually because each constituent participates in a biogeochemical cycle. (The best known is the hydrological cycle, which is associated with continual changes in the atmospheric concentration of water vapor.) At present, we are interfering with the carbon cycle by burning fossil fuels, and thus emitting carbon into the atmosphere. The oceans and the plants absorb a large fraction, but much remains in the atmosphere so that the concentration there is rising rapidly.

> *A thousand years ago, the northern Atlantic was so warm that Greenland had a large enough population for the pope to send a bishop.*

The ocean, atmosphere, and biosphere form a complex interacting system capable of generating fluctuations on its own. This is known as natural variability, in contrast to variability forced by daily and seasonal changes in sunlight, or by human-induced changes in the composition of the atmosphere. Daily changes in the weather, the best-known examples of natural variability, are as natural as the swings of a pendulum and would be present even if there were no variations in sunlight. Another natural fluctuation, with a much longer timescale of years rather than days, is the oscillation between El Niño and La Niña in the Pacific Ocean. From a strictly oceanic perspective, these phenomena are associated with changes in sea surface temperatures, in the currents, and so on, that are attributable to changes in the winds. Along the equator, those winds are intense during La Niña, weak during El Niño. Why do the winds change? From a meteorological perspective, the large

ALBEDO EFFECT: Snow-covered regions effectively cool the Earth by reflecting sunlight back into space, and hence changes in the range of snow cover can serve to amplify climate changes.

temperature contrast between the western and eastern equatorial Pacific during La Niña drives intense winds that weaken when the contrast weakens. This circular phenomenon—atmospheric changes are both the cause and consequence of oceanic changes—implies that El Niño and La Niña are consequences of interactions between the ocean and atmosphere.

We know a great deal about daily changes in weather because we have ample opportunities to study those changes. Over the past few decades, we learned a fair amount about El Niño, because that phenomenon occurred several times during that period. The past centuries and millennia were also characterized by naturally occurring fluctuations, but information about those climate fluctuations is scant, because of the lack of instrumental records. A thousand years ago, the northern Atlantic was so warm that Greenland had a large population, sufficiently large for the Pope to send a bishop.

That warm period was followed by the frigid Little Ice Age. Those changes were presumably aspects of natural variability, but as yet they are unexplained. Because we know very little about natural variability, it is not possible to determine whether a few unusually warm years, or a few intense hurricanes such as Katrina, or the unusually strong El Niño of 1997, indicate the onset of global warming. Scientists had to search carefully for distinctive patterns, for the "footprints" of global warming, before they could conclude in the 2007 IPCC report that humans activities are affecting the global climate.

March 1998: El Niño

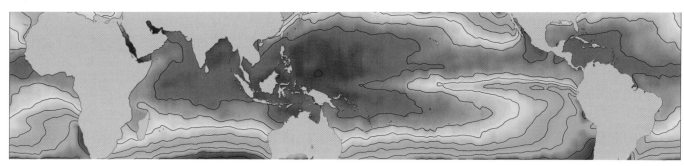

October 1988: La Niña

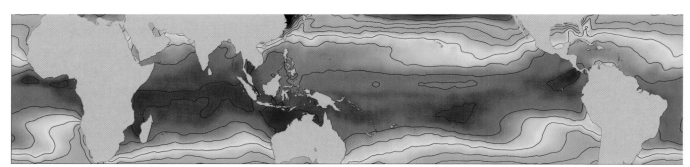

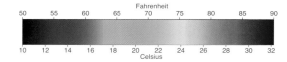

NATURAL FLUCTUATION: With a timescale of years rather than days, the oscillation between El Niño and La Niña in the Pacific Ocean governs weather patterns and storm activity.

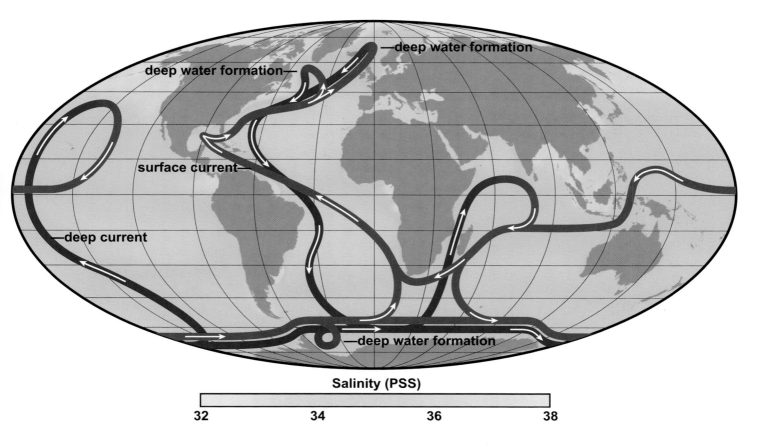

OCEAN CURRENTS: The warm surface currents (red) intertwine with the deep cold currents (blue), creating climate patterns across the Earth. (Robert Simons/NASA)

The composition of the atmosphere, which strongly influences climate, depends on biogeochemical cycles involving not only the ocean, atmosphere and biosphere, but also the solid Earth. *Terra firma* is anything but firm; its surface is composed of several slowly moving, nearly rigid plates, on some of which the continents float. This is the surface manifestation of motion deep in the interior of the Earth, where temperatures are very high because of the decay of radioactive material. Earthquakes are common along the plate boundaries, which feature tall mountains where the plates collide, or deep trenches where one plate dives (subducts) beneath another. In regions of subduction, volcanoes are common; that is why the Pacific rim is known as a "ring of fire." When they erupt, volcanoes emit carbon dioxide into the atmosphere. That gas interacts with water vapor to form an acid that erodes rocks, causing the removal of carbon dioxide from the atmosphere. Hence, the building of mountains—the creation of extensive rock surfaces—promotes the removal of carbon dioxide from

> *Some 65 million years ago, the Earth was so warm that there was no ice on the planet. Palm trees and crocodiles flourished in polar regions.*

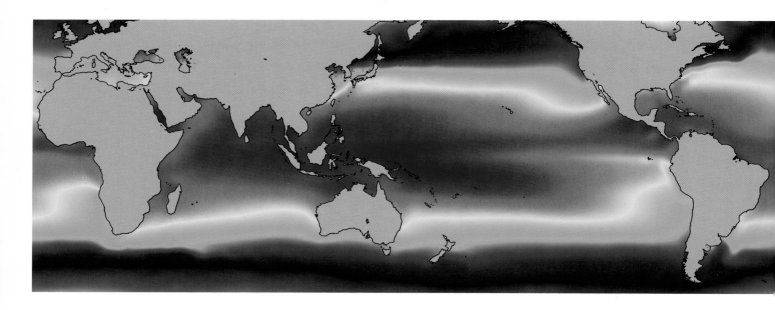

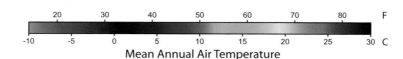

Mean Annual Air Temperature

TEMPERATURE MAP: The areas in dark red with the highest temperatures correlate to the Precipitation Map—regions with the highest precipitation are also the warmest.

the atmosphere. Continental drift therefore affects the atmospheric composition by bringing into play processes that increase, and others that decrease, the concentration of carbon dioxide. Volcanic eruptions contribute to the increase, the building mountains to the decrease. Continental drift affects climate in a more direct manner by changing the distribution of continents. At one time all the continents were together and formed a supercontinent, Pangea, with a northern part known as Laurasia, and a southern Gondwanaland that included the Antarctic continent. With the breakup that started around 250 million years ago, Africa and South America separated to form the Atlantic Ocean. India traveled northward until it collided with Asia, and started creating the Himalayas.

For those interested in global warming, what happened after the demise of the dinosaurs some 65 million years ago is of special interest. At that time, the Earth was far warmer than it is today, so warm that there was no ice on the planet. Palm trees and crocodiles flourished in polar regions, in part because the atmospheric concentration of carbon dioxide was much higher than it is today. Subsequently, the continued drifting of the continents, accompanied by decreases in the atmospheric concentration of carbon dioxide, contributed to the global cooling. (This period is known as the Cenozoic, the age of new animals, specifically mammals.) What caused the Ice Ages? Why did they start 3 million years ago? The answers involve slight changes in sunlight. Sunlight varies daily and seasonally because the Earth rotates on its tilted axis once a day, while orbiting the sun once a year. Additional variations over

> *Glaciers, because they are white, reflect sunlight. This reflection deprives the Earth of heat, lowers temperatures, and promotes the growth of glaciers.*

much longer periods of thousands of years are associated with slight oscillations of the tilt of the axis, which also precesses, while the orbit changes gradually, from a circle into an ellipse, and back to a circle. The moon and several planets cause these Milankovitch cycles, which have been present throughout the Earth's history. The climate fluctuations induced by these sunlight cycles were modest up to 3 million years ago, but then started amplifying. That amplification required positive feedbacks that translated the slight variations in sunlight into Ice Ages. The feedbacks were brought into play by the drifting of the continents. A complex and poorly understood interplay between the slow, erratic drifting of continents, and the regular variations in sunlight, caused the Ice Ages to be absent during some periods, and prominent during others, such as the present.

The global cooling associated with the drifting of the continents that started 60 million years ago inevitably led to the appearance of glaciers, first on Antarctica, then on northern continents around 3 million years ago. Glaciers, because they are white, reflect sunlight. This deprives the Earth of heat, lowers temperatures, and

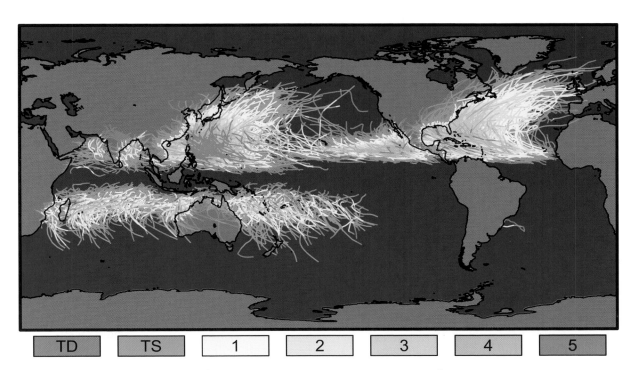

TD	TS	1	2	3	4	5

Saffir-Simpson Hurricane Intensity Scale

HURRICANE PATHS: A plot of the intensity and paths of hurricanes and typhoons. How global warming will affect the development and strength of storms is a subject of debate and study.

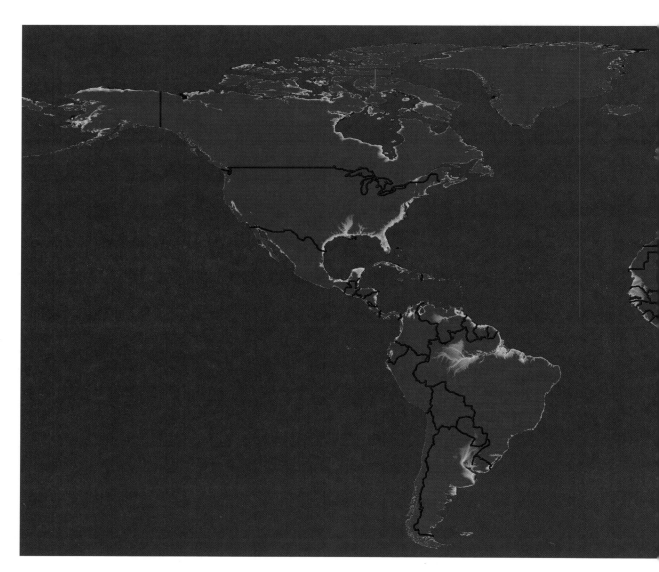

POTENTIAL FLOODING: This is a topographic map designed to emphasize regions near sea level that could potentially be vulnerable to sea level rise, though over centuries rather than decades.

promotes the growth of glaciers. Hence, the appearance of continental glaciers was one of the feedbacks that amplified the response to Milankovitch variations in sunlight. Trapped in those glaciers are bubbles of air that tell us about past changes in the atmospheric composition, past variations in the atmospheric composition of carbon dioxide. As yet it is not known why the concentration varied, or to what degree the variations contributed to the temperature fluctuations.

Solving the puzzle of the Ice Ages will be a major contribution to our ability to anticipate future climate changes, because the solution will tell us a great deal about the sensitivity of the Earth's climate to changes in the atmospheric concentration of carbon dioxide. In the meanwhile, familiarity with the data can give us a valuable perspective on global warming by giving us a geological context for our activities over the past century. From a geological perspective, the present is a special moment

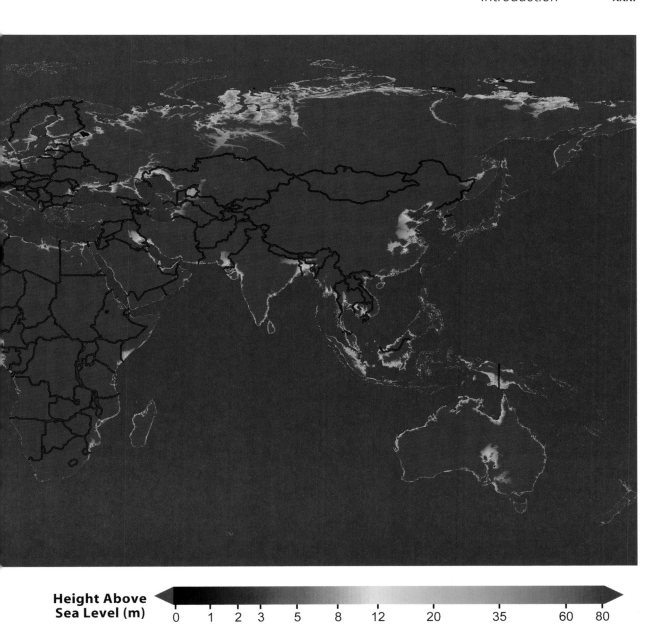

Height Above Sea Level (m)

0　1　2　3　5　8　12　20　35　60　80

in the history of our planet for at least two reasons. The first is that the Earth is currently in an era of high sensitivity to small disturbances. Starting approximately 3 million years ago, the Earth's response to slight variations in sunlight, the Milankovitch cycles, have included enormous climate fluctuations associated with recurrent Ice Ages. Only some of the feedbacks that are involved have been identified. The second reason why the present is special is that we are currently enjoying the temperates of one of the brief interglacial periods that separate prolonged Ice Ages.

The previous interglacial was more than 100,000 years ago but, at that time, we humans were few in numbers, and had very limited capabilities. We were ready when the current interglacial started, some 10,000 years ago, and proceeded to advance with astonishing rapidity, inventing agricultures, domesticating certain animals, developing cultures, and building cities. We developed so rapidly that we are now

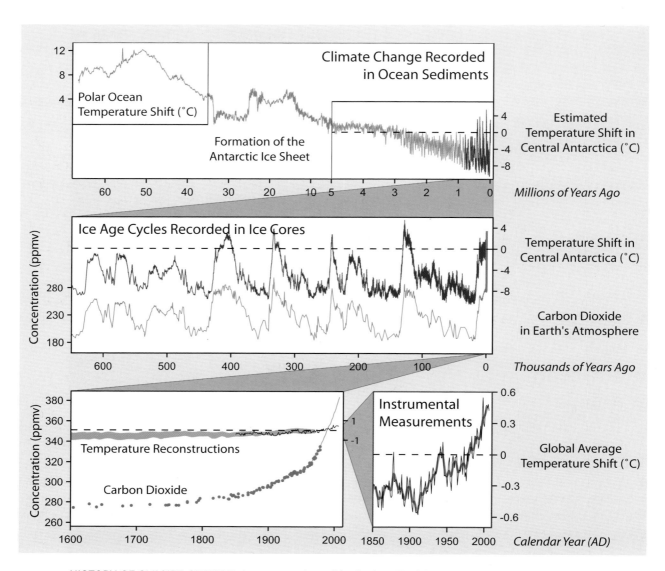

HISTORY OF CLIMATE CHANGE: A compound graphic depicts Earth's climate change across the millennia and centuries. Top: Global cooling over the past 60 million years. Middle: Recurrent Ice Ages over the past 600,000 years. Bottom: Rise in temperature and carbon dioxide over the past four centuries. The information in the top panel comes from cores drilled into the ocean floor, where sediments contain remains of primitive organisms that live near the ocean surface. The information in the middle panel comes from Antarctica, where the accumulated snowfall of hundreds of millennia created deep glaciers.

geologic agents, capable of interfering with the processes that make this a habitable planet. For more than a century, we have caused the atmospheric concentration of carbon dioxide to grow exponentially. This, surely, is a time for circumspection and caution.

MAPS AND PLOTS PREPARED BY ROBERT A. ROHDE
UNIVERSITY OF CALIFORNIA, BERKELEY

Pacific Ocean

THE PACIFIC OCEAN—named the "peaceful sea" by Ferdinand Magellan, a Portuguese explorer leading a Spanish expedition—is the largest ocean in the world, covering 65.3 million sq. mi. (169.2 million sq. km.), encompassing 32 percent of the total surface of the Earth, and holding 46 percent of the Earth's water. Altogether, there are 25,000 islands in the Pacific, the vast majority south of the equator, which bisects the ocean.

RISING SEA LEVELS

Global warming and climate change pose many real threats to the Pacific Ocean. The major focus of much attention around the world has been on the rising water levels, which is likely to inundate many of the low-lying Pacific Islands. Independent countries such as Fiji, Kiribati, the Federated States of Micronesia, Nauru, Palau, Samoa, and Tuvalu risk losing the vast majority of their land if the rising world temperature continues to raise the water level of the ocean. Atolls in French Polynesia and in Wallis and Futuna are also under threat. In addition to those places, all the countries in the Pacific have an increased risk of flooding, which could lead to permanent soil loss, as well as an increased risk of the prevalence of insect-borne diseases such as malaria and dengue fever as mosquitoes find further breeding grounds. The rising sea levels also threaten mangrove swamps in many areas, including off the northeastern coast of Australia, and in many Pacific Islands, with 13 percent of the world's mangrove swamps at risk of being lost.

For this reason, many of the countries in the Pacific have been at the forefront of urging countries around the word to embrace the Kyoto Protocol and limit carbon dioxide emissions. The Republic of Nauru, the country with the highest per capita rate of carbon dioxide emissions in the Pacific, went as far as adding a long addenda to the Kyoto Protocol, arguing that it did not feel that the protocol went far enough. Two U.S. territories in the Pacific, Guam and American Samoa, have considerable carbon dioxide emissions. The Solomon Islands, Papua New Guinea, and Vanuatu have, respectively, the lowest rates of carbon dioxide emissions in the Pacific, at rates similar to that of many African countries.

CHANGES TO MARINE LIFE

Other problems in the Pacific Ocean regarding global warming focus on the marine flora and fauna. The area most dramatically affected has been the bleaching of coral reefs around the Pacific, with studies by the International Ocean Institute of the University

Reefs in Fiji, in the Pacific Ocean. Global warming and climate change many cause a rise in sea level.

of the South Pacific in Fiji conducting surveys of coral reefs in the southwest Pacific as part of the International Coral Reef Initiative. In many cases, the damage to coral reefs has come from overpopulation, and through overexploitation through tourism, but even many reefs located in remote parts of the Pacific have experienced bleaching, showing that the damage can be ascribed as much to global warming as to other problems.

As well as coral reefs, there have been significant changes to the marine life, especially the fish in the Pacific. The most dramatic changes have been the reduction in the diversity of fish shoals, as well as the decline in the number of fish, the latter probably as much from overfishing as from global warming. However, there still remain large numbers of tuna fish and also some cluepoids in the central part of the Pacific Ocean, as well as sardines and jack mackerel along the coast of Chile, anchovy off the coast of Ecuador and Peru, mackerel and Saury off the Pacific coasts of Mexico and the United States, and sardine and salmon off the Pacific coast of Canada.

Some 4 percent of the ozone in the Earth's stratosphere is lost each decade, and a hole has appeared over Antarctica, leading to a higher risk of skin cancer from ultraviolet light in places such as Australia, New Zealand, Chile, and southern Argentina. Although there has been a great focus on their effects on humans, the ultraviolet rays have also been linked to the reduction in the plankton population in the southern part of the Pacific Ocean. The removal of much of the plankton has major effects on the food chain throughout the Pacific, especially on the whale population, which has been growing following a moratorium on commercial whaling in 1986, although Japan continues whaling for ostensibly "scientific" reasons.

One last major area of problems in the Pacific Ocean through global warming and climate change has been changes in the ocean currents, which have been caused by the rise in the temperature of the water. Although few Polynesians travel long distances in traditional canoes, as they did about 1,000 years ago during the populating of many of the islands, the currents are very important, not just for shipping, but also for the movement of marine life such as shoals of fish. The warmer temperature and changes in the current seem to have had major effects on the spawning process of some fish species, and this may be responsible for a decline in the population of certain fish.

Although there is a serious worry about global warming and its effects on the Pacific Ocean, one report in 1997 by scientists from the Lamont-Doherty Earth Observatory at Columbia University claims that the vast size of the Pacific Ocean has led to the dissipation of many of the effects of global warming and climate change, and might account for the fact that the world's temperature has only risen half the level of that in some projections.

SEE ALSO: Alliance of Small Island States (AOIS); Floods; Marine Mammals; Oceanic Changes; Sea Level, Rising.

BIBLIOGRAPHY. Hodaka Kawahata and Yoshio Awaya, *Global Climate Change and Response of Carbon Cycle in the Equatorial Pacific and Indian Oceans and Adjacent Land Masses* (Elsevier, 2006); John Morrison, Paul Geraghty, and Linda Crowl, eds., *Science of Pacific Island Peoples* (Institute of Pacific Studies, 1994); Eric Shibuya, "Roaring Mice Against the Tide: The South Pacific Islands and Agenda-Building on Global Warming," *Pacific Affairs* (v.69/4, 1996–97); Hans von Storch and Ann Smallegauge, *The Phase of the 30- to 60-Day Oscillation and the Genesis of Tropical Cyclones in the Western Pacific* (Max-Planck-Institut für Meteorologie, 1991); Clive Wilkinson, ed., *Status of Coral Reefs of the World* (Australian Institute of Marine Science, 2000).

JUSTIN CORFIELD
GEELONG GRAMMAR SCHOOL, AUSTRALIA

Pakistan

PAKISTAN, IN SOUTHERN Asia, is situated on the Arabian Sea. Pakistan shares borders with Afghanistan (to the north and northwest), Jammu and Kashmir (to the northeast), India (to the east and southeast) (796,095 sq. km.) and Iran (to the west). Pakistan has an area of 307,374 sq. mi., and supports a population of approximately 150 million. Pakistan's topography is varied, with mountains to the north and west, and includes four regions: northern plateau, Baluchistan plateau, southeast desert, and Indus plain. The Indus River, the major river in Pakistan, divides the southern part. Only 5 percent of Pakistan's land has forests, although trees are an essential resource for rural communities and wildlife. Extensive logging in northern areas has resulted in slope instability with the potential for soil erosion and water pollution. Overgrazing results in desertification and soil salinity.

Pakistan's climate is mostly dry, with extremes in elevation and temperature. In the mountain regions of the north and west, temperatures fall below freezing during winter; in the Indus Valley area, temperatures range between about 90–120 degrees F (32–49 degrees C) in summer. Precipitation (often in July and August) is scarce in most of the country, ranging from more than 20 in. annually in the Punjab region, to less than five in. in the arid southeast and southwest.

Pakistan already faces complex environmental problems of air, water, and soil pollution, as well as overuse of natural resources causing deforestation, desertification, and energy and water shortages. Ten percent of households lack access to improved water supply, and 38 percent lack improved sanitation. Air pollution and water shortages are common in the south, with increasing health problems. Warmer temperatures could increase the incidence of heat-related illnesses, lead to higher concentrations of ground-level ozone pollution causing respiratory illnesses (diminished lung function, asthma and respiratory inflammation especially in those persons already susceptible), and increase the risk of contracting certain infectious diseases from water contamination or disease-carrying vectors. Rising sea levels associated with global warming will cause problems on the Indus delta and coastal plains with the loss of land, displaced coastal villages, and population displacement. Flooding and storm surges associated with sea-level rise could increase the incidence of water-borne diseases.

Pakistan's environment minister, speaking at the United Nations, voiced concern for climate change impacts on the environment, the economy, and humans. Some problems already experienced include decreased agricultural productivity (arid land and decreasing water supply) and melting glaciers in the Himalayas in the northern part of Pakistan. The minister indicated Pakistan's major goals include water management, improving energy production, changing agricultural practices, tree planting for carbon sequestration, and establishing a committee for research, advising, and action. Like all preparedness measures, financial resources are necessary for development and implementation. While the future is a concern for Pakistan, solutions for current water quality and energy deficiencies are taking priority.

The University of Peshawar is home to the Department of Environmental Sciences (DES), formed by the university in collaboration with the Pakistan government's Environment and Urban Affairs Division in 1987. DES educates at the graduate level and postgraduate level in Environmental Sciences, Public Health Safety, Natural and Occupational Hazards, Applications of Remote Sensing to Environmental Monitoring and Hazard Mapping. The University of Peshawar uses geographic information systems for planning and natural resources management, performs research, and informs government policymakers in sustainable industrial growth, resource conservation, and environmental preservation.

SEE ALSO: Carbon Sequestration; Desertification; Salinity.

BIBLIOGRAPHY. "Climate Change Affecting Pakistan's Environment: Faisal Saleh," Daily Times, www.dailytimes.com.pk (September 26, 2007); Pakistan Department of Environmental Sciences, www.upesh.edu.pk (cited November 2007).

LYN MICHAUD
INDEPENDENT SCHOLAR

Palau

THE REPUBLIC OF Palau, formerly known as Caroline Islands, is a group of 200 islands located in the Pacific. In total, they have a land area of 177 sq.

mi. (459 sq. km.), with a population of 20,000 (2006 est.), and a population density of 111 people per sq. mi. (43 people per sq. km.). Because of the prosperity of Palau, it has one of the highest rates of per capita carbon dioxide emissions in the world, and is the country with the 15th largest rate of emissions. It was 15.3 metric tons per person in 1990, but was steadily reduced to 12.3 metric tons per person by 2003. Much of this comes from heavy use of electricity in air conditioning, and also, in spite of the size of the islands, relatively heavy use of private automobiles.

Palau is expected to suffer from rising water levels. This will probably lead to the permanent loss of some of the low-lying islands, with the others at risk from flooding. With the country's public water system badly contaminated and posing a serious health risk in January 2002, flooding could overwhelm the water supply for the country, and could cause the spread of insect-borne diseases such as malaria.

The government of Palau took part in the UN Framework Convention on Climate Change signed in Rio de Janeiro in May 1992, and it accepted the Kyoto Protocol to the UN Framework Convention on Climate Change on December 10, 1999, with it entering into force on February 16, 2005.

SEE ALSO: Alliance of Small Island States (AOIS); Floods; Salinity.

BIBLIOGRAPHY. A.H. Leibowitz, *Embattled Island: Palau's Struggle for Independence* (Praeger, 1996); P.H. Saville, *The Animal Health Status of Palau* (Secretariat of the Pacific Community, 1999).

ROBIN S. CORFIELD
INDEPENDENT SCHOLAR

Paleoclimates

PALEOCLIMATES ARE, LITERALLY, past climates. The term "paleo" is, however, normally reserved for those time periods that are prehistorical, although usage of the term to describe climate in the first millennium C.E. and earlier is not uncommon. Like reading a historical text, paleoclimates give glimpses into the var-

ied past of Earth's climate system, and understanding of paleoclimates may well be critical to understanding how Earth's climate will change in the future.

Unfortunately, understanding of this critical climate history does not come easily. Unlike present climate that is carefully recorded and catalogued by various observational and physical measurements, paleoclimate records must be extracted from Earth's geological history; in particular, the archives of sedimentary rocks and large ice sheets. These geological archives do not contain precise measurements of climatic variables such as temperature and precipitation, but rather record environmental responses to ambient climate conditions. As such, the climate indicators extracted from the geological record are known as proxy data.

PROXY DATA

An example of the use of proxy data would be conducting a North American winter snow survey by searching North American garages for snow shovels in the summer. In regions where snow shovels are common in garages, a researcher might reasonably interpret cold winter temperatures and the presence of snow. The shovels, though not actually snow, are a proxy for the occurrence of winter snow. In regions without snow shovels, winter temperatures are either warmer with no snow, or snow is removed in a different way. Because proxy data may be definitive in one sense, but ambiguous in another, using them to decipher the nature of

The Petrified Forest, in northeastern Arizona, has one of the best fossilized records of the Late Triassic period in the world.

paleoclimates requires a suite of data types, or a multi-proxy approach. Fortunately, paleoclimatologists have proven extremely innovative in their development of climate proxies, which range from preserved vegetation in ancient pack rat middens (nests) to the relative abundances of different stable oxygen isotopes (notably ^{16}O and ^{18}O) in the shells of small, ocean-dwelling plankton known as foraminifera.

Some of the oldest proxy data are those related to preserved vegetation. That vegetation may be actual fossils (permineralized plant material or traces of plant material) or simply desiccated (mummified) plant remains, some of which can be tens of millions of years old in cold climates. Vegetation is used as a climate indicator in two primary ways. The first is based on the observation that certain types of plants inhabit certain types of climates. Alders, for example, prefer wetter climates, while Ponderosa Pines thrive in drier conditions. As such, if a plant fossil is found in a given location and there is reasonable certainty that the plant actually grew there, climate in that location can be interpreted based on the known preferences of the identified vegetation.

This is relatively straightforward when the preserved plant material is from a known, living species such as an Alder or Ponderosa Pine. It becomes increasingly difficult further back in geological time when many plants are unknown. In these instances, the climate tolerances of the extinct plant's nearest living relative are used to interpret climate. The other common use of preserved vegetation as a climate indicator is through a statistical process known as leaf margin analysis. Statistical studies of large, modern data sets have indicated that the shape of a leaf's margin (smooth or toothed, elongated or rounded) has some correlation to the climatic conditions that the plant prefers. By performing the same analyses on ancient leaves, scientists can compare ancient leaf samples to the modern climate interpretation and make some assessment of paleoclimatic conditions.

While analyses of preserved vegetation continue to provide significant insight into paleoclimates, more recently, scientists have developed a suite of chemically based climate proxies that make use of the lengthy climate records stored in the sediments of the ocean floor. The most common of these chemical proxies is based on the stable isotopes of oxygen (in particular ^{16}O and ^{18}O, which are the most abundant).

Because different isotopes of a given element have different masses, they are relatively easier or harder to evaporate and, inversely, relatively harder or easier to precipitate. This means that water vapor will always have more ^{16}O than the liquid it evaporated from and is, therefore, lighter. Liquid precipitation will always have more ^{18}O than the vapor it condensed from and is, therefore, heavier. These processes can be used to decipher past climate.

For icehouse climates with large, continent-spanning ice sheets, the relative volume of water stored in large glaciers largely influences the oxygen isotopic concentration of global seawater. Because lighter isotopes evaporate preferentially, precipitation is almost always lighter than the global ocean. If precipitation falls as snow and is then trapped as glacial ice, the overall ocean isotopic composition will get heavier as more and more of the light isotope is locked up in growing ice sheets. If climate warms and ice sheets begin to melt, the lighter isotopes are returned to the ocean and the ocean isotopic composition becomes lighter. Thus, changes in the oxygen isotopic composition of ocean water can be tied to growing and shrinking ice sheets and, by extension, cooling and warming climate.

The primary source of information on past oceanic oxygen isotopic composition is the shells of ocean dwelling microorganisms, in particular, foraminifera. Foraminifera create their shells out of calcium carbonate ($CaCO_3$). The oxygen incorporated into the foraminifera shells comes from the ocean water and, therefore, reflects the ocean's oxygen isotopic composition. The exact isotopic ratio of the shell, however, is also influenced by the ambient temperature and, to a lesser and largely negligible extent, salinity, as well as the organic processes of the foraminifera. Empirically derived equations relate the oxygen isotopic composition of calcium carbonate shells to the ocean temperature and oxygen isotopic composition. If any two of those values are known, the third can be calculated. While the isotopic composition of ancient foraminifera shells can be measured in the laboratory, the overall oceanic oxygen isotopic composition and temperature in the past must either be estimated or derived from other proxies.

CLIMATE EXTREMES

This panoply of proxy climate indicators records a startling array of paleoclimates in Earth's history.

Climatic conditions on Earth have ranged from extreme icehouse conditions with, potentially, the entire planet covered in glaciers (a paleoclimate known as Snowball Earth) to extreme hothouse conditions, with atmospheric carbon dioxide concentrations as much as 20 times higher than those at present, and tropical forests extending nearly pole to pole. Earth's climate has also apparently resided everywhere in between these extremes and at times moved rapidly from one to another.

Whether it is because the current climate falls relatively in the middle of the climate spectrum, or because extremes are more likely to be preserved in the geological record, or because understanding of extremes may provide the greatest insight into the climate system as a whole, the extreme paleoclimate events are the most studied. In the realm of extreme warmth, there were the hothouse climates of the Cenomanian/Turonian boundary (90 million years ago) and the Early Eocene Climatic Optimum (52 million years ago), or nearer-term warm climates like the Miocene Climatic Optimum (14 million years ago), the mid-Pliocene warm period (3.5 million years ago) or the Altithermal of the middle Holocene (5,000 years ago).

At the other end of the spectrum lie the extreme cooling of Snowball Earth (630 million years ago), the rapid inception of large Antarctic ice sheets (35 million years ago), or the peak glaciation of the last glacial maximum (18,000 years ago). Equally fascinating, though even more difficult to quantify, are transient or abrupt climate changes such as Pleistocene Heinrich and Dansgaard-Oeschger events where circum North Atlantic temperature changed by as much as 9 degrees F (5 degrees C) in 30 or 40 years or the Initial Eocene Thermal Maximum, when temperatures in the Arctic Ocean reached 73 degrees F (23 degrees C) for 50,000 to 100,000 years. Climate extremes such as these in Earth's history, and incomplete explanations for them, help show that the current climate samples a very finite portion of Earth's climatic possibilities, and if scientists wish to have a solid understanding of what the climate of the future may hold, paleoclimates must first be understood.

SEE ALSO: Cenozoic Era; CLIMAP Project; Climatic Data, Proxy Records; Earth's Climate History; Ice Ages; Mesozoic Era; Paleozoic Era; Snowball Earth.

BIBLIOGRAPHY. R.S. Bradley, *Paleoclimatology: Reconstructing Climates of the Quaternary* (Harcourt Academic Press, 1999); T.J. Crowley and G.R. North, *Paleoclimatology* (Oxford University Press, 1995).

Jacob Sewall
Virginia Tech

Paleozoic Era

THE PALEOZOIC ERA is the earliest of three geologic eras of the Phanerozoic eon. This era spanned from roughly 542 million years ago to roughly 251 million years ago. The Paleozoic era is subdivided into six geologic periods: the Cambrian, Ordovician, Silurian, Devonian, Carboniferous, and Permian. The Paleozoic covers the time from the first appearance of abundant, hard-shelled fossils to the time when the continents were beginning to be dominated by large reptiles and modern plants. The oldest geological period was classically set at the first appearance of creatures known as trilobites and archeocyathids. The youngest geological period marks a major extinction event 300 million years ago, known as the Permian extinction.

At the start of the era, all life was confined to bacteria, algae, sponges, and a variety of enigmatic forms known collectively as the Ediacaran fauna. The Cambrian Explosion resulted in an exponential increase of life-forms. There is some evidence that simple life may already have invaded the land at the start of the Paleozoic, but substantial plants and animals did not take to the land until the Silurian and did not thrive until the Devonian. Although primitive vertebrates are known near the start of the Paleozoic, invertebrates were the dominant life-forms until the mid-Paleozoic. Fish populations exploded in the Devonian. During the late Paleozoic, great forests of primitive plants thrived on land forming the great coal beds of Europe and eastern North America. By the end of the era, the first large, sophisticated reptiles and the first modern plants had developed.

The Paleozoic era began shortly after the breakup of a supercontinent called Pannotia and at the end of a global ice age. During the early Paleozoic, the Earth's landmass was broken up into a number of relatively small continents. Toward the end of the era, the continents gathered together into a supercontinent called Pangaea, which included most of the Earth's land area.

The Early Cambrian climate was probably moderate at first, becoming warmer over the course of the Cambrian, as the second-greatest sustained sea level rise in the Phanerozoic got underway. Gondwana moved south with considerable speed. By the Ordovician period, most of West Gondwana (Africa and South America) lay directly over the South Pole. The Early Paleozoic climate was also strongly zonal. The climate became warmer, but the continental shelf marine environment became steadily colder. The Early Paleozoic ended, rather abruptly, with the short, but apparently severe, Late Ordovician Ice Age. This cold spell caused the second-greatest mass extinction of Phanerozoic time. The Middle Paleozoic was a time of considerable stability. Sea levels had dropped coincident with the Ice Age, but slowly recovered over the course of the Silurian and Devonian.

The slow merger of Baltica and Laurentia and the northward movement of bits and pieces of Gondwana created numerous new regions of relatively warm, shallow seafloor. The far southern continental margins of Antarctica and West Gondwana became increasingly less barren. The Devonian period (410 to 360 million years ago) resulted in diversifiaction of life on the land, including the first terrestrial vertebrates, the amphibians, and the first forests of trees. In the waters fish continued their diversification with the rise of the lobe-finned and ray-finned fish. The Devonian ended with a series of turnover pulses which killed off much of Middle Paleozoic vertebrate life, without noticeably reducing species diversity overall. Global cooling tied to Gondwanan glaciation has been proposed as the cause of the Devonian extinction, as it was also suspected of causing the terminal Ordovician extinction. Rocks in parts of Gondwana suggest a glacial event. The forms of marine life most affected by the extinction were the warm-water to tropical ones.

The Late Paleozoic consisted of the Carboniferous period (360 to 286 million years ago), also known as the Mississippian period. The period began with a spike in atmospheric oxygen, while carbon dioxide plummeted. This destabilized the climate and led to multiple ice age events during the Carboniferous. The supercontinent of Pangaea was assembled during this time, causing the uplift of seafloor as continental land masses collided to build the Appalachian and other mountains.

This created huge arid inland areas subject to temperature extremes. The Permian period spanned the time interval from 286 to 245 million years ago. During the Permian the assembly of Pangaea was completed and a whole host of new groups of organisms evolved.

The Permian ended in the greatest of the mass extinctions, where over 90 percent of all species were extinguished. With the assembly of Pangaea and resulting mountain building, many of the shallow seas retreated from the continents. The Permian saw the spread of conifers and cycads, two groups that would dominate the floras of the world until the Cretaceous period with the rise of the flowering plants. The end of the Permian, also the end of the Paleozoic era, was marked by the greatest extinction of the Phanerozoic eon. During the Permian extinction event over 95 percent of marine species went extinct, while 70 percent of terrestrial taxonomic families suffered the same fate. The fusulinid foraminiferans went completely extinct, as did the trilobites. The majority of extinctions seem to have occurred at low paleolatitudes, possibly suggesting some event involving the ocean. The exact cause of the terminal Permian extinction remains unknown; however, many theories have been hypothesized. Regardless, this event proved to be a massive and severe crisis for life. Many groups of organisms went extinct at that time. Surviving groups diversified during the Triassic period and gradually a more modern world developed.

SEE ALSO: Climatic Data, Proxy Records; Earth's Climate History; Ice Ages.

BIBLIOGRAPHY. I.P. Montañez, "CO_2-Forced Climate and Vegetation Instability during Late Paleozoic Deglaciation," *Science* (v.315, 2007); Paleozoic Era, www.palaeos.com (cited November 2007); University of California Museum of Paleontology, www.ucmp.berkeley.edu (cited November 2007).

FERNANDO HERRERA
UNIVERSITY OF CALIFORNIA, SAN DIEGO

Panama

LOCATED IN CENTRAL America, the Republic of Panama has a land area of 29,157 sq. mi. (78,200 sq. km.), with a population of 3,343,000 (2006 est.), and a population density of 111 people per sq. mi. (43 people per sq. km.). Only 7 percent of the land in the country is arable, the second lowest percentage in Central America, with 20 percent used for meadows and pasture, and 44 percent of the land forested. The level of carbon dioxide emissions in Panama was 1.3 metric tons per capita in 1990, rising to 2.3 metric tons per person in 2001, and then falling slightly to 1.9 metric tons per person by 2003. Most of this comes from liquid fuels, which make up 89 percent of all carbon dioxide emissions from the country, with cement manufacturing contributing 6 percent, and solid fuels (coal and charcoal) contributing another 3 percent.

The Caribbean coast of Panama has long had problems with hurricanes, but the rising water temperature in both the Caribbean Sea and the Pacific Ocean have led to increased worry over flooding, and has caused some bleaching of coral reefs in the Archipelago de Bocas del Toro off the northwest coast of the country. Although some 30 percent of the country has been set aside for conservation, the deforestation of many areas used for pasture has led to soil erosion, which has also helped contribute to the destruction of the mangrove swamps. There have also been effects on the wildlife in the pristine cloud forest on the Quetzal Trail around the Parque Nacional Volcán Barú, and worries about flooding, which in turn could lead to a spread of insect-borne diseases such as malaria and dengue fever.

The Panamanian government of Guillermo Endara took part in the UN Framework Convention on Climate Change signed in Rio de Janeiro in May 1992, and that of Ernesto Pérez Balladares signed the Kyoto Protocol to the UN Framework Convention on Climate Change on June 8, 1998. The Kyoto Protocol to the UN Framework Convention on Climate Change was ratified on March 5, 1999, and came into force on February 16, 2005.

SEE ALSO: Diseases; Floods; Hurricanes and Typhoons.

BIBLIOGRAPHY. "Dengue Fever in Costa Rica and Panama," *Epidemiological Bulletin* (v.15/2, 1994); "Dengue in Central America: The Epidemics of 2000," *Epidemiological Bulletin* (v.21/4, 2000); "Panama—Climate and Atmosphere," www.earthtrends.wri.org (cited October 2007).

JUSTIN CORFIELD
GEELONG GRAMMAR SCHOOL, AUSTRALIA

Papua New Guinea

LOCATED IN THE Pacific, on the eastern half of the island of New Guinea, Papua New Guinea has a land area of 178,703 sq. mi. (462,840 sq. km.), with a population of 6,331,000 (2006 est.), and a population density of 34 people per sq. mi. (13 people per sq. km.). In spite of having a tropical monsoon climate, only 0.1 percent of the land is arable, with 82 percent of the country forested. Conservationists and environmentalists condemn the rate of the timber logging industry as unsustainable.

Some 55 percent of the electricity in the country comes from fossil fuels, mainly liquid fuels, which account for 93 percent of Papua New Guinea's carbon dioxide emissions. The remainder of the electricity in the country comes from hydropower. Plans to harness the power of the Purari in the early 1970s were shelved, with the Ramu River used to generate hydropower from the 1980s.

The effect of climate change and global warming in Papua New Guinea has been significant. A rise in temperature has led to the bleaching of some coral reefs off the south coast of the country, which has exacerbated the problems from logging and the crown-of-thorns starfish. There has also been a decline in the fish stocks, both in diversity and numbers. In addition, some parts of the country have experienced flooding, which has led to an increase in the prevalence of insect-borne diseases such as malaria and dengue fever.

The Papua New Guinea government took part in the United Nations Framework Convention on Climate Change signed in Rio de Janeiro in May 1992. It signed the Kyoto Protocol to the UN Framework Convention on Climate Change on March 2, 1999, and ratified it on March 28, 2002, with it entering into force on February 16, 2005.

SEE ALSO: Deforestation; Floods; Forests.

BIBLIOGRAPHY. *Approaches to Environmental Planning in Papua New Guinea* (Office of Environment and Conservation, 1980); *Hydroelectric Potential of Papua New Guinea* (Australian Department of Housing and Construction for the Papua New Guinea Electricity Commission, 1974); "Papua New Guinea—Climate and Atmosphere," www.earthtrends.wri.org (cited October 2007); Eric Shibuya,

"Roaring Mice Against the Tide: The South Pacific Islands and Agenda-Building on Global Warming," *Pacific Affairs* (v.69/4, 1996–97).

ROBIN S. CORFIELD
INDEPENDENT SCHOLAR

Paraguay

THIS LANDLOCKED SOUTH American country has land borders with Argentina, Bolivia, and Brazil, and covers an area of 157,047 sq. mi. (406,752 sq. km.). It has a total population of 6,036,900 (2003 est.), with a population density of 38.4 people per sq. mi. (14.5 people per sq. km). However, the real population density for much of the country is higher, as the country is bisected by the Rio Paraguay (Paraguay River), and the vast majority of the population lives in the eastern half of Paraguay. Only 6 percent of the land is arable, although 55 percent of it is used for grazing animals.

The main environmental problem in the country is the lack of arable land. This has resulted in deforestation to create more farmland, although officially 35 percent of Paraguay remains forested. The western half of the country, the Chaco Desert, has extremely poor soil and is used for cattle grazing and ranching, but the cattle industry has faced many problems. Even in the more fertile eastern half of Paraguay, the soil is poor, and much of the land is also used for grazing cattle, for which there are 9,400 head per 1,000 people in the country.

The underdeveloped economy and the low standard of living have helped reduce the effect of the country on global warming, with Paraguay ranking 157th in the world for carbon dioxide emissions per capita of 0.5 metric tons of carbon dioxide per capita in 1990, rising to a peak of 0.9 per capita in 1998, and then falling back to 0.7 in 2000–03. Much of this was because of the low use of electricity by the poor in rural areas and in shantytowns, and the relatively low use of cars.

There has been a sharp increase in car usage in recent years that has coincided with a marked decline in public transport. The railway network that covered parts of eastern Paraguay closed down in the 1970s, and only operates a tourist train, lead-ing to an increase in road haulage, and the tram system in Asunción, the capital, was closed for general use in 1995. There have also been few environmental controls in the country. However, Paraguay has been active at the international level, as a member of the United Nations Framework Convention on Climate Change signed in Rio de Janeiro in 1992, and as one of the early countries to sign the Kyoto Protocol to the UN Framework Convention on Climate Change, which was done on August 25, 1998. Paraguay was the 13th country to ratify the Kyoto Protocol, which took place on August 27, 1999.

One of Paraguay's main contributions to reducing the threat of global warming was the construction of the Itaipu Dam, which started generating power in 1982. By 2000, it was supplying hydroelectric power to much of the country, making 93 percent of all electricity generation in Paraguay; and providing 20 percent of the electricity used by Brazil, generating significant income for the Paraguayan economy. Subsequently, work has begun on the Yacyretá Dam, located on the Paraguayan-Argentine border, which will provide electrical power for sale to Argentina.

SEE ALSO: Agriculture; Deforestation; Transportation.

BIBLIOGRAPHY. Patrick McGrath, "Paraguayan Powerhouse," *Geographical Magazine* (v.55, 1983); R.A. Nickson, "The Itaipu Hydro-Electric Project: The Paraguayan Perspective," *Bulletin of Latin American Research* (v.2/1, October 1982); G.D. Westley, "Electricity Demand in a Developing Country," *Review of Economics and Statistics* (v.66/3, 1984).

ROBIN S. CORFIELD
INDEPENDENT SCHOLAR

Penguins

FROM THE TROPICS to Antarctica, penguins depend on predictable regions of high ocean productivity where their prey aggregate. There are between 16 and 19 species of penguins, all generally restricted to the southern hemisphere, with the greatest species diversity found in New Zealand. Changes in precipitation, sea ice, ocean temperature and productivity, and prey distributions associated with

Changes in precipitation, sea ice, and ocean temperature associated with global warming are affecting penguins.

global climate warming are affecting penguins and changing their distribution and abundance. BirdLife International includes climate change among the threats for seven of 10 species of penguin listed as endangered or vulnerable.

The global climate signal is strongest in the Antarctic Peninsula, where air temperature has increased, glaciers have retreated, and ice shelves have collapsed. Adélie and Emperor penguins, the most ice-associated and southerly of the penguin species, are suffering more reproductive failures because of increases in rain and snow, early breakup of ice, and blocking of colony access by icebergs. Gentoo and Chinstap penguins, more northerly and less ice-tolerant, have extended their ranges farther south, but suffer more failures because of increased precipitation. Snow covers nest sites and rain soaks the down of chicks, which then freeze. King penguins, the second largest species, harvested during the whaling era for their oil, have increased and expanded their range northward.

Many of the temperate species of penguins are in decline because of human perturbations including harvest, accidental capture by fisheries, petroleum pollution, breeding habitat destruction, and climate variation that changes the distribution and abundance of their prey. Temperate penguins also suffer from increases in rainfall or air temperatures, as young chicks cannot regulate their body temperatures when their down is wet. Increased frequency of El Niño events associated with global warming reduced Galapagos penguins to about half of what they were in the 1970s. During El Niño, the water warms, ocean productivity declines, and penguins quit breeding, and under the strongest and longest events, penguins die. Peruvian penguins declined after the 1972 El Niño and have never recovered, in large part because the anchovy fishery, the second largest fishery in the world after Alaskan Pollack, harvests much of their prey to make fishmeal.

Patagonian penguins, the most common of the temperate penguins, have declined by 22 percent since 1987 at their largest breeding colony, in Argentina. They swim farther north during incubation than they did a decade ago, likely a reflection of shifts in their prey in response to climate change and reductions in prey abundance due to commercial fishing. African penguins are only 10–20 percent of their previous number, because humans are harvesting more of the ocean's produce than in the 1950s. African penguins are also shifting their breeding locations because of changes in prey distributions linked to climate. Breeding colonies in protected reserves are no longer near sources of food. Ocean productivity will likely continue to decline as climate warms, which is a problem for penguins, as they need areas of high productivity to survive. As global sentinels, penguins are showing that global climate change is creating new challenges for their populations.

SEE ALSO: Benguela Current; Detection of Climate Changes; El Niño and La Niña; Oceanic Changes; Sea Ice.

BIBLIOGRAPHY. BirdLife International, www.birdlife.org (cited July 2007); T.D. Williams, *The Penguins: Spheniscidae* (Oxford University Press, 1995).

P. Dee Boersma
Ginger A. Rebstock
University of Washington

Penn State University

PENNSYLVANIA STATE UNIVERSITY (also known as Penn State) is a large public multicampus university with its main campus located in University Park, Pennsylvania. The University has 24 campuses located throughout the U.S. state of Pennsylvania, including a virtual World Campus. The enrollment at Penn State is over 84,000 students, placing it among the 10 largest public universities in the United States. Penn State offers more than 160 majors. Pennsylvania State University ranks in the top five in the world in total number of citations in the area of global warming research. Penn State Institutes of Energy and the Environment is the central coordinating structure for energy and environmental research at Pennsylvania State University.

The concept of the Penn State Institutes of the Environment arose from intense interactions between Penn State's administration and faculty and remains a novel partnership between the two. It is organized under the Office of the Vice President for Research (OVPR) and is designed to position the environmental faculty to compete vigorously in this new interdisciplinary environmental science and engineering prototype. It facilitates environmental research, teaching, and outreach across eight colleges, including the University Park colleges of Agricultural Sciences, Earth and Mineral Sciences, Engineering, Health and Human Development, and the Liberal Arts, plus the Hershey College of Medicine and Harrisburg Capital College. The PSIEE promotes Penn State's interdisciplinary environmental enterprise through a wide variety of activities.

The PSIEE has been instrumental in assisting faculty and departments in the creation of world-class programs such as climate change, and hydrogen energy, as well as other areas.

Courses offered at Pennsylvania State University that focus on global warming and climate change include:

EARTH 103: EARTH IN THE FUTURE: PREDICTING CLIMATE CHANGE AND ITS IMPACTS OVER THE NEXT CENTURY.

The United States is actively working on national assessment of the impacts of the climate change predicted to occur over the next century. The U.S. National Assessment has developed three major documents: an overview written for Congress, a foundation document giving the sources of information and their interpretation, and a series of regional and sector (water, health, agriculture, forests, and coastlines) reports. These reports present an exceptional opportunity to connect advances in the natural sciences to society. The course has four major objectives: (1) to gain an understanding of climate science and of the possible scenarios of how climate may change in the future; (2) to analyze the linkages between climate and major human and natural systems (e.g., agriculture, human health, water, coastal ecosystems, and forests), necessary to assess the potential impacts of climate change; (3) to demonstrate that the impacts of climate change, and the way in which society responds, are dependent on factors such as age, economic capability, lifestyle (e.g., urban vs. rural), generational differences, and cultural differences; and (4) to understand the different types of responses that humans may have to climate change, including adaptations to change and possible mechanisms to mitigate the factors that are forcing change to occur.

GEOSC 320: GEOLOGY OF CLIMATE CHANGE

Geologic records provide a critical perspective on climate change, with implications for our behavior. Ice cores, ocean sediments, tree rings, and others reveal that agriculture and industry have arisen during a few thousand years of anomalously stable climate. Natural changes half as large as the entire difference between ice-age and modern conditions have occurred repeatedly in mere years, affecting hemispheric or broader regions. Such climate jumps have been linked to changes in greenhouse gases, but not driven by them. Students in this course will learn how records of recent climate changes are recovered, read, and dated, how the climate system works and has worked, and the causes of ice-age cycles and faster climate jumps.

BIOL 436: POPULATION ECOLOGY AND GLOBAL CLIMATE CHANGE

In this course, students investigate the factors shaping the characteristics of populations and their dynamics in time and space, with emphasis on the responses of populations to climatic fluctuation and global climate change. These concepts include the science of climate change, how temperature trends are estimated, the data used in assessment reports by the Intergovernmental Panel on Climate Change, large-scale climate systems such as the North Atlantic Oscillation and the El Niño Southern Oscillation, the basic characteristics of populations, how population densities are estimated, and the types of population data used in studies of population responses to climate change.

SEE ALSO: El Niño and La Niña; Greenhouse Effect.

BIBLIOGRAPHY. Pennsylvania State University, www.psu.edu (cited November 2007); Real Climate, www.realclimate.org (cited November 2007).

FERNANDO HERRERA
UNIVERSITY OF CALIFORNIA, SAN DIEGO

Pennsylvania

BECAUSE OF PENNSYLVANIA'S interior location in the northeast, it is in prime position to experience the many negative effects associated with global warming. By 2100, average summer temperatures in Pennsylvania could increase between 7–9 degrees F (4–5 degrees C). This temperature change could cause extreme cases of precipitation. Some parts of the state could experience up to a 50 percent increase in rainfall, while other areas face drought conditions.

In Philadelphia, by 2050, heat-related deaths during a typical summer could increase by 90 percent, from about 130 deaths per summer to more than 240. Currently, "red alert" air quality days happen about two days every summer in Pittsburgh. By the middle of the century, this number could rise to five days per summer. Also, ozone levels in the city are already above the Environmental Protection Agency's healthy standard at least 10 days out of the year. Global warming could cause this number to increase to 22 days in the near future, meaning more cases of respiratory diseases such as asthma. Loss of wildlife and habitat are also possible threats caused by global warming, which could mean a loss of tourism dollars.

Over the last century, the average temperature in Harrisburg, Pennsylvania, has increased 1.2 degrees F (0.6 degrees C), and precipitation has increased by up to 20 percent in many parts of the state. Over the next century, climate in Pennsylvania is expected to change even more. Precipitation is estimated to increase by about 10 percent in spring, by about 20 percent in winter and summer, and by as much as 50 percent in fall. The amount of precipitation on extreme wet or snowy days is also likely to increase, which would cause an increase in extremely hot days in summer because of the general warming trend. Although it is not clear how severe storms would change, an increase in the frequency and intensity of summer thunderstorms is possible.

Higher temperatures and increased frequency of heat waves may increase the number of heat-related deaths and the incidence of heat-related illnesses. Pennsylvania, with its irregular, intense heat waves, could be especially susceptible. Similar but smaller increases have been projected for Pittsburgh, from about 40 heat-related deaths to 60, or a 50 percent increase. Winter-related deaths could drop from 85 per winter in Philadelphia, to about 35 per winter if temperatures warm.

The complications that global warming could have on the major cities in the future will also be felt by the rural areas that make up a large portion of the state. Pennsylvania's farming and agriculture industries are vital to the state's economy, as well as the outdoor tourism market; both would be greatly debilitated by effects of global warming. In 2001, more than 4.5 million people spent nearly $3 billion on hunting, fishing, and wildlife viewing in Pennsylvania, which in turn supported 56,113 jobs in the state.

HEALTH VARIABLES

Climate change may also increase ground-level ozone levels. For example, high temperatures, strong sunlight, and stable air masses tend to increase urban ozone levels. If a warmed climate causes increased use of air conditioners, air pollutant emissions from power plants also will increase. A preliminary modeling study of the Midwest, which included the area around Pittsburgh, found that a 4 degrees F (2 degrees C) warming, with no other change in weather or emissions, could increase concentrations of ozone, a major component of smog, by as much as 8 percent. Currently, ground-level ozone concentrations exceed national

The Delaware River overflowing in 2005. Pennysylvania has some of the most intense flooding in the United States.

ozone health standards in several areas throughout the state. Ground-level ozone has been shown to heighten respiratory illnesses such as asthma, as well as other complications. In addition, ambient ozone reduces crop yields and is harmful to ecosystems.

Warming and other climate changes may cause an increase in disease-carrying insects, thus the potential for the spread of diseases such as malaria and dengue fever is increased. Mosquitoes flourish in many areas around Pennsylvania. Some can carry malaria, while others can carry encephalitis, which can be lethal or cause neurological damage. Incidents of Lyme disease, which is carried by ticks, have also increased in the northeast. If conditions become warmer and wetter, mosquito and tick populations could increase in Pennsylvania, increasing the risk of these types of diseases.

WATER RESOURCES

Pennsylvania's valuable water resources would also be affected by changes in precipitation, temperature, humidity, wind, and sunshine. Changes in stream flow tend to coincide with changes in precipitation. Water resources in drier climates are more sensitive to climate changes. Because evaporation often increases with the onset of a warmer climate, the result could be lower river flow and lower lake levels, especially in the summer. If this happens, groundwater will consequently be reduced. In addition, a rise in precipitation could lead to increased flooding. Pennsylvania's Susquehanna River drains much of

the eastern two-thirds of the state, and the Allegheny and the upper Ohio rivers drain most of the western third. A warmer climate would lead to earlier spring snowmelt, and could result in higher stream flows in winter and spring and lower stream flows in summer and fall. However, changes in rainfall also could have significant effects on stream flow and runoff. This alerts many Pennsylvanians because some of the most intense flooding on record in the United States has occurred in Pennsylvania.

SEE ALSO: Diseases; Floods; Penn State University; Rainfall Patterns.

BIBLIOGRAPHY. D.I. Benn and D.J.A. Evans, *Glaciers and Glaciations* (Wiley, 1998); Clean Air Council, www.cleanair.org (cited July 2007); "Climate Change and Pennsylvania," Environmental Protection Agency, www.epa.gov (cited July 2007); "Global Warming and Pennsylvania," National Wildlife Fund, www.nwf.org (cited July 2007); Johannes Oerlemans, *Glaciers and Climate Change* (Balkema, 2001); Penn Environment, www.pennenvironment.org (cited July 2007).

ARTHUR MATTHEW HOLST
WIDENER UNIVERSITY

Perfluorocarbons

THE PERFLUOROCARBONS ARE a group of chemically related greenhouse gases covered by the Kyoto Protocol. Although emissions of perfluorocarbons are low compared to many other pollutants, they are of great concern because the perfluorocarbons are extremely powerful greenhouse gases with very long atmospheric lifetimes. Furthermore, the release of man-made perfluorocarbons is on the rise, due to increasing aluminum and semiconductor chip manufacture. Annual releases of PFM, the most abundant perfluorocarbon, are the global warming equivalent to about 70 megatons of CO_2, roughly one two-hundredth of the amount of CO_2 released annually.

In the context of climate change, the most important perfluorocarbons are perfluoromethane (PFM) and perfluoroethane (PFE). Also of interest in a wider environmental context are the oxygenated perfluorocarbons, PFOS and PFOA. These latter compounds are highly soluble in water, and they are found in the ocean environment and in living tissues, but rarely in the atmosphere. Since 1980, the atmospheric concentration of PFM has risen by around 30 percent, despite reductions in emissions per ton from the aluminum industry, and is thought to have risen by around 70 percent since 1960. The atmospheric concentration of PFE has doubled from its concentration in 1980, and is believed to be more than 10 times higher than its 1960 value.

The perfluorocarbon molecules are strong absorbers of infrared radiation, and are therefore powerful greenhouse gases. Although the atmospheric concentration of perfluoromethane is around 100,000 times lower than CO_2, the radiative forcing due to this atmospheric loading of PFM is as much as one five-hundredth of the radiative forcing due to CO_2. (Radiative forcing is a measure of the global warming effect of a chemical at a given atmospheric concentration.) PFM is a much more powerful greenhouse gas than CO_2 as measured by its global warming potential. The global warming potentials of PFM and PFE are 7,390 and 12,200, respectively.

The PFCs are extremely environmentally stable; they are only very slowly destroyed by the action of sunlight and oxygen. The main pathway for removal of PFCs from the environment is via high temperature combustion processes when air is taken into vehicle engines or power station furnaces. This environmental stability arises from the molecules' chemical structure. The PFCs are related to simple hydrocarbons by replacement of all hydrogen atoms by fluorine. For example, the simplest hydrocarbon is methane, CH_4. The corresponding PFC is perfluoromethane, CF_4. The carbon-fluorine chemical bond is tremendously robust with respect to normal mechanisms by which the atmosphere cleans itself. Consequently, the atmospheric lifetimes of PFM and PFE are around 50,000 years and 10,000 years, respectively.

Natural PFM emissions from soils give rise to a background "clean air" concentration of about 40 pptv (parts per trillion by volume). Its concentration has been increasing throughout the latter half of the 20th century, to its current value of about 75 pptv due to industrial activity. Both PFM and PFE are produced as a by-product of the electrochemical extraction of aluminum from its ores. Over the past 20 years, the global aluminum industry has significantly improved its performance: currently, an average of 400 grams of PFM and about 40 grams of PFE are released per ton of aluminum produced, down by almost two-thirds since the 1980s. However, this per-ton emissions reduction has been somewhat offset by increases in aluminum production volumes. The semiconductor chip manufacturing industry is a major source of PFE and a secondary source of PFM. Consequently, the increase in chip manufacture is a major influence on the growing emissions of PFE.

Oxygenated perfluorocarbons, PFOS (perfluorooctane sulfonate) and PFOA (perfluorooctanoic acid) are thought to be harmful to human health. The U.S. Environmental Protection Agency regards PFOA as a "likely carcinogen"; it has been shown carcinogenic in rodents, as well as causing immune and reproductive system damage. PFOA and PFOS are released to the environment from manufacture and use of nonstick materials, fabric protectors, and fire-fighting foams. Due to their high water solubility and extremely long environmental lifetime, they are found in low concentrations in the blood of humans worldwide, and in many animals, including U.S. dolphins, Chinese pandas, and Arctic polar bears.

SEE ALSO: Hydrofluorocarbons; Intergovernmental Panel on Climate Change (IPCC); Kyoto Protocol.

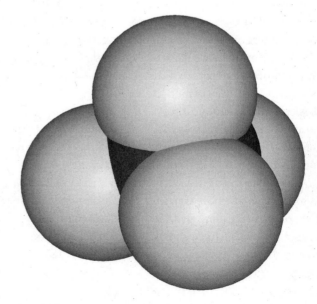

A tetrafluoromethane molecule: The most abundant atmospheric perfluorocarbon is tetrafluoromethane.

BIBLIOGRAPHY. Intergovernmental Panel on Climate Change, www.ipcc.ch (cited November 2007); M.A.K. Khalil, et al., "Atmospheric Perfluorocarbons," *Environmental Science and Technology* (v.37, 2003).

CHRISTOPHER J. ENNIS
CLEAN ENVIRONMENT MANAGEMENT CENTRE
UNIVERSITY OF TEESSIDE

Peru

PERU'S 0.5 PERCENT contribution of greenhouse gases to the world's atmosphere is small, compared to the impact on Peru expected as a result of climate change. Peru is ranked as the fourth country most impacted by climate change. El Niño has regularly affected the 386,102 sq. mi. (one million sq. km.) of Peruvian territory, with droughts in the Andean south, and floods in the northern Pacific coast. These impacts, however, are small compared to the impacts of climate change. The 125 mi. (200 km.)-long White Mountain Range, the world's largest ice-covered tropical range and Peru's main concentration of ice, has been losing volume in the last 50 years. The glaciers are melting, leading to glacier reduction, the formation or increase of glacial lakes, and changes in ecosystem composition. The glaciers of White Range Park have retreated 82 ft. (25 m.) in the last 50 years according to CONIDA, Peru's aerospace agency, and CONAM, the Peruvian environmental agency.

Glacier retreat has implications for downstream river flows. In rivers fed by glaciers, summer flows are supported by glacier melt (with the glacier contribution depending on the size of the glacier relative to basin area, as well as the rate of annual melt). If the glacier is in equilibrium, the amount of precipitation stored in winter is matched by melt during summer. However, as the glacier melts as a result of global warming, flows would be expected to increase during summer—as water is released from long-term storage—which may compensate for a reduction in precipitation. As the glacier gets smaller and the volume of melt reduces, summer flows will no longer be supported and will decline to below present levels. The duration of the period of increased flows will depend on glacier size and the rate at which the gla-cier melts; the smaller the glacier, the shorter lived the increase in flows and the sooner the onset of the reduction in summer flows. In 18 glaciers in the Peruvian Andes, mass balances since 1968 and satellite images show a reduction of more than 20 percent of the glacial surface, corresponding to 11,300 million cu. m. of ice, according to B. Morales-Arnao and INAGGA-CONAM.

In addition to the White Mountain Range, 19 mountain range glaciers exist in Peru, which according to a study elaborated by the Corporation For the Development of Santa in 1970, occupied 965 sq. mi. (2,500 sq. km.) These mountain ranges lodge 1,500 lagoon glaciers. The White Mountain Range is one of the most sensible thermometers to measure global warming, showing that in the last half-century the Peruvian mountains have lost hundreds of cubic meters of ice. Peru was hit in 1998 with the destruction of the Machu Picchu hydropower station, and rebuilding costs amounted to $160 million, according to RGS Ltd.

Glacial lakes are formed on the glacier terminus due to the climate change–induced glacier retreating processes. The majority of these glacial lakes are dammed by unstable moraines, which were formed by the glaciations of the cooler period of climate within the last millennium. Occasionally, the lake happens to burst and suddenly releases an enormous amount of its stored water, which causes serious floods downstream along the river channel. This phenomenon is known as glacial lake outburst flood (GLOF). Since 1702, more than 25 catastrophic GLOFs have occurred in the White Mountain Range. They come with little or no warning, and are made up of liquid mud that transports rolling stones and blocks of ice with the capacity to destroy cities and lives. The most serious GLOFs destroyed parts of the city of Huaraz in 1725 and 1941, as well as the GLOF from Lake Jancarurish in 1950. Additionally, two other destructive avalanches from the summit of North Huascarán destroyed, in 1962 and 1970, several villages and caused the deaths of more than 25,000 inhabitants.

SEE ALSO: Floods; Glaciers, Retreating; Glaciology; Peruvian Current.

BIBLIOGRAPHY. INAGGA-CONAM, *Vulnerabilidad de Recursos Hídricos de Alta Montaña* (1999); B. Morales-Arnao, "Estudios de ablación en la Cordillera Blanca," *Boletín* *del Instituto Nacional de Glaciología del Peru* (v.1/5, 1969); B. Morales-Arnao, "Estudio de la evolución de la lengua glaciar del Pucahiurca y de la laguna Safuna," *Boletín del Instituto Nacional de Glaciologiá del Perú* (v.1/6, 1969).

CARLOS SORIA
UNIVERSIDAD NACIONAL MAYOR DE SAN MARCOS

Peruvian Current

ORIGINATING IN THE frigid waters off the coast of Antarctica, the Peruvian Current moves north along the western coast of South America. When in reaches the continental shelf along South America, the current rises, carrying cold water with it to the surface of the Pacific Ocean. The prevailing winds of the South Pacific and Earth's rotation cause the Peruvian Current to rotate; the Coriolis Force causes the current to rotate clockwise. The Peruvian Current extends 125 mi. (201 km.) west from the coast of South America. As the current moves north through the coasts of Chile, Peru and Ecuador it splits into two masses where Cabo Blanco, Peru, meets the Gulf of Guyanquil. The main current turns west into the Pacific Ocean, while the remnant of the current moves along the coast of Ecuador. At that point, the second branch of the Peruvian Current also moves west, rejoining the main current near the Galapagos Islands.

The Peruvian Current is also known as the Humboldt Current after its discoverer, German scientist Alexander von Humboldt. The Peruvian Current affects Peru year round, and moderates the climate of Chile in spring and summer, when it displaces a subtropical center of high pressure. Ordinarily the coast of Chile would warm in spring and summer, but the onset of the Peruvian Current diminishes temperatures and forestalls any rain. The air that accompanies the current is dry, keeping the coast arid. Some weather stations along the Chilean coast have never recorded rainfall; others areas receive considerably less than 1 in. (2.5 cm.) of rain per year. The northern coast of Peru is dry from May to November, and receives light rain between December and April. Even though some areas of the coast are humid, rain nevertheless does not fall. The arid coastline supports few plants, and so sunlight either is absorbed by the land or radiates back into space. Rainfall along the southern coast of Ecuador totals 12 in. (30 cm.) per year,

though in the north, where the Peruvian Current weakens, rainfall increases tenfold. Some regions receive as many as 197 in. (500 cm.) of rain per year.

With Peru located near the equator, one might expect warm temperatures, but the Peruvian Current keeps the coast of Peru at 75 degrees F (24 degrees C). Lima varies from 70 degrees F (21 degrees C) in January and 50 degrees F (10 degrees C) in June. Areas inland from the current often record temperatures of 90 degrees F (32 degrees C). Periodically El Niño disrupts the Peruvian Current, bringing warm water from the tropical Pacific to the western coast of South America. Temperatures along the coast rise and rain falls on some parts of the coast.

As it cools the western coast of South America, the Peruvian Current creates a climate of unremitting dryness. Temperatures are moderate but rainfall is scant. Seabirds inhabit the western coast of South America but humans have only colonized the region in small numbers. The deserts of Chile are especially forbidding. Without rain, the land ceases to sustain plant life. In contrast to the sterility of the desert, life abounds in the ocean. The Peruvian Current carries plankton to the surface of the ocean, and fish feed on it in large numbers. Seabirds in turn feed on the fish. Despite creating an arid climate, the Peruvian Current teems with life.

SEE ALSO: Climate; Coriolis Force.

BIBLIOGRAPHY. Perry Cohen, *Geographical Aspects of the Peruvian Coastal Current* (Thesis Paper, 1950); Kirill Ya Kondratyev and Vladimir F. Krapivin, *Global Environmental Change: Modeling and Monitoring* (Springer, 2002).

CHRISTOPHER CUMO
INDEPENDENT SCHOLAR

Pew Center on Global Climate Change

IN 1998, WITH a grant from the Pew Charitable Trusts, and within the context of increasing concern over the impact of climate change, the Pew Center on Global Climate Change was established. It is a nonprofit, nonpartisan, independent organization. The role of the Pew Center is to build a bank of credible information about climate change and its impacts, and to try to provide solutions to the problems created by climate change. The Pew Center aims to provide an objective forum for research, and development of sensible solutions and policy suggestions.

Underpinning this mission is the commitment by Pew Center staff to educate policymakers and the general community about the causes and consequences of climate change. In so doing, the Pew Center has commissioned many significant reports on the impacts of climate change in various contexts, including domestic and international policy, economic and environmental impacts, and practical solutions. For example, reports have focused on the relationship between developing and developed worlds in relation to climate change policies. Many reports highlight the effects of climate change on marine ecosystems. The Pew reports offer policymakers and the public the opportunity to access and learn about detailed case studies from across the world on this topic. The Pew Center also hosts conferences and workshops on climate-related topics to stimulate engagement and interaction between business, government, and nongovernmental organizations. Staff from the Pew Center also regularly participate and attend key meetings on climate change issues, such as negotiations on the UN Framework Convention on Climate Change (UNFCC).

Within the policy center, staff at the Pew Center for Global Climate Change investigate how policies can be changed and developed within the areas of science and technology, market-based mechanisms, adaptation, international engagement, cross-sector policies, transportation, manufacturing agriculture, and energy production and use policies. Staff members also work within the United States at congressional and state levels to encourage policymakers, and politicians in these areas, to activate plans and legislate on climate change solutions and issues.

UNDERSTANDING CLIMATE CHANGE THROUGH INTERNATIONAL DIALOGUE WITH BUSINESS

The Pew Center's position on climate change is that there is strong scientific consensus that climate change is real, happening now, and is largely the result of human-induced activities. For example, the Pew Center cites the emissions of carbon dioxide and other greenhouse gases produced by industrial processes, fossil fuel combustion, and changes in land use as the primary cause of climate change. The result of these activities is that there will be an additional warming over the 21st century, and that there will be a global increase of 2.5 degrees F (1.4 degrees C) to a global average of 10.4 degrees F (5.7 degrees C). The Pew Center identifies a number of consequences from global warming, including sea-level rise, coastal inundation, beach erosion, flooding from coastal storms, changes in precipitation patterns, increased risk of droughts and floods, threats to biodiversity, and potential public and environmental health problems.

At an international level, the Pew Center is working to develop an international forum for agreement on climate change through the Climate Dialogue at Pocantico, established in 2005. This was a forum convened by the Pew Center on Global Climate Change to give senior policymakers the forum to informally discuss core issues with other policymakers and stakeholders in an international context. These meetings, held between July 2004 and September 2005, resulted in an agreement between the 25 international participants; the dialogue was then presented to the wider community as a platform for initiated change and options for promoting the climate change effort in an international arena. Participating countries and members include Argentina, Australia, Brazil, Canada, China, Germany, Japan, Malta, Mexico, Tuvalu, the United Kingdom, and the United States; senior executives from Alcoa, BP, DuPont, Eskom (South Africa), Exelon, Rio Tinto, and Toyota; and experts from the Pew Center, the Energy and Resources Institute (India), and the World Economic Forum.

The Pew Center has also established the Pew Center's Business Environmental Leadership Council (BELC). Business interests represented include diversified manufacturing, oil and gas, transportation, utilities, and chemicals. The BELC was created when the Pew Center was established, due to the belief that engaging business must be one of the cornerstones for effecting change in relation to addressing climate change impacts. This council is based on the understanding that businesses that are proactive in climate-proofing their business, and minimizing their impact, will have a future competitive advantage. To date, the BELC is the largest consortium of businesses working together to combat climate change, with a membership of 44 members. Together, the membership represents $2.8 trillion in market capitalization and over 3.8 million employees.

These companies are providing leadership by negotiating the establishment of emissions reduction objectives; through investment in new best practice and climate friendly technologies; and by supporting programs designed to achieve cost-effective emissions reductions. For example, 32 of the companies represented within the BELC have greenhouse gas reduction targets. Many of these companies are also implementing strategies in energy supply, energy demand solutions, process improvement, waste management practices, transportation, carbon sequestration and offsets solutions, and emissions trading and offsets.

Overall, the Pew Center advocates that the reduction of emissions of carbon dioxide and other greenhouse gases is the crucial first step to addressing this problem. To do so, it proposes a restructuring of the global economy away from its current reliance on fossil fuels to increased dependence and use of renewable and more efficient energies.

SEE ALSO: Carbon Emissions; Nongovernmental Organizations (NGOs); Policy, International.

BIBLIOGRAPHY. Pew Center on Global Climate Change, www.pewclimate.org (cited September 2007).

MELISSA NURSEY-BRAY
AUSTRALIAN MARITIME COLLEGE
ROB PALMER
RESEARCH STRATEGY TRAINING

Philippines

THE PHILIPPINES ARE located across 7,107 islands forming an archipelago in southeastern Asia, east of Vietnam. The country lies squarely in the typhoon belt, and is affected by an average of 15 storms (five to six of them direct strikes) annually. Most of the population of 91 million is at risk from landslides, flash flooding, and tsunamis. The Philippines face severe environmental problems, ranging from water pollution, soil degradation, deforestation, and loss of coral habitat, and are expected to suffer from rising sea levels and increased storm intensity.

The Philippines have already been feeling the impact of oscillations in Pacific sea temperatures known as El Niño and La Niña. In El Niño periods, the islands tend to see a decrease in rainfall, leading to droughts. During La Niña periods, rainfall can come in intense bursts, leading to landslides and flash flooding. Supertyphoons are also on the rise, with 2007 bringing a number of category 4 and category 5 cyclones into the Pacific basin.

Deforestation, both in the rainforest and the country's vast mangrove swamps, is another serious environmental issue for the Philippines. After a series of devastating landslides in the 1980s, the government instituted a ban on timber harvesting. Still, the country lost a third of its forest cover to logging 1990–2005. Deforestation has slowed to 2 percent annually, and the government has been more aggressive about pursuing illegal loggers.

With over 36,000 km. of shoreline, a rising sea level will have devastating consequences for the Philippines. Some models show a 100 cm. rise in ocean levels on the Philippine coast by 2080. In the heavily populated capital city of Manila, this rise would potentially displace up to 2.5 million people, and put millions more at risk for flooding from storm surges.

The Philippines is home to southeast Asia's second-largest coral reef system, covering 9,676 sq. mi. (25,060 sq. km.) As temperatures rise, coral bleaching events are expected to become more severe, leading to the eventual death of the reef. The impact will be felt in several ways: it will signal the end of an important habitat; it will destroy the livelihood of more than a million fishermen along the coast; it will reduce an important source of food from the Philippine market; and it will make the coast more vulnerable to storm surge and rising tides, as bleached reefs have been shown to be less of a buffer to sea level rise.

Carbon emissions rose 40 percent between 1990–98, but with CO_2 emissions at 1,000 metric tons per capita in 1998, the Philippines is not a significant contributor to global carbon emission totals. An estimated 73 percent of emissions come from liquid fuels, 18 percent from solid fuels, and 9 percent from cement manufacturing. Climate change has already cost the Philippines billions in lost agricultural production, reduced fish yields, storm damage, and a reduction in tourism. The government is taking steps to educate the public and to institute sustainable policies.

SEE ALSO: Deforestation; El Niño and La Niña; Hurricanes and Typhoons; Sea Level, Rising.

BIBLIOGRAPHY. "Climate Change Scenarios For the Philippines," Climatic Research Unit, University of East Anglia, www.cru.uea.ac.uk (cited October 2007); "Crisis or Opportunity: Climate Change in the Philippines," Greenpeace, www.greenpeace.org (cited October 2007); "Philippines Loses Billions to Climate Change—Philippines Today," Philippines Today, www.philippinestoday.net (cited October 2007).

HEATHER K. MICHON
INDEPENDENT SCHOLAR

Phillips, Norman

NORMAN PHILLIPS IS a theoretical meteorologist who pioneered the use of numerical methods for the prediction of weather and climate changes. His influential studies led to the first computer models of weather and climate, as well as to an understanding of the general circulation of the atmosphere, including the transports of heat and moisture that determine the Earth's climate. His 1955 model is generally regarded as a ground-breaking device that helped to win scientific skepticism in reproducing the patterns of wind and pressure of the entire atmosphere within a computer model.

Phillips received his B.S. from the University of Chicago in 1947 and his Ph.D. in 1951. He was the first to show, with a simple general circulation model, that weather prediction with numerical models was possible. The advent of numerical weather predictions in the 1950s also marked the transformation of weather forecasting from a highly individualistic effort to a cooperative task in which teams of experts developed complex computer programs. With the first digital computer in the 1950s, scientists tried to represent the complexity of the atmosphere and its circulation in numerical equations. Nineteenth- and early 20th-century mathematicians such as Vilhelm Bjerknes and Lewis Fry Richardson had failed to come up with adequate mathematical models. Through the 1950s, some leading meteorologists tried to replace Bjerknes and Richardson's numerical approach with methods

based on mathematical functions, working with simplified forms of the physics equations that described the entire global atmosphere. They succeeded in getting only partial mathematical models. These reproduced some features of atmospheric layers, but they could not show persuasively the features of the general circulation. Their suggested solutions contained instabilities as they could not account for eddies and other crucial features. Discouraged by such failures, scientists began to think that the real atmosphere was too complex to be described by a few lines of mathematics. The comment of such a leading climatologist as Bert Bolin is revealing of this skepticism. In 1952 Bolin argued that there was very little hope for the possibility of deducing a theory for the general circulation of the atmosphere from the complete hydrodynamic and thermodynamic equations. Yet, computers opened up new possibilities in the field, although the first digital specimens were extremely slow and often broke down.

Jule Charney was the first to devise a two-dimensional weather simulation. Dividing North America into a grid of cells, the computer started with real weather data for a particular day and then solved all the equations, working out how the air should respond to the differences in conditions between each pair of adjacent cells. It then stepped forward using a three-hour step and computed all the cells again. The system was slow to operate and it had imperfections, yet its completion paved the way for more researches to be carried out. It was Norman Phillips who sought to address the problems in Charney's model. The challenge for meteorologists now became the computation of the unchanging average of the weather given a set of unchanging conditions such as the physics of air and sunlight and the geography of mountains and oceans. This was a "boundary problem." A parallel problem that they had to face was that of the "initial value," where the operation of calculating how the system evolves from a particular set of conditions found at one moment becomes less accurate as the prediction moves forward in time.

Phillips was inspired by "dishpan" experiments carried out in Chicago, where patterns resembling weather had been modeled in a rotating pan of water that was heated at the edge. For Phillips this showed that "at least the gross features of the general circulation of the atmosphere can be predicted without having to specify the heating and cooling in great detail."

Phillips argued that if such an elementary laboratory system could effectively model a hemisphere of the atmosphere, a more advanced tool such as a computer should be able to do it as well. Although certainly more advanced than a dishpan, Phillips's computer was still quite primitive. Thus, his model had to be extremely simple. By mid-1955 Phillips had devised improved equations for a two-layer atmosphere. To avoid mathematical difficulties, his grid covered not a hemisphere but a cylinder, 17 cells high and 16 in. in circumference. The calculations allowed the representation of a plausible jet stream and the evolution of a realistic-looking weather disturbance over a period of a month.

This settled an old controversy over what procedures set up the pattern of circulation. The simulation-based approach became the generally accepted method to devise circulation models. For the first time scientists could visualize how giant eddies spinning through the atmosphere played a key role in moving energy from place to place. Phillips's model was quickly hailed as a "classic experiment," the first true general circulation model (GCM). Phillips used only six basic equations (PDEs) which have been since described as the "primitive equations". They are generally conceived of as the physical basis of climatology. These equations represent well-known physics of hydrodynamics. The model was able to reproduce the global flow patterns of the real atmosphere. Phillips was awarded the Benjamin Franklin Award in 2003.

SEE ALSO: Atmospheric General Circulation Models; Bolin, Bert; Computer Models; Richardson, Lewis Fry.

BIBLIOGRAPHY. Edward N. Lorenz, *The Global Circulation of the Atmosphere* (Princeton University Press, 2007); Spencer Weart, *The Discovery of Global Warming* (Harvard University Press, 2004).

LUCA PRONO
UNIVERSITY OF NOTTINGHAM

Phytoplankton

PHYTOPLANKTON ARE A group of free floating microscopic organisms predominantly classified as algae. Over 4,000 species of phytoplankton have been identified and this list is rapidly growing. Phytoplankton are mostly single cellular organisms and all are autotrophic (i.e., they contain photosynthetic pigments). These pigments allow phytoplankton to use the sun's energy to convert CO_2 and inorganic nutrients, through photosynthesis, into biological molecules such as proteins and carbohydrates. This process of creating new biological molecules is called primary production. Phytoplankton are the only primary producers in the open oceans, and thus form the basis of the food chain in over 70 percent of the world's surface area.

Specific aspects of primary production are often considered more carefully and these warrant definition. The Net Primary Production (NPP) refers to the amount of organic carbon available after respiration has been subtracted from the total amount of photosynthesis. It is an important term because this is the amount of carbon that is available to the rest of the foodweb, and is the upper limit on respiration. The Net Community Production (NCP) is the difference between net primary production and heterotrophic respiration. It is measured by gross changes in oxygen or biomass in a specific time. The New Production is the fraction of primary production driven by newly-available nitrogen. This is principally the nitrate and nitrite that becomes available when deep ocean waters are brought up into the euphotic zone, but could include sources from the atmosphere or river inputs.

The global distribution of phytoplankton throughout the world's oceans is not evenly distributed. This distribution is the result of growth being limited by the availability of nutrients. High concentrations of phytoplankton are seen in upwelling areas, such as the Benguela Current off the southwest coast of Africa.

PHYTOPLANKTON AND CLIMATE CHANGE

CO_2 does not limit phytoplankton growth. CO_2 is not thought to limit phytoplankton growth in any region of the ocean. Primary production, in general, is limited by the availability of inorganic nutrients. Therefore it is not expected that increased atmospheric concentrations of CO_2 will have a significant impact on the phytoplankton population in the oceans.

The "Biological Pump" Locks Away CO_2. The "Biological Pump" is the mechanism by which anthropogenic CO_2 is taken from the atmosphere and stored in the deep ocean, which transfers carbon from the surface to the deep ocean, across the barrier of the permanent thermocline. The pump is powered by phytoplankton fixing carbon and sinks to the deep ocean. This removes CO_2 from the surface ocean, causing more CO_2 to be drawn from the air to maintain equilibrium. Carbon is locked away from the atomosphere as a result of being taken below the permanent thermocline. This contrasts to primary productivity on land which builds plants (such as trees) and in turn animals to, locking CO_2 away from the atmosphere.

Although increasing levels of CO_2 do not have a fertilizing effect on the oceans, the longer growing season in temperate and polar regions is expected to lead to an increase in primary productivity.

BIBLIOGRAPHY. L. Bopp, et al., "Will Marine Dimethylsulfide Emissions Amplify or Alleviate Global Warming,?" *Can. J. Fish. Aquat. Sci.* (v. 61, 2004); C.M. Lalli and Timothy R Parsons, *Biological Oceanography, 2nd Edition,* (Oxford, 1997).

CARL PALMER
INDEPENDENT SCHOLAR

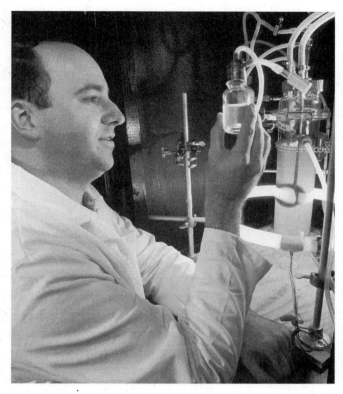

Microbiologist Kevin Schrader examines a culture of green algae (Selenastrum capricornutum).

Plants

THE ANCIENT GREEKS were the first to identify a relationship between climate and plants. Following this insight, other naturalists recognized that fossilized plants revealed the climate in prehistory. Plants colonized the land 410 million years ago, shaping the climate as they spread throughout the globe. By absorbing carbon dioxide, plants have the potential to cool the climate. By releasing water vapor into the atmosphere, plants have the potential to warm the climate by trapping heat, or to cool the climate by forming clouds. Rooted to the ground, plants must either adapt to the climate or die. Sudden climate changes threaten some species with extinction, whereas hardier species survive. From equator to pole, climate determines what plants grow at given latitudes.

HISTORY OF THE RELATIONSHIP BETWEEN CLIMATE AND PLANTS

Theophrastus, a pupil of Aristotle and the founder of botany, may have been the first to ponder the relationship between plants and climate. He understood that each species of plant is adapted to a particular climate, and that in a foreign climate a plant will not thrive and may not survive. Plants are thus an indicator of climate. The mangrove, for example, is an indictor of a climate wet enough to form swamps. With the work of Theophrastus, the promising synthesis of botany and the study of climate was in its infancy, but the Romans did not bring this synthesis to maturity. The Romans were a practical people, with no interest in the theoretical relationship between plants and climate. The Middle Ages were no better for the study of the relationship between plants and climate. The emphasis on theology undercut any progress in science.

In 1876, Norwegian botanist Axel Blytt revived Theophrastus's notion that plants are an indicator of climate. Working on prehistoric climates, Blytt identified fossils of trees that no longer grew in Denmark. From this observation, he posited that the climate in Denmark had once been suitable for these species of trees, but was no longer. The climate was not therefore a static, unchanging entity. Rather, the climate has changed over time. Working in a similar vein, Swedish geologist Ernst Von Post identified fossils of *Alnus*, a genus of tree, in the strata of rocks throughout Europe. *Alnus* is adapted to a warm wet climate, and Von Post tracked the tree fossils as they migrated from southern to northern Europe at the end of the Cenozoic Ice Age. As the glaciers retreated, *Alnus* rooted itself in the warm wetlands that followed the ice age.

PLANTS AND CLIMATE CHANGE

Fossilized pollen can likewise pinpoint changes in climate. The evergreen red beech is adapted to warm locales. The abundance of its pollen 8,000 years ago implies that this time correlates with the maximum temperature since the end of the Cenozoic Ice Age. On the other hand, the abundance of pollen from *Phyllocladus*, a shrub that is adapted to the cold, at 26,000 years ago and again at 20,000 years ago, indicates these years as the coldest during the ice age.

Photosynthetic algae evolved in the ocean as early as four billion years ago. From this beginning, the first plants colonized land 410 million years ago, before animals. By 360 million years ago, the beginning of the Carboniferous period, plants had spread throughout the planet, forming lush forests. The temperature and amount of carbon dioxide in the atmosphere were both higher than today. So lush was the growth of plants that when they died, they formed one layer upon another. Under heat and pressure, these many layers of plants formed the vast deposits of coal, natural gas, and petroleum that humans are now exploiting. The immensity of these deposits underscores the massive growth of plants during the Carboniferous period.

The fact that plants absorb carbon dioxide during photosynthesis has an important consequence for the climate. Carbon dioxide correlates with temperature. A high concentration of carbon dioxide correlates with high temperatures, whereas a low concentration of the gas correlates with low temperatures. This relationship holds true, because carbon dioxide is a greenhouse gas: it traps sunlight that reflects from Earth, preventing light, in the form of infrared radiation, from returning to space. In trapping sunlight, carbon dioxide traps heat, thereby increasing the temperature of the atmosphere. Plants lower the concentration of carbon dioxide by absorbing it during photosynthesis. When plants absorb carbon dioxide faster than Earth produces it through volcanism, the concentration of carbon dioxide diminishes, and temperatures decline. In this context, plants may have contributed to the onset of the Ice Ages by absorbing carbon dioxide.

The absorption of carbon dioxide is cyclical. In spring and summer, when plants grow vigorously, they absorb large amounts of carbon dioxide. In autumn, however, plant growth slows and in winter it stops. The decay of dead plants in autumn returns the carbon dioxide that they had absorbed while alive to the atmosphere. The Northern and Southern hemispheres contribute to the carbon dioxide cycle in opposite fashion, for when plants are growing vigorously in the Northern Hemisphere, they are dead and decaying in the Southern Hemisphere, and vice versa.

In another sense, the carbon dioxide cycle spans eons of time. In the Carboniferous period, lush forests absorbed prodigious amounts of carbon dioxide, storing it in their tissues as sugars. Upon the death of plants, their conversion to coal, natural gas, and petroleum locked up these vast amounts of carbon dioxide. Since the Industrial Revolution, humans have burned these fossil fuels for energy, and in the process liberated carbon dioxide, returning to the atmosphere the carbon dioxide that plants had absorbed during the Carboniferous period. The liberation of this carbon dioxide has increased global temperatures. Humans are burning fossil fuels constantly, leading climatologists to predict that humans will, by 2080, double the amount of carbon dioxide in the atmosphere, further increasing temperatures and perhaps leading to the flooding of coastal cities.

Plants absorb carbon dioxide through their stomata, pores on the leaves. At the same time, plants shed water through their stomata. Plants transpire as water vapor more than 90 percent of the water that they absorb through their roots. Through transpiration, plants change the climate, though not in a straightforward way. On the one hand, water vapor is a greenhouse gas, absorbing as heat the sunlight reflected from Earth.

The warmth of the Carboniferous period, a time of abundant plant growth, was due to the greenhouse effect. Along with carbon dioxide, the water vapor transpired by plants contributed to a warm climate. On the other hand, the water vapor that plants transpire forms clouds, which reflect 30 percent of sunlight back into space before it can heat the Earth. In their role in forming clouds, plants cool the climate.

One scientist predicted that a doubling of carbon dioxide concentration might cool, rather than heat Earth, because plants will, in transpiring water vapor, hasten the formation of clouds.

CLIMATE CHANGES AND THE GREENHOUSE EFFECT

The product of a long evolutionary history, plants are adapted to the climate in a way that humans are not. Humans fashion their material culture to suit the climate, or, when climate worsens, they migrate to a more hospitable locale. Plants, rooted to the ground, must adapt or die. Migration is not an option, though seed dispersal is a kind of intergenerational migration. Through dispersal, seeds travel roughly one kilometer per year, a rate too slow to adapt a plant to rapid climate change. As the climate deteriorates, some plant species die out, to be replaced by hardier species. In Neolithic and Bronze Age Denmark, for example, the climate was warmer and wetter than it is today.

The climate was warm and wet enough to sustain the growth of oak trees. In the Iron Age, however, temperatures dropped to current figures. Unable to cope with the decline in temperature and moisture, oak trees died out and were replaced by grasses and heather. In New Hampshire, the tree *Pinus strobus* was for centuries the most numerous tree in the region's deciduous forests. *Pinus* came through the fire of 1665 unscathed, but was unable to cope with climatic catastrophe. In 1921, a tornado, and in 1938 a hurricane, swept through the forests, felling large trees of several species. The catastrophes wiped out *Pinus*, ending centuries of its dominance, and leaving other species to reconstitute the forests.

These sudden changes in climate and flora are dramatic and easy to quantify. No less dramatic has been the effect of hydrofluorocarbons on climate and plants. Humans have released large quantities of hydrofluorocarbons into the atmosphere, where they have thinned the ozone layer. Consequently UV-B radiation, a type of ultraviolet light, penetrates the ozone layer in greater amounts than in the past. UV-B radiation damages half of all plant species. Damaged plants grow small leaves and short shoots and photosynthesize at a slow rate. These effects are magnified by the fact that several crops are among the plants sensitive to UV-B radiation. One study concluded that a 25 percent reduction in the ozone layer would halve soybean yields.

The tropics support lush vegetation, with more than 80 in. of rain per year, temperatures nearly uniform and warm year-round, and abundant sunshine. Trees have thin bark, for they don't need insulation against the cold or protection against water loss. Trees grow in layers, with those in the innermost layers able to survive without exposure to direct sunlight. Because sunlight does not penetrate to the rainforest floor, little vegetation grows along the ground. At higher latitudes north and south of the tropics, rainfall diminishes and temperatures vary year round. These climates have seasons, with vigorous plant growth in spring and summer and dormancy in autumn and winter.

To cope with a diminution in rain, plants in temperate climates have evolved small leaves to minimize the loss of water through transpiration. In areas that have a dry season, trees evolved the shedding of leaves to stop transpiration and, in other species of trees, the growth of needles rather than leaves to minimize transpiration. With their needles, the evergreens and conifers are adapted to short summers because they can carry out photosynthesis as soon as temperatures warm, whereas deciduous trees must regrow their leaves before they can photosynthesize. Evergreens and conifers grow where winter is cold enough to freeze the ground. Once the ground freezes, roots cannot absorb water, making winter a period of drought and favoring trees that can minimize transpiration in response to frost-induced drought.

In contrast to the rainforest, sunlight penetrates to the forest floor in temperate forests, permitting the growth of plants, often grasses, in abundance along the ground. As the dry season lengthens and rainfall diminishes still further, forest gives way to grassland. Grasses need less water than trees. Trees grow alongside grasses on the African savanna, but few trees grow on the Russian steppe. There, grasses are the dominant flora. At high latitudes, temperatures fall below 40 degrees F (4 degrees C) for six to nine months per year. Summers are brief, with temperatures above 50 degrees F (10 degrees C). Rainfall ranges between 10–40 in. (25–102 cm.) per year, with between 15–24 in. (38–61 cm.) typical. This climate favors the growth of coniferous forests. In addition to their needles, conifers have thick bark to retard water loss and to protect against the cold. Temperature separates coniferous forest from tundra. Where summer temperatures exceed

50 degrees F (10 degrees C), conifers predominate, but wherever summer temperatures fall below 50 degrees F (10 degrees C), tundra results. Grasses and sedges are the tundra flora. During the 50 or 60 days of summer, the sun melts a thin strip of ground. Free from the grip of ice and benefiting from the nearly continuous sunlight of summers in high latitudes, plants grow vigorously and then are dormant for the long, bitter winter.

EFFECTS OF CLIMATE CHANGE ON PLANTS

The increase in temperatures that is the likely outcome of the greenhouse effect will affect plants. By one estimate, a two or three degree F increase in temperature will raise crop yields in the temperate zone, though an increase above 3 degrees F (1.7 degrees C) will decrease yields. Any temperature increase will likely reduce yields in the tropics, where crops are already at their maximum heat tolerance.

The climate of the future is sure to affect plants. Despite predictions to the contrary, the increase in carbon dioxide will likely increase temperatures. One study suggests that a doubling of carbon dioxide in the atmosphere will triple the growth rate of plants and trees. Forests will grow more densely and will extend their range to higher latitudes. Plants will grow more vigorously on marginal land. Another study indicates that a doubling of carbon dioxide will shift temperate forests 310 to 621 m. (500 to 1,000 km.) north in the northern hemisphere, and south in the southern hemisphere. The concentration of carbon dioxide is likely to double by 2080, but trees are not likely to migrate so far so fast. Global warming therefore endangers temperate forests. As the climate warms, trees will advance north and south, taking over ground that had been tundra. As temperatures rise, dead plants will decay more rapidly, liberating still more carbon dioxide into the atmosphere.

The cutting down of forests will surely harm the plants that survive. The amount of rainfall will decrease as forests are cut down. With fewer forests, the rate of transpiration will diminish. Whereas forests absorb sunlight, bare ground reflects sunlight back into space. A 15 percent decline in rainfall would replace the forests of South America with grassland. A 30 percent decrease in rainfall would replace the forests of Zaire with grasses. A 70 percent decrease in rainfall would make the Amazon basin a desert.

The climate of the future may imperil many plant species, but as a kingdom, plants are resilient. They have survived the Ice Ages and the predation of herbivores in warm climates. Humans are fortunate that plants are so adaptable, for with their agriculture, humans are dependent on plants. Life would not exist without the diversity of plants. The most numerous form of terrestrial life during the Carboniferous period, plants occupy every biome. Even in deserts, their seeds lie dormant, awaiting the infrequent rains. Plants have adapted to every climate, from the tropics to frigid tundra. Even bodies of water are home to plants. The survival of plants depends on their ability to adapt to the climate of the future. The survival of the rest of the biota depends on the success of plants.

SEE ALSO: Climate; Cretaceous Era; Greenhouse Effect; Greenhouse Gases.

BIBLIOGRAPHY. Institute for Biospheric Research, *The Greening of Plant Earth: the Effects of Carbon Dioxide on the Biosphere* (Western Fuels Association, 1991); D.C. Money, *Climate, Soils and Vegetation* (University Tutorial Press, 1965); J.I.L. Morison and M.D. Morecroft, eds., *Plant Growth and Climate Change* (Blackwell Publishers, 2006); Jelte Rozema, Rien Aerts, and Hans Cornelissen, eds., *Plants and Climate Change* (Springer, 2006); F.I. Woodward, *Climate and Plant Distribution* (Cambridge University Press, 1987).

CHRISTOPHER CUMO
INDEPENDENT SCHOLAR

Pleistocene Era

THE INCREASING FREQUENCY and intensity of glacial-interglacial cycles toward the end of the Pliocene (1.806–5.332 million years ago) set the stage for the Pleistocene epoch (11.8 thousand years ago–1.806 million years ago), which is the final phase of the Quaternary period. Some argue that the lower Pleistocene boundary may be set too late because the general trend toward significant cooling and glaciation had begun in the mid-late Pliocene (2.75 million years ago). Hence, the term Plio-Pleistocene may be used to delineate this transitional phase between the two epochs.

Strong glacial-interglacial phases are the key climatic features that characterize the Pleistocene epoch and have shaped much of the modern landscape. Glacial stages may be referred to as ice ages, and are used to describe a period of extensive ice sheet presence in the polar, high latitude continental, and alpine regions. Glacial phases are synonymous with reduced global temperatures. Quaternary glacial-interglacial cycles occurred with a 41,000-year periodicity, starting in the late Pliocene (2.75 million years ago) to mid-Pleistocene (1.11 million years ago), followed by a 100,000 year cycle in the mid- to late-Pleistocene. The most intensely studied glacial stage during the Pleistocene is the last glacial maximum (21,000 years ago).

Marine fossil material and isotopic proxies were used to simulate sea surface temperatures, sea ice, continental ice sheets, and albedo during the last glacial maximum, with results indicating that high latitudes in the northern hemisphere cooled by 7–11 degrees F (4–6 degrees C), while simulated sea temperatures increased by 2–5 degrees F (1–3 degrees C) in the Pacific and Indian oceans. Most recent evidence suggests that with the exception of Central America and the Indo-Pacific, the climate was much drier than today, due to the combination of reduced evaporation, greater coverage of land surfaces by ice sheets, and wind anomalies.

Glaciation was most extensive in the northern hemisphere, with 2–2.5 mi.- (3–4 km.-) thick ice sheets covering Canada and parts of the northern United States, Greenland, northern Europe, Russia, and perhaps to a lesser extent, the Tibetan Plateau. In the southern hemisphere, the glaciation of Antarctica that began in the Pliocene continued through to the last glacial maximum, the Andes were highly glaciated, the Patagonian Ice Sheet covered much of southern Chile, and small glaciers formed in Africa, the Middle East, and southeast Asia, where simultaneously, deserts were expanding. Sea levels may have been up to 426.5 ft. (130 m.) lower than today. The hydrologic and geological consequences of the last glacial maximum and other glacial stages are still evident, particularly at the higher latitudes of the northern hemisphere, where the abundance of fresh water is effectively the result of glacial retreat and runoff. Remnants of Pleistocene glaciers also remain in high-altitude tropical localities such as on Mount Kilimanjaro and the Peruvian Andes, but these glaciers are quickly retreating.

The causes of the Pliocene-Pleistocene glacial-interglacial cyclicity are largely attributed to climate forcing caused by variations in the Earth's orbital parameters (Milankovitch cycles), but the sequence of events is difficult to establish. However, there is strong evidence that greenhouse gas levels fell at the start of glacials and rose during the interglacial retreat of the ice sheets. So far, eight glacial cycles have been identified from cores in Antarctica dating back to 740,000 years ago, but currently, it is the Vostok ice core dating back to 420,000 years ago that provides the clearest perspective on the link between greenhouse gases and sea surface temperatures over the last four glacial-interglacial cycles. CO_2 concentrations fall between 180–200 ppm during the coldest glacial periods, and 280–300 ppm during full interglacials, while methane concentrations were approximately 350 ppb during glacials, and roughly twice that amount during interglacials. Current thinking is that Pleistocene changes in greenhouse gas levels were probably caused by disturbance to the sources of these gases, of which the oceanic and terrestrial sources were most significant.

During the last glacial maximum, the presence of large ice sheets over the high latitudes of the Northern Hemisphere significantly reduced the amount of exposed vegetation, and combined with low atmospheric CO_2 and other regional climatic changes, created biomes and vegetation assemblages that have no modern analogue. The Laurentide Ice Sheet completely covered Canada and the northern United States, with taiga, desert, and grassland ecotones dominating the mid-latitudes. At this time, woodland and shrub communities were also present, but highly fragmented. An exception to this is the Canadian and Alaskan Pacific coasts, where the continuity of woody flora remains largely unchanged from the last glacial maximum.

Substantial winter cooling reduced the global extent of tropics and subtropics and caused local extinctions, but equable areas may have acted as regional refuges for species that otherwise would have disappeared. The expansion of more arid ecosystems is well documented from pollen data showing that grasslands and shrub ecotones spread into previously tropical areas such as the Amazonian Basin, equatorial Africa and southern Asia. The persistence of rainforests in central North America and Indonesia during the last glacial maximum can be attributed to the consistently high rainfall in these regions. By contrast, over half of central Australia was desert, with tropical grasslands lying in the north, and scrub-woodland vegetation dominating the eastern and western regions.

Substantial evidence exists to support the hypotheses that Pleistocene fauna was dually affected by the climatic oscillations of the early-mid Quaternary, and the hunting activity of ancestral humans. The disappearance of species that had evolved in colder climates, such as the woolly mammoth, woolly rhinoceros, and musk ox, is most consistent with the appearance of humans in North America. By contrast, the Pleistocene extinction of Eurasian megafauna was likely due to climate. The disappearance of African and South American mammals is unresolved, but current evidence points to the arrival of humans as a key factor.

SEE ALSO: Glaciology; Ice Ages; Pliocene Era; Quatenary Era.

BIBLIOGRAPHY. K.D. Alverson, R.S. Bradley, and T.F. Pederson, *Paleoclimate, Global Change and the Future* (Springer-Verlag, 2003); A.D. Barnosky, et al., "Assessing the Causes of Late Pleistocene Extinctions on the Continents," *Science* (v.306, 2004); J.A. Van Couvering, *The Pleistocene Boundary and the Beginning of the Quaternary* (Cambridge University Press, 1997).

JARMILA PITTERMANN
UNIVERSITY OF CALIFORNIA, BERKELEY

Pliocene Era

THE PLIOCENE EPOCH is the uppermost subdivision of the Tertiary period (65.5 to 2.588 million years ago), and represents a geological stage from about 1.806 to 5.332 million years ago. Although the Pliocene was generally warmer than the present, this epoch is characterized by pronounced climatic oscillations that ultimately led to the characteristic cooling of the late Quaternary glacial-interglacial cycles. Pliocene climate data are inferred from oxygen isotope, dust, microfossil, and in some cases pollen data from cores collected under the flag of the Ocean Drilling Program (ODP), as well as terrestrial deposits. These records have allowed climatologists to refine the absolute chronology of the Pliocene epoch, and provide a continuous climatic record of global ice volume, sea surface temperatures, aridity, and terrestrial vegetation patterns.

The first Pliocene cooling event is documented at 4.5 million years ago, and was followed by variable, but persistent reductions in temperature after 3.6 million years ago. A brief period of warmth followed until 3.5 million years ago, at which time a second cooling event took place. A well-characterized mid-Pliocene warm period dates to approximately 3.3 to 3.15 million years ago, and is followed by the return to progressive cooling that culminated in the arrival of early northern hemisphere glacial-interglacial cycles about 2.75 million years ago. Significant growth of ice sheets did not begin in Greenland and North America until approximately three million years ago, following the formation of the Isthmus of Panama. Many agree that this final Pliocene cooling period set the stage for strongly developed glacial events of the Pleistocene (1.8 million to 11,550 thousand years ago) and thus represents a climatic stage that is most relevant to the climates of late Tertiary and early Quaternary.

The contemporary significance of the mid-Pliocene warm period lies in its utility as a model for future scenarios of global warming. This is because continental distributions and climate-indicative plant taxa are thought to have been very similar to today. Members of the Goddard Institute for Space Studies (GISS) and the PRISM (Pliocene Research, Interpretations and Synoptic Mapping) group have exploited these paleofeatures in their efforts to model global Pliocene climate and vegetation distributions. Average mid-Pliocene global sea levels are modeled at 33 to 82 ft. (10 to 25 m.) higher than today, due to reduced Greenland and Antarctic ice cover, while sea surface temperatures were approximately 6.5 degrees F (3.6 degrees C) warmer than at present day. Mid-Pliocene climate simulations generally indicated increased surface air temperatures, particularly during the winter, and increased annual rainfall, evaporation, and soil moisture. Pollen records from land-based cores are less chronologically accurate, but consistent with a 7–18 degrees F (4–10 degree C) warmer northern hemisphere climate, coupled with higher continental moisture levels. This is especially evident in high latitude regions such as the Arctic.

The PRISM group has used fossil and pollen data to document vegetation patterns across the globe during the mid-Pliocene warm period. Their work indicates extensive conifer and mixed forests in the mid-Pliocene Arctic, and generally more northerly distributions of the mixed deciduous forests of eastern North America. Interior North America was likely to be moister, and warmer than today, with evidence of lakes in southeastern California, Arizona, and Utah. Northern Europe was warmer and wetter, with a greater abundance of swamps and wetland areas. Little information exists about Central and South America, but the limited numbers of pollen studies are consistent with GISS climate models suggesting a warmer, wetter climate, with a greater abundance of steppe and prairie vegetation. The Australian mid-Pliocene warm period is poorly documented, but it is thought to be wetter than today, with broader distributions of forest flora. Regions of Antarctica were significantly warmer than today, so increased exposure of soils supported the presence of mixed beech forests.

The cause of the mid-Pliocene warming is uncertain, but some combination of CO_2 increase and change in ocean heat transport may have been responsible. Carbon isotopic data from deep-sea microfossils, coupled with GISS climate models, support the increased strength of thermohaline circulation during the mid-Pliocene, particularly with respect to North Atlantic deep water production. However, simulations where CO_2 is the single variable show that the proposed, realistic patterns of mid-Pliocene oceanic heat transports would only have been possible at CO_2 levels greater than 1,200 ppm. There is no evidence supporting such elevated CO_2 excursions, but some workers suggest that even the predicted minor increases up to 380 ppm, in combination with altered ocean heat transport, may have been enough to catalyze mid-Pliocene warming.

Early Pliocene fauna was transitional, favoring grazers over browsers, as grasslands and savannas expanded in central North America and Africa, thereby replacing woodlands and their associated fauna. Charismatic Pliocene fauna included mammoths, mastodons, camels, and hippopotamus in the mid-Northern Hemisphere latitudes, while large turtles and marsupials were found in the southern hemisphere. Pliocene high-Arctic fauna was primarily Eurasian, characterized by now extinct species of beavers, badgers, deer, and caniids, the presence of which is consistent with mixed-evergreen forest vegetation. The Pliocene deposits of eastern North America revealed mostly Eurasian fauna, most notably new species red panda.

Pliocene Africa, prior to 2.8 million years ago, was wetter than today, as evidenced by deposits of mangrove swamps and tropical forests, which retreated southward as desertification intensified. The western Sahara desert likely formed after 2.8 million years ago. The Pliocene is a particularly important time for the evolution and diversification of hominids. The aridity-humidity cycles that were related to the late Pliocene glacials-interglacials in the northern hemisphere in the Pliocene climate of Africa may have shaped hominid evolution by creating cyclic opportunities for species extinction and innovation.

SEE ALSO: Global Warming; Pleistocene Era; Quartenary Era; Tertiary Climate.

BIBLIOGRAPHY. M.A. Chandler, D. Rind, and R.S. Thompson, "Joint Investigations of the Middle Pliocene Climate II: GISS GCM Northern Hemisphere Results," *Palaeogeography, Palaeoclimatology and Palaeoecology* (v.9, 1994); E.S. Vrba, et al., *Paleoclimate and Evolution, with Emphasis on Human Origins* (Yale University Press, 1995); J.H. Wrenn, J.-P. Suc, and S.A.G. Leroy, *The Pliocene: Time of Change* (Publishers Press, 1999).

JARMILA PITTERMANN
UNIVERSITY OF CALIFORNIA, BERKELEY

Poland

LOCATED IN EASTERN Europe, Poland has a land area of 120,728 sq. mi. (312,679 sq. km.), with a population of 38,125,500 (2006 est.), and a population density of 320 people per sq. mi. (122 people per sq. km.). Some 47 percent of the country is arable, with a further 13 percent used as pastures and meadows, and 29 percent is forested.

Regarding electricity generation in the country, 98.1 percent comes from fossil fuels, mainly coal that is mined in many parts of the country, with 1.5 percent from hydropower. As a result of this, even though Poland is less industrialized than many other European countries, it has a high per capita rate of carbon dioxide emissions—9.1 metric tons in 1990, falling slowly to 8.0 metric tons per person by 2004. Some 57 percent of all carbon dioxide emissions in the country come from the production of electricity, with 17 percent from manufacturing and construction, 11 percent from transportation, and 11 percent for residential purposes. The reliance on coal has meant that 76 percent of Poland's carbon dioxide emissions have been from solid fuels, with 15 percent from liquid fuels, and 7 percent from gaseous fuels.

The rising average temperatures in Poland as a result of global warming and climate change have caused hot summers in Lesser Poland, a region in the south of the country. Poland has been actively involved in various schemes to introduce carbon trading, and has even managed to reduce its own emissions rate, although it is hoping to cut back further. As a result, Poland has tried to follow a project that was developed by the Global Environment Facility, by which Mexico and Norway managed to reduce their power use through widespread introduction of compact fluorescent lamps in two major cities. In the case of Poland, this would also involve the conversion of coal-fired boilers to use gas. The main problem with this has been the political power of the coal-mining areas, which has hindered many attempts to reduce the dependence on coal.

The Polish government took part in the UN Framework Convention on Climate Change signed in Rio de Janeiro in May 1992. It signed the Kyoto Protocol to the UN Framework Convention on Climate Change on July 15, 1998, committing to a 3 percent reduction prior to ratification, which took place on December 13, 2002, with it entering into force on February 16, 2005.

SEE ALSO: Coal; European Union; Global Warming.

BIBLIOGRAPHY. Barbara Hicks, *Environmental Politics in Poland: A Social Movement between Regime and Opposition* (Columbia University Press, 1996); Gordon Hughes and Julia Bucknall, *Poland: Complying with E.U. Environmental Legislation* (World Bank Publications, 2000); Tim O'Riordan and Jill Jäger, eds., *Politics of Climate Change: A European Perspective* (Routledge, 1996); "Poland—Climate and Atmosphere," www.earthtrends.wri.org (cited October 2007); Farhana Yamin, ed., *Climate Change and Carbon Markets: A Handbook of Emissions Reduction Mechanisms* Earthscan, 2005).

JUSTIN CORFIELD
GEELONG GRAMMAR SCHOOL, AUSTRALIA

Polar Bears

THE POLAR BEAR (order Carnivora, family Ursidae) is the largest bear species and is thought to have evolved from brown bears, *Ursus arctos*, approximately one million years ago. There are 19 recognized populations, distributed in Canada, the United States (Alaska), Norway (Svalbard Islands), Denmark (Greenland), and Russia. The current estimated worldwide population is 20,000–25,000.

Polar bear territories can cover tens of thousands of sq. km. They live solitarily, but often congregate around food sources. Their diet consists primarily of ringed seals, but includes other seals, walruses, and beluga whales. They get the majority of their nutritional intake in the spring and summer from seal pups, which can be as much as 50 percent fat. Polar bears can swim for 37 mi. (60 km.) without resting, at speeds up to 6.2 mi. per hour (10 km. per hour). They possess several adaptations for a semi-aquatic existence, including partially webbed front paws, eyes adapted to see underwater, and a thick fat layer (eight to 12 cm.) that provides buoyancy and insulation in water.

Polar bears are also highly adapted to the Arctic climate. Arctic temperatures can drop to minus 49 degrees F (minus 45 degrees C) for days or weeks, but polar bears can withstand this due to their thick fat layer, a dense undercoat of fur with longer guard hairs, and black skin (which absorbs heat from sunlight). Other Arctic adaptations include white-seeming fur for camouflage (the hairs are actually colourless and hollow), small ears and tail (which reduce heat loss), large furry feet (which act like snowshoes), and a digestive system very effective at absorbing and storing fat.

Females become sexually mature at four to five years of age. Males may not mate successfully until they are 8 to 10 years old. Mating occurs from April to June and each male may mate with more than one female. The females have induced ovulation, mating multiple times causes the release of an egg. The implantation of the blastula (the fertilized egg after several cell divisions) is delayed until September/October, and in November/December the female excavates a den in the snow. She eventually gives birth in December or January. One to three cubs are born (two-thirds of cubs are twins) and they are nursed in the den until March/April, when they emerge. Cubs are weaned at the age of two or three years. During

A reduction of sea ice would result in the loss of polar bear habitat. It would also affect their access to food and to mates, and disrupt their migration routes. Many drowned polar bears have been reported, the result of larger areas of open water between ice floes.

this period, they learn important hunting skills from their mother. Occasionally, cubs are attacked and killed by males, so mothers are fiercely protective. Females breed after they wean the current cub(s), but will not reproduce at all if conditions are unfavorable. Cub mortality rates are high.

The most common cause of death for subadult bears is starvation, as they do not yet have a territory and must compete with larger bears, pushing many into marginal habitats. In adulthood, mortality drops sharply (to less than 5 percent annually); the most common natural cause of death in adults is attacks by other bears. Human activities and impacts that pose the greatest risk to polar bears include hunting, pollution, industrial development, and climate change.

HUMAN EFFECTS ON POLAR BEARS

Humans kill polar bears for aboriginal subsistence, sport, and defense of human life and property. In some areas, monitoring of polar bear kills is effective (in Norway and the United States), but in other areas there is little reliable information (Russia and Greenland). Concerns have recently been expressed

about the threat posed by trophy hunting (currently allowed only in Canada and Greenland). Quotas are often based on poor population data. Approximately 80 trophies are imported into the United States each year from Canada.

Arctic marine mammals can accumulate high concentrations of pollutants in their blubber. As top predators, polar bears accumulate the highest concentrations, and in some areas there is concern about pollution effects on the bears' health. Polar bears with abnormal genitalia and other defects have been recorded, and there have been suggestions that these are caused by exposure to certain contaminants. Development in the Arctic is also an issue. The Arctic is rich in natural resources, especially oil and gas. Exploration for and extraction of these resources involve construction that could potentially reduce habitat, produce pollution, or cause disturbance.

Because of their exclusively Arctic habitat and their charismatic nature, polar bears have become the "poster child" for the impacts of climate change. The primary concern is the loss of sea ice. This would remove essential habitat, for both the bears and their marine mammal prey. Increasing numbers of drowned polar bears have been reported, the result of increasing areas of open water between ice floes, presumably leading to overexertion during swimming. The loss of ice cover could also reduce the ability to access prey and mates, disrupt migration routes, and increase distances animals have to travel to find food (exacting an energetic cost). Many females also construct their birthing dens on ice. In the past 20 years, the proportion of dens located on sea ice has halved.

Habitat loss and prey reduction may force bears closer to human habitation to find food. Sightings of animals near Arctic towns and villages are occurring with greater frequency. This increases the likelihood of negative human/polar bear interactions. In addition, researchers have reported a significant decrease in polar bear body condition over the past 20 years, and such a decline is likely to have impacts on reproduction and survival.

Climate change may also cause chronic overheating in the highly cold-adapted bear. Finally, rising temperatures are likely to cause a shift in the distributions of other bear species; for example, brown bear populations may shift farther north and could compete, or hybridize, with polar bears.

CONSERVATION OF POLAR BEARS

In 1965, the "polar bear" nations met and agreed that each country should take whatever steps were necessary to conserve the species. Cubs, and females accompanied by cubs, should be protected throughout the year; each nation should, to the best of its ability, conduct research on polar bears within its territory; and each nation should exchange information on polar bears freely. This eventually led to the signing of the International Agreement on the Conservation of Polar Bears in May 1973 in Oslo, Norway. This agreement is currently the major international polar bear treaty.

In 1982, largely due to declines resulting from hunting pressure, the World Conservation Union (IUCN) listed the polar bear as Vulnerable. This rating was reduced to Lower Risk in 1996. However, in May 2006, polar bears were relisted as Vulnerable due to a predicted population reduction of more than 30 percent within three generations (45 years), and a decline in area of occupancy, extent of occurrence, and habitat quality resulting from climate change.

In the United States, polar bears are managed by the Fish and Wildlife Service and are protected by the Marine Mammal Protection Act (MMPA), which prohibits harassment, hunting, capture, or killing, or attempts to do any of these. There are exemptions for Alaska Native subsistence hunting, as well as scientific research and "incidental harassment" from activities such as oil and gas exploration. However, a controversial amendment in 1994 permitted the import of sport-hunted polar bear trophies into the United States from some Canadian populations. A legislative attempt is ongoing to repeal this amendment.

In February 2005, the Center for Biological Diversity and Greenpeace USA petitioned the U.S. government to list polar bears as threatened (that is, likely to become endangered within the foreseeable future throughout all or a significant portion of its range) under the U.S. Endangered Species Act. Such a listing would obligate the U.S. government to reduce anthropogenic impacts on polar bears and to devise a plan to aid their recovery. In December 2006, the U.S. Secretary of the Interior announced a proposal to list the polar bear as threatened. The deadline for deciding on the listing was January 2008.

SEE ALSO: Impacts of Global Warming; Marine Mammals; Sea Ice.

BIBLIOGRAPHY. A.E. Derocher, et al., "Polar Bears in a Warming Climate," *Integrative and Comparative Biology* (v.44, 2004); A.S. Fischbach, et al., "Landward and Eastward Shift of Alaskan Polar Bear Denning Associated with Recent Sea Ice Changes," *Polar Biology* (in press); Charles Monnett and J.S. Gleason, "Observations of Mortality Associated with Extended Open-Water Swimming by Polar Bears in the Alaskan Beaufort Sea," *Polar Biology* (v.29, 2006); M.E. Obbard, et al., "Temporal Trends in Body Condition of Southern Hudson Bay Polar Bears," *Climate Change Research Information Note* (v.3, 2006); Aqqalu Rosing-Asvid, "The Influence of Climate Variability on Polar Bear (*Ursus maritimus*) and Ringed Seal (*Pusa hispida*) Population Dynamics," *Canadian Journal of Zoology* (v.84/3, 2006); Ian Stirling and A.E. Derocher, "Possible Impacts of Climatic Warming on Polar Bears," *Arctic* (v.46/3, 1993); Ian Stirling and C.L. Parkinson, "Possible Effects of Climate Warming on Selected Populations of Polar Bears (*Ursus maritimus*) in the Canadian Arctic," *Arctic* (v.59/3, 2006); Ian Stirling, et al., "Long Term Trends in the Population

Ecology of Polar Bears in Western Hudson Bay in Relation to Climatic Change," *Arctic* (v.52/3, 1999).

E.C.M. Parsons
B.J. Milmoe
George Mason University
Naomi A. Rose
Humane Society International

Policy, International

INTERNATIONAL POLICY IS complex because of the necessity to consider the needs of each sovereign nation against economies, global environmental impact, and contributions to human-induced climate change. Much of the work toward international cooperation is done through the United Nations (an international organization with nearly all countries as members). The United Nations facilitates discussion and commitment, though it has limited authority or power. Members agree that something most be done about climate change and the environment; however, getting member nations to agree on solutions is harder. Conflicts and disagreements usually include protection of national interests by not turning authority over to the United Nations, wording of agreements that appears to not protect the interest of private industry or national interest, and economic differences between developing and developed countries.

CURRENT INTERNATIONAL POLICY

The UN Environment Programme (UNEP) was formed in 1972 with the specific purpose of encouraging collaboration on conservation and development. In 1977, the international community adopted a Plan of Action to Combat Desertification due to impacts on economy, society, and environment. In 1991, the UNEP concluded that even with some successes, land degradation due to desertification had increased, and at a 1992 conference proposed sustainable development and for the UN General Assembly to establish an Intergovernmental Negotiating Committee (INCD) to prepare a Convention to Combat Desertification, which was adopted in 1994, with entry into force on December 26, 1996. Over 179 countries were parties to the convention as of March 2002.

The Montreal Protocol on Substances that Deplete the Ozone Layer was agreed on September 16, 1987, and entered into force on January 1, 1989, and has been ratified by 191 countries. In meetings for the Montreal Protocol, the parties established adjustments and reductions of production and consumption of the controlled substances.

At the 1992 UN Conference on Environment and Development, member nations agreed to work together to formulate solutions. The Convention on Biological Diversity (CBD) was adopted at the conference for commitments to maintaining ecology balanced with economic development, though the United States disagreed with some document wording as not protecting biotechnology firms.

The Intergovernmental Panel on Climate Change (IPCC) was formed in 1988 by partnership with the World Meteorological Organization (WMO) and the UN Environment Programme to assess the risk of human-induced climate change, possible impacts of climate change, and provide options to deal with climate change. The First IPCC Assessment Report completed in 1990 provided the impetus for the formation of the UN Framework Convention on Climate Change (UNFCCC) by the UN General Assembly.

The UNFCCC is an international treaty joined by most countries and adopted in 1992, and entering into force in 1994. The UNFCCC provides the overall policy framework for addressing climate change and for coping with the impacts from inevitable temperature increases. From initial discussions, the idea came about of incorporating policy-based commitments into the international climate negotiations. The variety of options considered by the Intergovernmental Negotiating Committee was a system incorporating national assessment, along with implementation strategies and programs for reducing greenhouse gas emissions by considering their country's needs and responsibilities within the scope of international, regional, and local circumstances to determine what agenda items would be a priority and for the setting of objectives.

Developed countries committed under the initial negotiations to take the majority of responsibility for creating policy and actions to mitigate climate change and limit/reduce their output of emissions of greenhouse gases to 1990 levels by the year 2000. In debating further commitments, the participants discussed a variety of options, including policies and measures, as well as quantified emission targets. The 1995 Berlin Mandate, under negotiation, requested more significant quantification of emissions and stricter policy to enforce reduction objectives in a certain period of time. The various participants presented and supported policy approaches including across-the-board standards and taxation as well as options to choose from. Some proposals blended mandatory strict policy standards and taxation, along with a variety of options for reaching those goals.

Proposals from the European Union, the primary proponent of common and coordinated policy, listed specific actions including mandatory commitments, highly recommended commitments, and voluntary commitments. The United States and other participants favored target-based approaches to allow countries complete autonomy in choosing policies and measures.

By the late 1990s, working in voluntary partnerships was leading to reductions in conflict among stakeholders, new ideas on sustainable economic development and poverty alleviation, new thinking about the relationship between conservation areas and the communities in and around them, and more focused application of existing resources. An addition to the UNFCCC treaty is the Kyoto Protocol, drafted in 1997 by 160 nations, calling for the 38 industrialized countries releasing the most greenhouse gases to cut their emissions to levels five percent of 1990 levels by 2012 to achieve a worldwide reduction of greenhouse gases.

THE KYOTO PROTOCOL

The Kyoto Protocol entered into force on February 16, 2005, and is legally binding on countries that ratify the agreement. In initial negotiations, the United States voluntarily accepted a more ambitious target, promising to reduce emissions to 7 percent below 1990 levels; the European Union, which had wanted a much tougher treaty, committed to 8 percent; and Japan, to 6 percent. Since the adoption of the Kyoto Protocol, 175 parties have ratified it. The protocol sketched out the basic features of its mechanisms and compliance system, for example, but did not explain the all-important rules of how they would operate. The United States has not ratified the agreement because of wording and exclusions of developing countries from emissions reduction.

Though the United States has not ratified the Kyoto Protocol, the government has established a comprehensive policy to slow the growth of emissions with voluntary and incentive-based programs, strengthen institutional advancement of climate technologies and climate science, and enhance international cooperation. In February 2002, the U.S. government announced a comprehensive strategy to reduce the greenhouse gas intensity of the American economy by 18 percent 2002–12, preventing the release of more than 100 million metric tons of carbon-equivalent emissions to the atmosphere (annually) by 2012 and more than 500 million metric tons (cumulatively) 2002–12.

A core challenge in addressing global climate change is arriving at multilateral arrangements ensuring adequate effort by all major economies to moderate and reduce their greenhouse gas (GHG) emissions. Thus far, the multilateral effort has relied most heavily on a particular form of commitment-economy-wide emission targets. Developed countries agreed to voluntary targets under the UNFCCC, and most later agreed to binding targets under the Kyoto Protocol. Most developing countries, however, view quantified emission limits as a potential cap on their growth and are unlikely to accept binding targets in a post-2012 climate agreement.

At the World Conference on Disaster Reduction, The Hyogo Framework for action signed in January 2005 by 168 countries details steps to reduce the impact of natural hazards on populations. It assesses disaster capabilities and needs and incorporates risk reduction strategies and adaptations associated with existing climate variability and future climate change, including risks from geological hazards such as earthquakes and landslides. The result has been 40 countries adjusting national policies to give priority to disaster risk reduction.

The mission of the Millennium Development Goals is to alleviate poverty by integrating sustainable development into national policies and prevent the loss of natural resources. These goals allow countries to determine their own priorities. One way this is being done is with the creation of model forests to show how sustainability practices can have a positive impact on human life and to reverse environmental degradation with conservation and reforestation.

FUTURE OUTLOOK FOR INTERNATIONAL POLICY

Some international policies retain support over the long term, like the Montreal Protocol on Substances that Deplete the Ozone Layer and the Convention to Combat Desertification. The reduction in greenhouse gas emissions measures agreed to in the Kyoto Protocol end in 2012.

In determining future policy, the success of the target-based approach to reducing greenhouse gas emissions and its enforceability must be examined. The commitment is based on outcome, and coupled with emissions trading, provides economic benefits as well as freedom to apply reduction strategies in the most cost-effective manner. This works for industrialized nations, with already established economies. Developing countries may not accept imposed limits on emissions, because those same limits may limit the potential for growth. Developing countries already have multiple concerns to improve first, including environmental issues like water quality, air quality, poverty, and health issues like access to sanitation, nutritious food, and health services.

Ongoing negotiation and commitment will be needed to continue international cooperation on climate change policy. As discussed at the Climate Dialogue at Pocantico, flexibility seems to be the most important factor in gaining support by perhaps allowing countries to develop national policies in developing countries to reduce emissions on a more individual basis, instead of being limited to strict limits set by an economy-wide emission limit. Within this framework, nations would need to establish quantifiable measurements of emission impacts.

Since the Kyoto Protocol applied emission reduction levels to industrialized nations, gaining agreement to emission standards from developing countries may require some incentive-based policies. Researchers have made a variety of proposals, including the trading of carbon/emission credits as long as emissions can be quantified, control of conventional air pollutants, or improvement of agricultural productivity. Taking measures for limiting the extent of global warming from the greenhouse effect includes preventive measures (reducing emissions, enhancing carbon sequestration) and adaptive measures (construction to protect against the effects of climate change, improving water resources, and improving cultivation practices, or shifting crops to match the plants' ideal weather conditions for maximum production).

Also to be considered is a shift to previously underutilized or unused commodities like carbon taxes on all fossil fuels, such as gas, oil, coal, and the electricity generated from these sources. These taxes would shift the burden of emissions reduction to the consumer by inflating the price of using carbon-based energy and making the use of energy conservation measures and alternative resources more attractive and cost-effective, as well as increasing the demand for alternative sources. The trading of carbon credits (with each nation allocated a certain permissible level of carbon emissions and the ability to sell leftover allocations to nations who have exceeded their allocation limits), subsidizing non-carbon-based fuel providers instead of fossil fuel providers, and research and development expenditures promoting the commercialization of alternative technologies and promoting the transfer of technology to developing nations could also help.

Any measures taken to prevent global climate change will have economic effects, both positive and negative, on the economy, including production, employment, and investment. Continued scientific research on climate and atmosphere, as well as environmental education and continued voluntary dialogue between countries on the future of global environment and economic policies, can help lead to action and global policy development to overcome the limitations of determining economic policies.

Policy must also take into account the damages caused by the impacts of climate change. While climate models vary on temperature increase for each region, depending on the already present climate, potential risks include rising sea levels causing flooding, loss of coastal wetlands, beach erosion, saltwater contamination of drinking water, and damage/decreasing stability of low-lying property and infrastructure; possible increase in frequency and intensity of storms make flooding a possibility in many areas made worse by difficult drainage and causing other damage. Flooding and runoff could contaminate water supplies (with eroded soils and agricultural chemicals containing high concentrations of nitrates, pesticides, and soil nutrients); other damages could include decreased water supplies, population (both human and animal) displacement, changes in agriculture (with benefit to colder climates and loss to warmer climates), forest loss with persistent drought, and loss of trees unsuited to higher temperatures.

Changes within established ecosystems would affect wildlife, including breeding grounds of waterfowl and migratory birds. Human health risks include contracting certain infectious diseases from water contamination or disease-carrying vectors such as mosquitoes, ticks, and rodents. Warmer temperatures would increase the incidence of heat-related illnesses and lead to higher concentrations of ground-level ozone pollution, causing respiratory illnesses (diminished lung function, asthma, and respiratory inflammation). In addition, psychological factors related to higher stress could be prevalent. All of these potential impacts will also have an impact on the economy, and policy, international, national, or local, must take these possibilities into account, as some areas are more vulnerable than others; those with high emissions may not be the ones who suffer from the consequences.

Groundwork for emissions crediting is being demonstrated through the Clean Development Mechanism of the Kyoto Protocol. The projects already approved could generate 870 million tons of emission credits with the current estimated value of between $6 and $9 billion. This monetary value makes carbon crediting a viable source of income, especially for developing countries and those with carbon sequestration in forests and agricultural areas. This type of market enterprise would require independent verification. The limitation to this program might be creating disincentives for industrialized nations to reduce emissions with the option to purchase carbon credits or to not agree to the program at all if they will not be able to benefit from the income produced.

The quantification of standards thus far has been based on present levels and emissions assessments. For developing countries that have not had access to the vast amounts of energy used by industrialized nations, the strategy is not reduction of emissions, but limiting emissions to standards for consumption of energy against the development of renewable energy and improved social structures. The effectiveness of a policy is only as good as the results; other variables may be in action instead of just the policy.

As environmental awareness grows and more individuals choose climate-preserving strategies, those will have a cumulative effect on emissions. Options for telecommuting reduce driving, an economic downturn affects production, and service industries consume less energy than industrial production. Improved agricultural practices remove carbon from the air. In some cases, the effectiveness of the emissions reduction policy will be difficult to determine. The most vital part of international policy is raising awareness of the environment and what can be done to combat climate change and the effects of climate change.

SEE ALSO: Intergovernmental Panel on Climate Change (IPCC); Kyoto Protocol; Preparedness; United Nations Environment Program (UNEP); World Health Organization; World Meteorological Organization.

BIBLIOGRAPHY. International Development Research Centre, www.idrc.ca (cited May 2007); Pew Center on Global Climate Change, www.pewclimate.org (cited May 2007); United Nations, www.un.org (cited May 2007).

LYN MICHAUD
INDEPENDENT SCHOLAR

Policy, U.S.

U.S. GLOBAL WARMING and climate change policy is a hotly contested issue, one fraught with partisan bickering throughout the course of at least the last three presidential administrations. The official U.S. position has vacillated considerably over the last two decades, swinging from initial global leadership displayed during the very first climate change hearings in the U.S. Congress during summer 1988, to a mixed bag of sorts during the tenures of President George H.W. Bush and President Bill Clinton, to periods of obstruction and outright suppression of scientific studies during the early 21st century under President George W. Bush.

Despite this checkered past, though, a previously politically hamstrung United States is now making considerable advances in climate change policy. Thanks in large part to the federalist model of a national government that shares some power with its individual states, as well as local municipalities, and a fundamental separation of powers among executive, legislative, and judicial branches at the national level, notable changes are underway.

Many analysts now believe that the United States has reached a tipping point in terms of public awareness of climate change. Scientific consensus on both the rising global temperatures and anthropocentric roots of that shift, combined with concerns about energy insecurity and its ties to international terrorism, have pushed climate change discourse to the forefront, particularly in the context of what some now label the post–Hurricane Katrina effect, the growing recognition among Americans that they too are vulnerable to the vagaries of climate change.

No longer is this a problem just for their children or grandchildren to consider, or a problem that threatens primarily the developing world or small island states. According to a survey commissioned by the Yale Center for Environmental Law and Policy in March 2007, 83 percent of Americans believe global warming is a serious problem. Analysts increasingly believe that a reasonable debate about the regulation necessary to reduce greenhouse gas emissions is possible in the United States and that this regulation makes good economic sense, much in the model of the 1987 Montreal Protocol on Substances that Deplete the Ozone Layer.

Yet, when the Kyoto Protocol, the only global treaty with binding measures to address climate change, finally entered into force on February 16, 2005, the United States was one of only two Annex I industrialized countries (Australia being the other) that did not ratify the protocol. In total, 175 states of the world joined together without the world's leading economic engine, and its biggest polluter. United States resistance to the Kyoto Protocol rests firmly on this division between Annex I and Annex II countries, the developed and developing world, and the fact that Kyoto exempted Annex II states from any reductions in greenhouse gas emissions, at least until 2012.

While India is also a concern, American diplomats fear China most. China is the world's largest and most populous country, one that has ranked among the world's fastest-growing economies for two decades. China is also heavily coal-dependent and becoming ever more so, building an average of one new coal-fired plant a week. Its total energy-related carbon emissions have more than doubled since 1980, and it is widely regarded as on pace to pass the United States as the leading greenhouse gas emitter before 2020. Nevertheless, per capita emissions of carbon dioxide in China are still seven times less than that of the United States, according to the Sierra Club. With only 4 percent of the global population, Americans account for roughly 22 percent of the planet's greenhouse gases. Because China was such a late entry into an industrialization process that has created the climate change problem, the ethics of reducing greenhouse gas emission there are spotty. However, the country's economy and greenhouse gas emissions are growing at such a fast pace, reductions seem necessary.

The globalization argument as to dirty industry being able to relocate to China if it continues to be exempt from any global agreement is essentially a red herring. According to the nonpartisan Pew Center on Global Climate Change, the vast majority of U.S. greenhouse gases come from transportation, commercial, residential, and agricultural sectors that cannot leave the country. Only industry itself could potentially do so, which in 2004 accounted for 30 percent of U.S. greenhouse gases.

HISTORY OF CLIMATE CHANGE IN THE U.S.

The United States was an early leader in the field, dating at least to the aforementioned congressional climate change hearings during summer 1988. Experts such as National Aeronautics and Space Administration's (NASA) Jim Hansen first sounded alarms as to the severity of the climate change problem. Four years later, at the 1992 UN Conference on Environment and Development (UNCED) in Rio de Janeiro, Brazil, the United States continued to actively participate in climate deliberations on the international stage. President George H.W. Bush attended what was at the time the largest gathering of world leaders in history, signing the UN Framework Convention on Climate Change (UNFCCC), one of the five major agreements reached at what has since been referred to popularly as the Earth Summit.

But both President Bush's presence and signature were couched within political compromise. Bush feared mandatory limits on greenhouse gas emissions would severely hamper the U.S. economy, and only agreed to attend if the final document would go no further than suggesting voluntary limits by each country. His famous saying was that America's way of life was not up for negotiation. Popular opinion in the United States had yet to grasp the concept of sustainable development, the idea that economic development and environmental resource protection can go hand in hand.

Thus, the great irony of Rio is that a conference that was intended to firmly establish the 1987 Brundtland Report's definition of sustainable development, outlined in their publication *Our Common Future* as development that meets the needs of the present generation without sacrificing the needs of future generations, actually did the exact opposite. An entire cohort of American diplomats handcuffed themselves by bowing to the popular connotation that a tension existed between economic and environmental health. For the next decade and a half, American climate change politics would be shaped primarily by President Bush's language, one that insinuated a trade-off between environmental protection and economic development and completely ignored the underlying foundation of economic health, namely a healthy environment.

The following year, President Bill Clinton, after defeating President Bush in his reelection bid, was faced with precisely this type of obstacle in public sentiment. Despite placing perhaps the most qualified environmental politician of his generation in the number two slot of the Democratic ticket in 1992, President Clinton was unable to generate much political traction when it came to the climate change debate during the eight years in office with Vice President Al Gore, Jr. Two key examples bear out this point. The first is the attempt early in Clinton's first administration to institute a British thermal unit (BTU) tax, one that would raise taxes for a family income of $40,000 by approximately $17 a month and those making under $30,000 by none at all. This proposal failed in short order and reinforced the perception in American politics that the public will not pay to solve a long-forming, distant problem with uncertain consequences. There are simply too many more immediate concerns the average American faces on a day-to-day basis.

The second major example rests squarely upon the foundation of the international climate change debate, the Kyoto Protocol. In the lead-up to the United States signing that international treaty in December 1997, Clinton Administration officials took a central international role. With the preceding April 1995 Berlin Mandate, for instance, the United States was a passionate supporter of the argument for differentiated responsibilities, where the industrialized world would agree to restrict its emissions on

the average of 5 percent below its 1990 levels before the developing world would do so, much like the Montreal Protocol had divided the world between developed and developing world.

But well before Vice President Gore left for Kyoto in late 1997 to join the U.S. delegation led by former senator Tim Wirth (D-CO), the U.S. Senate passed the bipartisan Byrd-Hagel Resolution, which stated that body would only support a treaty that included all countries of the world. Touting what would come to be labeled global apartheid, the Byrd-Hagel Resolution demonstrated an overwhelming sentiment, passing by a vote of 95–0. Thus, despite signing Kyoto, Clinton never submitted the treaty for ratification in the Senate, a constitutional requirement that dictates any international treaty must receive two-thirds Senate support before becoming force of law.

THE GEORGE W. BUSH ADMINISTRATION

Such failures pale in comparison to the politicization of climate change during the successor administration of President George W. Bush. Over his two terms, borrowing in large part from the tobacco industry playbook, President Bush has overseen a White House that first vociferously discounted the science of climate change, and then turned to outright suppression of its evidence. Thus, while President Bush has developed a combination of modest bilateral and regional initiatives such as the Asia-Pacific Partnership on Clean Development and Climate, International Partnership for a Hydrogen Economy, and the Carbon Sequestration Leadership Forum, the lasting legacy of President Bush in terms of climate change will be a decidedly negative one.

He has ignored scientific consensus on climate change and its human-induced links in the form of the widely regarded assessments by the Intergovernmental Panel on Climate Change (IPCC). He has withdrawn the United States from Kyoto, citing the familiar refrain on lack of participation by China. And most egregious of all, he has obstructed and even deleted the dissemination of information from his own Environmental Protection Agency (EPA) and other governmental scientists. The most publicized of these was reported by the *New York Times* in its coverage of e-mails obtained by Greenpeace under the Freedom of Information Act, which depict the White House Council on Environmental Quality chief of staff Phil Cooney, a former oil industry lobbyist, as actually doctoring science by editing climate scientist reports to render them innocuous in 2002.

Despite these actions by the Bush Administration, separation of powers in the United States has allowed both the legislative and judicial branches to weigh in regarding climate change. In April 2007, for example, the U.S. Supreme Court, in the landmark 5–4 decision of *Massachusetts et al. v. EPA et al.*, sided with the Commonwealth of Massachusetts in its suit of an EPA that had heretofore ignored whether greenhouse gas emissions cause or contribute to climate change. The Court declared Massachusetts, as well as the supporting states of Connecticut, Illinois, Maine, New Jersey, New Mexico, New York, Rhode Island, Vermont, and California, had every right to request the EPA to regulate greenhouse gases.

Six months prior, the sweeping midterm elections of November 2006 saw a redistribution of party lines in the U.S. Congress, with Democrats gaining advantages in both the House and Senate (with the two Senate independents in Vermont and Connecticut, respectively, agreeing to caucus with the Democrats) for the first time since 1994. While that election was primarily driven by sentiment over the war in Iraq (and climate change debate in Congress is by no means divided solely along partisan lines, with western coal interests and mid-western automobile labor interests often influencing Democrats more than Republicans), the tangible results within congressional committees were nearly immediate, as at least five different greenhouse cap and trade bills began circulating in Congress when it returned to session in January 2007.

Most notably, the newly minted chairwoman of the Senate's Environment and Public Works Committee, Sen. Barbara Boxer (D-CA) replaced James Inhofe (R-OK), who held a grand total of five hearings on climate change in four years as head of that committee, including science-fiction writer Michael Crichton as his star witness. Senator Inhofe was also well known for labeling global warming as the greatest hoax ever perpetrated on the American people. Senator Boxer promptly held five hearings on climate change in her first three months at the helm and proposed what will likely become the centerpiece of a renewed national debate with Sen. Bernie Sanders (I-VT). This bill, formerly known as S-309 but more popularly referred to as the Global Warming Pollu-

tion Reduction Act, requires emissions reductions in the United States to the 1990 level by 2020, 27 percent below 1990 by 2030, 53 percent below 1990 by 2040, and 80 percent below 1990 by 2050.

CLIMATE POLICY: STATE AND REGIONAL LEVELS

The most significant advances in U.S. climate policy to date have actually occurred at the state level. It is in these laboratories of democracy where federalism has allowed a number of enlightened governors and their respective legislatures to forge ahead where the federal government has floundered. Where the United States is flexing its muscle most is in California under Republican governor Arnold Schwarzenegger. As the eighth-largest industrial engine on the planet, California's 2006 law targeting a 25 percent cut in carbon dioxide by 2020 is more than mere lip service to climate change.

Furthermore, as the first U.S. law imposing mandatory caps on carbon dioxide, it turned heads all the way up to the Potomac and beyond. This was not California's first foray into the climate change debate. Four years prior, in 2002, the state passed legislation creating vehicle emissions standards that required reductions of 22 percent in tailpipe greenhouse gas emissions from new vehicles by the 2012 model year and 30 percent by the 2016 model year. Governor Schwarzenegger and California are responding to the increasing loss of Sierra mountains snowpack, their primary source of drinking water, and believe they must compensate for a federal government that has dragged its heels on the issue of climate change for far too long. They also believe their economy is large enough to force others to sit up and take notice.

That logic appears to be on target. In February 2007, governors of Arizona, New Mexico, Oregon, and Washington joined California in signing an agreement establishing the Western Regional Climate Action Initiative, a joint effort to tackle climate change by reducing greenhouse gas emissions with a market-based system. This set-up mimics that of the Regional Greenhouse Gas Initiative (RGGI), which became the first mandatory U.S. cap-and-trade program for carbon dioxide in December 2005. Negotiated by the governors of seven northeastern and mid-Atlantic states (Connecticut, Delaware, Maine, New Hampshire, New Jersey, New York, and Vermont) with Maryland joining in April 2007, RGGI sets a cap on emissions of carbon dioxide from power plants at current levels in 2009, and then reducing emissions 10 percent by 2019.

U.S. BUSINESS INTERESTS

Aside from state and regional initiatives, the business community is making perhaps the most significant inroads regarding climate change policy. Echoing the logic and business acumen that drove DuPont Chemical from its position as a stiff opponent of domestic regulation of chlorofluorocarbons (CFCs) in the late 1970s to that of a staunch supporter of the globally-reaching Montreal Protocol in the mid-1980s, an incredibly diverse cross-section of corporations now stands in favor of mandatory limits on carbon dioxide emissions.

A business-NGO alliance, the U.S. Climate Change Action Partnership (USCAP), which touted 33 different members in September 2007, six of which are environmental non-governmental organizations (NGOs) and 27 of which are corporations, is a prime example. These businesses cover a broad spectrum of the economy, including several from the fossil-fuel industry and transportation sector as well as energy and electric power interests, including BP America Inc., Chrysler LLC, ConocoPhillips, Duke Energy, Ford Motor Company, General Electric, General Motors Corp., and Shell.

As Pacific Gas & Electric (PG&E) CEO Peter Darby explained in his congressional testimony in February 2007, these companies worry that a motley collection of different U.S. state regulations will emerge if no national direction is given, making their day-to-day operations much more complicated and expensive than they need to be. And they fear a United States that has abdicated its global leadership in this area risks handicapping their financial interests even further in the global marketplace. This is the driving rationale of USCAP and the reason its January 2007 publication, *A Call for Action*, calls for mandatory, market-driven cap and trade programs in carbon dioxide emissions. In short, USCAP outlines a proposal to first slow, then stop, and ultimately reverse emissions on the order of 60 to 80 percent below the current level by 2050.

CONCLUSION

Thus, on numerous levels, from both the governmental to the business sector, climate change policy in the United States has enjoyed a rebirth of late. The IPCC first noted a discernible human impact on climate change in 1995. In the 2005 aftermath of a devastating Hurricane Katrina and its utter destruction of the truly unique American city of New Orleans, Americans finally concurred. And according to NASA's chief climatologist Jim Hansen, this public opinion tipping point is about to encounter its scientific brethren. Hansen believes there are only 10 more years to act before it is too late, before the phenomenon of irreversibility enters the complex climate models. Others hope the time horizon to be a bit longer, but concur that we are talking about only a two degree C rise before the threat of irreversibility arises.

It is here that a third sector of democratic policymaking, that of civil society, becomes all the more relevant. Since the Frenchman Alexis de Tocqueville first visited the United States in the 1830s, democracy scholars have singled out the United States as a unique bastion of civil society. De Tocqueville's *De la démocratie en Amerique* (1835, 1840) found American individualism and its dominant moneymaking ethic supported the unusually high level of civic interactions outside official government structures. This is part of what handicapped the Bush and Clinton administrations in the 1990s and their efforts to face climate change policy seriously: Americans do not like to be told what they can and cannot do, especially when it limits earnings potential. But it is also what explains the burst of grassroots activity now underway regarding climate change. Americans see great financial loss if action is not taken soon.

Al Gore's 2007 Academy Award–winning film *An Inconvenient Truth* is but one piece in the puzzle galvanizing action. Environmental NGOs from the Sierra Club and Environmental Defense to the National Wildlife Federation and Natural Resources Defense Council to The Nature Conservancy and World Resources Center continue to beat their advocacy drums, and are now showing results. American public opinion surveys over the past decade show a decided maturation of understanding that climate change is real, is in large part human-induced, and, perhaps most notably, actually impacts their own lives in the United States.

SEE ALSO: Bush (George W.) Administration; California; Energy; Gore, Albert, Jr.; Hansen, James; Intergovernmental Panel on Climate Change (IPCC); Kyoto Protocol; Oil, Consumption of; Pew Center on Global Climate Change; Policy, International; Precautionary Principle; Public Awareness.

BIBLIOGRAPHY. Ross Gelbspan, *Boiling Point* (Basic Books, 2004); Intergovernmental Panel on Climate Change, *Fourth Assessment Report* (Cambridge University Press, 2007); National Assessment Synthesis Team, *Climate Change Impacts on the United States: The Potential Consequences of Climate Variability and Change* (U.S. Global Change Research Program, 2004); Public Broadcasting Service, *Frontline: Hot Politics* (April 24, 2007); B.G. Rabe, *Statehouse and Greenhouse: The Emerging Politics of American Climate Change Policy* (Brookings Institution Press, 2004); White House Office of the Press Secretary, "Climate Change Fact Sheet: The Bush Administration's Action on Global Climate Change" (May 18, 2005).

MICHAEL M. GUNTER
ROLLINS COLLEGE

Pollution, Air

THE ATMOSPHERE IS an important resource for the survival of all species on the planet, as a source of fresh air for breathing and as a protective layer against direct solar radiation. The Earth's atmosphere is composed of 78.084 percent nitrogen, 20.948 percent oxygen, 0.934 percent argon, 0.031 percent carbon dioxide, and 0.003 percent trace gases such as water vapor and air pollutants. The analysis of air bubbles trapped in ice cores provides evidence that the contents of so-called greenhouse gases, such as carbon dioxide, methane, nitrous oxide, ozone, sulfate, and carbonaceous aerosols, have significantly increased over the past 200 years. This historic change of the atmospheric composition is not fully understood, but it has roots in natural processes and human activity. As a result, both roles of the atmosphere are affected. First, the increase of greenhouse gases contributes to the increase in the amount of solar radiative energy trapped at the Earth's boundaries, which directly affects the planetary climate. Second, the composition of atmospheric air, particularly the air pollutants, strongly affects the human and environmental health.

The air pollutants are defined as substances that adversely affect humans, animals, plants and/or damage property. The air pollutant substances are gases, liquids, or solids, which are suspended in the atmosphere and emitted from different stationary or mobile sources. The pollution sources can be located in outdoor or indoor environments, and as a result the pollution levels are location dependent. Typical outdoor pollutants are particulate matter resulting from different combustion processes, including transportation. The gaseous pollutants include nitrogen oxides (NO_x), sulfur oxides (SO_x), and carbon monoxide (CO), also resulting from combustion processes. These primary pollutants can have chemical reactions in the atmosphere and create secondary air pollutants such as chemical substances forming smog. An example of naturally occurring pollutant is radon (Ra), a radioactive gas, which is released from the soil, and can be dangerous when trapped in poorly ventilated building basements. Indoor air quality is also becoming important because symptoms called sick building syndrome were correlated to the high levels of indoor air pollutants such as volatile organic compounds emitted from common building materials.

Many countries have established their regulations and standards for air pollution. The so-called first class standards define the maximum concentration levels of target pollutants, which are then attained through regulations passed by environmental protection agencies. The second-class standards provide the scales that define risk levels associated with outdoor activity, which are provided as advisory information to the public.

SEE ALSO: Coal; Diseases; Industrialization; Nuclear Power; Oil, Consumption of; Oil, Production of; Pollution, Land; Pollution, Water.

BIBLIOGRAPHY. Jeremy Colls, David M. Farrell, *Air Pollution.* (Spons Architecture Price Book, 2002); F. Patania, C. A. Brebbia, eds., *Air Pollution XI* (WIT Press, 2003).

JELENA SREBRIC
PENNSYLVANIA STATE UNIVERSITY

Pollution, Land

LAND POLLUTION IS the degradation of the land surface through misuse of the soil by poor agricultural practices, mineral exploitation, industrial waste dumping, and indiscriminate disposal of urban wastes. It includes visible waste and litter, as well as pollution of the soil. The contamination of land usually results from its commercial and industrial uses or from the spillage and dumping of waste, including landfill. These activities leave behind levels of trace metals, hydrocarbons and other compounds on the land, which have the potential to cause harm to people or the environment. The main human contributors to pollution are landfills. About half of the waste is disposed of in landfills. The gradual decomposition of landfill wastes over several decades also generates new environmental problems in the form of air pollutants. Trace organic gases are emitted from landfills, along with significant amounts of methane and carbon (IV) oxide, both of which are greenhouse gases. Garbage and other forms of waste arising from homes, municipalities, industries, and agricultural practices are the major sources of pollution on the land environment. The indiscriminate discharge of these wastes into the environment creates a filthy environment.

Unlike contaminated air and water, which directly affect human health, pollution of the land from the dumping or burial of solid wastes affects people less directly. The primary environmental concern is that a waste material in the soil may migrate into surface water or groundwater where it can be ingested and harm living organisms. Soil pollution is mainly due to chemicals in pesticides. Soil erosion and degradation are some of the problems facing the state of the land. The causes are losing six hectares of land every year, and losing 24 billion tons of topsoil. Globally, a minimum of 15 million acres of prime agricultural land is lost to overuse and mismanagement every year. Desertification is threatening about one-third of the world's land surface.

TYPES OF WASTE

Litter is waste material dumped in public places such as streets, parks, picnic areas, bus stops, and near shops. The accumulation of waste threatens the health of people in residential areas. Waste decays, encourages household pests, and turns urban areas into unsightly, dirty, and unhealthy places to live in. The following measures can be used to control land pollution. Anti-litter campaigns can educate people against littering, organic waste can be dumped in places far from residential areas, and inorganic materials such as metals, glass, paper, and plastic can be reclaimed and recycled.

One of the main factors influencing fast generation of municipal sewage and garbage and agricultural, commercial, and industrial wastes is population growth. The world human population has increased tremendously, and there has been phenomenal urban growth due to the migration of rural-area dwellers to urban areas. The larger the population, the larger the wastes generated and the greater the pollution. Pollution becomes even more pronounced when the population is crowded into a smaller space.

The sources of domestic wastes are garbage, rubbish, and ashes. Municipal wastes emanate from bulky wastes, street refuse, and dead animals. Municipal solid wastes are wastes collected by private or public authorities from domestic, commercial, and industrial sources. No two wastes are the same. The wastes generated within a municipality vary widely depending on the community and its level of commercial venture. The data on waste will depend on the level of sophistication of the waste management operation. Domestic waste from a house will vary from week to week and from season to season. Waste varies from socioeconomic groups and from country to country. In most cases, the number of refuse dumps decreases with increasing distance from the city center. Other factors that influence the distribution of solid waste dumps in cities are distance from main markets, positions of residential houses, commercial and industrial centers, and topographic characteristics of the city that determine accessibility by vehicles.

Commercial wastes are traceable to markets, stores, and shops, while industrial wastes are from factories, power plants, and treatment plants. Commercial, domestic, agricultural, and industrial activities generate vast amount of wastes, which include paper, food, metals, glass, wood, plastics, and dust. Effluents from domestic and industrial sources are also potential land pollutants. Many commercial houses and industries, especially in developing countries, do not have an organized method of disposing of their wastes. They are dumped indiscriminately, thus constituting a menace, and if they are toxic or in any way harmful, they become hazardous to the health of the public. Spillage of oil on land is a source of pollution. Land can also be polluted by the introduction of pesticides. Acid deposition also changes the integrity of the land. Contamination of land gives rise to impairment of the quality of groundwater, and impoverishment of soil to the

Land pollution occurs in many forms. Some of the sources of land pollution are agricultural, commercial, industrial, military, and from the general public. About half of the waste is disposed of in landfills.

extent of not supporting plant and animal life. Land pollution leads to the uptake of pollutants by plants, thereby introducing the pollutant to the food chain.

Garbage or trash is a component of municipal solid waste, which includes all of the wastes commonly generated in residences, commercial buildings, and institutional buildings. Municipal solid wastes consist of such things as paper, packaging, plastics, food wastes, glass, wood, and discarded appliances. Similar kinds of wastes generated by industrial facilities also are part of municipal solid wastes. The additional wastes generated by manufacturing processes, construction activities, mining and drilling operations, agriculture, and electric power production are referred to as industrial wastes. The environmental threats posed by municipal and industrial wastes are varied. Though defined as nonhazardous wastes, many of these wastes are capable of harming human health and the viability of other living species. They

contain discarded hazardous wastes like batteries, paints, solvents, and waste motor oil, items that add trace metals and organic compounds to the inventory of potential contaminants in soil.

Environmental pollution by industrial wastes has become a threat to the continued existence of plants, animals, and humans. Industrial pollution contains traces of quantities of raw materials, intermediate products, final products, coproducts, and by-products, and of any ancillary or processing chemicals used. They include detergents, solvents, cyanide, trace metals, mineral and organic acids, nitrogenous substances, fats, salts, bleaching agents, dyes, pigments, phenolic compounds, tanning agents, sulfide, and ammonia. Many of these substances are toxic. Because of the larger volumes of waste materials, landfills are the preferred method of waste disposal. The pollutants arising from a particular industry are different from those arising from another industry. The waste generated differs from industry to industry. The level of pollution arising from the industry depends on the nature and magnitude of its wastes.

Pollution by trace metals occurs largely from industries, trade wastes, agricultural wastes, and automobile exhausts. These wastes are large in magnitude and varied in types. They include large quantities of raw materials, by-products, coproducts, and final products. Mining is a major area where metal pollution occurs. Apart from natural occurrence such as erosion, metal pollution on land is a direct result of anthropogenic activities. The dumping of old or damaged vehicles on land occurs especially in developing countries. Also, the dumping of obsolete or dangerous military wastes on sites is another source of pollution. Apart from trace metals, the wastes contain organic materials, biological and chemical warfare explosives, pesticides, solid objects, and other materials peculiar to military operations. Trace metals in soil also can enter the food chain via uptake by plants and vegetation that are subsequently consumed by animals and humans, with deleterious consequences. Land disposal sites serve as breeding grounds for disease-carrying organisms.

Pollution from agricultural practices is due to animal wastes, materials eroded from farmlands, plant nutrients, vegetation, inorganic salts, and minerals resulting from irrigation and pesticides that farmers use on their farms to increase agricultural yield and fight pests and weeds. Agricultural wastes are made up of unwanted parts of crops during harvesting season. Examples are maize sheaves and cobs, maize stalks of guinea corn, millet and rice and their chaffs, yam vines, cassava stems, and yam and cassava peelings. Studies have shown that groundwater can be contaminated through seepage by leachate arising from solid wastes dumped on the ground. Land application of wastes is the most economical, practical, and environmentally sustainable method for managing agriculture wastes, especially animal wastes. Application of agricultural wastes to the land recycles valuable nutrients and organic matter into the system from which they originated. Land application can also be an effective component of management strategies for other organic wastes like food processing wastes.

Radioactive wastes are peculiar and dangerous. Their harmful effects on living organisms are induced by radiation, rather than by chemical mechanisms. They also remain dangerous for several years. Radioac-tive wastes are products of usage of nuclear energy. An example is the mining of uranium ore and its processing into nuclear fuel, which is used for electric power production. Power plants may also be radioactive. The environmental impacts of nuclear waste vary with the nature and form of the waste material. The most dangerous of these include the spent fuel from nuclear reactors, as well as the radioactive liquids and solids produced from any reprocessing of spent fuel. This high-level waste is characterized by the intensity of its radioactivity and long half-life. Death from exposure to intense radiation can occur, depending on the intensity and duration of the exposure. Human exposure can occur through inhalation of radioactive substances and ingestion of food containing radioactive materials.

WAYS TO MINIMIZE WASTE

The best way to avoid the environmental problems of solid waste disposal is to desist from generating wastes in the first instance. Pollution prevention programs aimed at this objective have become widespread. Recycling and reuse of materials are ways to avoid waste generation. At the residential level, recycling programs for newspapers, glass, and metal containers have been implemented. However, some municipal programs have been criticized for increasing environmental emissions of air pollutants from the fuel combustion.

The ultimate land disposal methods used for municipal solid wastes are land filling, land farming, and deep well injection. Land filling of solid wastes involves the controlled disposal of solid wastes on or in the upper layer of the Earth's mantle, which has been excavated to a depth of about 13 ft. (4 m.). When solid wastes are placed in sanitary landfills, biological, chemical, and physical processes occur. Biological decay of organic materials occurs by either aerobic or anaerobic processes, resulting in the evolution of gases or liquids. The chemical oxidation of waste materials occurs, dissolving and leaching of organic and inorganic materials by water and leachate moving through the fill also occur.

Land filling in moist climates produces large quantities of leachate that are toxic and of high organic strength and require treatment in wastewater plants. Land filling in dry climates produces localized air pollution problems. There is also movement of dissolved material by concentration gradients and osmosis. Initially, the organic material in the landfill undergoes aerobic decomposition due to some oxygen amount obtained in air trapped in the landfill. Within a few days, the oxygen content is exhausted and long-term decomposition occurs under aerobic conditions. The anaerobic conversion of organic compounds occurs in the transformation of high molecular weight compounds catalyzed by enzymes in soil bacteria into compounds suitable for use as a source. However, landfill sites cause soil and groundwater contamination if not properly operated. Additional environmental problems with landfill are odors, litter, scavengers, and rat infestation.

Solid wastes are those wastes from human and animal activities. In the domestic environment, the solid wastes include paper, plastics, food wastes, and ash. Improper management of solid wastes has direct adverse effects on health. Solid wastes may contain human pathogens, animal pathogens, and soil pathogens. Inadequate storage of such wastes provides a breeding ground for vermin, flies, and cockroaches, which may act as passive vectors in disease transmission. The pathogens that can cause fecal-related diseases are viruses, bacteria, protozoa, and helminths. As proper waste management involves recycling, reuse, transformation, and disposal, it is relevant to know the physical, chemical, energy, and biological properties of wastes. The physical properties that are relevant include density, moisture content, particle size distribution, field capacity, hydraulic conductivity, and shear strength. Chemical analyses required are proximate analysis, ultimate analysis, and energy content analysis. The important elements in waste energy transformation are carbon, hydrogen, oxygen, nitrogen, and sulfur.

Only 2 percent of waste is actually recycled. Solid waste recycling implies recovery of a component of waste for use in a manner different from its initial function. Recycling consists of recovering from waste the matter of which a product was made and reintroducing it into the production cycle for reproduction of the same item. Composting after decomposition by aerobic bacteria mostly readily recycles garbage, grass, and organic matter. Composting may be defined as the decomposition of moist, solid, organic matter by the use of aerobic microorganisms under controlled conditions. The end product of the decomposition is a sanitary, nuisance-free, humus-like material that can be used as soil conditioner and as partial replacement for fertilizer. In a typical operation, the municipal wastes are presorted to remove noncombustible materials and those that might have salvage value such as paper, cardboard, rags, metals, and glass. Refuse is then shredded and stacked in long piles where it degrades to humus much as it would in soil. Usually, the decomposed material contains less than 1 percent of each of the three primary fertilizer nutrients. The final step is grinding and bagging for ultimate sale as soil conditioner.

Plants die because of land pollution. Crops are affected, as they do not mature and grow well. There are three ways that people pollute the land: littering all over the land, improper garbage disposal, and dumping of chemical fluids on the land. It is not uncommon to see people throw the trash on the road while in the car. Every day, people are polluting the land. Because of pollution, people do not only affect the cleanliness of the land, but also destroy the beauty and increase avenues of contracting diseases. These negative tendencies have effects on tourism potentials of nations as tourists are turned off. Tourists won't like to take risks in an unsafe environment because of pollution. Mosquitoes live in littered empty cans. Thus, the threat of mosquito bite is imminent in a polluted land. A greater proportion of land pollution is instigated and carried out by man. Governments of nations should be alive to their responsibilities of providing a safe and secured world environment to its people.

SEE ALSO: Diseases; Nuclear Power; Pollution, Air; Pollution, Water.

BIBLIOGRAPHY. C.M.A. Ademoroti, *Environmental Chemistry and Toxicology* (Foludex Press, 1996); G. Kiely, *Environmental Engineering* (McGraw-Hill International Editions, 1998); E.S. Rubin and C.I. Davidson, *Introduction to Engineering & the Environment* (McGraw-Hill, 2001).

AKAN BASSEY WILLIAMS
COVENANT UNIVERSITY

Pollution, Water

WATER POLLUTION USUALLY describes the introduction or presence of harmful or objectionable substance in water in magnitude sufficient to alter the quality indices of natural water. It also connotes the presence of polluting substances in rivers, lakes, bays, seas, streams, underground water, or oceans in levels capable of resulting in measurable degradation of the water quality or usefulness. For example, if water contains too much contamination as a result of certain harmful chemical compounds or microorganisms, it could be rendered unsafe in its existing state for an intended purpose. This could be described as water pollution. In most cases, water pollution may arise from the use to which the water has been put. Although some kind of water pollution can occur through natural processes, it is mostly caused by human activities.

Water pollution has many sources and characteristics. These sources can be categorized into point and nonpoint sources. Point sources of water pollution are direct discharges to a single point, or simply stationary location discharges. Examples include discharges from sewage treatment plants, power plants, factories, ships, injection wells, and some manufacturing or industrial sources. Nonpoint sources of water pollution are more diffused across a broad area and their contamination is traceable to a single discharge point. Examples of nonpoint sources include mining activities and agricultural and urban runoffs. Water pollution arising from nonpoint sources accounts for the majority of contaminants in streams, rivers, bays, underground water, and seas.

A water pollutant is any biological, chemical, or physical substance if when present in water at excessive concentrations has the capability of altering the chemical, physical, biological, and radiological integrity of water, thereby reducing its usefulness to living organisms, including man. Although there are many sources of pollutants in our waters, the primary sources of water-polluting substances come from sewage, agricultural runoffs, oil spills, industrial wastewaters, land drainage, and domestic wastes. The major categories of common water pollutants include heavy metals, pathogens, nutrients, acids, organic chemicals, and radioactivity. Many of these substances are toxic and are capable of interacting additively or synergistically or antagonistically to give varying responses in aquatic ecosystems and in humans. However, the influence of a pollutant in natural waters varies according to the polluting substance, the local environmental conditions, and the organism involved.

MAJOR TYPES OF WATER POLLUTANTS

Heavy metals are toxic and include many metal pollutants that could have potentially harmful effects on human health and aquatic ecosystems. Common examples include cadmium, nickel, arsenic, lead, vanadium, mercury, and selenium. Typical sources of metal pollutants include wastes from domestic, industry, agriculture, urban, and mining drains. Acids are inorganic water pollutants caused by industrial discharges, especially sulfur dioxide from industrial power plants, drainage from mines, wastes from industry, and aerial acid deposition. Acids have the potential of causing harm to aquatic ecosystems via the mobilization of toxic heavy metal pollutants.

Organic chemicals such as insecticides, herbicides, petroleum hydrocarbons, detergents, and a range of volatile organic compounds such as solvents discharged into aquatic ecosystems have the potential of altering the integrity of natural waters. This variety of chemicals regarded as water pollutants arises from agricultural use of pesticides, especially insecticides and herbicides, industrial wastes, marine oil spillage, and domestic wastes. They are potentially harmful to human health and aquatic organisms. Nutrients arising from sewage and agricultural use of fertilizers may cause eutrophication in aquatic ecosystems.

Although nutrients are elements essential for the growth of living organisms, human-caused contaminations can greatly enhance the presence of nutrients (especially nitrogen and phosphorus compounds), leading to anthropogenic or cultural eutrophication. Continuous nutrient loading to aquatic systems could ultimately increase the phytoplankton population, resulting in algal bloom, by providing more food for the algae than is normally available. Nutrients may affect human health. Excessive algal population in water has the potential of unbalancing the food chain, discoloration of water, and reduction in the quantity of light radiation that is available to aquatic life. However, when the algae dies, the rotting algae could produce a strong, unpleasant smell and the remains could be toxic to aquatic fauna and flora. This process could also result in depletion of dissolved oxygen in water.

There are several sources of water pollutants, and these are domestic and industrial wastewaters, agricultural runoff water, and other nonpoint sources. Domestic wastes commonly carry organic matter, microbiological contaminants, and sometimes physical and chemical pollutants. Industrial wastes contain mostly chemical and radioactive pollutants, while agricultural run-off water may carry mainly nutrients, pesticides, and heavy metals. Moreover, water pollution can be broadly classified into different types and these include microbiological, chemical, physical, and thermal water pollution.

Biological hazards associated with water pollution include disease-causing (pathogenic) microorganisms, like parasites, bacteria, and viruses. People exposed to biologically contaminated waters can become sick from drinking, washing, or swimming. Disease-causing pathogens commonly associated with fecal contamination of water include *Shigella dysenteriae*, *Salmonella typhi*, *Salmonella paratyphi*, *Vibrio cholerae*, *Entamoeba histolytica* and poliomyelitis virus responsible for causing bacterial dysentery, typhoid fever, paratyphoid fever, cholera, amoebic dysentery, and infantile paralysis, respectively. Also, the consumption of microbe-contaminated seafood, especially shellfish, could lead to outbreaks of food poisoning.

CHEMICAL WATER POLLUTION

Chemical form of water pollution includes the presence of a wide range of chemicals from industry, such as lead, arsenic, nitrates, radioactive substances, metals and solvents, and even chemicals which are formed from the breakdown of natural wastes (ammonia, for instance). Effluents from chemical industries and oil pollution from accidental crude spillage are categorized under chemical form of water pollution. In aquatic systems, these chemicals are poisonous to fish and other aquatic life. Chemical pollutants can be generally categorized into persistent (degrade slowly) and nonpersistent (degradable) substances.

Nonpersistent pollutants include domestic wastes, fertilizers, and some classes of industrial wastes. These polluting substances can be broken down into simple nonpolluting molecules or compounds such as carbon dioxide, and nitrogen by chemical or biological processes. Persistent water pollution is the most rapidly growing type of pollution, and includes polluting substances that degrade or do not grade or cannot be broken down at all. These pollutants tend to remain in aquatic environments for a long period of time. Common persistent chemical pollutants include some pesticides (such as dieldrin, heptachlor, and DDT), petroleum products, polychlorinated biphenyls (PCBs),

chlorophenols, dioxins, polycyclic aromatic hydrocarbons (PAHs), radionuclides, and heavy metals. Toxic metals discharged in effluent can be accumulated in seafood, especially fish and shellfish such as prawns, cockles, mussels, and oysters, to levels in excess of public health limiting levels, therefore posing serious health concerns to people who eat them.

Pesticides used in agriculture and around the home, especially those used for controlling insects (insecticides) and weeds (herbicides), are another type of toxic chemical. These chemicals are used to kill unwanted animals and plants, and may be collected by rainwater runoff and carried into streams, lakes, bays, rivers, and seas, especially if these substances are applied in excessive quantities. Some of these chemicals are biodegradable and may quickly decay into harmless or less harmful forms, while others are nonbiodegradable and can persist in the environment for a long time. When animals consume plants that have been treated with certain nonbiodegradable toxicants (NBTs), such as dichlorodiphenyltrichloroethane (DDT) and chlordane, these chemicals are absorbed into the tissues or organs of the animals and can accumulate over time. When other animals feed on these contaminated animals, the chemicals are passed up the food chain. Some of these can accumulate in fish and shellfish and poison people, animals, and birds that eat them. Materials like detergents and oils float and spoil the appearance of a water body, as well as being toxic; and many chemical pollutants have unpleasant odors.

PHYSICAL WATER POLLUTION

A common form of physical water pollution is thermal pollution. This includes warm water from cooling towers, floating debris, foam, and garbage. In highly industrialized areas of the world, power plants are used in generating electricity, where warmer water generated in the process is generally released back to the environment. In nuclear plants, water is used in large quantity to cool reactors. The discharge of high-temperature water into a natural body of water can affect the downstream habitats, therefore altering the ecological balance. It can lead to cultural eutrophication, thereby promoting algal bloom. This development has the potential of threatening certain fish species, as well as disturbing the chemistry of the receiving water body.

Heat may also affect man's legitimate use of water for fishing. Another common and widespread type of thermal pollution is the unsafe removal of vegetations that should naturally keep streams and small lakes cool. Natural vegetations, mainly trees and other tall plants, are usually seen around streams and sizable water bodies and they block direct sunlight from heating and thereby increasing the surface temperatures of these waters. People often remove this shading vegetation in order to harvest wood from the trees, to make room for crops, or to construct buildings, roads, and other structures. When these vegetations are removed and the aquatic ecosystems are left uncovered, the water temperature could increase by as much as 18 degrees F (10 degrees C).

Many wastes are biodegradable, that is, they can be broken down and used as food by microorganisms like bacteria. Biodegradable wastes may be preferable to nonbiodegradable ones, because they will be broken down and not remain in the environment for a very long time. However, too much biodegradable material can cause the serious problem of oxygen depletion in receiving waters. Like fish, aerobic bacteria that live in water use oxygen gas, which is dissolved in the water when they feed. Invariably, the oxygen is not very soluble in water. Even when the water is saturated with dissolved oxygen, it contains only about 1/25 the concentration that is present in air. So if there are too many nutrients in the water, the bacteria that are consuming it can easily use up all of the dissolved oxygen, leaving none for the fish, which will die of suffocation. Once the oxygen is depleted, other bacteria that do not need dissolved oxygen take over. But while aerobic microorganisms convert the nitrogen, sulfur, and carbon compounds that are present in the wastewater into odorless, oxygenated forms like nitrates, sulfates, and carbonates, these anaerobic microorganisms produce toxic and smelly ammonia, amines, and sulfides, and flammable methane.

Nutrients are major chemical pollutants and they include nitrates and phosphates found in sewage, fertilizers, and detergents. Although phosphorus and nitrogen are essential elements necessary for plant growth, in excess levels nutrients overstimulate the growth of aquatic plants and algae. When discharged into rivers, streams, lakes, and estuaries, they cause nuisance growth of aquatic weeds, as well as blooms of algae, which are microscopic plants. Excessive growth of these organisms can clog navigable waters, deplete dissolved oxygen as they decompose, and block light

from penetrating deeper waters. Weeds can make a lake unsuitable for swimming and boating. Algae and weeds die and become biodegradable material. If the water is used as a drinking-water source, algae can clog filters and impart unpleasant tastes and odors to the finished water. It can also impair respiration by fish and aquatic invertebrates, which could lead to a decrease in animal and plant diversity.

Suspended solids originate from eroded stream banks, construction, and logging sites. They are a form of physical water pollution. These pollutants are also referred to as particulate matter because they contain particles of much larger size which remain suspended in the water column. Although they may be kept in suspension by turbulence, once in the receiving water, they will eventually settle out and form silt or mud at the bottom. As these sediments enter the rivers, lakes, and streams, they tend to decrease the depth of the body of water. If there is a lot of biodegradable organic material in the sediment, it will become anaerobic and contribute to the formation of algal bloom. Toxic materials can also accumulate in the sediment and affect the organisms that live there, and can build up in fish that feed on them, and so be passed up the food chain, causing problems along the food web. Also, some of the particulate matter may be coated with grease, which is lighter than water, and float to the top, creating an aesthetic nuisance.

CONCLUSION

The pollution of water resources can have serious and wide-ranging effects on the environment and human health. The immediate effects of water pollution can be seen in water bodies and the animal and plant life that inhabits them. Pollution poisons and deforms fish and other animals, unbalances ecosystems, and causes a reduction in biodiversity. Ultimately, these effects take their toll on human life. Drinking-water sources become contaminated, causing sickness and disease. Pollutants accumulate in food, making it dangerous or inedible. The presence of these toxic substances in food and water can also lead to reproductive problems and neurological disorders. The effects of water pollution are varied. They include poisonous drinking water, poisonous food animals (due to these organisms having accumulated toxins from the environment over their life spans), unbalanced river and lake ecosystems that can no longer support full biological diversity, deforestation from acid rain, and many other effects. These effects are, of course, specific to the various contaminants.

SEE ALSO: Industrialization; Pollution, Air; Pollution, Land.

BIBLIOGRAPHY. Scott Brennan and J.H. Withgott, *Environment: The Science Behind the Stories* (Pearson Education, Inc., 2004); Bill Freedman, *The Ecological Effects of Pollution, Disturbance, and Other Stresses* (Academic Press Limited, 1995); Kiely Gerard, ed., *Environmental Engineering* (McGraw-Hill International (UK) Ltd., 1997); R.M. Harrison, *Pollution: Causes, Effects and Control* (Royal Society of Chemistry, 1994); Miroslav Radojevic and V.N. Baskin, *Practical Environmental Analysis* (Royal Society of Chemistry, 1999).

NSIKAK BENSON
COVENANT UNIVERSITY

Population

CLIMATIC EVENTS HAVE had an important impact on the geographical distribution of human populations in the past. Nowadays, the growing consensus among the scientific community on the reality of human-induced global warming has raised concern that millions of people could be displaced.

A LONG HISTORY OF LINKAGE BETWEEN CLIMATE AND POPULATIONS

Population geography has acknowledged for many years the role played by climatic factors in explaining the history of population and the emergence of cities. Thus, for mankind, the passage across the Bering Strait to America 13,000 years ago was possible due to the low sea levels of the Ice Age, while the Medieval Climate Optimum which lasted between the 8th and 13th centuries stimulated the population of Polynesia by making navigation relatively easy, thanks to regular winds and clear skies. The desertification of the Sahara and the Arabian Peninsula also played an important part in the densification of the population on the banks of the Nile and consequently contributed to the birth of ancient Egyptian civilization.

More recently, the droughts of the 1930s in the plains of the American Dust Bowl (parts of Kansas, Oklahoma, Texas, New Mexico, and Colorado) forced thousands of migrants toward California, and those that struck the Sahel between 1969 and 1974 displaced millions of African farmers and nomads.

Notwithstanding the present media focus on global warming, the amount of systematic research on climate and populations remains limited. There is much vagueness surrounding the concepts employed, the underlying mechanisms involved, the number of persons affected, and the geographical zones concerned. Climatic factors are rarely the sole cause of migration, and the economic, social, and political situation of the zone under threat can, depending on the case, increase or decrease the flow of migrants.

The use by numerous authors of the term "climate refugee" has also led to confusion, because it evokes the juridical status recognized by the UN Convention of 1951 to political refugees. The High Commissioner for Refugees, aware of a risk of confusion between political and nonpolitical refugees, has always treated with the utmost prudence the idea of including environmental motivations in the international definition of refugees, even if this category of the population is deemed a part of the protective mandate toward displaced persons.

THE CONSEQUENCES OF GLOBAL WARMING

Alarming predictions of greater resource scarcity, desertification, risks of droughts and floods, and rising sea levels that could drive many millions of people to migrate appeared in the review on the economic consequences of global warming delivered to the British government by Sir Nicholas Stern at the end of November 2006. While it is extremely difficult to elaborate scientific predictions by combining climate and migration models, the expected consequences of climate change can be compared to past experiences so as to establish a list of the populations most at risk and the possible resulting emigration flows. Three consequences of climate warming forecast in the latest report of the Intergovernmental Panel on Climate Change (IPCC) for the end of the 21st century appear to be the most threatening potential causes of migrations:

1. The increase in strength of tropical hurricanes and in the frequency of heavy rains and flooding due

to the augmentation of evaporation correlative to temperature increase.

2. The growth in the number of droughts, with evaporation contributing to a decrease in soil humidity, often associated with food shortages.

3. The increase in sea levels resulting from both water expansion and melting ice.

While the first two consequences are the direct result of sudden natural disasters, the third is a long-term process, which has very different possible implications in terms of migrations.

HURRICANES, TORRENTIAL RAINS, AND FLOODS

The consequences of hurricanes and floods on population displacement are among the easiest to identify in that they manifest themselves in a brutal and direct manner. While the number of persons affected by flooding worldwide (106 million yearly between 2000 and 2005 according to the International Disaster Database) and by hurricanes (38 million) is known, the total number of people threatened by an eventual increase of this kind of disaster is, however, very difficult to estimate. No climate model is able to predict with accuracy whether or not the affected zones will be densely populated and whether the damage wrought will have tragic consequences.

Apart from this difficulty of forecasting, the studies carried out after such events tend to relativize their effects in terms of migration in general, and long-term migration in particular. Living mainly in poor countries, the victims have little mobility, and the majority of the displaced return as soon as possible to reconstruct their homes in the disaster zone. The results from numerous researches conducted worldwide on the subject tend to confirm this point with remarkable regularity. On a global level, the general conclusion therefore is that the potential of hurricanes and torrential rains to provoke long-term and long-distance migrations remains limited.

DROUGHT AND DESERTIFICATION

The latest report of the Intergovernmental Panel on Climate Change predicts increased water shortages in Africa (74 to 250 million people affected in 2020) and Asia. Case studies, however, bring to light a contrasting picture of the consequences for migrations of these kinds of evolutions. The effect of a lack of drinking and irrigation water on migration is actually less sudden than that of hurricanes and floods, and it only generates progressive departures. On one hand, there are many well-known cases of mass population departures following droughts, in particular in Africa (Sahel, Ethiopia) with an impressive figure of one million displaced persons during the drought in Niger in 1985, but also in South America (Argentina, Brazil), in the Middle East (Syria, Iran), in central Asia, and in southern Asia.

On the other hand, many researchers strongly relativize the possible direct link existing between drought and emigration by highlighting the fact that the latter, in general, is the last resort when all other survival strategies have been exhausted. For example, during the 1994 drought in Bangladesh, only 0.4 percent of households had to resort to emigration. Other researchers hold views similar to that of Nobel Prize winner for Economics, Amartya Sen, in remarking that famines are, in general, only marginally the direct result of environmental factors, but much rather political ones and add that this also holds for migrations. In certain contexts, the effect can even be inversed. This was the case in Mali during the drought of the mid-1980s: a reduction in international emigration was observed due to the lack of available means to finance the journey. Forecasts of increased migrations linked to drought-related phenomena remain hazardous. Consequently, it would be difficult to put a figure on the magnitude of populations at risk and the eventual migrations arising from global warming–induced droughts.

RISING SEA LEVELS

While the first two climatic hazards mentioned do not foreshadow massive population displacements due to climate change, the potential for migration when linked to an increase in sea level is considerable. Contrary to hurricanes, rain, and drought, this phenomenon is virtually irreversible and manifests itself over a long period of time. This could make migration the only possible option for the population affected. The localization of the consequences of rising sea levels is a relatively easy task because the configuration of coastlines, their altitude, and population are well known and thus easy to map. Hence, it is possible to calculate, on a global scale, the number of persons living in low-elevation coastal zones and threatened by either rising water levels, higher tides, or farther-reaching waves. Low-elevation coastal zones are defined as situated at an altitude of less than 33 ft. (10 m.). Even though these zones account for only 2.2 percent of dry land, they are presently home for 10.5 percent of the world's population, some 600 million people, of whom more than two-thirds live in Asia, and nearly a half in the poorest countries of the world.

It would be an exaggeration, however, to consider that these hundreds of millions of people are all potential migrants in a near future. The latest report of the IPCC evokes, of course, the possible melting of Greenland ice cover and the consequent 23 ft. (7 m.) rise in sea level, but this would occur over several thousand years. Of more concern is the scenario of thermic expansion of the oceans. According to a future CO_2 emission estimate based on continuing economic growth, but with a moderation of fossil fuel use, there would be an increase of 1 to 2.6 ft. (0.3 to 0.8 m.) of the oceans by 2300. On this basis, it seems reasonable to consider populations living at an altitude of less than one meter as being directly vulnerable by the next century. A study commissioned within the framework of the Stern Report gives a considerable figure of 146 million people for this group. Mainly situated in the major rivers' deltas and estuaries, the flood zones are particularly populated in south Asia (Indus and Ganges-Brahmaputra) and east Asia (Mekong, Yangtze, Pearl River). These two regions account for 75 percent of the population at risk. Certain Pacific states such as Tuvalu or Kiribati are, in the shortterm, among the most threatened, as they are situated only centimeters above water. Although far less populated, they nevertheless have inhabitants numbering several thousand persons.

The increase in sea levels is the greatest direct threat for numerous populations. Contrary to hurricanes and droughts, the localization of potential victims is ascertainable. If no measure of moderation is taken and if no effort is made to protect the groups at risk, then they will have no alternative but to emigrate.

CONCLUSION

Climate changes can generate migration flows. Global warming could, in particular, lead to major forced displacements. The latter will result principally from rising sea levels, but will only progressively manifest themselves over the forthcoming centuries, with the exception of the flooding of certain islands. The increase in droughts and meteorological disasters predicted by climatic models will also have impacts in terms of migrations, but these will remain regional and shortterm, and are at present difficult to estimate.

Existing research shows that due to the number of factors involved, no climatic hazards inevitably result in migrations. Many authors note that even if disasters become more frequent in the future, political efforts and measures of protection will be able to lessen the need to emigrate, provided that the necessary financial means are made available. Even rising sea levels could be partially counteracted by the erection of dikes or the filling in of threatened zones. The question of what kind of international system of burden-sharing and protection to put in place to face these challenges remains unanswered, and is all the more important because of the clear responsibility of rich countries for global warming. Bangladesh, for example, contributes only 0.14 percent of global CO_2 emission, but counts hundreds of thousands of people at risk of increased flooding.

SEE ALSO: Alliance of Small Island States (AOIS); Bangladesh; Climate Change, Effects; Desertification; Developing Countries; Drought; Economics, Impact from Climate Change; Floods; Impacts of Global Warming; Refugees, Environmental.

BIBLIOGRAPHY. Stephen Castles, "Environmental Change and Forced Migration: Making Sense of the Debate," *New Issues in Refugee Research, UNHCR Working Paper* (v.70, 2002); Nicholas Stern, *The Economics of Climate Change* (Cambridge University Press, 2007); J.D. Unruh, M.S. Krol, and Nurit Kliot, *Environmental Change and Its Implications for Population Migration* (Kluwer Academic Publisher, 2004).

ETIENNE PIGUET
UNIVERSITY OF NEUCHÂTEL

Portugal

AS REPORTED IN a major collaborative international research project, named SIAM II, Portugal is one of the European countries that are expected to suffer from the most extreme consequences of global warming and climatic change. These consequences are expected to entail three major geological effects in Portugal. First, due to the rise in sea levels, studies predict an increase in the erosion of Portugal's coastal areas. Second, scientists anticipate increased levels of rain precipitation and the concomitant occurrence of floods that will carry high social and economic costs. Finally, in dry areas of the country, studies point to a higher probability for the incidence of forest fires. Studies show that the tendency for increasingly hotter summers in Portugal has accelerated in the past few decades. Research and analysis based on data collected from 1931 to 2000 in Portugal demonstrates that the six hottest summers occurred in the last 12 years.

The Portuguese population is slowly becoming aware of the relationship between global warming and climate change within the borders of their nation, and this awareness is becoming stronger. One of the reasons for increased knowledge concerning the impact of global warming on Portugal has been the occurrence of major forest fires, as well as fluvial floods. In this context, the mass media and government officials have played a crucial role in sensitizing and educating the Portuguese population about the risks associated with global warming. Another reason is the fact that an increasingly large percentage of the Portuguese population is university educated and, in this context too, they are exposed to recent and important national as well as international research concerning the issue.

Due to Portugal's relatively late entry into the European Union in 1986, and its previous situation of economic isolation and considerable underdevelopment in terms of European standards, Portugal is still highly dependent on importing energy derived from nonrenewable sources. According to the Statistical Office of the European Union (Eurostat), Portugal is 99.4 percent dependent on such imported energy. Because of the economic weight that this carries, it is oftentimes difficult for enterprises to develop or invest in alternative and more ecological forms of energy supply. For this reason, it is feared that Portugal may face difficulties in reaching Kyoto established goals.

SEE ALSO: Deforestation; Energy; Floods.

BIBLIOGRAPHY. F.D. Santos, K. Foibes, and R. Moita, eds., *2002 Climate Change in Portugal: Scenarios, Impacts and Adaptation Measures—SIAM Project* (Gradiva, 2002).

KATJA NEVES-GRACA
CONCORDIA UNIVERSITY

Precambrian Era

THE PRECAMBRIAN ERA, or Supereon, refers to the geological time comprising the eons that came before the Phanerozoic eon. This time spans from the formation of Earth around 4.5 billion years ago to the evolution of abundant macroscopic hard-shelled animals, which marked the beginning of the Cambrian era, the first period of the first era of the Phanerozoic eon. The Precambrian era encompasses 86 percent of the Earth's history, however very little is known about this time period. In fact, the few fossil discoveries from this period were recently made in the late 20th century. Precambrian time can be further divided into three large eons, the Hadean, Archean, and Proterozoic eons.

The Precambrian's oldest eon, the Hadean (4.5 to 3.9 billion years ago), predates most of the geologic record. During the Hadean, the solar system formed out of gas and dust, the sun began to emit light and heat, and Earth took shape. Meteors and other galactic debris showered the planet over the first half-billion years, making it entirely uninhabitable. Planet Earth was very hot during its initial formation. As the earth began to cool and its mass increased, its gravitational field strengthened. This attracted meteorites and other debris, which continued to bombard the planet for at least another 500 million years, producing enough energy and heat to vaporize any water or melt any rock that may be present. Iron continued to sink to form the Earth's core, while silicon, magnesium, and aluminum gradually rose toward the surface. Gases released from magma

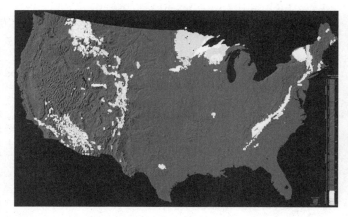

Highlighted in light areas are Precambrian rocks. They formed between 560 million years and 2.6 billion years ago.

inside the Earth escaped through cracks in the surface and began to collect in the early atmosphere. The likely presence of methane and ammonia among the gases made for conditions that would be highly toxic to life as we know it. Because there was little to no free oxygen, no protective ozone layers existed and damaging ultraviolet rays showered the Earth at full strength. As the meteorite bombardment finally slowed, Earth was able to cool, and its surface harden as a crust, rocks and continental plates began to form. Water began to condense in the atmosphere, resulting in torrential rainfall. After several hundred million years of falling rain, great oceans were formed. By about 3,900 billion years ago, Earth's environment had been transformed from a highly unstable state into a more hospitable place. This marked the beginning of the Archean eon (3.9 to 2.5 billion years ago). It was early in the Archaean eon that life first appeared on Earth.

The climate of the late Precambrian time, the Proterozoic eon (2.5 billion years ago to 543 million years ago) was typically cold with glaciations spreading over much of the earth. One of the most important events of the Proterozoic was the gathering of oxygen in the Earth's atmosphere. Though oxygen was undoubtedly released by photosynthesis well back in Archean times, it could not build up to any significant degree until chemical sinks unoxidized sulfur and iron had been filled. The first advanced single-celled and multicellular life roughly coincides with the oxygen accumulation. It was also during this period that the first symbiotic relationships between mitochondria and chloroplasts and their hosts evolved. At this time the continents were bunched up in to a single supercontinent known as Rodinia. It broke up starting around 750 million years ago, and as continental fragments reached the North and South poles they likely contributed to the great Ice Ages. In the latest Proterozoic era a new supercontinent called Pannotia came together. A number of glacial periods have been identified going as far back as the Huronian epoch, roughly 2,200 million years ago. The best studied is the Sturtian-Varangian glaciation, around 600 million years ago, which may have brought glacial conditions all the way to the equator, resulting in a "Snowball Earth". This theory states that the continents and oceans were covered in ice approximately 600 million years ago. The Earth may have remained in this frozen form, but it was rescued by the release of volcanic gases. While the Earth was in a deep freeze, chemical cycles were halted; as a result, carbon dioxide accumulated in the atmosphere causing an extreme greenhouse effect. After 10 million years of deep freeze, the Earth thawed in only a few hundred years. These dramatic events may have caused the explosion of lifeforms seen in Cambrian fossils.

At this point, at the start of the Cambrian period, Earth had taken on its current form in the life-filled oceans and oxygenated atmosphere. Coevolution of biosphere and lithosphere over billions of years led to this point. Anaerobes and oxygen-breathers had evolved complementary chemical cycles, and biogenic carbonates entered the plate-tectonic cycle of the crust and upper mantle with new efficiency.

SEE ALSO: Climate Change, Effects; Climate Cycles; Snowball Earth.

BIBLIOGRAPHY. J.C. Walker, "Precambrian Evolution of the Climate System," *Global Planet Change* (v.82, 1990); K. Zahnle, "A Constant Daylength during the Precambrian Era?" *Precambrian Research* (v.37, 1987).

Fernando Herrera
University of California, San Diego

Precautionary Principle

THE UPSHOT OF precaution is that it is better to be safe than sorry when there are severe or irreversible consequences. It has been a very important notion in environmental and public health policy. It has been advocated in several issues ranging from climate change to genetic engineering to phase-out of persistent organic pollutants. The invocation of precaution has been particularly controversial when there are significant business interests at stake. The problem with simply asserting precaution whenever a technology, policy, or action involves possible negative outcomes is that it often poses significant challenges in evaluating the public versus private trade-offs involved.

This use of precaution is often invoked when outcomes are uncertain. The notion of uncertainty is used to characterize how well future events or scientific truths can be predicted or known. It is used in both social and natural science disciplines, from mathematics to philosophy, to risk assessment and public policy. If probability is a measure of likelihood, then uncertainty is a measure of how well the probability is known. Uncertainty can be classified into known and unknown probabilities. Events with known probabilities are referred to as events with statistical uncertainties. Events with unknown probabilities are often called events with true uncertainty.

Uncertainty in the context of the environment mainly refers to scientific uncertainty. Here, science generates truths through the testing of hypotheses. But often, the affirmation of hypotheses involves a certain degree of uncertainty due to the method or research design. Scientists often use the benchmark of 95 percent certainty when deciding whether or not cause and effect have been correctly identified. Scientists often report confidence limits based on research design and sampling error in their studies to account for uncertainty.

The precautionary principle is often invoked under uncertain circumstances, particularly when the consequences are irreversible or permanent. This differs from the choice that scientists make when deciding what to do under conditions of uncertainty. Typically, scientists are interested in avoiding false negatives, because science is epistemologically conservative. Scientists do not want to suggest something as truth when in fact it may not be. In public or environmental policy, however, because the consequences are not epistemological but are ethical, there is desire to avoid false positives and be ethically conservative.

In public and environmental policy, it is important to understand how to make decisions in the absence of perfect information. Knowing the degree of uncertainty is particularly important when questions about risk arise. Risk assessment is a policy approach that deals with uncertainty. Risk assessment is widely used by the Environmental Protection Agency (EPA), but mainly focuses on known probabilities. Because of difficulties with codifying the precautionary principle into policy, the EPA has yet to include true uncertainty in environmental policy.

CLASSES OF SCIENTIFIC UNCERTAINTY

Scientist Schrader-Frechette describes four classes of scientific uncertainty dealt with by scientists and policymakers: framing uncertainty, modeling uncertainty, statistical uncertainty, and decision-theoretic uncertainty. In framing uncertainty, scientists often use a two-value frame to accept or reject a hypothesis. Frechette argues that in public policy, it is more appropriate to adopt a three-value frame that creates a category to deal with situations where significant uncertainty and serious consequences suggest adopting the precautionary principle. Modeling uncertainties involve those involved in the prediction of future scenarios. These are highly speculative, despite claims to be verified and validated models.

In public and environmental policy, statistical uncertainty should be dealt with in such a way that highlights the difference between epistemological consequences and ethical ones. When faced with decision-theoretic uncertainty, scientists are forced to distinguish between using expected value rules and the minimax rule. The former argues that decisions should be based on the expected value, while the latter seeks to prevent the worst-case scenario. More recently, Bayesian statistics has been used to help evaluate data under conditions of uncertainty by updating the probabilities as new data come to view. A Bayesian approach involves the introduction of prior knowledge into statistical models.

There are many environmental policy debates where questions about uncertainty are raised. In debates about global climate change, for example, scientists typically agree that there is significant uncertainty in the projection of future climate change models. Climate change skeptics, to discredit climate change science, often highlight uncertainty. In debates about genetic engineering, uncertainty about the prediction of how transgenic organisms will behave in the environment, or uncertainty about how markets will react to the adoption of transgenic organisms, is cited as a reason to invoke the precautionary principle. In debates about nuclear waste disposal at Yucca Mountain, uncertainty about how the storage facility will perform in the long term is cited as reason to question the suitability of nuclear waste repository.

SEE ALSO: Climatic Data, Nature of the Data; Environmental Protection Agency; Measurement and Assessment.

BIBLIOGRAPHY. Daniel Bodansky, "Scientific Uncertainty and the Precautionary Principle," *Environment* (v.33/7, 1991); John Lemons, ed., *Scientific Uncertainty and Environmental Problem Solving* (Blackwell Science Press,

1996); Allison MacFarlane and R.C. Ewing, *Uncertainty Underground: Yucca Mountain and the Nation's High-Level Nuclear Waste* (MIT Press, 2006).

Dustin Mulvaney
University of California, Santa Cruz

Precipitation

PRECIPITATION IS THE primary factor that controls the hydrologic cycle. It takes different forms, such as rain, snow, hail, sleet, drizzle, dew, and fog. It supplies most of the fresh water on the Earth. Most precipitation starts from space as snow, as the upper space is cooler. If the temperature of the surface closer to the ground is below 32 degrees F (0 degrees C), then the precipitation falls on the ground in the form of snow. If the ground and closer surface temperature is above 32 degrees F (0 degrees C), the precipitation takes the form of rain. When the air at the ground is below freezing, the raindrops can freeze while heating the ground, and that is known as freezing rain. When a dust particle in the atmosphere attracts a moisture drop, hail is formed. Drizzle consists of very small raindrops, 1/1000 of a normal raindrop size. Sleet is a type of precipitation between rain and snow, but very distinct from hail. Dew is another form of precipitation that can be seen in the early morning on colder days. Water vapor in the atmosphere condenses on the surface of exposed objects at a greater rate than that at which it can evaporate, developing dew. Fog as such is not precipitation, but is considered one because of its low-altitude occurrence. This consists of a cloud in contact with the ground, and produces water droplets when intercepted with vegetation or other exposed objects. If the precipitation evaporates before reaching the ground, it is then known as virga.

The 1996 Hurricane Fran, shown in a weather sattelite image, caused about $5 billion in damages in North Carolina alone. Greater precipitation, including hurricanes, typhoons, ocean depressions, and floods, are being experienced globally.

HOW PRECIPITATION OCCURS

Dynamic and adiabatic cooling causes precipitation. Due to this process, condensation of water vapor occurs and then falls to the Earth as rainfall, snowfall, or as other forms of precipitation. Vertical air motion is the leading factor of all rainfall from clouds after condensation. Rising air in the tropics and mid-latitudes (40 to 60 degrees N and S latitudes) causes more precipitation and descending air patterns in the subtropics (20 to 30 degrees N and S latitudes) and in the poles causes less. Precipitation is classified into different categories based on the conditions that generate vertical air motion: convective, orographic, and cyclonic.

Convective precipitation is mostly seen in the tropics. Heating up of the air at groundlevel, then moving upward and mixing with the water vapor in the atmosphere, with dynamic cooling in space, causes precipitation to fall on Earth. This is called convective precipitation. Orographic precipitation occurs due to the interception of moisture-laden air or clouds by mountain ranges. Cyclonic precipitation is caused by movement of moist air masses from high-pressure regions to low-pressure regions. In the hydrologic cycle, the total volume of precipitation onto land is measured at 42,471 sq. mi. (110,000 sq. km.), while the total volume of precipitation on the ocean surface is 176,834 sq. mi. (458,000 sq. km).

Precipitation has greater ecological, geographical, and regional impact due to its characteristics, such as relative amount, seasonal timing, and most importantly the size and intensity. Low-intensity and well-distributed seasonal precipitation is good for agriculture. High-intensity, long-duration precipitation creates more problems than good. Precipitation is governed primarily by atmospheric water vapor, but its variation depends upon other climatic factors such as temperature, wind, and atmospheric pressure at different locations and season. The amount of water vapor is very high in the atmosphere closer to water bodies. Therefore, coastal areas always have heavier precipitation (high-intensity, long-duration) than inland areas. Thunderstorms, hurricanes, typhoons, cyclones, blizzards, and hailstorms are high-intensity precipitations with damaging ability.

It is essential to understand the precipitation distribution process and its temporal and spatial variation for water resources planning and management for the betterment of the society. Precipitation distribution mechanisms include interception by vegetation, filling in depression storage, infiltration to soil and ground water, surface detention, and overland or surface flow or runoff. Low-intensity, short-duration storms are good for land and vegetation because most of the water either stays in depression storage or infiltrates to soil and groundwater for groundwater recharge. But high-intensity, long-duration storms are detrimental because most of the precipitation (rainfall) is wasted as runoff. The excessive runoff erodes soil, creates floods, landslides, and other damaging effects. Blizzards are examples of such high-intensity, long-duration precipitation. A heavy amount of snowfall accompanied by high wind creates problems. Many high-intensity, long-duration storms are currently being encountered, believed to be due to global warming.

EFFECTS OF GLOBAL WARMING ON PRECIPITATION

Global warming is the effect of increasing atmospheric concentrations of greenhouse gases. The near surface of the Earth has warmed by nearly 1 degree F (0.6 degrees C) during the 20th century. It may continue in this century, warming the globe further, so that a warmer ocean surface would result. There would be increased evaporation rate and subsequent increase in the other components of the hydrologic cycle, like water vapor in the atmosphere and consequent higher precipitation amounts. Computer simulation models found that a global warming by 7.2 degrees F (4 degrees C) is expected to increase global precipitation by about 10 percent and that rainfall intensity will increase. Scientists using models found that the upper tropospheric water vapor amount will increase by 15 percent with each degree of atmospheric temperature rise. The global water vapor amount will increase by 7 percent with each degree of atmospheric temperature rise.

Another major downside of global warming is less snowfall throughout the world. As the surface temperature is rising, raindrops, in many parts of the Earth, cannot take the shape of snow before reaching the ground. Snow is better for the land than rain because it helps in water conservation. Rainfall becomes runoff to oceans and is a waste if not conserved artificially through dams, earthen embankments, and soil and water conservation structures, whereas snow remains on the ground and melts slowly to release water for agriculture and other consumption. Slow melting of snow also facilitates more accumulation of soil moisture and ground water recharging. Agriculture in the northern United States and southern Canada depends upon soil moisture conserved by snowpack on the agricultural land. With less snowfall, due to increase in surface temperature, there may be less snow deposition on agricultural land, and agriculture would suffer.

The irony of global warming is the occurence of more rain resulting in less water. Due to warmer temperatures in the spring, snowpacks in the mountains are melting unseasonably and at a rapid rate. Many perennial rivers in the world are experiencing shortages of flowing water in summer months and groundwater recharging has lessened. The water from a quick snowmelt cannot be arrested in reservoirs due to lack of space. Therefore, the systems failing to hold the entire season of runoff would face challenges to meet the water demand for agriculture and other purposes. Increased quick melting of snowpacks (due to a rise in surface temperature) in the northwestern United States and India causes spring and summer floods.

It is estimated that there could be a 15 to 30 percent reduction of water available for human consumption from California's Sierra Nevada mountains. Agriculture in the Canadian prairies could be the worst affected, due to spring water runoff. Sufficient water may not be available for subsequent crop seasons. Many rivers in the world will experience water shortages in dry seasons because during dry months, most of the perennial rivers get water from snow and glacier melting. Glacial retreats are clearly visible in the Poles, Greenland, and Antarctica. This occurs due to an insufficient supply of water to glaciers from precipitation as compared to the loss of water from melting and sublimation.

Global warming would cause more precipitation on the Earth, but would create more detrimental effects to agriculture, ecology, and, above all, society. The wetter areas of Earth would become wetter, and the drier areas may become even drier.

SEE ALSO: Climate Change, Effects; Global Warming; Rain; Rainfall Patterns; Weather.

BIBILIOGRAPHY. American Geophysical Union, "Water Vapor in Climate System," www.agu.org (cited August 2007); R.R. Britt, "The Irony of Global Warming: More Rain, Less Water," *Live Science* (November 16, 2005); Thomas Dunne and L.B. Leopold, *Water in Environmental Planning* (W.H. Freeman and Company, 1978); P. Groisman, et al., "Changes in the Probability of Heavy Precipitation: Important Indicators of Climatic Change," *Climatic Change* (v.42, 1999); J. Huang and H.M. van den Dool, "Monthly Precipitation–Temperature Relations and Temperature Prediction over the United States," *Journal of Climate* (v.6, 1993); M. Hulme, "Estimating Global Changes in Precipitation," *Weather* (v.50, 1995); U.S. Geologic Survey, "Glacier Retreat in Glacier National Park, Montana," www.usgs.gov (cited August 2007).

Sudhanshu Sekhar Panda
Gainesville State College

Preparedness

PREPAREDNESS FOR GLOBAL warming and climate change requires multilevel planning, at the international, national, local, and individual levels to deal with the direct effects of climate changes in temperature, precipitation, wind, storm patterns, sealevel, as well as the indirect strain on world resources leading to migration, famine, and conflicts.

While mitigation strategies like reducing greenhouse gas emissions are a start, they do not take precedence over readiness to respond to natural disaster emergencies associated with the impact of climate change (intense storms, flooding, wildfires, public health, or the necessity to deal with future environmental pressures, the decreasing longevity of infrastructure with roads, bridges, waterworks, buildings, and facilities requiring earlier replacement, repair or modifications to remain safe for use. Adaptive measures in the form of physiological, social, and cultural measures will allow people to live throughout the world.

ASSESSMENT OF IMPACT

In order to prepare for climate change, assessments must be completed and available for decision-making processes. On the international level, the International

Governmental Panel on Climate Change encourages collaboration among scientists, allied professionals, and policymakers, providing a forum for the collection of research material and posing questions to spur planning. The take-away message is evaluation to determine global policy toward a common goal while addressing implications and increasing cooperation between governments.

On the national and local levels, any assessment of the extent of impact climate changes (higher temperatures, rising sea levels, changing weather patterns will have on human health, ecosystem diversity and productivity, agricultural production, water supply, sanitation, infrastructure) must factor in the associated costs and benefits in developing sustainability plans, preparation for emergency situations, and adaptive measures.

Each area is different, with varying susceptibilities, policies, institutions, and social/cultural structures. National and local assessments to be most effective must evaluate the capacity to handle emergency measures associated with natural disasters and identifying susceptible areas of the environment and resources including agriculture, fisheries, forestry, fauna, human health, water supply and sanitation, infrastructure and construction, land use in hazard-prone areas (flood plain, islands), and disaster management.

Challenges to assessment for impact include factors not related to climate change that will impact the areas being considered at risk because of climate change include the dynamics of society and economy (demographic trends, agricultural management, improving and new technologies, cultural preferences, opportunities for employment, availability, and changes in transportation). Complex dynamics in human relationships with the environment, self-reliance, growing population, and increased urbanization may produce very different impacts between urban and rural areas.

IMPACTS TO CONSIDER

When making assessments, a variety of elements need to be considered that cross boundaries of social structures and practices as well as environmental conditions. Severe weather destruction has highlighted areas that lack preparedness. The current global situation for preparedness is still lacking. Severe weather causes great devastation. Even in a highly developed country, as seen in the aftermath of the 2005 Hurricane Katrina hitting New Orleans and other areas on the U.S. Gulf Coast, problems were experienced with infrastructure, the failure of levees to hold back storm surge water, and mobilizing emergency supplies.

Planning for impacts from rising global temperatures must take into account a wide range of factors including the impact on human health, comfort, lifestyle, food production, economic activity, and residential and migration patterns including tourism.

Housing or shelter must be sufficient to meet the needs of an emergency situation or for migration from an area no longer suitable for living or to meet expectations for increased tourism to climates that so far have not been inundated by travelers. In addition, supportive measures must be able to supply water, sanitation, communication, energy, transport, and industry; and social and cultural services including health services, education, police protection, recreational services, parks, and museums for areas with an increased population.

With rising temperatures, weather patterns could alter in frequency and seasonality of precipitation, weather-related events could increase, including droughts, floods, severe tropical storms with associated storm surge flooding and wildfires from increased temperatures and drought conditions. With a rise in evaporation and precipitation rates, water availability and quality could be affected, with lower groundwater levels, decreased surface area and water levels of many lakes or inland waterways, as well as altering natural habitats. Potential impacts range from loss of property, effects on housing and street/road conditions, effects on construction materials, stress on sewage systems and potential overflow from excess storm water and drainage failures.

Rising sea levels could displace residents of delta regions (Nile, Ganges, Yangtze, Mississippi) from homes and livelihoods. Island nations could become uninhabitable. Coastal erosion on gradually sloping coasts by encroaching water could affect densely populated cities (New York, London) and important seaports.

Cooler climates may see an increase in agriculture and warmer climates may see a decrease in agriculture. Feeding the world's population means an inherent dependence on agricultural production, which is highly sensitive to climate change. Land degradation may produce either abandonment of the land or require changes in cultivation practices to improve yields and restoration of soil. Possible changes in fertileness of the land could increase or decrease food production capability for the agricultural country and as an export to other world regions.

Health impacts include temperature stress from either extreme heat or cold, especially among high-risk groups (children, the elderly, and those with already compromised health); air pollution exposure with increased incidence of respiratory disease; chemical pollution; water quality or water shortages from precipitation changes, flooding, or in some areas already a lack of safe water; and vector-borne diseases; lack of physical or economic access to health services or insufficient capacity of health services. In the event of disaster refugees or migration, sanitary facilities and housing could become quickly overburdened, enhancing the spread of communicable diseases.

Climate change could modify supplies and consumption patterns. These impacts would vary by region based on cost of various types of food and fiber. Changing availability of resources might lead to changed diets, production patterns, and employment levels. Major impact could be felt by the energy, transport, and industrial sectors, with increasing need balanced against dwindling supply.

Impact on physical and social environment would also vary by region and could include loss of housing (from wildfires, flooding, mud slides); loss of living resources (water, energy supplies, food, or employment); loss of social and cultural resources (cultural properties, neighborhood or community networks); decline in living standards (conditions caused by mandatory evacuation, contamination of water supply); total loss of livelihood following land degradation (erosion of top-soil, over-cultivation or deforestation); or a major natural disaster like flooding or drought. In some areas, physical and social environments could improve. Communication technologies (cell phones, computers, and fax machines) could have a positive impact including increased potential for decentralization of the population by enabling many professional and technical people to perform work in homes far removed from major metropolitan areas.

POSSIBLE SOLUTIONS

The Hyogo Framework for Action signed in January 2005 by 168 countries details steps to reduce the impact of natural hazards on populations. The result has been 40 countries adjusting national policies to give priority to disaster risk reduction.

Examples of disaster planning paying off in action include Jamaica in 2004. A community disaster response team issued early warning by megaphone, and used risk maps and equipment assembled by the Red Cross for successful evacuation of all area residents. Hurricane awareness is taught in schools in Cuba along with practice drills. The Citizens' Disaster Response Center in the Philippines helps to create disaster management plans and provides emergency response. These examples indicate effectiveness and indicate the need for global, regional, national, and local early warning systems to alert populations of impending disasters.

While the previous examples are for disasters, the same ideas can be used to implement climate change preparedness, of which natural disasters are one element. One caveat for planning is the necessity to include feedback loops to allow for changes as new information becomes available or conditions are modified.

The challenge is the predictive factor (strategies planning far into the future, 10 years, 25 years, 50 years, or 100 years) in the presence of current needs of access to clean water management, production of energy allowing for cost-effectiveness and to allow developing countries to utilize cheap fossil fuels available to them, and supplying enough food for populations with limited agriculture due to already stressed conditions. To meet food supply needs far into the future, research to develop new varieties of wheat, corn, and soybeans for resistance to drought and heat and still produce good yields and strategies for future irrigation sources should begin now, building new wells and reservoirs instead of waiting until water shortages persist. Centralized stockpiles of grain could provide for increases in food needs or to supply areas with crop devastation.

With changing weather patterns, coastal areas should prepare for and be able to demonstrate effective evacuation procedures to deal with rising sea levels and more severe hurricanes, as well as to assess the risks of new construction in low-lying areas. Global, regional, national, and local early warning systems to alert populations of impending disasters should be developed, implemented, and tested.

Water supplies, both in rural areas and in municipal water infrastructure, should be set up for equitable availability and pricing. Some areas have no access to fresh water and must use desalinization systems to make clean water from seawater.

Environmental policy should strengthen institutional and legislative environmental framework at national and regional levels, including environmental authorities, and incorporating new laws, as well as implementing environmental control standards for air quality, including reduced emissions of greenhouse gases, reducing ground level ozone and particulates from stationary and transport sources, and more tightly enforcing pollution and land use regulations.

Emphasis should be placed on nature and biodiversity such as management of protected areas and a national biodiversity strategy and provide for better management of all natural resources, including forests, fishing, soils, water, air, wildlife. Altered habitats will cause animal migration. Barriers standing in their way could be removed by setting up migration corridors to connect natural areas and allow them to migrate safely. In the event these measures aren't enough, wildlife managers may have to capture and move certain species.

Reducing health impacts from climate change can be acted upon early by preventing the onset of disease from environmental disturbances, in an otherwise unaffected population (such as supply bed nets to all members of a population at risk of exposure to encroaching malaria, taking precautions against mosquito bites to prevent West Nile virus, early weather watch warning systems). Surveillance systems could be improved in sensitive geographic areas with the potential for epidemics under certain climatic conditions, including those bordering areas of current distribution of vector-borne diseases (plague from prairie dogs, hanta virus from rats, malaria from mosquitoes). Vaccination programs could be intensified and pesticides used for vector control. Drugs for prophylaxis and treatment could be stockpiled.

In certain areas of the world, irrespective of climate change, breakdowns in public health measures have been responsible for many recent outbreaks of disease. In those areas, climate change would add to the health burden; current and future health problems related to the environment share similar causes of eco-

nomic factors and access issues. All areas must have sufficient trained staff to handle healthcare issues, as well as technology for diagnosing and treatment and medications or medical interventions required.

Construction practices and advancing technology make possible new building techniques, including floating houses to weather storm surge and flooding in hazard areas. Current building material (including roads) may not be able to withstand future climate change conditions and will require replacement or restoration. Public reconstruction to ensure safe bridges and other infrastructure is necessary. In some areas, drainage improvements will need to be made for water and sewage.

Adaptation refers to actions taken to lessen the impacts of the anticipated changes in climate. The ultimate goal of adaptation interventions is the reduction, with the least cost, of disruptions in living standards, of suffering from diseases, injuries, or disabilities, and of destruction of habitats, ecosystems, and species (both plant and animal).

AWARENESS RAISING

To make any plan for preparedness work, the information collected in assessments must reach all relevant stakeholders in the process of building and reaching a consensus on adaptation strategy. Providing environmental information to the public is essential to any preparedness method; if people don't know what to do in the event of an emergency or steps to take to prevent or adapt to changing conditions, the plan will fail. Awareness campaigns directed at the general public and to those persons who are directly affected should be made available through all forms of media and should be easily understood.

On the individual level, certain precautions can be taken to prepare for natural disasters, and some associated climate change, including extreme heat or winter weather, flooding, hurricanes, landslides and mudslides, tornadoes, tsunamis, and wildfires. Taking preparedness actions helps people deal with disasters in an effective manner. Having emergency supplies of water, food, and first aid prepared and easily available allows an individual or families to either evacuate quickly or shelter in place as appropriate. Having a plan of action for where to go, how to reconnect with family members, and dealing with finances lessens the stress involved with disasters.

SEE ALSO: Intergovernmental Panel on Climate Change (IPCC); Public Awareness; World Health Organization.

BIBLIOGRAPHY. Daniel Sitarz, ed., *Agenda 21: The Earth Summit Strategy to Save our Planet* (EarthPress, 1993); University of Wyoming, "Continued Global Warming Could Destroy Existing Climates and Create New Ones," www.uwyo.edu (cited March 2007); WHO, "Early Human Health Effects of Climate Change and Stratospheric Ozone Depletion in Europe" www.who.int (April 9, 1993); Worldwatch Institute, *Vital Signs 2006–2007* (W.W. Norton & Co., 2006).

LYN MICHAUD
INDEPENDENT SCHOLAR

Princeton University

PRINCETON UNIVERSITY IS a private coeducational research university located in Princeton, New Jersey. It is one of eight colleges and universities that belong to the Ivy League.

Originally founded at Elizabeth, New Jersey, in 1746 as the College of New Jersey, it relocated to Princeton in 1756 and was renamed "Princeton University" in 1896. Princeton was the fourth institution of higher education in the United States to conduct classes. Princeton has never had any official religious affiliation, rare among American universities of its age. At one time, it had close ties to the Presbyterian Church, but today it is nonsectarian and makes no religious demands on its students. The university has ties with the Institute for Advanced Study, Princeton Theological Seminary, and the Westminster Choir College of Rider University.

Princeton has traditionally focused on undergraduate education and academic research, though in recent decades it has increased its focus on graduate education and offers a large number of professional master's degrees and Ph.D. programs in a range of subjects. The Princeton University Library holds over six million books. Among many others, areas of research include anthropology, geophysics, entomology, and robotics, while the Forestall Campus has special facilities for the study of plasma physics and meteorology.

The Atmospheric and Oceanic Sciences Program is a unique collaboration between a renowned academic institution, Princeton University, and a world-class climate research laboratory, the Geophysical Fluid Dynamics Laboratory (GFDL) of the National Oceanic and Atmospheric Administration. The program hosts graduate students, postdoctoral researchers, and visiting senior researchers, as well as permanent research staff and faculty. The highly flexible graduate program offers students opportunities for research and courses in a wide variety of disciplines related to climate, including the physics and dynamics of the atmosphere and ocean, atmosphere and ocean chemistry and biological processes, global climate change, and paleoclimate. Through the in Science, Technology, and Environmental Policy (STEP) program at the Woodrow Wilson School of Public and International Affairs students can explore climate- [and air pollution] related policy. Students benefit from an unusually low student-to-faculty ratio and access to GFDL's supercomputing resources.

The Program in Atmospheric and Oceanic Sciences (AOS) offers graduate study under the sponsorship of the Department of Geosciences. An understanding of the complex behavior of the atmosphere and oceans requires a balanced effort in theoretical analysis, numerical modeling, laboratory experiments, and analysis of observations. The AOS program benefits from the research capabilities of the Geophysical Fluid Dynamics Laboratory (GFDL) of the National Oceanic and Atmospheric Administration, which is located on the Forrestal Campus. GFDL has a major in-house supercomputer facility to which students have direct access for their research. Many GFDL scientists are active in the AOS program as lecturers with the rank of assistant through full professor. The geosciences department, with its activities in physical and chemical oceanography, paleoceanography, and paleoclimatology, collaborates with GFDL in providing an academic program of courses and seminars.

The program is internationally recognized for its development of models of atmospheric and oceanic circulation and climate, particularly studies related to global warming. Additionally, it is world renowned for its development of earth system models, specifically as related to the global carbon cycle, and for its training of graduate students and postdocs. The student to postdoc to faculty ratio is typically about 1:1:1, which provides graduate students and postdocs with a highly stimulating environment for learning and carrying out their research.

Courses offered by the department which focus on climate change include:

Introduction to Physical Oceanography: Study of the oceans as a major influence on the atmosphere and the world environment. The theoretical and observational bases of our understanding of ocean circulation and the oceans' properties.

Introduction to Atmospheric Science: Atmospheric composition and thermodynamics including effects of water. Simple radiative transfer, elementary circulation models, phenomenological description of atmospheric motions, structure of the troposphere, stratosphere, mesosphere, and thermosphere, chemistry of ozone, and comparison with atmospheres on other planets.

Atmospheric Radiative Transfer: The structure and composition of terrestrial atmospheres. The fundamental aspects of electromagnetic radiation, absorption and emission by atmospheric gases, optical extinction by particles, the roles of atmospheric species in the Earth's radiative energy balance, the perturbation of climate due to natural and anthropogenic causes, and satellite observations of climate systems are also studied.

Atmospheric Chemistry: Natural gas phase and heterogeneous chemistry in the troposphere and stratosphere, with a focus on elementary chemical kinetics; photolysis processes; oxygen, hydrogen, and nitrogen chemistry; transport of atmospheric trace species; tropospheric hydrocarbon chemistry and stratospheric halogen chemistry; stratospheric ozone destruction; local and regional air pollution; and chemistry-climate interactions are studied.

Atmospheric Thermodynamics and Convection: The thermodynamics of water-air systems. The course gives an overview of atmospheric energy sources and sinks. Planetary boundary layers, closure theories for atmospheric turbulence, cumulus convection, interactions between cumulus convection and large-scale atmospheric flows, cloud-convection-radiation interactions and their role in the climate system, and parameterization of boundary layers and convection in atmospheric general circulation models are also studied.

Introduction to Geophysical Fluid Dynamics: Physical principles fundamental to the theoretical, observational, and experimental study of the atmosphere and oceans; the equations of motion for rotating fluids; hydrostatic and Boussinesq approximations; circulation theorem; and conservation of potential vorticity; scale analysis, geostrophic wind, thermal wind, quasigeostrophic system; and geophysical boundary layers.

Atmospheric and Oceanic Wave Dynamics: Observational evidence of atmospheric and oceanic waves; laboratory simulation; surface and internal gravity waves; dispersion characteristics; kinetic energy spectrum; critical layer; forced resonance; and instabilities.

Physical Oceanography: Response of the ocean to transient and steady winds and buoyancy forcing. A hierarchy of models from simple analytical to realistic numerical models is used to study the role of the waves, convection, instabilities, and other physical processes in the circulation of the oceans.

Numerical Prediction of the Atmosphere and Ocean: Barotropic and multilevel dynamic models; coordinate systems and boundary conditions; finite difference equations and their energetics; spectral methods; water vapor and its condensation processes; orography, cumulus convection, subgrid-scale transfer, and boundary layer processes; meteorological and oceanographic data assimilation; dynamic initialization; verification and predictability; and probabilistic forecasts.

Current Topics in Dynamic Meteorology: An introduction to topics of current interest in the dynamics of large-scale atmospheric flow. Possible topics include wave-mean flow interaction and nonacceleration theorems, critical levels, quasigeostrophic instabilities, topographically and thermally forced stationary waves, theories for stratospheric sudden warmings and the quasi-biennial oscillation of the equatorial stratosphere, and quasi-geostrophic turbulence.

Weather and Climate Dynamics: An examination of various components of the Earth's climate system. Dynamics and physical interpretation of principal tropospheric circulation systems, including stationary and transient phenomena observed in middle and low latitudes. Reviews of phenomena of topical interest, such as El Niño, seasonal climate anomalies, and natural and anthropogenic climate changes.

SEE ALSO: Climate Change, Effects; Global Warming; Oceanography.

BIBLIOGRAPHY. Geophysical Fluid Dynamics Laboratory, www.gfdl.noaa.gov (cited November 2007); Princeton University, www.princeton.edu (cited November 2007).

FERNANDO HERRERA
UNIVERSITY OF CALIFORNIA, SAN DIEGO

Public Awareness

PUBLIC AWARENESS IN the United States of the issue of global warming increased from about one-third in the early 1980s to near 100 percent 25 years later. By 2007, climate change was featured in the media almost daily. Awareness does not necessarily imply acceptance; although polls indicate that over half of Americans consider climate change to be real, there remains widespread public uncertainty about the degree to which human activities are involved, and to what extent CO_2 emissions need to be curtailed. There also remain widespread misconceptions about the meaning of global warming, and likely effects.

Public acceptance of human-induced climate change as a real phenomenon has lagged well behind the scientific consensus. In the mid-1970s, the popular media widely reported that the Earth was cooling and may be entering the next glacial interval, accelerated by light reflected off atmospheric particulates from pollution. The reports were based on the ideas of several scientists espoused primarily outside peer-reviewed literature. By the late 1970s, scientific consensus emerged from early generation global climate models that the warming influence of greenhouse gases was stronger than the cooling influence of particulates and insolation change. Scientific evidence that the climate was warming first received major coverage in a 1981 front-page article in the *New York Times*. Considerable advances in scientific understanding of current and past climate change occurred in the 1980s; this received enhanced public recognition with the 1988 congressional testimony by climatologists that coincided with a record hot summer.

As calls for government controls to reduce greenhouse gases increased, climate change discussions and media coverage of it grew politicized. In the 1990s, media, in efforts to offer "balanced" reporting, covered a small number of climate change skeptics in roughly equal proportion to the scientific consensus that climate is warming, which had grown to close to 100 percent in peer-reviewed scientific literature. The public was thereby given the impression that a considerable scientific controversy still existed. Debate about U.S. participation in the international Kyoto Accord in late 1997 and again in 2001 further increased politicization of the issue.

Several events in the mid-2000s swung U.S. public opinion from simple awareness of the issue to greater acceptance that global warming was happening. During this time, skeptics also changed stances, from whether climate change was happening to whether humans were causing observed changes. The severe hurricane season of 2005 (in particular, Hurricanes Katrina and Rita) centered U.S. public attention on potential domestic human and financial costs of climate change. Al Gore's 2006 documentary, *An Inconvenient Truth*, one of the most watched documentaries of all time, stimulated a groundswell of activity to further increase awareness, though to some degree maintaining the politicization of the issue. Several very warm years globally during the 2000s also helped give climate change greater reality to a broader geographic segment of U.S. citizens accustomed to hearing about warming in other areas of the world.

FACTORS THAT MAKE PUBLIC UNDERSTANDING OF CLIMATE CHANGE DIFFICULT

Scale: It is difficult for most people to grasp: scales of space the size of the Earth and its atmosphere; scales of time that include analyzing data from thousands of years in the past and up to decades or centuries into the future; and scales of human influence that involve billions of people, each contributing some quantity of CO_2 to the atmosphere.

Complexity: Climate is a complex system that is difficult for any one person, even specialists, to understand in total; in global warming, some places cool, which is why climate change is now the preferred term; spatial and temporal variability in weather systems means that even places that are warming on average may be occasionally unusually cold; in some places climate change is manifested more by precipitation change than temperature change.

Models and uncertainty: Computer models of climate are complex sets of mathematical equations that are "black boxes" for most of the public, and even many scientists; it can be confusing that different results occur for different researchers' models, as is evaluating the probability of events decades in the future, contingent on human actions in the meantime.

Personal reality: In many places climate change may not be readily evident from casual neighborhood observations; changes at the poles may seem personally irrelevant; warming may sound attractive in cooler climates.

Personal beliefs: Spiritual or philosophical beliefs may indicate that the Earth does not change, that humans need not concern themselves with personal influence on the Earth, or that near-term spiritual events on Earth will render changes to the Earth meaningless.

EDUCATION

Many efforts have developed in recent years to go beyond increasing public awareness, to educate the public on the nature of climate change and what can be done about it. The former is especially the purview of science education and the latter of personal and governmental action. Few K-12 school curricula have climate change as a major topic, though building block concepts such as greenhouse warming may be found in Earth and environmental courses at the high school level. Professional development for teachers at the state and national level may be needed in large numbers in the coming decades. Nonformal educational groups such as Scouts and 4-H are alternative settings for introducing climate education to youth. Major research partnerships such as Global Learning and Observations to Benefit the Environment (GLOBE) provide opportunities to collect data for scientific research that is relevant to climate change. Many colleges and universities are developing campus sustainability models and associated student action groups and courses.

Some organizations of informal education, such as museums and science centers, are in a good position to present public climate change education to a wide range of age groups. The International Polar Year (2007–09) provided an opportunity to present outreach associated with research on polar processes, including polar warming. Many grassroots groups have started in towns and cities across the United States to provide information on climate change to local citizens, particularly on how to take action to reduce CO_2 emissions. Among the most influential means of informal education remain radio and television documentaries on the topic.

SEE ALSO: *An Inconvenient Truth*; Education; Gore, Albert, Jr.; Media, Internet; Media, TV; United States.

BIBLIOGRAPHY. Al Gore, *An Inconvenient Truth* (Rodale, 2006); K.R. Stamm, Fiona Clark, and P.R. Eblacas, "Mass Communication and Public Understanding of Environmental Problems: The Case of Global Warming," *Public Understanding of Science* (v.9, 2000); Spencer Weart, "The Public and Climate Change," American Institute of Physics, www.aip.org (cited September 2007).

Robert M. Ross
Warren D. Allmon
Paleontological Research Institution

Qatar

LOCATED IN THE Persian Gulf, the State of Qatar has a land area of 4,416 sq. mi. (11,437 sq. km.), with a population of 841,000 (2006 est.), and a population density of 192 people per sq. mi. (74 people per sq. km.). With massive prosperity from the petroleum industry and from the production of natural gas, some 90 percent of Qatar's population lives in urban areas. Only one percent of the country is arable, with a further five percent used for meadows and pasture, with the country having no forests or woodland.

Qatar has the highest rate of carbon dioxide emissions per capita, and one of the highest in the world as far back as 1950. In 1990, it was measured at 22.5 metric tons per person, rising dramatically to 37.4 metric tons per person, dramatically increasing to 55.3 metric tons in 1992. It rose steadily, reaching 69.2 percent in 2004, considerably more than the second highest per capita emitter, Kuwait, which reached a peak of 38 metric tons in 2004. This high level comes from the use of natural gas for electricity generation, all electricity coming from the use of fossil fuels. Electricity usage remains high, with heavy use of air conditioning, and also the running of a large desalination plant.

Gaseous fuels account for 80 percent of all carbon dioxide emissions, with liquid fuels making up a further 19 percent. The generation of electricity contributes 28 percent of all carbon dioxide emissions, with other energy industries making up 32 percent, and manufacturing and construction another 32 percent. In spite of its small size, transportation accounts for 8 percent of the emissions, a result of a wealthy economy that has a large private ownership of automobiles, and gasoline is cheap.

The Qatar government took part in the UN Framework Convention on Climate Change signed in Rio de Janeiro in May 1992. They accepted the Kyoto Protocol to the UN Framework Convention on Climate Change on January 11, 2005, with it entering into force on April 11, 2005.

SEE ALSO: Automobiles; Oil, Consumption of; Oil, Production of.

BIBLIOGRAPHY. "Qatar—Climate and Atmosphere," www.earthtrends.wri.org (cited October 2007); M.A. Weaver, "Revolution from the Top Down," *National Geographic* (v.203/3, 2003).

ROBIN S. CORFIELD
INDEPENDENT SCHOLAR

Quaternary Era

THE QUATERNARY PERIOD, the most recent geologic interval, represents the last 1.8 million years of time. Its most striking feature is that the Earth had cold polar regions, which led to periodic development of continental glaciers. The prolonged Ice Ages, comprising the main part of the Quaternary interval, ended 10,000 years ago, when the continental glaciers had melted. Compared to the many cold periods (glacial intervals) of the Quaternary, the last 10,000 years of the present interglacial has been comparatively warm.

The changes in climate over the past 150 years have shown a varied history. The world endured a historic cool period during the late 1800s (the Little Ice Age), followed by a warm period of the 1930s (the Dust Bowl years), and since about 2000, the climate has been more variable, reaching extremes in warmth at high latitudes. Reports from Vikings show that 1,000 years ago the climate of Greenland and Iceland was warmer than today (the Medieval warming). While these historic shifts in temperature are relatively moderate, much larger changes in temperature on Earth have occurred in the geologic past.

PROXY DATA

Using the changing oxygen isotope ratios from ice cores and marine sediment cores, scientists have discovered global changes, recorded synchronously over a wide range of latitudes. One long climate proxy record was taken in the Antarctic, the Vostok core. From it, the inferred temperature over a long interval is based on the temperature-sensitive ratio of oxygen isotopes, ^{18}O to ^{16}O. In the Vostok core, isotopes of oxygen have been used to develop Earth temperature histories extending over 400,000 years. Trapped gas bubbles record the history of atmospheric CO_2 concentrations for over this period (data from the National Climatic Data Center, Asheville, North Carolina). Because the isotope ^{18}O is heavier than ^{16}O, the proportions of each vary depending on the climate region. In alpine areas such as in the Alps of Switzerland and in the polar regions, ^{18}O is more abundant, while in the lowlands of the middle and low latitudes ^{16}O prevails in the atmosphere. By calibrating these, scientists can use the proportions of oxygen isotopes taken from ancient sediments or ice cores as an index of average annual temperature.

The changing proportions of isotopes can be matched and dated with a geomagnetic signal of polar reversals (the Earth's poles changed in magnetic signals), a known time scale based on isotopically dated magnetic signals that are worldwide. Another source of long climate records are the deep-sea sediment cores from which oxygen isotopes can be extracted from calcareous plankton (foraminifera). These data carry the paleoclimatic records back through more than 60 million years through the Tertiary period and the time of the last dinosaurs.

The Ice Ages, which comprised the main part the last million years, was a globally cold period with increasingly variable swings of climate. The glacial pattern continued over long intervals, with only a few comparatively short warm or interglacial periods. The last major glaciation came in two parts; in the United States, these are called the early Wisconsin (80,000–28,000 years before present, oxygen isotope stage 4) and the Full Glacial or late Wisconsin (23,000–15,000 years ago, oxygen isotope stage 2); between these a somewhat warmer middle Wisconsin period occurred 28,000–23,000 years ago. The maximum of the last major glaciation occurred about 18,000–15,000 years ago (called the Full Glacial). After 15,000 years ago, a global warming began, and continental ice sheets melted by about 10,000 years ago. The period after 10,000 years ago, or postglacial, is called the Holocene or Recent period representing the present interglacial.

The period of the last glaciation (Full Glacial) brought continental ice down to the middle latitudes in both hemispheres. In Europe, ice from the Scandinavian highlands spread southward over the Netherlands and mountain glaciers covered the Alps. Ice stood over parts of northern Siberia, Greenland, and parts of Alaska. Permafrost ice developed underground in Siberia, northern Canada, and Alaska as much as 300 ft. (91 m.) thick. Equivalent ice expansions occurred in the Southern Hemisphere.

As temperatures warmed between 15,000 and 9,000 years ago, the average annual temperatures at mid-latitudes increased by about degrees 11 F (6 degrees C). At Lamont National Observatory, scientists estimated that the difference in solar insolation between the Full Glacial and the mid postglacial was about 8 percent.

An overall trend within the Quaternary is apparent. In the first half, the amount of variance from year to year was fairly low, while in the last half of

A glacier in New Zealand. The prolonged Ice Ages, comprising the main part of the Quaternary interval, ended 10,000 years ago. The continental glaciers melted during a comparatively warm cold period, but some have remained.

the Quaternary, variance became more and more extreme. Looking back over records of the past million years, geologists estimate that there were at least 40 cold or glacial intervals, and these were of varying length and were not regular in occurrence. In the last 400,000 years of the Ice Ages, proxy data indicate that there were four extreme cool periods (glacials) interspersed with five variable warm periods (interglacials). The last major interglacial (spanning the interval of about 80,000–122,000 years before present) was similar in warmth to the present interglacial climate, but the previous interglacials were definitely cooler that the present one.

CAUSES

The main climatic changes of the Quaternary are linked to the orbital position of Earth in relation to the sun. Astronomer Milutin Milankovitch proposed the orbital theory that is established by a variety of evidence, starting with tree ring variation. The precession of the equinox, the obliquity of the Earth's orbit around the sun, and the eccentricity of the Earth's orbit all contribute to the level of insolation received by Earth and are therefore the cause of geologic shifts in climate.

Before the Quaternary was the warm so-called Tertiary period that brought tropical conditions to the mid-latitudes, and extensive temperate forests grew in the Northern Hemisphere. The north and south polar areas were at least 36 degrees F (20 degrees C) warmer than today, perhaps around 41 degrees F (5 degrees C) average annual temperature. Orbital causes for this part of Earth's history have not been specified, but undoubtedly were important factors.

IMPACTS

Global changes in climate forced major changes in plant and animal distributions, especially in the high and mid-latitudes. During the Full Glacial, remains of trees that now grow in the northern boreal forest were found in Tennessee where their fossils were associated with deciduous hardwoods and even cypress-swamp types, creating strange mixtures of genera. Plants and animals that had dispersed to lower latitudes and in areas of southerly environments during glacials dispersed northward again during the postglacial warming. Thus, the Quaternary period records some of the more extreme climate changes known in Earth's history.

SEE ALSO: Ice Ages; Little Ice Age; Milankovitch, Milutin; Orbital Parameters, Eccentricity; Orbital Parameters, Obliquity; Orbital Parameters, Precession.

BIBLIOGRAPHY. H.R. Delcourt and P.A. Delcourt, *Quaternary Ecology: A Paleoecological Perspective* (Chapman & Hall, 1998); Alan Graham, *Late Cretaceous and Cenozoic History of North American Vegetation* (Oxford University Press, 1999); P.D. Moore, W.G. Chaloner, and P. Scott, *Global Environmental Change* (Blackwell Science, 1996).

ESTELLA B. LEOPOLD
UNIVERSITY OF WASHINGTON

Radiation, Absorption

RADIATION IS ENERGY transmitted by electromagnetic waves. Electromagnetic waves travel at the speed of light (when passing through a vacuum) and have a characteristic wavelength, λ, which is inversely proportional to their frequency, ν, by

$$\lambda \ (m) = c \ (m \ s\text{-}1) \ / \ \nu \ (s\text{-}1),$$

where c is the speed of light. Electromagnetic radiation is conceptualized in contemporary theory both as a wave and as a stream of particles called photons (this dual approach is referred to as wave-particle duality). The energy of any photon, E, of radiation is inversely proportional to the wavelength by

$$E = h \ \nu,$$

where h is Planck's constant. This relationship allows us to order electromagnetic waves from high energy/ short wavelength (for example, x-rays), to low energy/ long wavelength (for example, radio waves). The resulting progression is referred to as the electromagnetic spectrum (Figure 1). The visible region of the electromagnetic spectrum is bound by infrared (IR) radiation on the lower energy side of the visible region (around

Figure 1: The electromagnetic spectrum

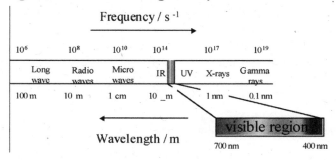

1 μm to 1 mm in wavelength), and by UV radiation (UV) on the higher energy side (from 400 nm to 1 nm). Microwave radiation is slightly lower in energy than IR, with a wavelength of around 1 cm.

All objects both emit and absorb radiation. Although all objects emit radiation at all wavelengths, the frequency of maximum emission, λmax, is proportional to the temperature of the object by Wien's law

$$\lambda max = \alpha \ / \ T,$$

where α is a constant equal to 2897 μm K. This implies that hotter objects emit higher energy radiation, as would be expected from everyday experiences. From Wien's law, the surface temperature of the sun can be calculated based on its emission peak at ~0.5 μm (green light) to be around 5800 K. The average temperature of the Earth's surface is around 18 degrees C (290 K) which corresponds to a peak emission at around 10 μm, in the infrared to microwave region.

There are three basic modes of motion: translational (movement through space), rotational, and vibrational. These are important, because along with electronic energy, they are the ways in which gas molecules can store energy. Quantum theory dictates that energy levels are discrete, not continuous; this implies that molecules will only absorb discrete frequencies of radiation that correspond to the gap between a high and lower energy state. UV radiation corresponds to the gap in energy between electronic energy levels in a molecule. When a molecule absorbs UV radiation, it may be promoted to an electronically excited state. In the general, this will make the bonds holding the atoms together weaker and may help facilitate reactions or the breakup of molecules. For example, the reactions that complete the Chapman mechanism in the stratosphere:

(i) $O2 + hv \rightarrow O + O$ ($\lambda < 240$ nm)
(ii) $O+O2 + M \rightarrow O3 + M$
(iii) $O3 + hv \rightarrow O2 + O*$ ($\lambda < 320$ nm)
(iv) $O* + M \rightarrow O + M$

The Chapman mechanism is chemically a null cycle, but it is important in the context of life, as it prevents most of the high-energy radiation from below 320 nm reaching the Earth's surface. IR and microwave radiation, being lower in energy than UV, correspond to the gaps between rotational and vibrational energy levels, respectively. Quantum theory dictates that molecules interact with IR/microwave radiation only when two conditions (selection rules) are met:

1. The energy of the radiation corresponds to the energy gap between two of the discrete energy levels in the molecule and,
2. The resulting motion results in a change in the dipole moment (electron distribution) of the molecule.

This implies that atmospheric components that are symmetrical molecules, such as N_2 and O_2 that have an even electron distribution and cannot be rotated, bent, or stretched in such as way as to create one, do not absorb microwave and IR radiation. Molecules possessing a dipole, which can be altered by bending and stretching, absorb radiation. Atmospheric components that fill this requirement include CO_2, CH_4, H_2O and CFC's, for example, the asymmetric stretch of CO_2. These gases are collectively known as greenhouse gases. Radiation from the sun is received at the Earth's surface, mainly in the UV and visible region, with other frequencies cut out by the atmosphere and electromagnetic field.

SEE ALSO: Chemistry; Radiation, Infrared; Radiation, Long Wave; Radiation, Short Wave; Radiation, Ultraviolet.

BIBLIOGRAPHY. Thormod Henrikson and David H. Maillie, *Radiation and Health* (Taylor & Francis, 2002); James R. Mahan, *Radiation Heat Transfer* (Wiley & Sons, 2002).

CARL PALMER
INDEPENDENT SCHOLAR

Radiation, Infrared

INFRARED RADIATION IS the part of the electromagnetic spectrum popularly (but not entirely accurately) conceptualized as heat. The IR region covers wavelengths that span nearly three orders of magnitude; it is conventional therefore to break this down into further subgroups (Table 1).

Table 1: Conventional breakdown of IR radiation

Name	Wavelength	Comments
Near-infrared (NIR)	0.75 – 1.4 μm	Absorbed by water and commonly used in fiber-optic technology.
Short Wavelength Infrared (SWIR)	1.4 – 3 μm	Strongly absorbed by water and used in long-range telecommunications.
Mid wavelength Infrared (MWIR)	3-8 μm	Used in heat-seeking missile technology.
Long Wave Infrared / Far IR (LWIR)	8-1000 μm	Not absorbed by water, therefore used for thermal-image sensors.

The main application of IR radiation in relation to climate change is in the fields of meteorology and climatology. Satellite measurements of IR radiation received from the Earth can be used to derive cloud types and heights; these are used for weather forecasting, but knowledge of the number and type of clouds present is useful in calculating the Earth's radiative budget. It is possible to use the IR radiation returned to space from the earth to measure land and sea surface temperature, both of which are important parameters in calculating the Earth's heat budget and in monitoring global climate change. Satellite measurements of IR are also used to determine cloud height and rates of convection.

IR radiation also finds a variety of industrial, military, other scientific, and domestic applications, for example:

1. Night vision—Night vision devices use a photon muliplier to amplify the signal from the available ambient light which is then augmented with the IR radiation.
2. Thermal imaging—non-contact, non-destructive technique that generates a false-color thermal image of a subject finding a wide range of industries.
3. Heating—IR lamps can be used for heating; examples include for frozen aircraft wings and for patio heaters.
4. Spectroscopy—IR spectroscopy (also called rotational spectroscopy) is a technique used by chemists for the identification of molecules and for elution of chemical structure.

SEE ALSO: Radiation, Absorption; Radiation, Microwave.

BIBLIOGRAPHY. Guy Guelachvili and K. Ramamohan Rao, *Handbook of Infrared Standards* (Elsevier Science & Technology, 1986); Thormod Henrikson and David H. Maillie, *Radiation and Health* (Taylor & Francis, 2002).

CARL PALMER
INDEPENDENT SCHOLAR

Radiation, Long Wave

LONG WAVE (OR longwave) radiation is the part of the electromagnetic spectrum emitted at spectral wavelengths generally greater than one micrometer (μm). Types of long wave radiation include infrared, microwave, and radio waves. Emittance of radiation is a function of temperature, and objects giving off long wave radiation are colder than those radiating at short wavelengths. For example, the sun (approximately 5800 K) radiates primarily in the short wave part of the spectrum (especially visible light from 0.4 to 0.7 micrometers), whereas the Earth (approximately 290 K) emits radiation at much larger wavelengths. Climatologically, long wave radiation generally refers to radiation emitted by the Earth-atmosphere system (also called terrestrial radiation), largely at wavelengths of 5–15 μm. Long wave radiation emitted by the Earth's surface and atmosphere falls primarily within the thermal infrared ("below the red") region of the electromagnetic spectrum. It can be sensed through the sensation of heat.

In the Earth-atmosphere system, short wave radiation from the sun is absorbed and converted to long wave radiation. Various components of the Earth-atmosphere system absorb the incoming short wave radiation; among those are the Earth's surface, gas and dust molecules in the atmosphere, and clouds. Long wave radiation is then reradiated from those components, after which it is referred to as outgoing long wave radiation or counterradiation. Counterradiation may be reabsorbed (and reradiated) by those very same components that initially absorbed short wave radiation. This process is behind the greenhouse effect.

Globally, the Earth's atmosphere is relatively transparent to radiation between 8 and 15 μm. This atmospheric window allows much long wave radiation to be lost to space. However, the window may be closed locally by the presence of large amounts of water vapor or clouds. Additionally, increasing amounts of greenhouse gases can also potentially close this window. Thus, the role of long wave radiation in the greenhouse effect is fundamental. In the absence of an atmosphere containing long wave–absorbing greenhouse gases (for example, water vapor, or carbon dioxide), the Earth's average temperature would be approximately 0 degrees F, (minus 18 degrees C or 255 K). However, due to the efficiency with which greenhouse gases reabsorb counterradiation, the Earth's average temperature is 59 degrees F (15 degrees C, or 288 K).

Water vapor and carbon dioxide are the most abundant greenhouse gases by volume. They are particularly effective at absorbing counterradiation. The amount of water vapor in the atmosphere is a direct response to temperature. Other long wave–absorbing gases can be produced by human activity. Carbon dioxide and certain other trace gases (such as methane, nitrous oxide, and chlorofluorocarbons) can potentially upset the long wave exchanges involved in the Earth's energy balance. Such deviations can lead to changes in the average global temperature. Potential consequences of an escalating amount of these atmospheric gases include a growing proportion of outgoing long wave radiation being "trapped" in the atmosphere, leading to an increase in the Earth's temperature.

Clouds serve as very effective absorbers of outgoing long wave radiation. This, in turn, affects surface and near-surface temperatures. The presence of clouds in a nighttime sky results in warmer temperatures than a cloudless one. In the absence of incoming short wave radiation, only long wave exchanges occur at night. With clouds trapping much of the outgoing long wave, substantial amounts are redirected back toward the surface. This increases the near-surface temperatures. Similarly, the lack of clouds leads to greater amounts of long wave radiation escaping the Earth-atmosphere system to space.

SEE ALSO: Albedo; Radiation, Microwave; Radiation, Short Wave; Radiative Feedbacks.

BIBLIOGRAPHY. C.D. Ahrens, *Meteorology Today* (Thompson Brooks/Cole, 2007); T.R. Oke, *Boundary Layer Climates* (Routledge, 1987); P.J. Robinson and Ann Henderson-Sellers, *Contemporary Climatology* (Prentice Hall, 1999).

PETRA A. ZIMMERMANN
BALL STATE UNIVERSITY

Radiation, Microwave

THE EXISTENCE OF microwaves was first postulated by James Clerk Maxwell in 1864 and confirmed by the experiments of Heinrich Hertz some 20 years later. Microwaves are subdivided into categories as listed in Table 1.

Table 1: Microwave frequency bands as defined by the Radio Association of Great Britain

Designation	Wavelength *
L band	15 cm – 30 cm
S band	8 cm – 15 cm
C band	3.75 cm – 8 cm
X band	2.50 cm – 3.76 cm
Ku band	1.7 cm – 2.50 cm
K band	1.1cm – 1.7 cm
Ka band	0.75 cm – 1.1 cm
Q band	0.6 cm – 1 cm
U band	0.5 cm – 0.75 cm
V band	0.4 cm – 0.6 cm
E band	0.33 cm – 0.5 cm
W band	0.3 cm – 0.4 cm
F band	0.2 cm – 0.33 cm
D band	0.27 cm – 0.18 cm

*Note the definition is originally in units of GHz and has been converted here to cm for comparison to the other measurements. This has inevitably led to some rounding.

Microwaves are most well known in popular culture for their role in heating food (in a microwave oven), and for their use in mobile telecommunications such as mobile phones and wireless networking. Microwave ovens work by passing S-band radiation though food, which excites water, sugar, and fat molecules. These molecules in turn reradiate in the infrared to heat the food. With the rapid expansion of mobile phone technology, many studies have been done to look at the effects of microwave radiation on human health. The studies show mixed results however, and the general conclusion is that only doses high enough to heat up tissue are likely to have negative impacts.

SEE ALSO: Radiation, Absorption; Radiation, Ultraviolet; Technology.

BIBLIOGRAPHY. Thormod Henrikson and David H. Maillie, *Radiation and Health* (Taylor & Francis, 2002); James R. Mahan, *Radiation Heat Transfer: A Statistical Approach* (Wiley & Sons, 2002) Monika Willert-Porada, *Advances in Microwave and Radio Frequency Processing: 8th Inter-*

national Conference on Microwave and High-Frequency Heating (Springer-Verlag, 2006).

CARL PALMER
INDEPENDENT SCHOLAR

Radiation, Short Wave

RADIATION TRAVELING IN waves shorter than one micrometer (μm) is characterized as short wave, and includes gamma rays, x-rays, ultraviolet light, and visible light. Climatologically, short wave radiation commonly refers to the incoming radiation from the sun. There is an inverse relationship between the temperature of an object and the wavelengths at which it primarily emits. Because the sun is a hot object (approximately 5800 K), it emits radiation at short wavelengths. Since shorter wavelengths carry more energy than longer ones, they are more intense. Most of the short wave emitted by the sun is in the visible region of the electromagnetic spectrum, which spans from 0.4 μm (violet) to 0.7 μm (red). The sun's wavelength of maximum emission is found at 0.5 μm.

The amount of emitted solar radiation that reaches the Earth decreases inversely with the square of the distance between the Earth and sun, by the inverse square law. The total of short wave radiation that reaches the top of the Earth-atmosphere system is called the solar constant. Although the amount varies slightly throughout the year, due to the elliptical nature of the Earth's orbit around the sun, the solar constant averages around 1370 W m^{-2}.

Because the eccentricity of the Earth's orbit varies between nearly zero to five percent (with a periodicity of 110,000 years), the amount of short wave received increases or decreases over time from present values. The current orbit is nearly circular, leading to little seasonal variation of incoming short wave radiation. However, a more highly elliptical orbit would render a difference of up to 30 percent between aphelion (maximum Earth-sun distance) and perihelion (minimum Earth-sun distance). Incoming solar radiation can take several avenues once it enters the atmosphere. Over a year, 30 percent of total short wave radiation is reflected back to space, either by gas molecules or other particles in the atmosphere, by clouds, or by the Earth's surface; this is the Earth's albedo (the proportion of radiation that is reflected from a surface). The atmosphere and clouds absorb an additional 20 percent of short wave radiation. Approximately 50 percent strikes the surface, where it is absorbed. These values may vary locally and at shorter time scales.

Some of the short wave radiation reaching the Earth's surface arrives directly, unimpeded by clouds or atmospheric constituents; this is direct radiation. Some is scattered about the atmosphere and arrives at the surface indirectly. This is diffuse radiation. Diffuse radiation is a product of scattering, which occurs when short wave radiation strikes small particles in the atmosphere, including gas molecules. Upon impact, the radiation is scattered omnidirectionally. Some short wave radiation will be scattered toward the surface. Energy received at the Earth's surface is the total amount of direct and diffuse radiation.

The presence and type of clouds in the atmosphere reduces the amount of short wave radiation reaching the Earth's surface. Thin clouds, such as cirrus, have lower albedos than thick ones, such as cumulus. Once short wave solar radiation reaches the surface and is absorbed, it is converted to long wave forms of radiation.

Short wave radiation impacting the Earth also includes ultraviolet light. These wavelengths are shorter than visible radiation, so ultraviolet light carries more energy than visible light. Ultraviolet radiation is classed into three categories: UV-A (0.32–0.40 μm), UV-B (0.29–0.32 μm), and UV-C (0.20–0.29 μm). Excessive exposure to UV-A and UV-B radiation has been linked to skin cancers and skin damage. UV-C radiation is largely absorbed by stratospheric ozone. Absorption of ultraviolet radiation breaks ozone down into atomic and molecular oxygen. The presence of stratospheric ozone helps protect the Earth's surface from the damaging effects of this form of short wave radiation.

SEE ALSO: Radiation, Ultraviolet; Sunlight.

BIBLIOGRAPHY. C.D. Ahrens, *Meteorology Today* (Thompson Brooks/Cole, 2007); D.L. Hartmann, *Global Physical Climatology* (Academic Press, 1994); R.B. Stull, *Meteorology for Scientists and Engineers* (Brooks/Cole, 1999).

PETRA A. ZIMMERMANN
BALL STATE UNIVERSITY

Radiation, Ultraviolet

ULTRAVIOLET RADIATION WAS discovered as a result of the observation that silver salts darken on exposure to sunlight. In 1801, the German physicist Johann Wilhelm Ritter first observed that invisible electromagnetic radiation was responsible for this darkening. These rays eventually became known collectively as UV, so named as this radiation is immediately beyond violet in the electromagnetic spectrum. This implies that UV is more energetic than the visible light. Conventionally, UV radiation is broken down into further subdivisions as shown in Table 1:

Table 1: Conventional subdivisions of UV radiation

Name	Wavelength	Comments
1. Near	400 nm – 200 nm	Referred to as 'Blacklight'
UV-A	400 nm – 320 nm	Strongly absorbed by O_3
UV-B	320 nm – 280 nm	Strongly absorbed by O_3
UV-C	290 nm – 200 nm	Strongly absorbed by O_2
2. Far	200 nm – 31 nm	Strongly absorbed by O_2
3. Extreme	31 nm – 1 nm	

In humans, UV radiation is important for health. UV-B has health benefits, as it is responsible for the production of vitamin D. A deficiency of vitamin D is thought to lead to a range of cancers and also to osteomalacia (the adult equivalent of rickets), with symptoms ranging from painful bones to brittleness and fractures.

Exposure to UV radiation also causes the skin to release a pigment (melanin), giving the skin a darker color that is regarded as healthy in most Western cultures. Melanin provides some protection again the more harmful effects of UV exposure.

Exposure to UV radiation also has a range of negative health effects. The most widely publicized of these is the link between exposure to UV and skin cancer. UV radiation is strongly absorbed by DNA, the cellular molecule responsibly for the transfer of hereditary information. The absorption of UV radiation causes chemical bonds in the DNA to be broken and reformed in the wrong order. This can lead to mutations and cancerous growths. UV radiation is also harmful to the eyes, leading to short-term uncomfortable conditions such as arc eye or to more serious conditions such as cataracts. There may also be a link between excessive UV exposure and poor immune response. UV radiation also finds a range of industrial and domestic applications; these include:

1. Astronomy: Many hot objects in the universe emit large amounts of UV radiation and are therefore better observed in the UV region. However, as the atmosphere absorbs a lot of this UV, these observations are generally only made from space.
2. Spectrophotometry: A widely used technique in analytical chemistry to determine chemical structure.
3. Analyzing minerals: Many minerals glow characteristic colors under a UV lamp, aiding identification.
4. Sterilization of surface and drinking water: UV radiation is effective at killing pathogens and is used to sterilize critical workspaces (such as biochemistry labs) as an alternative to chlorination. Methods to sterilize water based on using UV from sunlight may provide a carbon neutral solution to domestic water treatment.

SEE ALSO: Chemistry; Geography; Radiation, Absorption; Radiation, Infrared.

BIBLIOGRAPHY. W.B. Grant, "An Estimate of Premature Cancer Mortality in the U.S. Due to Inadequate Doses of Solar Ultraviolet-B Radiation," *Cancer* (v.94/6, 2002); Thormod Henrikson and David H. Maillie, *Radiation and Health* (Taylor & Francis, 2002); World Health Organization, www.who.int (cited November 2007).

CARL PALMER
INDEPENDENT SCHOLAR

Radiative Feedbacks

WHILE MOST OF the feedbacks at play in the Earth's ecosystem are subject to the closed system of matter, radiative feedbacks are those that deal with the open system in which the sun's energy is transferred to the Earth and then absorbed or reflected back, to varying degrees. Because the amount of absorbed or reflected solar energy is the principal component in the planet's temperature (the other factors then determine what happens to that heat, but the solar energy provides the initial quantity), these feedbacks are key to understanding and modeling global climate. Feedbacks are relationships found in complex systems, in which the output (or result) of the system is returned to the input. For instance, as the summers get hotter, people run their air conditioners longer, releasing more chlorofluorocarbons, accelerating global warming, and causing hotter summers. That is a simple example of a positive feedback often used in schools. The radiative feedbacks in question involve principally clouds, ice, and water vapor.

A critical feedback in global warming is the albedo-ice feedback. Though the melting of the polar ice caps is often associated with rising sea levels, it's only one of the important effects, and simply the most vivid to illustrate. The more complicated effect is in the change it enacts on how much solar energy is retained by the Earth. Ice is far more reflective than land or liquid water, and as the polar ice melts, the area of the surface that it occupied is replaced by one of those two things. As a result, more sunlight is absorbed by the surface instead of being reflected, which then warms the poles further, melting more ice.

Clouds are also critical. Clouds act as a sort of imperfect barrier, absorbing a limited amount of heat moving in either direction, heat emanating from the Earth into space, and sunlight shining on the Earth. Models disagree on whether their overall feedback effect on global warming will prove positive or negative. On the one hand, the evaporative feedback in general accelerates global warming: as the temperature increases, the capacity of the air to hold moisture increases exponentially, so as water evaporates into the warm air, it is able to stay there. Since water vapor is a greenhouse gas, as it accumulates, it causes the temperature to get warmer and warmer. On the other hand, sufficiently great cloud cover shields the Earth from solar radiation, and cloudy days aren't as hot. High-enough clouds can reflect sunlight back down on the Earth, however, more than balancing out the sunlight they have blocked. Water vapor also retains heat better than the atmosphere does; a wetter atmosphere loses heat more slowly. As more and more water evaporates, more and more heat from the sun can be retained, which makes for a warmer Earth, and greater evaporation.

It is possible that the effects of positive radiative feedbacks on global warming have been muted until recently by global dimming, the gradual reduction of the Earth's irradiance (emitted radiation, including heat and light) in the second half of the 20th century, which has reversed in the 21st century. Global dimming is probably caused by the coalescing of water vapor around anthropogenic particles in the air, the product of pollution and other industrial activities. The resulting water droplets form a little differently than they would otherwise, and make more highly reflective clouds. The Clean Air Act and other anti-pollution efforts are presumed to be responsible for ending global dimming. The apparent recent increase in global warming may be because that dimming had been reducing global warming's effects by limiting the amount of sunlight entering the system.

SEE ALSO: Albedo; Atmospheric Absorption of Solar Radiation; Biogeochemical Feedbacks; Climate Feedbacks; Cloud Feedback; Dynamical Feedbacks; Evaporation Feedbacks; Ice Albedo Feedback; Radiation, Absorption; Sunlight.

BIBLIOGRAPHY. K.O. Buesseler, et al., "Revisiting Carbon Flux Through the Ocean's Twilight Zone," *Science* (v.316/5824, 2007); J.P. Kenneth, et al., *Methane Hydrates in Quaternary Climate Change: The Clathrate Gun Hypothesis* (American Geophysical Union, 2003); G.A. Meehl, et al., "How Much More Global Warming and Sea Level Rise," *Science* (v.307/5716, 2005); M. Torn and J. Harte, "Missing Feedbacks, Asymmetric Uncertainties, and the Underestimation of Future Warming," *Geophysical Research Letters* (v.33/10, 2006).

BILL KTE'PI
INDEPENDENT SCHOLAR

Rain

RAIN IS COMPOSED of water droplets formed from vapor that has condensed in the Earth's atmosphere. Regular rainfall is vital to the functioning of the biosphere, and even minor fluctuations in the amount of rain can have substantial impact on the ecosystem. Climatologists believe that rainfall patterns are going to change dramatically as the atmosphere heats up in the coming decades, with potentially damaging consequences.

HOW RAIN FORMS

Precipitation forms under three main types of conditions: convective uplift, frontal uplift, and orographic uplift. Convective uplift is common in the equatorial tropics and during the summer months. As the sun heats up the Earth's surface, the humid, warm air rises into the atmosphere, cools, forms tall, unstable cumulonimbus clouds, and then releases its moisture in a quick downpour.

Frontal uplifts are the result of cold and warm fronts colliding. This can take different forms. Fronts are transition zones between air masses of different temperatures. When warm air collides with cooler air along a warm front, the warm air gently rides up over the cooler air. If cold air collides with warm air along a cold front, the colder, denser air pushes up the warm air rapidly. Depending on the particular conditions, rains from frontal events can be a steady, soaking rain that last for several hours, or severe weather events that spawn quick downpours, high winds, and tornadoes. Orographic uplift, sometimes called "relief rain," occurs when a warm air mass meets a geographical barrier (usually a mountain range) that pushes the warm air up into the atmosphere, where it cools and condenses.

No matter what the type, the development of rain requires the presence of extremely small particles called condensation nuclei to form droplets. These particles, trapped in cloud formations, can be composed of anything from dust and soot to sea salt and phytoplankton, and at 0.02 mm. in diameter, are just 1/100th the size of a typical raindrop. Condensation nuclei give water vapor in the clouds something to coalesce around, thus increasing the size of the individual molecules of water vapor into droplets heavy enough to fall out of the clouds.

The speed with which raindrops fall depends primarily on their weight: for example, a 0.5 mm. raindrop would fall at around 6.5 ft. (2 m.) per second, while a five mm. drop would fall at 29.5 ft. (9 m.) per second. The average raindrop is around two mm. in diameter. The largest raindrop ever recorded was 10 mm. in diameter, but those are rare; drops larger than five mm. tend to be unstable, often breaking into smaller droplets as they fall. Raindrops are not shaped like teardrops: rather, they are spherical, or flattened on the top, with larger drops sometimes shaped liked parachutes.

RAINFALL AND GLOBAL WARMING

Rainfall is vital to the creation of food and water on Earth. With less than 1 percent of all the planet's fresh water available for human and animal consumption, the regular recharging of groundwater and reservoirs from rain and snow is critical to keep fresh water sources flowing, and all plant life requires at least some water to grow. Regular, moderate precipitation is critical to the production of most of the 2,000 food crops under cultivation. Because of the importance of rainfall to the environment, climate experts have been trying to determine the possible impact of global warming on rainfall distribution around the world.

In 2007, a group of researchers constructed a computer modeling program drawing on data collected 1925–99, and, taking into account potential future increases in levels of greenhouse gases and solar activities, made projections well into the next century. Their study found the precipitation in areas lying between 40 and 70 degrees N latitude is likely to increase at a rate of 2.44 in. (62 mm.) per decade, and in areas lying between zero and 30 degrees S latitude is likely to increase 3.23 in. (82 mm.) per decade. However, the area lying between zero and 30 degrees N latitude is likely to see precipitation decreases of 3.86 in. (98 mm.) per decade. This means that regions toward the poles, including Europe, Canada, and parts of Latin America, will become wetter, while already parched regions will become much dryer.

This could be catastrophic for the Sahel region in Africa. A 2,400-mi. (3862 km.) long belt stretching from Senegal on the west coast to Eritrea on the east, the Sahel is the boundary line between the Sahara Desert and the more fertile sub-Saharan region to the south. The area is home to an estimated 50 million people, 62 percent

A patch of rain moving through Tobago in the Caribbean. Climate models predict more heavy rain events, separated by prolonged periods of dry weather. In a global warming scenario, the warmer air in the atmosphere holds more water.

of whom live in poverty. Since the 1970s, the Sahel has been struck by longer and more frequent droughts, with overall precipitation decreasing by around 20 percent during the period. Modeling indicates that in the coming century, rainfall averages in this region could drop by another 30 percent. The death toll from previous droughts has numbered in the millions; in the future, the area may become uninhabitable.

Areas where rainfall is expected to increase will likely see new patterns of precipitation as well. The overall amounts of rain may not change dramatically from year to year, but climate models predict more heavy rain events, separated by prolonged periods of dry weather. Much of this will be due to the heating of the atmosphere: warmer air holds more water, raising the potential for a quick release of a large volume of water. Air pollution will also play a role, as more particulates in the atmosphere give this increased amount of water vapor more condensation nuclei around which to coalesce.

Heavy rains are a potential threat to human life and property, raising the probability of events such as flash floods and landslides. Research has also found that frequent heavy rains combined with long dry spells quickly erode the vitality of a region's biomass, and in the long term contribute to a loss of soil quality and overall biodiversity.

Climate experts predict there could be some initial benefits to plant life in areas of increased rainfall. Scientists have seen a rise in the amounts of reactive nitrogen in the atmosphere, which they believe stems from a combination of rising rates of nitrogen oxides from the burning of fossil fuels and ammonia (NH_3) from agricultural sources. This reactive nitrogen is carried to the ground by rain, where it stimulates plant growth, particularly in the forests. Increasing the vitality of the forest will increase the terrestrial carbon sink, as trees and plants absorb harmful carbon dioxide from the atmosphere as a process of photosynthesis. The amount of reactive nitrogen has increased fourfold since the beginning of the Industrial Revolution, and will likely to continue to grow in coming years. Scientists are not sure at what point there is too much of this substance available, when its risks to the environment begin to outweigh its benefits.

SEE ALSO: Floods; Hurricanes and Typhoons; Precipitation; Rainfall Patterns; Thunderstorms.

BIBLIOGRAPHY. Environmental Protection Agency, www. epa.gov (cited November 2007); N. Gillett, et al., "Detection of Human Influence on Twentieth-Century Precipitation Trends," *Nature* (v.448, 2007); P. Green, et al., "Global Water Resources: Vulnerability From Climate Change and Population Growth," *Science* (v.289, 2000); Intergovernmental Panel on Climate Change, www.ipcc.ch (cited November 10, 2007); W. Moomaw, "Energy, Industry and Nitrogen: Strategies for Decreasing Reactive Nitrogen Emissions," *AMBIO: A Journal of the Human Environment* (v.31, 2002); M. Ritter, "Atmospheric Moisture," University of Wisconsin, www.uwsp.edu (cited November 2007).

HEATHER K. MICHON
INDEPENDENT SCHOLAR

Rainfall Patterns

CHANGE IN RAINFALL pattern is a consequence of global warming. The world's agriculture, especially third world agriculture, depends upon the seasonal rainfall pattern. Recent erratic changes in rainfall pattern lead toward low agriculture production, thus creating food insecurity for an ever-increasing world population. Flood, drought, and famine are the consequences of these changing patterns.

Rainfall pattern means the distribution of rain geographically, temporally, and seasonally. The tropics receive more rainfall than deserts. Cooler places like the poles receive no rainfall, as it is converted to snow before it falls to the ground. Rainfall happens more in a particular time of a year, during a rainy season. In other seasons, rainfall is scant. Therefore, agriculture (rain-fed), worldwide, is planned based on rainfall's natural pattern. Water storages, irrigation networks, and urban water supply systems are designed according to the average annual rainfall. If it rains a lot on a continuous basis for a longer time, there is possibility of flood and subsequent disaster to the infrastructure. No rainfall or little rainfall for a longer period (years) in an inhabited area could lead to drought and famine.

Global warming leads to a near-term collapse of the ocean's thermohaline circulation. Thermohaline circulation is a global ocean circulation pattern that distributes water and heat both vertically, through the water column, and horizontally across the globe. Due to this collapse of thermohaline circulation, warm surface waters move from the tropics to the North Atlantic and extra-warm water surfaces in the Pacific Ocean surrounding the equator. Thus, Western Europe, some parts of Asia, and many parts of the Americas get warmer than normal, and some parts of Europe get cooler rapidly. El Niño and La Niña are examples of this. This latest deviant trend generates dramatic weather impacts, such as rapid cooling in some parts of the world, and greatly diminished rainfall in agricultural and urban areas. UNESCO and other studies found that changes in rainfall pattern could be attributed to the shifts in global wind pattern. These shifts are due to the changes in the ocean surface temperature. Effect of human activity on the surface vegetation is also causing rainfall pattern variation. Widespread deforestation in parts of Africa and Asia is causing scarce rainfall and subsequent drought.

The world's scientific community commonly accepts that global warming has affected rainfall patterns. More precipitation is happening in northern Europe, Canada, and northern Russia, but less in swaths of sub-Saharan Africa, southern India, and southeast Asia. A Canadian research team found with 75 years (1925–99) of rainfall data analyzed through 14 powerful computer models that the Northern Hemisphere's midlatitude (a region of 40 to 70 degrees N) received increased precipitation over the years. It corroborates with the change in thermohaline circulation. The models also showed that in contrast, the Northern Hemisphere's tropics and subtropics (a region between the equator and 30 degrees N) became drier, while the Southern Hemisphere's similar region became wetter. This study was conducted for rainfall patterns over land.

Researchers claim that a natural pattern of rainfall is good for plant growth, as variable rainfall patterns lead to lower amounts of water in the subsurface level of the soil (in the upper 30 cm.). Variable rainfall patterns also cause plant diversity in a particular land. That means that weeds grow rapidly with variable rainfall. The significance of these changes is evident from recent large-scale failure of the crops, rangelands, and water-supply systems in the world. Mass starvation in Sahelian Africa is the stark proof of this.

Some might argue that changes in rainfall patterns are unfounded due to lack of instrumental records for a large time period. However, studies using indirect methods have proved that global warming, in fact, is causing serious rainfall pattern variability. Tree-ring analysis for predicting rainfall amounts in previous years (hundreds of years back) is one study, which proved that rainfall pattern variability is extensive in recent years.

If this trend continues, environmental managers need to make new decisions about the management of water and land. They need to accurately understand the interannual variability of rainfall and a possibility of runs of dry and wet years, which may cause important changes in runoff, sedimentation, soil erosion, or changes in communities of vegetation and animals, and of the viability of large water resource developments. Rainfall pattern variability would certainly cause mass human migration.

SEE ALSO: Rain; Thermohaline Circulation; Thunderstorms.

BIBILIOGRAPHY. Thomas Dunne and L.B. Leopold, *Water in Environmental Planning* (W.H. Freeman and Co., 1978); Cheryl Dybas, "Increase in Rainfall Variability Related to Global Climate Change," Earth Observatory, NASA (December 12, 2002); "Global Warming has Already Changed World's Rainfall Patterns," CommonDreams.org (cited July 2007); UNESCO, "Changes in Climate," *Arid Zone Research* (v.20, 1963); Warren Viessman, Jr., and G.L. Lewis, *Introduction to Hydrology* (Prentice Hall, 2003).

SUDHANSHU SEKHAR PANDA
GAINESVILLE STATE COLLEGE

Refugees, Environmental

IN THE LAST 10 years, the issue of environmental refugees has emerged as a pressing issue. Most refugees are fleeing from natural disasters such as the Asian tsunami in 2004, or as a result of the impacts of global climate change, such as sea level rise. As the executive director of the United Nations Environment Programme (UNEP) noted in 1989, "as many as 50 million people could become environmental refugees if the world does not support sustainable development." Since then, many studies have considered this topic, with Norman Myers (one of the leading thinkers in this field) estimating that environmental refugees will soon become the largest group of involuntary migrants.

One of the difficulties in managing the issue of environmental refugees is their classification. The 1951 Convention relating to the Status of Refugees classifies a refugee as a person who, "owing to a well-founded fear of being persecuted for reasons of race, religion, nationality, membership of a particular social group, or political opinion, is outside the country of his nationality, and is unable to or, owing to such fear, is unwilling to avail himself of the protection of that country." Refugees who are outside their country for environmental reasons do not strictly speaking fit within this category. The first definition of environmental refugees came from UNEP researcher Essam El-Hinnawi in 1985: "environmental refugees are those people who have been forced to leave their traditional habitat, temporarily or permanently, because of a marked environmental disruption (natural and/or triggered by people) that jeopardized their existence and/or seriously affected the quality of their life [sic]. By 'environmental disruption' in this definition is meant any physical, chemical, and/or biological changes in the ecosystem (or resource base) that render it,

Many are still living as refugees from the 2002 eruption of the Mount Nyiragongo volcano in the Congo.

temporarily or permanently, unsuitable to support human life." Environmental refugees are different from environmental migrants, in that they do not have any choice about their situation.

Today, there are at least 25 million environmental refugees. There are 22 million people in the world that classify as refugees under the traditional definition. Environmental refugees number one person to every 225 worldwide. A further 900 million people may also become environmental refugees because they live in marginal environments, or are driven into marginal environments for political, economic, social, cultural, legal, and institutional reasons.

EXAMPLES OF ENVIRONMENTAL REFUGEES

There are a number of reasons to become an environmental refugee. First, it may happen as a result of a natural disaster. Natural disasters include floods, cyclones, earthquakes, tsunamis, or any other major event that will make the lived environment temporarily or permanently inhabitable. One example is the eruptions of the Soufriere Hills Volcano on the Caribbean Island of Montserrat in 1995–98. As a result of these eruptions, 7,000 residents were forced to evacuate. Second, individuals or whole groups of people might become environmental refugees due to the appropriation of habitat or land by external parties, hence dispossessing and permanently displacing people.

For example, the building of the Three Gorges Dam in China has displaced up to 850,000 people and overall has the potential to displace up to 1.3 million people by 2009. Third, people may become environmental refugees as a result of an ongoing deterioration of their land and seas. Desertification is a good example as is sea level rise of the type of activity that ultimately creates environmental refugees of people in their own homelands. Movement from one area to another occurs as families and settlements find it harder to sustain livelihoods. This is the group that most often finds it hard to get support, as it is hardest for this group to be recognized as having refugee status.

There are many examples of this situation across the world due to climate change. For example, water shortages caused by climate change will cause huge reductions in agriculture and people's way of life. Tropical forests are estimated to lose another 40 to 50 percent of their cover due to climate change, and in turn this will dispossess many millions of people living within and dependent upon it for survival. Up to 500 million people could experience absolute shortages in fuel wood supply as a result of climate change impacts. Other (preliminary) estimates highlight that the total of people at risk of sea level rise in Bangladesh could be 26 million, in Egypt 12 million, in China 73 million, in India 20 million, and elsewhere 31 million, making an aggregate total of 162 million. At least another 50 million people are at risk through increased droughts, desertification, and related climate disruptions.

RESPONSES

One response to the problem is to ensure that environmental refugees have a formal classification within the 1951 Convention relating to the Status of Refugees. Another is to address the root causes of the environmental problems motivating relocation of people across the world. Promoting policies of and achieving sustainable development programs is a good first step to help achieve this goal. Nations must also develop policy to address what support structures and frameworks they are going to put in place to deal with and acknowledge environmental refugees in their countries. Implementing specific projects in countries most affected is another pathway nations could take to support each other on this issue, including the relief of foreign debt and the granting of foreign aid that will help ameliorate this problem. For policymakers and governments, while the issue of how to deal with refugees is politically contentious, the plight of environmental refugees merits special attention because, due to climate change, any country may be vulnerable and need support in the future.

SEE ALSO: Climate Change, Effects; China; Cyclones; Floods; Policy, International; Policy, U.S.; Preparedness; Risk; Tsunamis; United Nations Environment Programme (UNEP); Volcanism.

BIBLIOGRAPHY. Diane Bates, "Environmental Refugees? Classifying Human Migrations Caused by Environmental Change," *Population and Environment* (v.23/5, 2002); Derek Bell, "Environmental Refugees: What Refugees: What Rights? Which Duties?" *Res Publica* (v.10, 2004);

Tim Dyson, "On Development, Demography and Climate Change: The End of the World as We Know it?" *Population and Environment* (v.27/2, 2005); E. El-Hinnawi, *Environmental Refugees* (UN Environmental Programme, 1985); FoE, *The Citizens Guide to Climate Refugees* (Friends of the Earth International, 2007); Lori Hunter, "Migration and Environmental Hazards," *Population and Environment* (v.26/4, 2005); Y. Lou, "Immigration Policy Adjusted in Three Gorges," *Beijing Review* (v.43, 2000); Norman Myers, "Environmental Refugees," *Population and Environment* (v.19/2, 1997); R. Ramlogan, "Environmental Refugees: A Review," *Environmental Conservation* (v.23, 1996); M. Tolba, "Our Biological Heritage under Siege," *Bioscience* (v.39, 1989).

Melissa Nursey-Bray
Australian Maritime College
Rob Palmer
Research Strategy Training

Regulation

IN THE CONTEXT of law, the United Nations Framework Convention on Climate Change (UNFCCC) and the complementary agreement of the Montreal Protocol on Substances That Deplete the Ozone Layer are legal regulatory instruments. The UNFCCC Article 4 section 1 governs scientific, technological, technical, socioeconomic initiatives in the research, observation, and development of data on anthropogenic emissions into the atmosphere. To effect the desired regulation of the amount of greenhouse gases released into the atmosphere requires both a legal rule of what is or is not permitted, and the inclusion of regulatory features in equipment and processes or activity to control the emissions of the relevant substances into the atmosphere.

The atmosphere and other planetary features operate according to laws that have been articulated in scientific terms. These scientific laws are anthropocentric in their perspective and can be used to describe the conditions under which human life continues on this planet. The dominant conditions of the laws according to which the planet dictates or regulates human life and activity are related to but distinct from the regulatory operations of neoclassical human law. Anthropocentric law is the law made by persons for persons and their property. It is the actions of persons and their property that have consequences for global warming and climate change. Alternatively, global warming and climate change occur at the intersection of the laws of the planet usually perceived in the context of scientific laws and the anthropocentric laws of human behavior.

LEGAL REGULATION

In a neoclassical utilitarian framework, the regulator is typically a natural or legal person or an entity with the power to regulate within its domain. The difficulty in the context of global issues such as planetary warming and climate change is the absence of a single regulator with the jurisdiction for the global atmosphere or the climate and hence no one with authority to regulate the global commons. Since local climate is not independent or separable from global atmosphere and does not observe legal jurisdictions, regulating local climate change or warming locally without some provision for the larger global dimensions would be an exercise in futility. Further, in the dominant utilitarian approach, the global physical features such as the atmosphere and the climate are a common resource beyond human or state jurisdiction and available for use by all who have access to it.

Within the context of law or legal systems, regulation can refer either to a regulatory regime being a set of laws about some action, or a subset of detailed technical rules that are to be applied in particular physical contexts. In the latter sense, regulations are a set of physical specifications that are to be applied to particular products or processes. Regulatory regimes as part of legal systems in particular jurisdictions differ in the international and the domestic national contexts. Details are particular to the applicable legal system, and may not have comparable features in other legal systems.

The recognition of national sovereignty results in the distinction between the international and the domestic state jurisdictions. In the international context, states may agree to be bound at their own discretion as sovereign entities and not to be compliant with the regulation of a global regulator. Thus, the climate or the atmosphere can only be legally regulated within the jurisdictions of the various sovereign states' legal systems and by agreement with each other. Agreements such as the UNFCCC and the more specific Kyoto Protocol are indispensable in achieving global action. Such agreements to regulate anthropogenic emissions into the atmosphere must be implemented within the jurisdiction of sovereign states by domestic actions and procedures.

In implementing an international agreement in the domestic realm, a state issues laws and enforces them in order to achieve compliance with the international and domestic objectives. In doing so, the state would typically rely on scientific opinion. For instance, a scientific opinion stating that emissions from automobiles include greenhouse gases that should be regulated could lead to applicable regulations. The legal regulation of vehicles may then include a legal rule or a law or legislation that requires the greenhouse gases in tailpipe emissions of an operating automobile to be within certain parameters. This rule could then be accompanied by related regulations that set out in detail the technical specifications as to how the determination of the emissions is to be made and the concentrations of such gases and other acceptable limits of the emissions and may further specify variations by type and model of automobile. The technology in the vehicle itself may then involve systems such as computers or other mechanical systems that adjust or regulate the operation of the equipment to obtain the desired result of compliance with the legal limits.

For the most part, legal regulations are formulated by taking into account the available technical capabilities as well as society's expectations. The Kyoto Protocol, for example, sets legally binding target limits to the emissions into the atmosphere. Some people find the targets inadequate for achieving the stated objective of the UNFCCC and expect further reductions, and others, citing technical and financial reasons, consider these targets unattainable.

FUTURE INTERESTS

A significant feature of the UNFCCC and the Kyoto Protocol is the attempt to regulate the present human behavior in view of the future. This forward concern contrasts with typical legal regimes that usually codify past practice by referring to existing technology and social expectations based on past experience. In the case of climate change, the legal regulation seeks to operate closely with the mechanical or technological aspects to influence future changes.

For those who prefer the typical legal regime, regulating the future is resisted, as the future is not predictable. For others, the scientific understanding of inertia in the trends of climate changes dictates that setting and meeting targets for the future is essential to prevent foreseeable results. To put it in the negative, waiting for the climate to change before regulating current conduct would be meaningless as it would be too late then to make a difference to the outcome.

In law, predictability and foreseeability are distinct and important, with significant implications in practice. Consider the requirement for safety features in a vehicle. One group may take the view that installing safety features would be an unacceptably expensive proposition to provide for an unpredictable future event such as that a particular vehicle will be involved in a particular collision, which may not occur. Another may prefer to consider that it is foreseeable that a vehicle could be involved in the same particular collision and prefer to install features that will enable the avoidance of that outcome. In the climate change context, for the latter, waiting until climate change is predictable before regulating contributing factors may be compared to installing and using safety features on a vehicle after it is predictably involved in a collision. Since the consequences of climate change have been compared second only to those of global nuclear war, waiting for predictability in outcomes of climate change is waiting for the nuclear weapons to detonate before doing something to stop it.

PROBABILITY AND UNCERTAINTY

Predictability and foreseeability also involve different roles for the concepts of probability and risk in technological and scientific and legal regulation. In formulating laws, scientific probability may be characterized as uncertainty, and without sufficient foundation for concern about the atmospheric conditions. Here, scientific uncertainty leads to an absence of regulation.

On the other hand, science can and does operate on probabilities or risk and in that sense does include scientific uncertainty without diminishing the development of appropriate regulatory regimes. For instance, in economic and business decision-making context, risk management is an established approach in many segments of society. Insurance and investment products in the marketplace are examples of socially valued items that are designed specifically by reference to uncertainty and are foundational features of a market economy and have been legally regulated since the last century.

INFORMATION AND REGULATION

Since regulation is a result of social and political actions based on information in which perception is a key component, changing perceptions correspond to changing behavior and regulatory activities. For instance, there is a perceived shift in the discussion from global warming to a broader concern for climate change as including issues other than increases in temperature. Responding to this shift in perception, the U.S. Environmental Protection Agency (EPA) has renamed its website, which also indicates its self-understanding of its regulatory function.

Changing perceptions involve social and political changes and may also lead more directly to changes in regulatory regimes. In a democratic state, legislation is made by acts of a corresponding legislature and its authorized bodies that together formulate the regulatory regime being the law and its regulations.

MODELS AND REGULATION

The perception of the complexity of the atmosphere and climate change has grown over time. Scientific models used to study these phenomena have evolved with a corresponding increasing rapidity as research continues. Weather has played a strategic role in conflict throughout history. The value of information about the weather was appreciated when it was critical in the success of the Allied Forces D-Day invasion of Europe in 1944. After that, meteorology gained in its value as a matter of strategic interest to security forces. It was the incidental observations of meteorologists that led to the development of climate change as a research topic in its own right.

Since meteorologists observed a correspondence between atmospheric CO_2 levels and mean global temperatures, scientific information has been critical in the modeling and understanding of the interaction between human activities and the climate. That the climate does change has always been observed, as has the impact of human activities on the atmosphere, if only at a local scale. The issue now is that of scale in the question of global human activities and global atmosphere and climate changes. It is the position of the Nobel Peace Prize–winning institution known as the Intergovernmental Panel on Climate Change (IPCC), which provides peer-reviewed scientific information on climate change, that anthropogenic emissions are having an impact on global climate patterns.

Established in 1988 prior to the Toronto Conference, the IPCC provides peer-reviewed scientific information to the public which may be used in formulating policies and regulations. Emerging regulations rely on the acceptance of scientific opinions such as that expressed by the IPCC. The IPCC has concluded that without any change in practices, and based on current models, the current trend of increasing greenhouse gas emissions may increase the mean global temperature by 1.4 to 5.8 degrees C by 2100 compared to the temperature in 1990.

In its 1989 *Report to Congress: The Potential Effects of Global Climate Change on the United States,* the U.S. EPA Office of Policy, Planning and Evaluation notes that its report relies on scenarios that are based on information at a certain point in time, and assumes that carbon dioxide levels will have doubled and the climate will have stopped changing, neither of which are realities. This increasing rate of change has already appeared in the complexity of scientific modeling. The capability of regulating such increasing rates of change in the traditional manner has been limited. At the same time, some debate the desirability of regulating changes.

The variety of chemicals now produced and used and the resulting atmospheric emissions add to the complexity of atmospheric models. Contributors to the greenhouse effect include emissions of atmospheric CO_2, methane, CFC/NO_x, and SO_2. Increasing amounts of these gases and particulate matter in the atmosphere is linked to the warming trend in mean global temperatures. The links are still being researched, as the effect of each of the atmospheric components is different. For instance, CFCs and CO_2 remain in the atmosphere up to 100 years, while methane breaks down in 10 years, and SO_2 in a week. They also occur in different proportions and the chemical reactions they induce are different.

Commercial usage of CFCs has declined, and so has SO_2, which also causes acid rain. Each of these emissions and related products and processes are all susceptible to regulation.

THE UNFCCC AND KYOTO

The UNFCCC and the Kyoto Protocol is the principal international agreement relevant to climate change to which more than 140 countries forming a substantial proportion of the global community have signed on. The consensus expressed in the Kyoto Protocol includes the agreement to legally binding obligations on the signatories. Domestic acceptability of this international agreement is an ongoing concern in some states.

Article 1 of the UNFCCC provides definitions of terms and phrases. These definitions indicate some of the complexity and scope of the discussions and indicate the difficulty of regulating the activities necessary to achieve the objective. For instance, "Climate system" means the totality of the atmosphere, hydrosphere, biosphere, and geosphere and their interactions would include most if not the entire human environment on the planet. "Climate change" is similarly comprehensive in referring to climatic change attributed directly or indirectly to human activity that alters the composition of the global atmosphere and which is in addition to natural climate variability observed over comparable time periods. "Emissions" means the release of greenhouse gases and/or their precursors into the atmosphere over a specified area and period of time. All of which would require regulatory systems in law as well as in the applicable science, technology, and economics of human activities.

The convention's reference to the climate system as a shared resource indicates the continuing dominance of the utilitarian approach to the planet and recognizes the anthropogenic source of changes to the system. The observation of historically unprecedented anthropogenic changes to the system as a whole then also raises corresponding requirements for unprecedented solutions and strategies. In this endeavor, information and cooperation at the same global scale become critical. The Kyoto Protocol established legally binding levels of greenhouse gas emissions on the signatory nations. In doing so, the protocol's regulatory features advance the stated objective of the UNFCCC.

SOVEREIGNTY AND REGULATION

The format of the UNFCCC as a framework convention with protocols provides a foundation for a diversity of regulatory contexts and a variety of rates of change. This design relies on implementing regulatory regimes in sovereign states. Depending on the form and structure of each nation, the regulatory regime within a country will also vary. For example, in the context of the United States, which is a system of federated states, federal law, state law, and more local jurisdictions, all become pertinent to the implementation of international agreements.

Within each sovereign state, the specific form of the regulatory regime will also vary according to the state's legal system and internal rules. In a federal system, the implementation will require the appropriate jurisdiction to enact relevant laws. In a common law jurisdiction such as Canada, the federal and provincial jurisdictions are a constitutionally defined matter. Within each such jurisdiction, there can be further delegated authority that may need to act. For example, it would not be unusual to establish and authorize municipal authorities and regional organizations to fulfill obligations.

The legal regulatory regimes that are applicable to climate change are as complex and diverse as the human activities that influence climate change. These human activities may be broadly differentiated into social, technological, and economic spheres. The range of these activities was reflected in the development of the Kyoto Protocol, where delegates represented every organization with a vested interest in the outcome of the agreement. These included governmental and nongovernmental entities.

The divergent interests in this type of global agreement make discussions complex and differences more apparent. The motivation to achieve the desired objective and the capability of delivering the commitments vary. The use of a trading system to meet commitments was a significant victory for the United States and the overall effective commitment was to a reduction of greenhouse gas emissions to 5.2 percent below 1990 levels. The scientific consensus in contrast recommends a stabilization of the composition of the atmosphere. To achieve this scientifically recommended stabilization would require a reduction of about 50 percent below 1990 levels and an immediate stop to deforestation.

The emerging understanding of the complexity of the climate system and changing human behavior has led to a corresponding shift in the priority of various issues. For example, previous concern for the stratospheric ozone layer has declined while that for particulate matter has increased. Previously unperceived activities such as contrails from aviation activities have gained attention as possible anthropogenic contributors to the changes.

Complexity also results from the variety in the matters subject to regulation and the capacity of previous legal categories to accommodate emerging issues. For instance, in federal systems of national jurisdiction like Canada, transboundary transport of energy supplies would be federal jurisdiction, with the registration of vehicles and local deployment of the energy within the local jurisdiction. A single issue of the supply and use of energy then spans multiple jurisdictions even within a single sovereign state.

Again, if energy is considered as including electricity or gas and their use in commercial or domestic buildings, the related issue of the energy efficiency of building structures becomes relevant. To achieve the regulation of the buildings then involves the formulation of building codes which may depend on the materials used in construction and thereby involve international supplies, federal, and local jurisdictions, including land use regulations.

Land use provisions also relate to the preservation of habitats, which then determine the continuity of species. Changing species distributions impacts emissions to the atmosphere and human activities and therefore also become relevant to regulating climate change. Regulating the activities impacting on migratory species such as birds, energy transmission, and land use that usually modifies species habitat is all relevant to climate change.

Valuation of species is also related to consumer products and services. Consumer behavior and its modification through regulation and appropriately targeted municipal and private market services become imperative. Taxation systems can and do influence choices and have an effect on anthropogenic emissions. Other means of contributing to the amelioration of the climate change problem would be through actual regulatory measures coupled with market mechanisms. The product design and inventories and other similar information may be critically associated with the development of market mechanisms that include cap and trade systems.

Non-state entities also generate rules that are not typically included in the legal sense of the term regulation, but would be regulatory within the entity. For example, a business may take note of international and state laws or lack thereof in formulating its own internal rules and procedures. Where such entities are multinational, these internal rules may also be applicable in multiple state jurisdictions. Trade or industry groups and other associations can also formulate rules and regulations. Independent industry associations have also formulated rules or standards for environmental management systems along with procedures to certify those that comply with such standards.

The enterprise of modifying the climate changes has been inadvertent until recently, and the shift toward regulating the change intentionally has been growing in strength. Early attempts at regulating the climate involved the use of chemicals such as silver iodide to seed clouds to induce rainfall. For some, the ability to regulate the climate was also conceivably a strategic weapon that could be deployed in conflict.

In the United Kingdom, early air pollution issues arose with the regulation of the alkali industry with the Alkali Act of 1863, 1874, and consolidated in the Alkali Works Regulation Act of 1906. The Alkali Inspectorate later became Her Majesty's Inspectorate of Pollution and now amalgamated into the Environmental Agency. With separate legislation for smoke abatement in 1926 and the Clean Air Act in 1956 enacted following the death of 12,000 people from the London smog, regulating atmospheric gases and particulate emissions is not new. However, it was only in 1974 that a coherent regulatory structure was set up with the Health and Safety at Work Act and the Control of Pollution Act.

In the United States, regulation has proceeded unevenly between the federal and state levels and between states. For instance, California has taken a much more aggressive approach to regulating greenhouse gases than the federal government and some other states. Regional cooperation has also resulted in some cooperative regulatory regimes. The unevenness of the regulations regionally and globally has begun to have impacts also on the economic context with flows of products and processes being modified with differing features and specifications in response to local regulatory circumstances.

With the consumption of energy a dominant feature of human activity that results in the emission of greenhouse gases, the regulation of such usage has gained prominence. The installation of smart meters, which provide detailed information on the usage patterns for electricity consumption, along with the possibility of controlling energy distribution patterns by time of day, geographic location, and land use has begun to appear in places like Toronto, Canada. Thus, climate change-related concerns are beginning to be included in urban and regional planning regulations.

Additionally, the emergence of new viable renewable energy alternatives with reduced atmospheric impacts is also part of the trend to new regulations. In some jurisdictions, it is now possible for locally generated wind and solar power to be sold to the statewide grid operated by the usual utility companies. The tax regimes are also being considered legitimate means of regulating greenhouse gas emissions. From taxing the products that lead to the emissions to industries that directly emit the greenhouse gases, the complexity of the human response to climate change is growing.

CONCLUSION

In the context of climate change, regulation proceeds from an anthropogenic understanding and utilizes sociopolitical actions to implement technological and legal regulations as they occur in the totality of the climate system. From the understanding of the climate system as defined in Article 1 of the UNFCCC, very little human activity is irrelevant to the climate change and regulating the changes in the climate system is critical to the future of this planet.

SEE ALSO: Intergovernmental Panel on Climate Change (IPCC); Kyoto Mechanisms; Kyoto Protocol; Policy, International; Policy, U.S.

BIBLIOGRAPHY. M.D. Adler, "Corrective Justice and Liability for Global Warming," *University of Pennsylvania Law Review* (v.155/6, 2007); Terry Davies, *The Potential Effects of Global Climate Change on the United States Report to Congress* (U.S. EPA Office of Policy, Planning, and Evaluation, 1989); Nicholas DiMascio, "Credit where Credit is Due: The Legal Treatment of Early Greenhouse Gas Emissions Reductions," *Duke Law Journal* (v.56/6, 2007); B.D.A. Reifsnyder, "An U.S. Perspective on Environmental Regulation: The Larger Context–Transnational Cooperation," *Canada-United States Law Journal* (v.18, 1992); H. Thompson, Jr., "Tragically Difficult: The Obstacles to Governing the Commons," *Environmental Law* (v.30/2, 2000); M.C. Trexler and L.H. Kosloff, "Global Warming, Climate-Change Mitigation, and the Birth of a Regulatory Regime," *Environmental Law Reporter* (v.27/1, 1997).

LESTER DE SOUZA
INDEPENDENT SCHOLAR

Religion

RELIGION IS A universal and varied phenomenon. While there have been those who have rejected all religion and espouse atheism, they have been for practical purposes a very small minority until recently. The attempts since the Enlightenment in Western Europe in the 1700s to foster a belief in atheism has met with little success, except among educated elites in the more advanced industrialized countries. During the time of Communist domination in the Soviet Union or in Communist China, as well as other Communist countries, the attempts to suppress religion have almost completely failed.

ANCIENT RELIGIONS

For much of humanity, especially among tribal peoples, spirits are viewed as the animating part of nature. There are believed to be spirits in rocks, trees, plants, animals, or the forces of nature. Placating these spirits, which can be deadly, is a part of the acts of worship of many different groups of people. This may involve sacrifices to quiet a volcano, or the sacrifice of young men and women to the water god at the bottom of a well in order to bring the rains. Many of the religions of the Middle East in ancient times were fertility religions. Baal worship was done with cultic sexual practices that were believed to affect nature with fertility.

Among many peoples around the world, shamans, witch doctors, or medicine men have been the chief spiritual leaders. Shamans, who would beat drums and were transformed into a dancing bear or a seal or some other totemic animal, led the Inuit of northern Canada. The shaman, in a transformed and trance-like state, would then go on a spiritual journey where he would meet spirits in the spirit world who would be able to give him (or sometimes her) knowledge of the causes of sickness, the absence of the fish, or the failure of crops among people living in more temperate societies. The shaman might also wrestle with evil spirits that were causing nature to war on humans.

Other ancient religions, such as those of the Greeks, Romans, and Chinese, believed that dramatic events in nature were messages from the gods or spirit ancestors. For the Chinese, natural events could be signs that the ruling dynasty had lost the mandate of heaven. Signs such as droughts, floods, earthquakes, fires, or other natural disasters, especially if accompanied by events such as an unexpected arrival of a comet, were among the signs that said the ruling dynasty should be replaced. Virtually all of the movements that changed dynasties in Chinese history were religiously inspired.

The Chinese philosophy of Daoism (Taoism) founded by Lao Tzu, author of the Tao Te Ching, taught a philosophy of living in accord with nature. The sage, or just the common man who sought wisdom, would live wisely if he or even she went with the flow of nature. There was therefore a rhythm to life that led people to live according to nature.

CONTEMPORARY RELIGIONS

The Europeans who came to North America were farmers and settlers, rather than adventurers or traders, like many of the other Europeans who settled in what became Latin America. Their theology was based upon the Bible, which taught in Genesis that people should multiply and fill the Earth. In addition, they were to subdue the Earth and make a garden of it. Obedience to these commands may have been fulfilled too well. However, the land and water practices of Europeans had developed independently of Christian influences for centuries before their conversion. The Europeans had effectively deforested much of Europe and killed off a variety of species before the arrival of Christianity. What Christianity taught was stewardship.

It was like the stewardship practiced in ancient Israel. The religion of the Israelites included farming land owned by a family that would be handed down to the succeeding generations. This meant that the owners needed to practice careful husbandry that preserved the land and its resources for continued use over the generations. The idea of "development" by commercial developers common in the United States today would have been completely alien. In the modern State of Israel, there has been an ongoing project of afforestation, of identifying ancient practices and using them to make the Negev region, which is very arid, to bloom. The religious beliefs of earlier times have pushed the greening of the land since the beginning of the Zionist movement to return to the land of Israel. Oddly enough, those Christians who are most supportive of the return of the Jews to Israel are often not inclined to conservation.

Those groups that are usually evangelicals and fundamentalists who accept the end of history with the return of Jesus Christ in his Second Coming are those who believe that this divine event will happen very soon. To them, it is important to use Earth's resources before Christ returns, or else they will be wasted. The theological doctrine they commonly subscribe to is premillennialist. This, however, has historically been a minority view. Most Christians have been amillennialists, who believe that the Second Coming will be a complete surprise.

Therefore, the appropriate ethic for the eschaton (the period of waiting or the period of the last days) is to practice good stewardship of the resources of nature. The remaining theological position is postmillennialism. It is a very optimistic teaching about the unfolding of the events of the last days. The belief is that most people will become Christians and that the world will make steady progress until virtually near the end of the age. Most adherents discarded this optimistic view after the two world wars.

Religion has a powerful effect on people in a variety of ways. It can satisfy spiritual needs and organize energies to accomplish grand projects. The building of the pyramids, and other numerous architectural monuments in the ancient and medieval world, were driven by religious faith. Today, most of the peoples of the world belong to one of the major religions of Hinduism, Buddhism, Jainism, Sikhism, Judaism, Islam, or Christianity. Those who still follow Confucianism or Daoism are mostly confined to Taiwan or to other overseas Chinese communities. Among these and many smaller religious groups, there is developing a great concern for the present state of the globe.

Jainism has since its beginnings over 2,500 years ago practiced reverence for life or *ahimsa*. For the Jains, all living things are not to be harmed, because this will add karma to the soul of those who kill other living creatures, such as animals or even insects. This has made merchants of Jains, rather than butchers or farmers. Their ethic is one that requires urban living tolerance of wildlife if it invades the urban area.

For example, there are both Christian ethicists who teach in seminaries and lay people who work with environmental issues who are seeking to develop principles of a Christian environmental ethic. The goal is then to apply the Christian ethical principles to agriculture, to natural resources, and to the environment that go beyond traditional principles of stewardship. Ethical debates about environmental issues such as global warming are about problems that are current or that are likely to arise. The problems are likely to be caused by technological advances that allow for exploitation of resources in ways that are very productive, but also destructive of future utilization. The issues that arise concern agreement over what are the environmental facts, the nature of the problem, and the appropriate solution(s).

At the core of these disagreements are different values and beliefs related to nature itself, and the use and management of nature by people. A growing number of scholars in the humanities, social sciences, and physical and biological sciences are emphasizing that the problems causing environmental issues are fundamentally interrelated with ethical issues.

SEE ALSO: Climate Change, Effects; Conservation; Education; Ethics.

BIBLIOGRAPHY. C.E. Deane-Drummond, *Ethics of Nature* (John Wiley & Sons, 2004); David Edward, Edward Cooper, and S.P. James, *Buddhism, Virtue and Environment* (Ashgate Publishing, 2005); R.C. Foltz, ed., *Islam and Ecology: A Bestowed Trust* (Harvard Center for the Study of World Religions, 2003); Stephanie Kaza and Kenneth Kraft, eds., *Dhrama Rain: Sources of Buddhist Environmentalism* (Shambhala, 2000); Michael Northcott, *Environment and Christian Ethics* (Cambridge University Press, 1996); J.M. Sleeth, *Serve God, Save the Planet: A Christian Call to Action* (Zondervan, 2007); F.H. Van Dyke, et al., *Redeeming Creation: The Biblical Basis for Environmental Stewardship* (Intervarsity Press, 1996).

ANDREW WASKEY
DALTON STATE COLLEGE

Renewable Energy Policy Project (REPP)

FOUNDED IN 1995, the Renewable Energy Policy Project (REPP) is based in Washington, D.C. The organization researches strategies to make renewable sources competitive in energy markets and to stabilize carbon emissions. REPP supports reindustrialization through the use of renewable technologies. It demonstrates that solar, wind, biomass, and other renewable sources can provide energy services at or below the cost of nonrenewables when structural barriers are removed. REPP works directly with states and firms to help them develop their renewable portfolio. The organization also provides expert information to consumers to improve energy efficiency and guide their transition to alternative energy options. To promote sales of renewable energy products and services, REPP created a buyer's guide and consumer directory for approximately 5,000 businesses.

REPP was initiated with support from the Energy Foundation and the U.S. Department of Energy. While financial support is determined on an annual or project-by-project basis, major donors have included the Oak Foundation, SURDNA Foundation, Turner Foundation, Bancker-Willimas Foundation, Joyce-Mertz-Gilmore Foundation, National Renewable Energy Lab, and the U.S. Environmental Protection Agency. REPP's board of directors includes leaders from renewable energy businesses, the financial sector, environmental advocacy groups, regulators, government officials, and multilateral development institutions.

A core issue for REPP is assisting the United States to stabilize its carbon emissions using renewable energies. This goal would have a significant impact on the world's carbon balance given that the United States produces approximately two-sevenths of global carbon emissions. Half of this, or one of seven total global wedges, comes from electricity. REPP's use of the wedge concept builds from the work of Stephen Pacala and Robert Socolow. Based on REPP's research, they suggest that mitigation of one wedge can be implemented with the annual production of 18,500 Megawatts of electricity from renewables.

REPP works with the renewable energy–component manufacturing sector to make it more transparent, facilitate entry, improve production, and promote growth. For example, wind energy requires rotors, towers, and generators. All the components for wind turbines can be produced domestically, and 25 states already have firms active in this manufacturing. REPP attempts to refine the supply chain to determine the exact specifications and avoid any potential bottlenecks as the use of wind power rapidly expands. This type of expertise has been underdeveloped and has slowed domestic manufacturing of renewable technologies. REPP also advocates for production tax credits for renewable energy projects.

REPP recently completed a state analysis for California, Michigan, Ohio, Wisconsin, Pennsylvania, Nevada, and Arizona. The REPP state reports provide an explanation of how manufacturing potential is calculated, and offer detailed analysis showing for a state, region, and county the potential for each of the 43 industrial codes (NAICS codes) that comprise the major component parts for the major renewable energy technologies. REPP provides states with a web-based product to present analysis in a graphical manner.

A total of 16 states have created Renewable Portfolio Standards that mandate that they generate a percent of their electricity from renewable sources. REPP analyzes the experience of these states and tracks industries to determine whether costs have declined. In addition to stimulating demand for renewables, REPP state reports identify the specific firms that could benefit from an existing or proposed national program.

REPP has linked social and economic development to ecological concerns. The staff argues that renewables take advantage of resources, such as biomass or wind, that are currently underutilized or wasted. This can provide local economic support, labor benefits, and job creation. As REPP promotes wind energy development in North Carolina, the organization advocates for a variety of direct and indirect benefits to local communities. An important element of REPP's work is promoting public awareness of success stories, such as a Nevada bill in 2003 that provided economic incentive to participants using photovoltaics. The first Solar Energy Systems Demonstration Program resulting from the bill was created in conjunction with the Washoe Tribe. REPP acquired the Center for Renewable Energy and Sustainable Technology (CREST) in 1999. CREST circulates policy briefs, fact sheets, and testimonials from

The Renewable Energy Policy Project researches strategies to make renewable sources competitive in energy markets. Solar, wind, biomass, and other renewable sources can provide energy services at or below the cost of nonrenewable sources.

public officials. It hosts popular renewable energy discussion groups with topics such as green building, stoves, photovoltaic use, and bioenergy.

SEE ALSO: Alternative Energy, Overview; Carbon Emissions; Department of Energy, U.S.

BIBLIOGRAPHY. Stephen Pacala and Robert Socolow, "Stabilization Wedges: Solving the Climate Problem for the Next 50 Years with Current Technologies," *Science* (v.305, 2004); Renewable Energy Policy Project, www.repp.org (cited June 2007); George Sterzinger and Jerry Stevens, *Component Manufacturing: Michigan's Future in the Renewable Energy Industry* (Renewable Energy Policy Project, 2006).

MARY FINLEY-BROOK
UNIVERSITY OF RICHMOND

Resources

A RESOURCE IS any item or substance that is in scarce supply and has some value. Resources are normally considered to be physical items, such as oil and natural gas. However, it is also possible to consider humans resources, since they are finite in number and are perishable under current technological conditions. Resources, when used in the context of computer or virtual environments, meanwhile, are inherently intangible in nature, although the hardware that produces them is not.

It is customary, when considering resources, to distinguish between those that are renewable and those that are nonrenewable. Resources such as oil are consumed in use and are, therefore, nonrenewable. However, in a number of other cases, it is possible to recreate or recycle some resources either in the original form or, at least, some components of the original. Glass and plastic bottles may, to some extent, be recycled into different forms and so value is created from spent resources that appear to be valueless. Considerable effort has been expended in determining which resources may be recycled or recreated in this way and, in some countries, it has led to significant social change as people become accustomed to considering the issues involved and sorting out recyclable household waste. The process is mirrored at the industrial

level, too, especially when economic incentives are provided to encourage this behavior.

Improved technology has also provided two other means of increasing the stock of resources, or at least minimizing their depletion. The first is to employ more commonly occurring resources for more rare ones. This substitution may be seen in the prevalence of plastic bags, which are dispensed with alacrity at many retail outlets. More recently, the negative impact of those plastic bags has prompted the search for other materials that would be more environmentally friendly (more biodegradable). The process of beneficiation, on the other hand, is one in which technology enables the gathering or exploitation of resources which were previously considered to be too difficult or expensive to obtain. The search for coal deeper underground or in mines located underwater is an example of this, while the continuing demand for oil and the ability to extract it means that sources previously ignored have become of considerable strategic importance.

For example, water is considered a renewable resource, because once used, it can be returned to the circulatory system that returns it to use via evaporation and precipitation. However, the modern world has seen a growth in populations and demand for water, together with climate change, that has demonstrated the extent to which water resources are in fact insufficient for future use, given current trends for demand. It is possible to characterize the Middle Eastern wars between Israel and neighbors as the result of fighting for scarce water resources, while the conflict in Darfur in Sudan has been characterized as resulting from nomadic movements of people searching for water.

MORAL AUTHORITY TO USE AND DEPLETE RESOURCES

Most human societies have developed with a religious basis that justifies mankind's prerogative to use the resources of the Earth for its own benefit. Christianity, Judaism, and Islam, for example, have similar roots in a tradition that states a divine provenance for the world and the entire universe and the passing of responsibility to shepherding the world to humanity. Certain variations in scripture explain dietary rules for the different religions, and these have led to different uses of the land and the resources of the world. The same is true of those now rare religions that are believed to offer stewardship of the world to one specific group of people. The animism of the Mongols, for example, in common with that of certain other steppe peoples, was used to help justify the destruction of resources, including people, not immediately wanted or needed by khans and other leaders.

Buddhism stresses the endless cycle of birth, rebirth, and suffering in which souls are reincarnated in a variety of forms through the ages. Since souls could inhabit not only animate but inanimate objects, then it benefits people to take care of those items appropriately. They may be used in moderation, but not abused and used excessively. Other religious beliefs also confer upon humanity the right to use natural resources, but with certain limitations. The same is true of some moral creeds that have an environmental basis. Proponents of the Gaia hypothesis, for example, hold the resources of the Earth to be central to the successful existence of nature; consequently, husbanding of those resources is a central part of the successful functioning of society.

Belief systems based on nonreligious bases have not always been so favorable to the environment. Communism, for example, appropriates the resources of the Earth for the betterment of society, and has little to say about conservation of those resources. The impact on the environment by the Russian and Chinese Communist parties has been among the most severe in the world. Similar levels of exploitation of resources, such as pollution and overlogging, for example, are also witnessed when private-sector, free-market interests have been able to gain access to resources. The *Tragedy of the Commons* by William Foster Lloyd framed the potential problem of a laissez-faire approach to the management of nature. The presence of democracy in a country, accompanied by fair and transparent policing of the laws, is one of the best means of ensuring that overexploitation of resources does not take place. The Indian economist Amartya Sen, who observed that no famine had ever occurred in a functioning democracy, originally noted this concept.

THE DIMINUTION OF NATURAL RESOURCES

The history of cod fishing in the Atlantic Ocean is a graphic example of the abundance of resources available in past centuries and the way in which those resources have been enormously diminished within the last century. For hundreds and in some cases thousands of years, the ability of man to harvest resources, renewable resources in any case, was exceeded by the fertility of nature in replacing them. The development of industrialized harvesting techniques succeeded not only in depleting the stocks of the fish, but also seriously damaged the environment, including the ecology, in which the cod thrived. The diminution of this resource, in common with so many others, has been so severe that it is not possible to recreate a satisfactory understanding of the amount of the resource previously available. This makes it extremely difficult to identify means of returning to the status quo before overexploitation and, hence, it is not very likely that such a state could ever be attained. It would take an event as severe in its impact as World War II, which effectively prevented deep sea fishing in the Atlantic Ocean altogether for several years, for fish stocks to be replenished to any meaningful degree.

Overfishing has probably already destroyed ocean ecologies beyond repair, and the same is true of the logging of hardwood trees in the former rainforests of Thailand and Burma. Arguments persist over whether the production of oil and natural gas has yet peaked, or is at its maximum now, but the existing oil is not going to be replenished. Human society must prepare to live in a world in which many of the resources on which it had previously relied are no longer available.

RESOURCE ALLOCATION

Given that resources are, by definition, finite and scarce in nature, then there must be some mechanism to allocate different shares to different sets of people. Allowing everyone who has an interest in resource exploitation to do so as freely as desired will lead to disastrous depletion of the resource. Consequently, the basis of allocation must be determined.

In mature, democratic societies, coalitions of interests will help to set the agenda by which resources are allocated. This can be quite efficient in determining the share that each set of interests will gain, but has proved to be less successful in setting the amount of resource that may be allocated on a sustainable basis. Democratic debate must, consequently, be supplemented with a technical limitation determining overall exploitation in order to be viable. This is a superior approach to those that rely on market power (where resources go to those who can afford them and are denied to the poor), since these suffer from equity issues and, more relevantly, from the stress inflicted by inequality and the high probability that it will lead to social unrest and ultimately rebellion.

Systems that reward the rich at the expense of the poor rely, therefore, on the ability and willingness of the former to mobilize the threat of armed violence against the latter. Even so, social systems of this sort still rely upon the labor of the masses to produce goods and services to facilitate the lifestyle of the rich. Sequestering oxygen or water, therefore, which are essential for life, will provide short and possibly medium-term gains for the rich, but the system is not sustainable over the long term because it will lead to the deaths of so many of the poor. This in itself might not threaten the survival of the system, but the reduction in production capacity will do so.

Irrespective of the means by which resource allocation is managed, it must be supplemented by attempts to determine the presence or creation of substitutes. Resource scarcity inevitably leads to inequality, and this reduces social stability. The promise of suitable substitutes at some stage in the future helps to alleviate the pressures that this builds.

SEE ALSO: Alternative Energy, Overview; Ethics; Oil, Consumption of; Oil, Production of; Religion.

BIBLIOGRAPHY. M.T. Klare, *Resource Wars: The New Landscape of Global Conflict* (Holt, 2002); Mark Kurlansky, *Cod: A Biography of the Fish that Changed the World* (Penguin, 1998); W.F. Lloyd, *Lectures on Population, Value, Poor-Laws and Rent* (Augustus M. Kelley, 1968 [1837]); William Marsden, *Stupid to the Last Drop: How Alberta is Bringing Environmental Armageddon to Canada (And Doesn't Seem to Care)* (Knopf Canada, 2007); Amartya Sen, *Development as Freedom* (Anchor, 2000); Vandana Shiva, *Water Wars: Privatization, Pollution, and Profit* (South End Press, 2002).

JOHN WALSH
SHINAWATRA UNIVERSITY

Resources for the Future (RFF)

RESOURCES FOR THE Future (RFF) is a nonprofit, nonpartisan organization with headquarters in Washington, D.C. Founded in 1952 under the Truman Administration, RFF initially had a domestic focus, but has since shifted to include international affairs. With a variety of outlets for data dissemination, RFF provides intellectual leadership in environmental economics. Research methods are based in the social sciences and quantitative economic analysis, including cost-benefit trade-offs, valuations, and risk assessments. RFF scholars compile core knowledge on a range of environmental topics, with the goal of contributing to scholarship, teaching, debate, and decision making. One major division of research is energy, electricity, and climate change.

RFF was the first think tank in the United States devoted exclusively to environmental issues. The impetus for RFF came from William Paley, who had formerly chaired a presidential commission charged with examining whether the United States was becoming overly dependent on foreign natural resources and commodities. Today, the RFF board of directors consists of members of the business community, former state officials, academics, and leaders of environmental advocacy organizations. It is increasingly inclusive in terms of nationality, race, and gender.

By 2006, RFF had operating revenue of $10.6 million, of which nearly 70 percent came from individual contributions and private foundations, as well as the 25 percent that is generated from government grants. The rest was withdrawn from a reserve fund valued at over $35 million that was created to support the organization's operations.

RFF has approximately 40 staff researchers composed of senior fellows, fellows, resident scholars, research assistants, and associates. In addition, RFF hosts visiting scholars from academia and the policy community. RFF scholars share their findings through seminars and conferences, congressional testimony, and global media. They publish in external peer-reviewed journals and several RFF publications, including discussion papers, reports, issue briefs, and *Resources* magazine. The online *Weathervane* is a guide to global climate policy. *RFF Connection* is an electronic newsletter that provides updates on events, research, and publications. RFF Press offers hundreds of titles on environmental issues written by the organization's staff and outside experts.

The focus of RFF's research has shifted to include global concerns, although U.S. policy innovation and implementation maintain importance. International research topics include environmental governance in the European Union, the UN-based negotiations that produced the UNFCCC and the Kyoto Protocol, and the U.S. proposal for a technology transfer agreement. RFF scholars also examine the related topics of climate stabilization and air pollution control in rapidly developing countries such as China and India.

RFF's Climate and Technology Program analyzes and critiques options for U.S. policy and the role of technology development and deployment in combating carbon emissions. In 2007, RFF scholars authored a series of background reports related to the design of federal climate policy. They provide stakeholders and policymakers with an understanding of policy options, from which effective mandatory federal policy might be crafted. Researchers have estimated the costs of emissions abatement, calculated the benefits of mitigating climate change impacts, assessed the effect of the choice of discount rate for long-term policies, and characterized uncertainty in such analyses.

RFF is certified as a U.S. General Services Administration (GSA) Management, Organizational and Business Improvement Services (MOBIS) contractor for consulting, survey, and facilitation services. MOBIS contractors assist the federal government to respond to new mandates and evolving practices. RFF researchers are also analyzing proposed actions and evaluating current efforts of state and local governments as well as the business sector.

RFF provides ongoing support to many state and nongovernmental organizations, including the Intergovernmental Panel on Climate Change. For decades, the RFF Seminar Series has provided the Washington community with a weekly forum in which scholars, journalists, advocates, and policymakers interact. RFF internships and doctoral and postdoctoral fellowships train and support future leaders and scientists.

SEE ALSO: Policy, International; Policy, U.S.; Sustainability.

BIBLIOGRAPHY. Ray Kopp and Billy Pizer, *Assessing U.S. Climate Policy Options: A Report Summarizing Work at RFF as part of the Inter-Industry U.S. Climate Policy Forum,* http://www.rff.org/rff/Publications/CPF_AssessingUSClimatePolicyOptions.cfm; Resources for the Future, www.rff.org (cited June 2007).

MARY FINLEY-BROOK
UNIVERSITY OF RICHMOND

Revelle, Roger (1909–91)

AN EARLY PREDICTOR of global warming, Roger Revelle helped to start the scientific debate on the issue in the late 1950s. He challenged the accepted notion that global warming was countered by the absorption of carbon dioxide from the oceans. Revelle discovered that the particular chemistry of sea water hinders such absorption. Because of the respect that he earned among the scientific community, Revelle was regarded as a spokesperson for science whose advice on as diverse matters as world population, agricultural policies, education, and the preservation of the environment were held in high esteem.

Born in Seattle, Washington, on March 7, 1909, Revelle was raised in Pasadena, California, and soon stood out as a gifted student during his academic career. In 1925, Revelle enrolled at Pomona College with an interest in journalism, but later switched to geology as his major field of study. In 1928, Revelle met Ellen Virginia Clark, a student at the neighboring Scripps College and a grandniece of Scripps College founder Ellen Browning Scripps. The couple married in 1931.

Revelle obtained his bachelor's degree from Pomona in 1929, and then entered the University of California–Berkley to pursue his studies in geology. In 1931, his professor George Davis Louderback recommended him for a research assistantship in oceanography at the Scripps Institute of Oceanography in

Roger Revelle lead navy research in oceanography. His carbon dioxide research led to his predictions of global warming.

La Jolla, California. While at Scripps, Revelle took part in several expeditions on the *Scripps*, the institute's small research vessel. He was also a guest on ships of the U.S. Coast and Geodetic Survey and the U.S. Navy. In 1936, Revelle completed his dissertation, "Marine Bottom Samples Collected in the Pacific Ocean by the *Carnegie* on its Seventh Cruise," and was awarded his Ph.D. He was immediately hired as an oceanography instructor at Scripps, under the directorship of Harald Sverdrup.

During World War II, Revelle served in the U.S. Navy as the commander of the oceanographic section of the Bureau of Ships and became head of their geophysics branch in 1946. His reputation and influence in the navy quickly grew during the war and enabled him to substantially influence the navy research program in oceanography. Revelle received an official commendation for this work from Secretary of the Navy James Forrestal after the war. After the conflict, he returned to Scripps in 1948 and directed it from 1951 to 1964. The scientist was involved in supervision of the first postwar atomic test on Bikini Atoll, Operation Crossroads. He led the oceanographic and geophysical components of the operation. His task was to study the diffusion of radioactive wastes and the environmental effects of the bomb at Bikini.

During his directorship at Scripps, Revelle was also appointed to other prestigious positions such as chairman of the Panel on Oceanography of the U.S. National Committee on the International Geophysical Year (IGY). Scripps, which Revelle was constantly expanding thanks to his administrative skills, was initially designated as a participant in the IGY, but was later promoted to the main center in the Atmospheric Carbon Dioxide Program. As a result, Revelle's interest in the general carbon cycle and the solubility of calcium carbonate grew and he began a systematic research which engaged him for the rest of his life. The result of this interest was a famous 1957 article published in *Tellus*, a European meteorology and oceanography journal, which Revelle coauthored with Hans Suess, one of the founders of radiocarbon dating. The article demonstrated that carbon dioxide had increased in the air as a result of the use of fossil fuels.

Following this discovery, in 1963, Revelle took a leave of absence from Scripps and committed himself to public policy. Revelle was among the first in the scientific community to bring the subject of rising levels of carbon dioxide to the attention of the public as a member of the President's Science Advisory Committee Panel on Environmental Pollution in 1965. The committee, chaired by Revelle, published the first authoritative U.S. government report which officially stated that carbon dioxide from fossil fuels was a potential global threat. Revelle also founded the Center for Population Studies at Harvard University, and spent more than a decade as director. His primary interests were applications of science and technology to world hunger. In 1976, Revelle returned to the University of California–San Diego where he received the title of Professor of Science and Public Policy and joined the Department of Political Science.

As the chair of the National Academy of Sciences Energy and Climate Panel in 1977, Revelle concluded that about 40 percent of the anthropogenic carbon dioxide has remained in the atmosphere. It was produced two-thirds from fossil fuel, and one-third from the clearing of forests. In his role of spokesperson against the dangers of global warming, Revelle influenced public opinion on the carbon dioxide issue thanks to a widely read article published in *Scientific American* in August 1982. His research emphasized the rise in global sea level and the melting of glaciers and ice sheets caused by the thermal increase of the warming surface waters. Through his international scientific contacts, Revelle circulated his research findings and fostered debates about his findings and the threatening environmental and social effects of increased atmospheric carbon dioxide. The scientist was an early advocate of governmental policy and action.

Revelle was a respected member of many academic, scientific, and government committees. He was science adviser to the secretary of the interior, president of the American Association for the Advancement of Science, and a member of the NASA Advisory Council. In November 1990, Roger Revelle received the National Medal of Science from President George Bush.

SEE ALSO: Carbon Cycle; Carbon Dioxide; Carbon Emissions; Global Warming; Navy, U.S.; Scripps Institute of Oceanography; Sea Level, Rising; Sverdrup, Harald Ulrik.

BIBLIOGRAPHY. Judith Morgan and Neil Morgan, "Roger Revelle: A Profile." www.repositories.cdlib.org/cgi/view

content.cgi?article=1001&context=sio/arch (cited November 2007); Spencer Weart, *The Discovery of Global Warming* (Harvard University Press, 2004).

LUCA PRONO
UNIVERSITY OF NOTTINGHAM

Rhode Island

RHODE ISLAND HAS an area of 1,545 sq. mi. (4,001 sq. m.) with an average elevation of 200 ft. (61 m.) above sea level. The land is divided into two regions: the coastal lowlands (made up of sandy beaches, rocky cliffs, lagoons, saltwater ponds and low plains) covering more than half of Rhode Island; and the Eastern New England Uplands, a region of small valleys, rolling hills, lakes, reservoirs and ponds.

Rhode Island's humid continental climate is a little milder than in most of New England, with the extremes of winter cold and summer heat moderated by the Atlantic Ocean and Narragansett Bay. Providence's annual average temperature is 51 degrees F (11 degrees C), with January temperatures averaging 29 degrees F (minus 1.5 degrees C), and July temperatures averaging 73 degrees

In 1969, massive storm surges from Hurricane Carol raged through the Rhode Island Yacht Club.

F (23 degrees C), with an annual precipitation of 46 in (117 cm.). Temperatures are moderated by warm winds off Narragansett Bay, but extremes do occur. The highest temperature recorded in the state was 104 degrees F (40 degrees C) on August 2, 1975, and the lowest temperature recorded in the state was minus 23 degrees F (minus 31 degrees C) on January 11, 1942. Rhode Island's soil is fertile, but very rocky, and about 60 percent of Rhode Island's land is covered by forest.

IMPACT OF CLIMATE CHANGE

Beaches do not stay put, waves and current move the sand and change the coastline, and this damage will likely increase as sea levels rise and the intensity of storms increases. The average temperature in Providence has risen about 3.3 degrees F (1.8 degrees C) over the last century. Coastal erosion and storm surges have already damaged many of the state's tidal flats and dunes, including those on Block Island and throughout the Rhode Island Refuge Complex. Rhode Island's 400 mi. (644 km.) of coastline is home to the bulk of the state's residents. The beaches along the south shore have already been severely damaged by hurricanes and storm surges. Many of Rhode Island's lakes and waterways are freezing for shorter periods of time, reducing traditional outdoor recreation opportunities such as hockey and ice-skating. Rhode Island's current ozone levels exceed national health standards, and the entire state is rated as having a "serious" problem attaining safe levels. Warmer weather could increase concentrations of ground-level ozone, which is known to aggravate respiratory problems.

Based on energy consumption data from the Energy Information Administration, Rhode Island's total CO_2 emissions from fossil fuel combustion in million metric tons for 2004 was 10.95, made up of contributions by source from commercial, 1.20; industrial, 0.62; residential, 2.80; transportation, 4.38; and electric power, 1.96.

Rhode Island adopted a "renewable portfolio standard" that calls for 16 percent of the state's energy to come from clean, renewable sources like solar and wind by 2020, and a greenhouse gas reduction target is to meet 1990 levels of six greenhouse gases by 2010 and below 1990 levels by 2020. Rhode Island joined the Climate Registry, a voluntary national initiative to track, verify, and report greenhouse gas emissions, with acceptance of data from state agencies, corporations, and educational institutions beginning in January

2008 and joined all the states in New England (as well as others in the mid-Atlantic area) in the Regional Greenhouse Gas Initiative (RGGI), the first multiple-states, market-based mandatory cap-and-trade program to reduce heat-trapping emissions from power plants. The Environmental Council of Rhode Island and more than 70 other state groups and businesses are part of the Rhode Island Climate Coalition, which is working to support the state's climate action plan.

SEE ALSO: Hurricanes and Typhoons; Sea Level, Rising.

BIBLIOGRAPHY. "Global Warming Will Hit U.S. Northeast Hard Unless Action Taken Now; Long-term Severity Depends on Near-term Choices, Scientists Say," Union of Concerned Scientists, www.ucsusa.org (July 11, 2007); Barry Keim, "Current Climate of the New England Region," *New England Regional Assessment* (June 1999); National Wildlife Federation, www.nwf.org (cited November 2007).

LYN MICHAUD
INDEPENDENT SCHOLAR

Richardson, Lewis Fry (1881–1953)

LEWIS FRY RICHARDSON was an innovative Britsh mathematician, physicist, and psychologist who first tried to apply mathematical concepts to weather forecast. Although his method for weather forecasting was not entirely successful during his lifetime, it was rediscovered with the advent of computers and formed the basis for computer-based weather forecast. The recorded change over a given distance of temperature and wind (gradient) is named the Richardson number after him.

Lewis Fry Richardson was born into a wealthy Quaker family in Newcastle-Upon-Tyne on October 11, 1881. His mother, Catherine Fry, was the daughter of corn merchants, and his father, David Richardson, came from a family of tanners, a profession that he took up himself. Lewis was the youngest of a large family of seven children. He attended Newcastle Preparatory School where he already showed his predilection for math, particularly the study of Euclid. Then in 1894 he went to Bootham School in

York, an elite Quaker institution established in 1823. It was here that Richardson first combined his interest for math with science and meteorology in particular. One of his teachers, Edmund Clark, was in fact an expert in meteorology and greatly influenced Richardson. The institution also reinforced Lewis's pacifism, a value that had been taught to him by his parents and a fundamental tenet of Quakerism which led him to difficult career choices in his maturity. After leaving Bootham in 1898, Richardson spent two years in Newcastle at the Durham College of Science where he studied mathematics, physics, chemistry, botany, and zoology. Richardson completed his education at King's College, Cambridge, from which he graduated with a First Class degree in the Natural Science Tripos in 1903.

After graduation, Richardson was employed at many different posts. He worked in the National Physical Laboratory (1903–04, 1907–09) and the Meteorological Office (1913–16), and he was hired as a university lecturer at University College Aberystwyth (1905–06) and Manchester College of Technology (1912–13). In addition he was a chemist with National Peat Industries (1906–07) and directed the physical and chemical laboratory of the Sunbeam Lamp Company (1909–12). He married Dorothy Garnett in 1909 and although they had no children of their own, they adopted two sons and a daughter.

Richardson was working for the Meteorological Office as superintendent of the Eskdalemuir Observatory at the outbreak of World War I in 1914. Because of his Quaker beliefs, he declared himself a conscientious objector and could not, therefore, be drafted into the military. This choice implied that he would never be able to qualify for university posts. While Richardson was not involved in military operations, from 1916 to 1919 he served in the Friends Ambulance Unit, attached to the 16th French Infantry Division, where his work earned him praise. After the war, Richardson returned to his position in the Meteorological Office, but had to resign from it in 1920 when the Meteorological Office became part of the Air Ministry. His pacifist beliefs could not allow him to continue to work for an institution which was part of the military. Richardson then went back to teaching. From 1920 to 1929 he headed the Physics Department at Westminster Training College, and from 1929 to 1940, he was principal of Paisley College of Technology and School of Art in Scotland. He retired in 1940 at the age of 59 to concentrate on research.

Richardson had a lifelong interest in the application of mathematics to meteorology. He was the first to apply the mathematical method of finite differences to the prediction of the weather in his study *Weather Prediction by Numerical Process* (1922). His method of finite differences was designed to solve differential equations, arising in his work on the flow of water in peat for the National Peat Industries. As these methods allowed him to obtain highly accurate solutions, he decided to apply them to solve the problems of the dynamics of the atmosphere encountered while working for the Meteorological Office. The initial conditions were defined through observations from weather stations, and would then be used to solve the equations. Finally, a prediction of the weather could be made. Richardson's remarkable insight was ahead of its time since the time taken for the necessary hand calculations in a pre-computer age took too long. Even with a large group of people working to solve the equations, the solution could not be found in time to be useful to predict the weather. Richardson himself admitted that it would take 60,000 people to have the prediction of tomorrow's weather before the weather actually arrived. In spite of this flaw, Richardson's work pioneered present day weather forecasting.

Throughout his life, Richardson published extensively on the application of mathematics to the weather and contributed to the theory of diffusion, specifically regarding eddy-diffusion in the atmosphere. For his scientific achievements, he was elected to the Royal Society in 1926. His deeply-rooted interest in pacifism led him to apply mathematics to the study of wars and military conflicts. His results were published in three major books: *Generalized Foreign Politics* (1939), *Arms and Insecurity* (1949), and *Statistics of Deadly Quarrels* (1950). Richardson used mathematics to challenge the assumption that war was a rational national policy in the interests of a nation. He gave systems of differential equations governing the interactions between countries. Starting with the armament of two nations, Richardson constructed an idealized system of equations calculating the rate of a nation's military buildup as directly proportional to the amount of arms its rival has and also to the disputes toward the enemy. This rate is, instead, nega-tively proportional to the amount of arms it already has itself. Richardson died on September 30, 1953, in Kilmun, Argyll, Scotland.

SEE ALSO: Climate Models; Weather.

BIBLIOGRAPHY. O.M. Ashford, *Prophet—or Professor? The Life and Work of Lewis Fry Richardson* (Adam Hilger, 1985); T.W. Korner, *The Pleasures of Counting* (Cambridge University Press, 1996); P. Lynch, *The Emergence of Numerical Weather Prediction* (Cambridge University Press, 2006); Spencer Weart, *The Discovery of Global Warming* (Harvard University Press, 2004).

LUCA PRONO
UNIVERSITY OF NOTTINGHAM

Risk

RISK IS A concept that captures the probability and, in some instances, the potential severity of the occurrence of a negative outcome (that is, being exposed to a hazard). There is much discussion surrounding the various risks associated with global warming and climate change, such as those related to the environment, ecosystem, human health, and the world economy. In this regard, various experts have used risk analysis to assess, manage, and communicate these associated risks.

Global warming occurs as a result of the accumulation of greenhouse gases in the atmosphere. These greenhouse gases occur both naturally (such as water vapor, carbon dioxide, methane, and ozone) and as a result of human activity (for example, from the burning of fossil fuels, deforestation, and the use of chlorofluorocarbons and fertilizers). The latter has been the focus of a 2007 report from the Intergovernmental Panel on Climate Change (IPCC). This has led to a great deal of discussion surrounding the various policy implications that lie ahead.

There are several environmental risks associated with global warming and climate change—some of which have been noted by researchers worldwide as already occurring, and others that have been forecast. Climate change affects countries differently depending partly on their geographical location.

The record high temperatures documented over the past two decades have resulted in early ice thaw on rivers and lakes. Furthermore, the rates of sea-level rise are expected to continue to increase as a result of both the thermal expansion of the oceans and the partial melting of mountain glaciers and the Antarctic and Greenland ice caps. It has been reported that increasing temperatures have also been the cause of many extreme weather events such as heat waves, droughts, and wildfires.

Moreover, changes to ocean temperatures and wind patterns have resulted in more frequent and intense rain and ice storms, floods, and some natural disasters such as hurricanes and typhoons. An increase in the frequency and severity of disasters can lead to secondary effects such as massive mudslides—as was the case with Hurricane Mitch in 1998. They can also have far-reaching effects that can result in a loss of livelihood, displacement, as well as local and global migration, particularly for those living in communities in the most vulnerable areas (such as low-lying coastal areas and estuaries, alpine regions, and tropical and subtropical population centers). Additionally, these risks can be exacerbated for marginalized groups as well as those living in more vulnerable communities (for example, regions that are poverty laden or more crowded).

HUMAN HEALTH AND DISEASE RISKS

Extreme weather events can have disastrous effects on physical and mental health as well as environmental health. The occurrence of droughts can lead to problems associated with water availability and quality (for example, people sharing water with livestock). Similarly, heat-related effects such as exhaustion, cramps, heart attacks, stroke, and even death are possible outcomes as a result of heat waves. Furthermore, excessive rainfalls and flooding are associated with the risk of injuries and death (as from drowning), as well as the spread of various water-borne diseases (via fecal-contaminated waterways and drinking supplies), and exposure to toxic pollutants (from nearby industrial sites and municipal sewage—as was the case with the Elbe flood which took place in 2002 in central Europe). The variation of risks associated with the transmission of infectious diseases as a result of extreme weather events have also been documented; however, their relation to global warming and climate change have not yet been conclusively reported in the literature.

The risks associated with the transmission of infectious diseases are dependent upon the kind(s) of weather event(s) that have occurred. As such, the reproduction and survival of disease-carrying vectors such as mosquitoes could be impaired by heavy rainfalls (such as flushing larvae from pooled water) or heightened by changes in climate and rainfall patterns. For instance, changes in climate have allowed vector-borne diseases such as malaria and dengue fever to survive in otherwise inhospitable areas (in higher elevations). Other vector-borne diseases (such as cholera, Ross River virus, and West Nile) and food-borne diseases (like the proliferation of bacteria in contaminated foods) are also at risk of occurring as a result of higher temperatures.

As noted by the Canadian Lung Association, climate change and the effects of it can lead to air quality problems such as those resulting from the increased burning of fossil fuels as a direct result of rising temperatures (for example, increased use of air conditioners, refrigerators, and freezers); increases in forest fires; and increased mold growth as a result of elevated levels of precipitation. Associated health effects such as asthma and allergies as well as other respiratory-related morbidity and mortality are also of concern. The Canadian Lung Association provides a more comprehensive explanation of the connection between climate change, air quality, and respiratory health.

Global warming and climate change can impact both the balance and health of the ecosystem, which in turn puts human health at risk. A report by the UN Environment Programme provides an overview of the relationship between climate change and ultraviolet radiation, ozone depletion, terrestrial and aquatic ecosystems, and biogeochemical cycles. Changes and losses related to biodiversity (disruption in ecosystems and species extinction) are also of concern. In a 2007 article, Frederic Jiguet and colleagues note a number of studies that have shown that habitat degradation is taking place; particular reference is made to research on plants, butterflies, beetles, mammals, bumblebees, birds, coral reefs, and coral-dwelling fishes—however, there are a number of other habitats which may also be at risk. Moreover, based on findings from the first comprehensive assessment of extinction risk, the Natural Resources Defense Council notes, "more than one million species could be committed to extinction by 2050 if global warming pollution is not curtailed."

Diversity in species is important, because it aids in ecosystem services/functions (maintaining soil fertility and pollinating plants and crops); a change in ecosystem services/functions can have far-reaching implications (it can affect agro-ecosystems, marine systems, and fresh water, as well as the transmission of vector-borne diseases). Furthermore, there are also risks associated with nutrition, which is dependent on the state of agricultural output (such as changes in food productivity and associated pests which are involved in the transmission of diseases) as well as other food sources (like fisheries and mammals); this can have consequences on ecosystem and human health in the immediate area(s) and globally.

ECONOMIC RISKS

As far as businesses are concerned, environmental risks are usually understood in terms of costs, managed by regulatory compliance, potential liability, and pollution release mitigation. Supply chain risks might occur, whereby suppliers may eventually pass the costs pertaining to carbon-related and more energy efficient alternatives (technology advancement) to customers. Also, businesses that generate significant carbon emissions could face similar litigation risks (lawsuits) as those experienced by tobacco, asbestos, and pharmaceutical industries that in turn may also put the company's reputation at risk.

Organizations that seek a competitive advantage in light of global warming and climate change may use a Strengths, Weaknesses, Opportunities, and Threats (S.W.O.T.) analysis to examine their organization's strengths and weaknesses in relation to the opportunities and threats posed by the environment in order to transform the various business-related threats into opportunities. For instance, organizations that are able to identify and implement new market opportunities for climate-friendly products and services may fare better than other organizations that are unable to achieve this. Additionally, organizations may want to take measures such as quantifying their carbon footprint (the amount of CO_2 created by business operations) in order to show consumers that they are aware of their role in climate change—and then take measures to correct the identified shortcomings. This could lead to the development of new and profitable products that are also environmentally friendly, as well as an increase in consumer loyalty.

There are other far-reaching economic implications outside the immediate business arena. For instance, the costs associated with insurance (both for the customer and for insurance companies) may affect those living in more vulnerable areas. Moreover, it has been argued that the impacts of climate change are not evenly distributed; for instance, developing countries are not only at a geographical disadvantage (with high rainfall variability), but their economic livelihood is heavily dependent on their agricultural output (which is at risk with climate change). Furthermore, the implementation of more stringent regulations as well as the adoption of better and safer technological alternatives and advancements may not seem like a feasible option economically for developing countries.

There are various types of risks that need to be considered with respect to global warming and climate change. Risk analysis is a process that considers the scientific, social, cultural, economic, and political issues that shape the identification, evaluation, decision making, and policy implementation concerning risk. As such, risk analysis encompasses the assessment, management, and communication of risks, each in turn having their own framework. In the context of environmental health, Annalee Yassi and colleagues have provided a comprehensive list of hazards which arise from both natural and anthropogenic (caused or induced by human activity) sources and explain the processes involved in the assessment and management of these risks—some of the hazards identified are applicable to global warming and climate change.

The risk assessment framework most frequently used in relation to environmental and human health follows the steps first identified by Lawrence, E. McCrae in 1983; these steps include problem formulation, hazard identification, dose-response relationships, exposure assessment, and risk characterization. Once characterized, risks are sometimes ranked by organizations through the use of a risk matrix. Risk matrices are made up of rows and columns that denote the severity or impact of the hazard and the likelihood or probability of its occurrence, where the former is usually more subjective, and the latter is relatively objective.

Accurately predicting (through the use of modeling and other forecasting techniques) and assessing (via the five aforementioned steps) the environmental, ecosystem, and human health risks related to global warming and climate change is challenging because of the level of complexity and uncertainty involved. These factors can result for various reasons; among these are: incomplete, insufficient or inaccurate data; gaps and errors in observation; measurement error(s); lack of knowledge (inadequate or conflicting modeling, or unknowns pertaining to feedback effects); variation in assumptions; unforeseen circumstances; variability in natural conditions, exposures, and human activity; and contributions of natural and human-induced changes.

Both risk assessment and cost-benefit analysis (a tool used to determine whether or not costs outweigh benefits—the applicability of this tool has been a widely debated issue in the face of climate policy) usually drive risk management. As such, risk management is the process through which a regulatory agency decides which action(s) to take and which regulation(s) and policies to implement based on risk assessment estimate(s) and cost-benefit analyses. With respect to global warming and climate change, there are three risk management strategies that are frequently discussed in the literature—these are mitigation, adaptation, and geoengineering.

Mitigation strategies are those that are concerned with taking actions that are aimed at reducing the extent of global warming and climate change. For instance, mitigation strategies might include adopting measures that reduce anthropogenic sources of climate change (for example, limiting greenhouse gas emissions relating to fossil-fuel combustion, or deforestation). On the other hand, adaptation strategies are aimed at decreasing vulnerability to global warming and climate change. Adaptation solutions include insulating buildings (for heat-related illness); installing window screens (for vector-borne diseases); and constructing strong sea walls (for health and extreme weather events). A third strategy is geoengineering; this is essentially when large-scale manipulation of the environment takes place in an attempt to correct climate change. An example of geoengineering would include the (proposed) manipulation of the Earth's global energy balance by blocking a percentage of sunlight (via the use of superfine reflective mesh, or orbiting mirrors) in order to offset the doubling of carbon dioxide. However, approaches such as these carry great risks and are therefore largely debated.

Good risk communication includes the two-way exchange of information, concerns, and preferences between decision makers and the public in a manner where the mutual understanding of risks is achieved. The risks associated with global warming and climate change are numerous and have far-reaching implications; it is therefore imperative that all aspects of risk are carefully considered.

SEE ALSO: Climate Change, Effects; Diseases; Ecosystems; Health; Impacts of Global Warming.

BIBLIOGRAPHY. Commission on Life Sciences National Research Council, *Risk Assessment in the Federal Government: Managing the Process* (National Academy Press, 1983); C.D. Klaassen, *Casarett & Doull's Toxicology: The Basic Science of Poisons* (McGraw-Hill Companies, Inc., 2001); J.J. McCarthy, et al., *Climate Change 2001: Impacts, Adaptation, and Vulnerability* (Cambridge University Press, 2001); Peter Wright, et al., *Strategic Management Concepts* (Prentice Hall, 1998); Annalee Yassi, et al., *Basic Environmental Health* (Oxford University Press, 2001).

ANN NOVOGRADEC
YORK UNIVERSITY

Romania

LOCATED IN EASTERN Europe, Romania has a land area of 92,043 sq. mi. (238,392 sq. km.), with a population of 21,438,000 (2006 est.), and a population density of 236 people per sq. mi. (93 people per sq. km.). Some 41 percent of the land of Romania is arable, with a further 21 percent used for meadows and pasture, and 28 percent is forested. For the generation of electricity in the country, 53 percent comes from fossil fuels, with 37 percent from hydropower, and 10 percent from nuclear power, with a small amount of electricity exported, and an even smaller part imported. The heavy use of hydropower has resulted in Romania having one of the lowest per capita rates of carbon dioxide emissions: 6.7 metric tons per person in 1990, falling to 3.8 metric tons per person in 1999, and rising slowly to 4.16 metric tons in 2004.

The generation of electricity contributed to 49 percent of the country's carbon dioxide emissions, with

Bucura Lake in the Retezat mountains in Romania. Global warming has increased the risk of flooding.

24 percent from manufacturing and construction, 11 percent from transportation, 8 percent from residential uses, and the remaining 8 percent from nonelectricity energy industries. The source of the emissions comes from gaseous fuels (33 percent), liquid fuels (32 percent), and solid fuels (31 percent), with 4 percent from cement manufacturing.

Global warming has caused increased risks of flooding in parts of the country. It is also thought to be the major reason for the heat wave in the summer of 2007, which resulted in shortages of water in some parts of the country, and the deaths of some people, especially in Bucharest and other urban centers. The highest temperature ever recorded in Romania was, however, in the 1950s, when 45 degrees C was registered at the city of Calafat in the far south of the country. The hot summer of 2007 followed an extremely warm winter, especially in January and February 2007, during which the Romanian Soccer Federation even announced that if the trend of warm winters continued, they would change their calendar, which brought the problem to national attention.

The Romanian government of Ion Iliescu took part in the UN Framework Convention on Climate Change signed in Rio de Janeiro in May 1992. They signed the Kyoto Protocol to the UN Framework Convention on Climate Change on January 5, 1999, and ratified it on March 19, 2001, with it entering into force on February 16, 2005; the government committing itself to a reduction of emissions by 1.2 percent as

a stage toward ratification, and by another 8 percent by 2012. The Romanian government has long signaled its interest in emissions trading.

SEE ALSO: Drought; Emissions, Trading; Floods.

BIBLIOGRAPHY. N.P. Peritore, *Third World Environmentalism: Case Studies from the Global South* (University of Florida Press, 1999); "Romania—Climate and Atmosphere," www.earthtrends.wri.org (cited October 2007); Farhana Yamin, ed., *Climate Change and Carbon Markets: A Handbook of Emissions Reduction Mechanisms* (Earthscan, 2005).

ROBIN S. CORFIELD
INDEPENDENT SCHOLAR

Rossby, Carl-Gustav (1898–1957)

CARL-GUSTAV ROSSBY WAS a Swedish-American meteorologist whose innovations in the study of large-scale air movement and introduction of the equations describing atmospheric motion were largely responsible for the rapid development of meteorology as a science. Rossby explained the large-scale motions of the atmosphere in terms of fluid mechanics and was one of the first scientists to notice the problem of global warming.

Rossby was born on December 28, 1898, in Stockholm, Sweden. When he was 20, he moved to Bergen, Norway, to study under pioneering atmospheric scientist Vilhelm Bjerknes at the Geophysical Institute. At that time, Bjerknes and his so-called Bergen School were making great progress in laying the foundations of meteorology as a science with their breakthroughs in the polar front theory and air mass analysis. The center was the world's leading center of meteorological research. The young Rossby contributed his brilliant ideas to the development of the group's projects. Because of the impact of Bjerknes's guidance, Rossby, who had previously been interested in studying mathematics and astronomy, committed himself to meteorology.

Rossby moved to the United States in 1926, where he worked in Washington, D.C. He was employed as a fellow of the American-Scandinavian Foundation

Carl-Gustav Rossby with a rotating tank used to study atmospheric motion. Rossby explained atmospheric motion.

for Research at the U.S. Weather Bureau to explain the innovations of the polar front theory. While at the Weather Bureau Rossby established the first weather service for civil aviation. The Weather Bureau was not a stimulating context for Rossby, who, in 1928, became professor and head of the first department of meteorology in the United States at the Massachusetts Institute of Technology, Cambridge. At MIT, he made important contributions to the understanding of heat exchange in air masses and atmospheric turbulence. He also investigated oceanography to study the relationships between ocean currents and their effects on the atmosphere.

Rossby was given American citizenship in 1938. The following year, he became assistant chief of the Weather Bureau. In that capacity, Rossby was responsible for research and education and began his studies of the general circulation of the atmosphere. In 1941, he became chairman of the department of meteorology at the University of Chicago. Rossby carried out pioneering work on the upper atmosphere, proving how it affects the long-term weather conditions of the lower air masses. Measurements recorded with instrumented balloons had demonstrated that in high latitudes in the upper atmosphere there is a circumpolar westerly wind, which overlies the system of cyclones and anticyclones lower down. In 1940 Rossby developed the theory of wave movement in the polar jet stream. He demonstrated that long sinusoidal waves of large amplitude, now known as Rossby waves, are generated by perturbations caused in the westerlies by variations in velocity with latitude. Rossby also showed the importance of the strength of the circumpolar westerlies in determining global weather. When these are weak, cold polar air will sweep south, but when they are strong, they cause the normal sequence of cyclones and anticyclones. Rossby worked on mathematical models for weather prediction and introduced the Rossby equations, which, with the introduction of digital computers in the 1950s, were of fundamental importance to forecast the weather. During World War II, Rossby was in charge of training military meteorologists, and, at the end of the war, hired many of them to work in his department at the University of Chicago. Rossby served as president of the American Meteorological Society for 1944 and 1945, and laid the foundations for the Society's first scientific journal, the *Journal of Meteorology*.

Rossby and his Chicago Group were able to compile weather charts over periods of five to 30 days to extract the general features, and tried to analyze these using basic hydrodynamic principles. The group made radical simplifying suppositions, ignoring essential but transitory weather effects like the movements of water vapor and the dissipation of wind energy. Still, they began to conceptualize how large-scale features of the general circulation might arise from simple dynamical principles.

In 1950 Rossby returned to Sweden, but continued to visit the United States. In his home country, he worked with the Institute of Meteorology, which he founded in connection with the University of Stockholm. From 1954 to 1957 he was instrumental in arousing interest in atmospheric chemistry and the interaction of airborne chemicals with the land and the sea. On December 17, 1956, Rossby appeared on the cover of *Time* magazine and was praised for his key role in raising meteorology to the status of science. The piece also referred to a theory that Rossby was developing as a result of his interest in atmospheric chemistry. According to Rossby, the world's climate might be altered by solar heat trapped in the atmosphere due to a buildup of carbon dioxide. This was one of the first insights into the problem of global warming and paved the way for many researches in the field. Rossby was unable to fully develop this insight as he died on August 19, 1957, just nine months after the *Time* article.

The Carl-Gustav Rossby Research Medal is the highest award for atmospheric science presented by the American Meteorological Society for outstanding contributions to the understanding of the structure or behavior of the atmosphere. Rossby himself was the second recipient of this prestigious award when it was still called Award for Extraordinary Scientific Achievement.

SEE ALSO: American Meteorological Society; Global Warming; Waves, Rossby.

BIBLIOGRAPHY. David Laskin, "The Weatherman & the Millionaire: How Carl-Gustaf Rossby and Harry F. Guggenheim Revolutionized Aviation and Meteorology in America," *Weatherwise* (v.58/4, 2005); Spencer Weart, *The Discovery of Global Warming* (Harvard University Press, 2004).

LUCA PRONO
UNIVERSITY OF NOTTINGHAM

Royal Dutch/Shell Group

THE ROYAL DUTCH/SHELL group is a major contributor to the release of greenhouse gases, and has been among the leading oil companies to publicly embrace the need for sustainable development, including the need to address climate change. It is by most measures the world's second largest oil company, with over 100,000 employees, operations in over 130 countries, 2006 production of nearly 3.5 million barrels of oil equivalent per day, and proven reserves of nearly 8.5 billion barrels of oil equivalent. Shell's 2006 income was $26.3 billion on revenue of $318 billion.

HISTORY

Shell Transport began in 1833 with a British shopkeeper importing oriental shells, leading to an export/import business importing oil. Royal Dutch Petroleum Company began producing petroleum in the Dutch East Indies. A partnership was formed in 1907, expanded rapidly, and was the main fuel supplier to the British in World War I, and the world's leading oil company by 1930. During this period, it also began developing its global network of service stations. Demand for petroleum exploded after World War II. During the 1960s, Shell strengthened its presence in the Middle East, and discovered reserves in the North Sea. The 1973 oil crisis led Shell to diversify into other energy sources such as coal and nuclear power, with little economic success. Shell also acquired 50 percent of an Australian solar energy company, and began producing renewable softwoods that could be used for paper, construction, and fuel. Shell is the world's leading biofuels distributor.

After oil prices collapsed in 1986, Shell invested in research and development that led to huge improvements in drilling techniques, and began some of its most challenging offshore exploration. During the 1990s, high oil prices allowed Shell to further develop biomass technologies. Since 2000, Shell's greatest expansion has been in China. In 2005, the old partnership was dissolved, and one company was created, Royal Dutch Shell.

SUSTAINABLE DEVELOPMENT

The company experienced two major public image setbacks in 1995 related to sustainable development. After the British government approved Shell's plans to decommission the Brent Spar oil storage platform by sinking it in the North Sea, Greenpeace claimed this would create large amounts of pollution. Shell argued this would create the least environmental damage. Greenpeace activists protested, boarding the rig as it was being towed to the disposal area. Widespread media coverage followed, and resulted in huge negative publicity for Shell, spawning boycotts and even a firebombing of a Shell station. Shell eventually reversed its decision and towed the rig to port in Norway for dismantling. Dismantling revealed that claims of a negligible amount of oil in the rig were accurate, and Greenpeace later issued an apology.

That fall, Shell experienced another public relations fiasco. A wholly-owned subsidiary, Shell Nigeria, is the major international oil-producing company in Nigeria, with joint ventures with the Nigerian state-owned oil company and other multinational oil companies. Significant revenues from oil production go to the central government, with little benefit accruing to people in oil producing regions. Author Ken Saro-Wiwa, a member of the Ogoni tribe from the Niger Delta, led protests against the government for not using oil revenues to benefit the Ogoni, many Ogoni leaders for complicity with the central government, and Shell for substantial pollution from exploration and pipeline spills and gas flaring and for seeking security assistance from Nigerian military forces. Violence broke out, in which several Ogoni chiefs were killed, and Saro-Wiwa and eight of his associates were arrested, tried, and found guilty of murder.

International human rights activists regarded the charges as groundless, and the trial as unfair. Activists called on the Nigerian government to commute the sentences, and on Shell to use its influence to this end. Shell's CEO and the Shell Nigeria managing director appealed for clemency on humanitarian grounds, but Saro-Wiwa and his associates were executed 10 days after the verdict. Shell was heavily criticized for not doing more to attempt to influence the government to free Saro-Wiwa.

These events prompted Shell to expand its attention to sustainable development. The company has a Social Responsibility Committee that directs its sustainable development policies and performance, and it annually produces an extensive sustainability report stating that sustainable development is part of the duties of every manager at Shell. Every one of Shell's businesses is responsible for complying with corporate sustainable development policies and achieving unit-specific targets in this area.

FUEL ALTERNATIVES

Shell expresses a commitment to help meet the energy challenge by providing more secure and responsible energy. Shell is developing more environmentally friendly fossil fuel technologies like gas to liquids (GTL), which turns natural gas into cleaner-burning fuels. While they don't receive the same financial support, Shell also supports several renewable energies. Shell was the first energy company to build demonstration hydrogen filling stations in Asia, Europe, and the United States, and is one of the world's leading distributors of biofuels. However, Shell has not yet provided the financial investment needed for hydrogen expansion, and the hydrogen is derived from fossil fuels.

In 2006, Shell sold over 3.5 billion liters of biofuels, mainly in the United States and Brazil, enough to avoid over 3.5 million tons of CO_2 production. Shell believes that "first generation" biofuels are unreliable,

requiring too much acreage to be planted to feedstocks, thus putting strain on the environment and food supply. Shell has therefore invested in "second generation" biofuels, such as the production of ethanol from straw rather than corn. Shell claims this second-generation biofuel could cut well-to-wheel CO_2 production by 90 percent, compared with conventional gasoline. In early 2007, Iogen, acquired by Shell in 2002, was one of six companies selected to receive funding under the U.S. Department of Energy's cellulosic ethanol program.

Shell invested in CHOREN Industries to create the first demonstration-scale biomass-to-liquids (BTL) plant, scheduled to come online in late 2007. This process relies on the use of a woody feedstock, gasifies it, and then uses the Shell Middle Distillates Synthesis (SMDS) process to convert the gas into a high-quality fuel identical to GTL that can be blended with diesel fuel. If used at 100 percent concentration, it could also cut well-to-wheels CO2 production by 90 percent compared with conventional diesel. Shell also has small investments in solar and wind power. Currently, their financial impact on Shell is very small, and they are not seen as offering substantial room for growth.

GREENHOUSE GAS REDUCTION

In 1997, Shell started managing its CO_2 output with the goal of reducing total greenhouse gas emissions. In 2006, Shell facilities emitted 98 million tons of greenhouse gases, about seven million lower than in 2005 and more than 20 percent below 1990 levels. Yet, Shell has target emissions limits of only 5 percent below the 1990 level for 2010. Shell claims its standards are very aggressive, but agrees they may need to be reconsidered if they have already been met. Most of Shell's reductions came from ending the venting of natural gas. Most of its anticipated reductions will continue to come from ending continuous flaring and increasing energy efficiency.

In 2006, Shell missed its annual Energy Intensity target, as it had underestimated how much extra energy would be required to produce more environmentally friendly low-sulfur fuels, and because of unplanned equipment shutdowns at several facilities that required extra energy to restart. In 2007, Shell launched a new energy efficiency effort that will make up for part of the increase. Its plan is to continue efforts to end continuous flaring by 2008, except in Nigeria where the target is 2009. Further greenhouse gas reductions are planned from energy efficiency improvements at refineries and chemicals plants.

Shell also states it is trying to reduce its customers' CO_2 emissions. Shell's customers emit six to seven times (750 million tons of CO_2 in a typical year) more CO_2 consuming Shell products than Shell does producing them. Shell promotes the use of natural gas, which emits less CO_2 than coal, and is a more profitable part of Shell's business than coal. Shell has also patented a coal gasification technology that can reduce CO_2 emissions by up to 15 percent, compared to conventional coal-fired power plants.

Shell also actively supports governments in designing and implementing effective CO_2 trading schemes. They are part of the UN Partnership for Clean Fuels and Vehicles and the World Bank Clean Air Initiative in Asia to provide cleaner fuel and improve air quality in the developing world. Shell is also a member of the U.S. Climate Action Partnership created in 2007 by over 30 companies and environmental groups.

Shell's annual sustainability report is audited by an external review committee, which has praised Shell for its leadership and transparency in reporting, but has questioned whether the speed with which Shell is acting to tackle climate change is consistent with the urgent nature of the challenge. The committee specifically noted the expected rise in future emissions, the lack of published targets after 2010, how Shell will achieve future greenhouse gas reductions after it stops flaring, and the absence of adequate research and development fund allocation information to assess Shell's commitment to develop renewable energy sources and to greenhouse gas mitigation.

SEE ALSO: BP; Oil, Consumption of; Oil, Production of.

BIBLIOGRAPHY. Remember Ken Saro-Wiwa, www.remembersarowiwa.com (cited September 2007); Shell, www.shell.com (cited September 2007); U.S. Climate Action Partnership, www.us-cap.org (cited September 2007).

GORDON RANDS
TYLER SAYERS
WESTERN ILLINOIS UNIVERSITY

Royal Meteorological Society

THE ROYAL METEOROLOGICAL Society is a British charity whose mission is to "continue to be a world-leading professional and learned society in the field of meteorology. It will encourage and facilitate collaboration with organizations that are active in Earth Systems Sciences. It will serve its professional and amateur members and the wider community by undertaking activities that support the advancement of meteorological science, its applications and its understanding." A Council and its committees are responsible for running the society, within the constraints of the Royal Charter.

The Council comprises a total of 21 officers and ordinary members of council elected at the annual general meeting. The president, elected for a two-year term, is supported by a vice-president for Scotland and three other vice-presidents, the treasurer, general secretary, four journal editors, four main committee chairmen and Ordinary Members of Council. The Council convenes five times a year to consider applications for membership and supervise the running of the Society through its Honorary Officers, Committees and permanent staff. The work of the Council is largely organized by the recommendations made by its Committees. The society staff are based at the society's headquarters in Reading where committee meetings are normally held. The society's patron is HRH The Prince of Wales. Its membership in 2006 consisted of more than 3,000 members worldwide.

The Royal Meteorological Society is the national British society for all those individuals whose profession or interests are in any way connected with meteorology or related subjects. It controls the national qualifications of the profession and, under its royal charter, follows its mission to advance meteorological science. The society intends the terms meteorological science in their broadest meaning which includes its day-to-day application in weather forecasting and in disciplines such as agriculture, aviation, hydrology, marine transport and oceanography, as well as in the areas of climatology, climate change, and the interaction between the atmosphere and the oceans. The society publishes the results of new research and provides support both for researchers and professional meteorologists and also for those whose work is connected to the weather or climate and for those who have a general interest in environment and the weather. The membership of the society thus includes a variety of figures: professional scientists, practitioners, and weather enthusiasts. Associate fellows may be any age and do not require any specific qualification in meteorology. Fellows normally require a formal qualification in a subject related to meteorology plus five years experience and must be nominated by two other fellows. The society has a number of regular publications: the monthly magazine *Weather*, the *Quarterly Journal of the Royal Meteorological Society*, *Meteorological Applications*, and the *International Journal of Climatology and Atmospheric Science Letters*.

The society was established in April 1850 with the name of the British Meteorological Society, and was later incorporated by Royal Charter in 1866, when its name was changed to the Meteorological Society. The privilege of adding 'Royal' to the title was granted by Her Majesty Queen Victoria in 1883. In 1921, the society merged with the Scottish Meteorological Society.

In 1995, the Royal Meteorological Society developed a set of atmospheric dispersion modeling guidelines to encourage good practice in the use of mathematical atmospheric dispersion models, stressing the importance of selecting the most suitable modeling procedures and of fully documenting and reporting the results of modeling assessments. The 1995 guidelines provided broad general principles of good practice for modeling studies applying across a wide range of modeling situations. The UK Atmospheric Dispersion Modeling Liaison Committee (ADMLC) commissioned an upgrading of the 1995 guidelines to take into account the new developments in modeling techniques. The updated guidelines were completed and published in 2004.

The Royal Meteorological Society has specific resources on global warming for teachers and for educational purposes. The society has acknowledged that global warming is taking place and that it is the result of human activity. Yet, one of its former presidents, Chris Collier, and one of its leading researchers, Paul Hardaker, complained in 2007 about catastrophism and the "Hollywoodization" of weather and climate that only work to create confusion in the public mind. They argue for a more sober explanation of the uncertainties about possible future changes in the Earth's climate so as not to undermine scientists' credibility.

According to both Collier and Hardaker, several organizations, including the American Association for the Advancement of Science (AAAS), have overplayed the evidence that the phenomenon of global warming is causing short-term devastating impacts.

SEE ALSO: United Kingdom; Weather.

BIBLIOGRAPHY. Royal Meteorological Society, www.rmets.org (cited November 2007); Michael A. Toman and Brent Sohngen, *Climate Change* (Ashgate Publishing, 2004).

LUCA PRONO
UNIVERSITY OF NOTTINGHAM

Russia

SITUATED IN BOTH Europe and Asia, the Russian Federation has a land area of 6,592,800 sq. mi. (17,075,400 sq. km.), with a population of 142,499,000 (2006 est.), and a population density of 21.8 people per sq. mi. (8.3 people per sq. km.). Moscow, the capital and the largest city, has a population of 10,654,000, with a density of 25,022 per sq. mi. (9,644 per sq. km.). The second largest city, Saint Petersburg, has a population of 3,990,267.

In spite of the vast size of Russia, some 8 percent of the land is arable, with a further 4 percent assigned as meadows and pasture, and 46 percent of the country is forested, including vast expanses of the Siberian tundra. The per capita rate of greenhouse gas emissions from the Russian Federation was 13.4 metric tons in 1992, falling steadily to 9.9 metric tons per person by 2002, partially as the economy in the country was struggling, and then rising to 10.5 metric tons per person by 2004. Because of its climate, and the need for extensive heating, 60 percent of the country's carbon dioxide emissions come from the production of electricity, of which 66.1 percent is generated from fossil fuels, 18.9 percent from hydropower, and 14.7 percent from nuclear power. This heavy use of fossil fuels is largely because of the abundance of coal, and also the availability of locally extracted petroleum in parts of the country. However, by 1998, the use of natural gas from Siberia had become much more important, with gaseous fuels making up 48 percent of all carbon dioxide emissions, solid fuels making up 26 percent, and liquid fuels 24 percent. In terms of the sector producing the emissions, with the bulk created in electricity production, 14 percent came from manufacturing and construction, 13 percent from transportation, and 10 percent from residential use.

The coal-mining areas of western Russia still remain important, politically, but the discovery of the Siberian gas fields around Omsk and other cities has led to the Russian Federation's exportation of gas to neighboring countries, and also some parts of Western Europe. Russia has suffered some of the effects of global warming and climate change, with the melting of the Arctic sea ice creating major problems for the northern parts of the country. Satellite measurements of the Arctic Ocean have revealed that the area of perennial ice cover has fallen by about 7 percent per decade since 1978. There has also been the melting of some of the Siberian permafrost, which some Russians have welcomed, as it has the potential to open up more arable land. However, it has also caused damage to hundreds of buildings in cities of Yakutsk and Norilsk, with the average temperature of the permanently frozen ground at Yakutsk having warmed by 2.7 degrees F (1.5 degrees C) between 1968 and 1998. Lake Baikal has also experienced a shorter freezing period in the last century, with winter freezing taking place 11 days later than had been the case, and the spring ice breaking up some five days earlier. There has been a similar problem in the Caucasus Mountains, where half of all the glacial ice there has disappeared in the last 100 years.

Although the amount of arable land has increased, and some people have managed to survive more easily in an otherwise hostile climate in Siberia, there has been a threat to some of the native fauna and flora. In Khabarovsk, in the far eastern part of the Russian Federation, drought and high winds led to forest fires in 1998, which destroyed 3.7 million acres (1.5 million hectares) of coniferous forest, the taiga, and threatened nature reserves where the last remaining Amur tigers live.

The Russian government of Boris Yeltsin took part in the United Nations Framework Convention on Climate Change signed in Rio de Janeiro in May 1992. The Russian government of Aleksandr Rutskoy accepted the Kyoto Protocol to the UN Framework Convention on Climate Change on March 11, 1999, but the new government of Vladimir Putin was initially against the ratification of it, voicing opposition in 2003. However, Putin decided to support it, and the Russian Federation ratified it on November 18, 2004, with it entering into force on February 16, 2005. This immediately changed the entire situation as regards the Kyoto Protocol, making Russia a central player in international climate policies. The Russian ratification means that countries contributing to more than 55 percent of the world's carbon dioxide emissions in 1990 had ratified the treaty, and as a result, the Kyoto Protocol became international law on February 16, 2005, 90 days after the Russian ratification. To try to reduce its greenhouse gas emissions, the Russian government has embarked on a process of expanding its hydroelectric power generation, with the Boguchan Dam scheduled to be completed in 2012.

SEE ALSO: Glaciers, Retreating; Kyoto Protocol; Natural Gas.

BIBLIOGRAPHY. Alain Bernard, et al., "Russia's Role in the Kyoto Protocol, Global Change," Massachusetts Institute of Technology, www.mit.edu (cited October 2007); Mike Edwards, "Soviet Pollution," *National Geographic* (v.186/2, 1994); Robert Henson, *The Rough Guide to Climate Change* (Rough Guide, 2006); "Russia—Climate and Atmosphere," www.earthtrends.wri.org (cited October 2007); Farhana Yamin, ed., *Climate Change and Carbon Markets* (Earthscan, 2005).

JUSTIN CORFIELD
GEELONG GRAMMAR SCHOOL, AUSTRALIA

Rwanda

THE REPUBLIC OF Rwanda, located in central Africa, has a land area of 10,169 sq. mi. (26,798 sq. km.), with a population of 9,725,000 (2006 est.), and a population density of 829 people per sq. mi. (320 people per sq. km.), the second highest in Africa, and the highest on the African mainland. Some 35 percent of the country is arable land, with 18 percent used for meadows and pasture, and 10 percent of the country remains forested.

A fisherman in Lake Kivu struggles with his boat. The lake has declined in size due to increased temperatures.

In terms of its per capita carbon dioxide emissions, Rwanda had less than 0.1 metric tons per person in 1990, with 0.07 percent in 2003, making it the 10th lowest country in terms of per capita emissions. With less than 1 percent of this coming from cement manufacturing, the remaining 99 percent is the result of the use of liquid fuels. This is because 97 percent of the country's electricity production comes from hydropower, with only 3 percent from fossil fuels.

The effects of global warming and climate change in the country have been rising average temperatures, with a decline in the size of Lake Kivu, and also less irrigation water in the marshlands around the River Kagera and from the other lakes, Lake Rwanye, Lake Ihema, and Lake Mugesera. This also represents a threat to the rainforests where the mountain gorillas live, and which are also a home for 43 species of reptiles (including many frogs) and 31 species of amphibians.

The Rwanda government took part in the United Nations Framework Convention on Climate Change signed in Rio de Janeiro in May 1992, which it ratified in 1998, and in 2001 ratified the Vienna Convention. It accepted the Kyoto Protocol to the UN Framework Convention on Climate Change on July 22, 2004, with it entering into force on February 16, 2005.

SEE ALSO: Deforestation; Forests; Species Extinction.

BIBLIOGRAPHY. Theo Schilderman, "Rwanda: Low-Cost Construction in Kigali," (*Minar*, v.38, 1991); Geordie Torr, "...To See the Gorillas?" (*Geographical*, v.78/11, 2006); World Resources Institute, "Rwanda—Climate and Atmosphere," www.earthtrends.wri.org (cited October 2007).

ROBIN S. CORFIELD
INDEPENDENT SCHOLAR

Saint Kitts and Nevis

LOCATED IN THE West Indies, Saint Kitts and Nevis are two islands in the Leeward Islands chain. They have a land area of 68 sq. mi. (168.4 sq. km.) for Saint Kitts and 35.9 sq. mi. (93.2 sq. km.) for Nevis. They have an overall population of 50,000 (2006 est.) and a population density of 426 people per sq. mi. (164 people per sq. km.). Some 75 percent of the population lives on the island of Saint Kitts, with 49 percent of the overall population living in urban areas, even though the capital—Basseterre—has a population of only 14,000. Some 22 percent of the land is arable, with another 3 percent used for meadow and pasture.

Traditionally, the economy of Saint Kitts, and especially Nevis, was centered on the sugar industry, with sugar cane growing well on the volcanic slopes of both islands. Although demand for sugar for food has declined, its use in ethanol, in an effort made to reduce the world's dependence on gasoline, has become more important starting in the 1990s. Although Saint Kitts and Nevis do face trouble from Caribbean hurricanes, they face less of a problem than many other islands in the West Indies with regard to rising water levels. However, rising water temperatures are expected to have a major effect on the population of sea turtles and also on many other types of marine life that live around the black coral off the shores of Saint Kitts. This is expected to have an effect on the growing tourism industry, which now makes up some 12 percent of the country's economy. The effect of increased development on both Saint Kitts and Nevis has resulted in an increase in per capita carbon dioxide emissions, up from 1.6 metric tons per person in 1990 to 3 metric tons per person in 2003.

The government of Kennedy A. Simmonds took part in the United Nations Framework Convention on Climate Change, signed in Rio de Janeiro in May 1992, but the government of Denzil Douglas, which came to power in 1995, has so far not expressed any position on the Kyoto Protocol to the UN Framework Convention on Climate Change.

SEE ALSO: Greenhouse Gases; Kyoto Protocol.

BIBLIOGRAPHY. Sonita Barrett, *Sugar Industry of St. Kitts and Nevis—The Post War Experience* (University of the West Indies, 1985); Ariel Lugo, "Development, Forestry and Environmental Quality in the Eastern Caribbean," *Sustainable Development and Environmental Management of Small Islands* (UNESCO, 1990); USA International Business Publications, *Saint Kitts And Nevis Ecology & Nature*

Protection Handbook (USA International Business Publications, 2005).

ROBIN S. CORFIELD
INDEPENDENT SCHOLAR

Saint Lucia

THE ISLAND OF Saint Lucia, located in the Windward Islands and surrounded by the Caribbean Sea and the Atlantic Ocean, has a land area of 239 sq. mi. (616 sq. km.), with a population of 165,000 (2006 est.) and a population density of 774 people per sq. mi. (298 people per sq. km.), making it the 39th most densely populated country in the world. About a third of the population lives in the capital, Castries, on the sheltered western coast, with 8 percent of the land being arable and a further 5 percent used for meadows and pasture. Some 41 percent of the exports of the country come from bananas.

Saint Lucia has a number of offshore coral reefs, the most well known being those near Soufrière, in the southeast of the island. Just south of these are other coral reefs that also attract many tourists. There are worries about their preservation, with signs of coral bleaching resulting from the increase in water temperatures. There is extensive public transport on the island, maintained by private bus companies. As with many other developing economies, Saint Lucia has seen a rise in carbon dioxide emissions per capita, going from 1.2 metric tons per person in 1990 to 2.2 metric tons in 1996 before falling to 1.4 metric tons in 1998, but rising again to between 1.9 and 2.1 metric tons per person since then. All the carbon dioxide emissions in the country come from liquid fuel, and all electricity production in the country comes from fossil fuels.

The government of John Compton took part in the United Nations Framework Convention on Climate Change, signed in Rio de Janeiro in May 1992, and ratified the Vienna Convention; two years later, Saint Lucia was represented at the Global Conference on the Sustainable Development of Small Island Developing States held in Barbados. The Saint Lucia government of Kenny Anthony signed the Kyoto Protocol to the UN Framework Convention on Climate Change on March 16, 1998, with the country ratifying it on August 20, 2003, and it entering into force on February 16, 2005.

Saint Lucia has seen a rise in carbon dioxide emissions per capita, from 1.2 metric tons per person to 2.1, since 1990.

SEE ALSO: Climate Change, Effects; Transportation.

BIBLIOGRAPHY. P.I. Gomes, ed., *Rural Development in the Caribbean* (C. Hurst, 1985); *St. Lucia Industrial Development* (National Development Corporation, 1973); "Vulnerable Caribbean Nations Prepare for Global Warming," *Environment News Service*, http://ens-newswire.com/ens/jun2001/2001-06-04-02.asp (cited September 2007).

JUSTIN CORFIELD
GEELONG GRAMMAR SCHOOL, AUSTRALIA

Saint Vincent and the Grenadines

SAINT VINCENT AND the Grenadines are part of the Windward Islands chain, surrounded by the Caribbean Sea and the Atlantic Ocean, and have a land area of 150 sq. mi. (388 sq. km.), of which the island of Saint Vincent itself accounts for 107 sq. mi. (240 sq. km.). The country has a population of 120,000 (2006 est.) and a population density of 798 people per sq. mi. (307 people per sq. km.), with Kingstown, the nation's capital, having a population of about 16,000.

Some 10 percent of the land is arable, much of which is used for the growing of coconuts and bananas, and a further 5 percent is used for meadows and pasture. Traditionally, few tourists visit Saint Vincent.

Some of the islands of the Grenadines are low lying, and the country faces major problems with the rising water level; its coral reefs are also threatened by the rising temperature of the water. The flooding feared in some parts of the islands threatens to increase the prevalence of insect-borne diseases such as malaria and dengue fever and also threatens low-lying parts of the country. Global warming has also been blamed for the deaths of many fish in 1999 off the coast of Saint Vincent and other nearby islands. The carbon dioxide emissions per capita in Saint Vincent remain very low, being 0.7 metric tons per person in 1990, although this number has risen significantly since then, reaching 1.6 metric tons in 2003. The carbon dioxide emissions come entirely from liquid fuels, with fossil fuels making up 73.2 percent of the electricity production in the country and the remainder—26.8 percent—coming from hydropower.

The government of James Fitz-Allen Mitchell took part in the United Nations Framework Convention on Climate Change, signed in Rio de Janeiro in May 1992, and two years later, Saint Vincent was represented at the Global Conference on the Sustainable Development of Small Island Developing States held in Barbados. In 1996, the government ratified the Vienna Convention, and on March 16, 1998, the Saint Vincent government signed the Kyoto Protocol to the UN Framework Convention on Climate Change, with the country ratifying it on August 20, 2003, and it entering into force on February 16, 2005.

SEE ALSO: Climate Change, Effects; Floods; Kyoto Protocol.

BIBLIOGRAPHY. Environment News Service, "Vulnerable Caribbean Nations Prepare for Global Warming," http://ens-newswire.com/ens/jun2001/2001-06-04-02.asp (cited September 2007); World Resources Institute, "St. Vincent and the Grenadines—Climate and Atmosphere," http://earthtrends.wri.org (cited October 2007).

ROBIN S. CORFIELD
INDEPENDENT SCHOLAR

Salinity

TWO ATTRIBUTES OF the oceans, temperature and salinity, determine the density of seawater, and the differences in density between the water masses in the world's oceans causes the water to flow in thermohaline circulation, thereby producing the greatest oceanic current on the planet.

Salinity is the distinct taste of seawater and is the result of the presence of dissolved salts (more than 85 dissolved constituents), among which chloride (Cl) and sodium (Na), the elements of common table salt, are the most abundant. The term *salinity* refers to the content of these dissolved salts and has been defined as grams of dissolved salts per kilogram of seawater. Salinity has been expressed as parts per thousand (‰ or ppt) and, more recently, by practical salinity units (psu).

On average, a kilogram of seawater has 35 grams of dissolved salts, so its salinity content is 35‰, or 35 psu. The accuracy of most laboratory salinometers (see below) is about 0.001 psu. Thus, only those components with a concentration over 0.001‰ will contribute to such salinity estimates. Only 15 of the dissolved salts have concentrations above that limit.

A key observational result (known as the principle of Dittmar, after William Dittmar, a Scottish professor of chemistry; the principle of Maury, after Matthew Fontaine Maury, an American astronomer, oceanographer, and geologist; or the hypothesis of Forchhammer, after Johan Georg Forchhammer, a Danish mineralogist and geologist) is that the relative concentration between some of these most abundant salts is virtually constant over much of the World Ocean. This finding indicates that the physical characterization of seawater is given by its temperature, pressure, and a single number reflecting the concentration of the most abundant components. Salinity is that number.

MEASURING SALINITY

In 1902, an international commission defined salinity as the total amount of solid material, in grams, contained in 1 kg. of seawater when all the carbonate has been converted to oxide, the bromide and iodine replaced by chlorine, and all organic matter completely oxidized. With this definition in hand, the commission estimated the salinity of several seawater samples and the fixed relationship between salinity (S) and chlorinity:

$$S(\text{‰}) = 003 + 1.805 \ Cl \ (\text{‰})$$

This was known as Knudsen's equation, after Martin Hans Christian Knudsen, a Danish physicist (1871–1949), and was redefined in 1969 as

$$S(\text{‰}) = 1.80655 \ Cl \ (\text{‰}).$$

Defining salinity in terms of chlorinity alleviates the practical difficulties of measuring salinity through evaporating water samples to dryness. For calibration purposes, artificial water with salinity almost equal to 35‰, known as Copenhagen water, is manufactured to serve as a reference. Copenhagen water has a chlorinity of 19.381‰. This approach requires the chemical titration of water samples usually obtained by Nansen bottles, named after Fridtjof Bedel-Jarlsberg Nansen, a Norwegian explorer and scientist (1861–1930), which are self-closing containers that collect water from different depths.

Pure water is a poor electrical conductor. However, the presence of dissolved salts greatly increases its conductivity, which, in fact, is a function of pressure, temperature, and the degree of ionization of the dissolved salts. In the second half of the 20th century, technical improvements in the measurement of the electrical conductivity of seawater led to the development of the so-called salinometers. Conductive salinometers measure the ratio between the conductivity of the sample against that of a reference sample of known salinity. Researchers using conductivity salinometers in the beginning were giving a salinity value of 35‰ to any sample having the same conductivity as the Copenhagen water, even though the mass of salt per kilogram of water was not guaranteed to be the same in both cases. This was because conductivity depends on the degree of ionization of the dissolved salts and not on the absolute mass of salt. In 1978, the salinity scale was redefined in terms of the conductivity ratio, K_{15}, between any given sample and the reference solution:

$$S(\text{psu}) = 0.0080 - 0.1692 \ K_{15}^{1/2} + 25.3851 \ K_{15} + 14.0941 \ K_{15}^{3/2} \\ - 7.0261 \ K_{15}^{2} + 2.7081 \ K_{15}^{5/2}$$
$$K_{15} = C(S,15,0)/C(KCl,15,0),$$

where $C(S,15,0)$ is the conductivity of the water sample at a temperature of 15°C and atmospheric pressure, and $C(KCl,15,0)$ is the conductivity of a standard solution that contains 32.4356 g. potassium chloride (KCl) at the same temperature and pressure. The practical salinity unit is thus defined as a ratio of conductivities and has no physical units.

An alternative to conductivity salinometers are refractive salinometers, based on the fact that the speed of light through a medium depends on its density. In a refractive salinometer, a drop of sample water is placed on a prism. Because the water and the prism have different densities, light passing through the system is refracted at an angle that depends on the density (i.e., the temperature and salinity). On average, conductive salinometers have a precision of about 0.001 psu, whereas laboratory refractive salinometers have a precision of 0.06 psu, and handheld refractometers have a precision of 0.2 psu. Today's standard instrument for measuring both temperature and salinity is the Conductivity-Temperature-Depth profiler, which allows a quasi-continuous vertical sampling and a precision of 0.005 psu. Based on the same principle, conductivity/temperature instruments are mounted on autonomous profilers (e.g., Argo floats) and the thermosalinographs that use near-surface water intakes of ships to continuously measure temperature and salinity.

A promising approach for remote sensing of the salinity is the microwave radiometry measuring the emissivity or brightness temperature of the sea surface, because the dielectric constant of seawater depends on temperature and salinity. The largest sensitivity of the surface emissivity to salinity has been observed in the L-band (1.40–1.43 GHz). The Soil Moisture and Ocean Salinity of the European Space Agency and the Aquarius-SAC/D mission of NASA-Argentine Space Agency are the first two space missions designed to provide global, synoptic estimates of the sea surface salinity with an accuracy of about 0.1 psu every 30 days with a 100- to 200-km. spatial resolution.

PROCESSES AFFECTING SALINITY

Since the *Challenger* expedition in 1877, when the chemical composition of seawater was first reported, no changes in the composition of seawater have been observed. Thus, it can be supposed that for the timescales pertinent to climate change, viz., decadal to centennial, salinity behaves as a conservative tracer.

Thus, its time evolution is given by the three-dimensional transport of salinity by advection (as water parcels carry properties) or diffusion (tendency to smooth salinity gradients even in still water). At the surface of the oceans (up to a depth where the turbulent action of the wind is balanced by the laminarity of the stable ocean stratification), salinity concentrations are also modified by the dilution/concentration resulting from mass fluxes through the air-sea interface such as evaporation and precipitation, river runoffs, and the thawing/freezing of ice caps. In open oceans, the lowest values of salinity (below 30 psu) are found at high latitudes and at the mouth of the largest rivers. The highest salinities are found in the subtropics (over 35 psu), where evaporation dominates. In the tropics, which tend to be regions of strong precipitation, salinity is around 34 psu.

SALINITY AND CLIMATE

The range of temperatures on Earth allows water to be present as a solid (ice) in ice caps and glaciers; liquid (water) in oceans, groundwater, lakes, and rivers; and gas (water vapor) in the atmosphere. The idealized path of a water molecule from one phase to the other is known as the hydrological cycle. The residence times, that is, the average time that the molecules spend in each phase, range from a few days (water vapor in the atmosphere), to several months (seasonal snow cover, rivers), to the thousands of years (oceans and groundwater). Changes in the hydrological cycle affecting precipitation, evaporation, ice cap thawing, and river runoff have the potential to change the salinity of the oceans. The reverse is also true, and salinity changes may have an imprint in the hydrological cycle after thousands of years.

The mechanisms by which salinity affects the hydrological cycle are numerous. Because of its role in density variations, salinity gradients contribute to ocean currents transporting heat, salt, microorganisms, and nutrients across the oceans. In regions of strong precipitation, a layer of low salinity may isolate the uppermost surface of the ocean from the cold ocean below, forming the so-called barrier layer, which blocks the wind-stirring effects that cool the surface by mixing heat downward. This manifests as warmer sea surface temperatures, modifying the surface temperature gradients that drive surface winds. In the equatorial Pacific Ocean, such a phenomenon is of importance in the El Niño–La Niña cycles. Sim-

ilar salinity effects also occur in the tropical Indian and Atlantic oceans and have potential feedbacks to the hydrological cycles in the region. Tropical surface anomalies may be advected to the deep convection regions, modulating the thermohaline circulation. One of the largest ocean climate events recorded in the Atlantic Ocean is the Great Salinity anomaly, which lasted from 1968 to 1982. A salinity anomaly propagated over thousands of kilometers reached the Labrador Sea and perturbed the thermohaline circulation intensity. The origin and evolution of these anomalies is still not fully understood because of the historical lack of salinity observations, and studies of the mechanisms by which these salinity anomalies evolve are usually based on ocean and climate models.

SEE ALSO: Climate Change, Effects; El Niño and La Niña; Hydrological Cycle.

BIBLIOGRAPHY. W.J. Emery and R.E. Thomson, *Data Analysis Methods in Physical Oceanography* (Elsevier, 2001); M. N. Hill, *The Sea. Volume 2: The Composition of Sea-Water* (Krieger Publishing, 1982); S. Levitus, R. Burgett, and T.P. Boyer, *World Ocean Atlas 1994. Volume 3: Salinity* (U.S. Department of Commerce, 1994); J.P. Peixoto and A.H. Oort, *Physics of Climate* (American Institute of Physics Press, 1991).

JOAQUIM BALLABRERA
INSTITUT DE CIÈNCIES DEL MAR
CONSEJO SUPERIOR DE
INVESTIGACIONES CIENTÍFICAS

Samoa

SAMOA, KNOWN UNTIL 1997 as Western Samoa, has a land area of 1,093 sq. mi. (2,831 sq. km.), with a population of 187,000 (2006 est.) and a population density of 169 people per sq. mi. (65 people per sq. km.). With the economy dominated by subsistence agriculture, 19 percent of the country is arable, and 47 percent is forested.

From 1990 until 2003, the per capita carbon dioxide emissions from the country have been fairly stable, at between 0.7 and 0.8 metric tons per person. These emissions come entirely from liquid fuels, which are

Samoa is at risk of serious land loss because of global warming and climate change.

produced from transportation, electricity generation, and the running of small household and factory generators. Fossil fuels—petroleum—generate 59.2 percent of the electricity, with the remainder coming from hydropower.

Samoa is at risk of serious land loss because of global warming and climate change. Indeed, Upolu and Savaí, the two major islands in the country, have both experienced the loss of about 1.5 ft. (0.46 m.) of shore each year for the last 90 years. The only two other inhabited islands in Samoa—Apolina and Manono—have also suffered land loss, and some of the uninhabited islands are expected to become completely submerged if the water levels continue to rise.

The Samoan government ratified the Vienna Convention in 1992 and took part in the United Nations Framework Convention on Climate Change, signed in Rio de Janeiro in May 1992, ratifying it in 1994. It signed the Kyoto Protocol to the UN Framework Convention on Climate Change on March 16, 1998, which was ratified on November 27, 2000, and entered into force on February 16, 2005.

SEE ALSO: Climate Change, Effects; Floods.

BIBLIOGRAPHY. R.C. Kay et al., *Assessment of Coastal Vulnerability and Resilience to Sea-Level Rise and Climate Change—Case Study: Upolu, Western Samoa: Phase 1: Concepts and Approach* (South Pacific Regional Environment Programme, 1993); "Samoa—Climate and Atmosphere," http://earthtrends.wri.org (cited October 2007); Paul Smitz, *Samoan Islands and Tonga* (Lonely Planet, 2006).

Robin S. Corfield
Independent Scholar

San Marino

THE LAND-LOCKED REPUBLIC, entirely surrounded by Italy, has a land area of 23.5 sq. mi. (61 sq. km.), with a population of 31,000 (2006 est.) and a population density of 1,198 people per sq. mi. (461 people per sq. km.), the 20th highest density in the world. It is a very prosperous country, with gross domestic product per capita being US$34,600. As a result, it makes heavy use of electricity—air conditioning in the hot summers and heating for the winter, as well as regular domestic and business use. All electricity for San Marino is supplied by Italy, which produces some 80 percent of its electricity from fossil fuels.

Because of its geographical position, there are few data available for San Marino for greenhouse gas emissions, with Sanmarinesi emissions usually included under Italy, which has had carbon dioxide emissions per capita ranging from 6.9 metric tons per person in 1990, rising to 7.7 metric tons per person by 2003. With tourism being the major source of income, most tourists come to the country in buses or by train from the nearby Italian town of Borgo Maggiore. The Sanmarinesi government took part in the United Nations Framework Convention on Climate Change, signed in Rio de Janeiro in May 1992, but has, so far, not expressed any policy position on the Kyoto Protocol to the UN Framework Convention on Climate Change.

SEE ALSO: Climate Change, Effects; Tourism.

BIBLIOGRAPHY. Central Intelligence Agency, "San Marino," https://www.cia.gov/library/publications/the-world-factbook/geos/sm.html; Andrea Suzzi Valli, "Generalità sullo studio fitosociologico della vegetazione boschiva nella Repubblica di San Marino," *Studi Sammarinesi* (v. 1, 1984).

Justin Corfield
Geelong Grammar School, Australia

São Tomé and Principe

THE CENTRAL AFRICAN country of São Tomé and Principe, formerly a Portuguese colony, is located on two islands, São Tome and Principe, in the Atlantic Ocean. Together they have a land area of 372 sq. mi. (964 sq. km.), with a population of 158,000 (2006 est.), and a population density of 454 people per sq. mi. (171 people per sq. km.). About 42 percent of the population live in urban areas. Only 2 percent of the land is arable, with a further 1 percent used for meadows and pasture, with 75 percent of the country being forested.

Until the recent discovery of oil, the country has been poor, with the carbon dioxide emissions in the country being 0.6 metric tons per capita from 1990 until 2003 when the emission level was recorded as 0.62 metric tons per person. For the electricity production in the country, 58.8 percent comes from hydropower, with 41.2 percent from fossil fuels. The entire carbon dioxide emission comes from liquid fuels, being from car emissions, and also from small gas-driven generators. Gasoline prices in the country have been relatively low, but there is a very poor public transport system.

Both the islands of São Tomé and Principe might suffer flooding if global warming and climate change continue to raise the level of the Atlantic Ocean. The rising temperature of the water might also affect the country's fishing industry, which centers on catching shrimp and tuna. The São Tomé government took part in the United Nations Framework Convention on Climate Change signed in Rio de Janeiro in May 1992, and ratified the Vienna Convention in 2001. The government has so far expressed no opinion on the Kyoto Protocol to the UN Framework Convention on Climate Change.

SEE ALSO: Carbon Dioxide; Climate Change, Effects; Floods; Kyoto Protocol.

BIBLIOGRAPHY. P.J. Jones, J.P. Burlison, and A. Tye, *Conservacao dos Ecossistemas Florestais na Republica Democratica de Sao Tomé e Príncipe* (Uniao Internacional para a Conservacao da Natureza e dos Recursos Naturais, 1991); World Resources Institute, "Sao Tome & Principe—Climate and Atmosphere," http://earthtrends.wri.org (cited October 2007).

ROBIN S. CORFIELD
INDEPENDENT SCHOLAR

Saudi Arabia

COVERING MOST OF the Arabian Peninsula, the Kingdom of Saudi Arabia has a land area of 829,996 sq. mi. (2,149,690 sq. km.), with a population of 24,735,000 (2006 est.) and a population density of 29 people per sq. mi. (11 people per sq. km.). Riyadh, the capital and the largest city, has a population of 4,193,000 and has a population density of 3,891 per sq. mi. (1,500 per sq. km.). Some 2 percent of Saudi Arabia is arable land, with a further 56 percent used for meadows and pasture. With a high standard of living coming from the petroleum industry, and fossil fuels being used for all the country's electricity production, Saudi Arabia has a high rate of per capita carbon dioxide emissions, rising from 12.1 metric tons per person in 1990 to 18.4 metric tons in 1993.

Massive sandstorms over Saudi Arabia, seen from space. A rise in global temperatures is likely to increase desertification.

In 1998, emissions had fallen to 10.9 metric tons per person but rose to 13.4 metric tons in 2004.

Some 64 percent of the carbon dioxide emissions in the country come from liquid fuels, and 32 percent come from gaseous fuels. Electricity generation accounts for 30 percent of carbon dioxide emissions, with other energy industries contributing another 30 percent. Some 22 percent of carbon dioxide emissions come from manufacturing and construction, with 16 percent from transportation—Saudi Arabia has a high level of private ownership of automobiles and a limited public transport service.

Global warming and climate change, leading to a rise in the temperature, is likely to destroy more arable land, leading to increasing desertification. As the country has no perennial rivers or permanent water bodies, it has established power-hungry desalination plants. It has also drawn up plans for desert reclamation and irrigation schemes to try to achieve self-sufficiency in basic foods.

The Saudi Arabian government took part in the United Nations Framework Convention on Climate Change, signed in Rio de Janeiro in May 1992. It accepted the Kyoto Protocol to the UNñ Framework Convention on Climate Change on January 31, 2005, with it entering into force on May 1, 2005.

SEE ALSO: Climate Change, Effects; Deserts.

BIBLIOGRAPHY. Hussein A. Amery, "Water Wars in the Middle East: A Looming Threat," *The Geographical Journal* (v.168/4, 2002); "Saudi Arabia—Climate and Atmosphere," http://earthtrends.wri.org (cited October 2007); Frank Viviano, "Saudi Arabia on Edge," *National Geographic* (v.204/4, October 2003).

Robin S. Corfield
Independent Scholar

Schneider, Stephen H. (1945–)

AMERICAN CLIMATOLOGIST AND professor in the Department of Biological Sciences, Senior Fellow at the Center for Environment Science and Policy of the Institute for International Studies, and professor by courtesy in the Department of Civil and Environmental Engineering at Stanford University since September 1992, Schneider has been an outspoken advocate of the global warming theory since the 1980s and has helped draw public attention to the issue of climate change. He has argued for sharp reductions of greenhouse gas emissions to combat the phenomenon. Schneider has also served as an adviser to different U.S. administrations and federal agencies since the presidency of Richard Nixon. His research includes modeling of the atmosphere, climate change, and the relationship of biological systems to global climate change.

Schneider was born on February 11, 1945, in New York. He received his Ph.D. in mechanical engineering and plasma physics from Columbia University in 1971. He investigated the role of greenhouse gases and suspended particulate material on climate as a postdoctoral fellow at NASA's Goddard Institute for Space Studies. He was appointed a postdoctoral fellow at the National Center for Atmospheric Research in 1972 and was a member of their scientific staff from 1973 to 1996, where he cofounded the Climate Project. Although Schneider emerged as a public figure in the global warming debate in the 1980s, his interest in the subject dates back to the early 1970s, when he coauthored an article in *Science* ("Atmospheric Carbon Dioxide and Aerosols: Effects

Stephen H. Schneider at the Center for Environment Science and Policy of the Institute for International Studies.

of Large Increases on Global Climate"), which examined the competing effects of cooling from aerosols and warming from CO_2. The paper, however, predicted that carbon dioxide would only have a minor role and warned about a large possible decrease of the Earth's temperature. In the late 1970s, Schneider modified his position stating that, at that time, it was not possible to be certain whether the climate was cooling or warming. In *The Genesis Strategy: Climate and Global Survival* (1976), he wrote that "consensus among scientists today would hold that a global increase in atmospheric aerosols would probably result in a cooling of the climate; however, a smaller but growing fraction of the current evidence suggests that it may have a warming effect."

It was with his 1989 book, *Global Warming: Are We Entering the Greenhouse Century?*, that Schneider became a main figure in scientific debates about human effects on the environment. In clear language exempt from academic jargon, Schneider argued his case for global warming. The burning of fossil fuels, he claimed, causes a buildup of greenhouse gases in the atmosphere. Those gases trap the solar energy reradiated by the earth that would otherwise escape into space. This phenomenon could have devastating effects such as droughts, more frequent and more powerful tropical storms, and a rise in sea level. Schneider believes that temperature change appears less noticeable than it would otherwise be thanks to the capacity of the oceans to absorb heat. He is persuaded of the necessity to improve climate models that will take fully into account the interactions between atmosphere and oceans.

Schneider has always tried to involve the public in his research and has frequently appeared in media events relating to environment. He has tried to popularize complex scientific ideas to make them more available to the larger public. This has earned him praise as well as criticism, as several of his colleagues have charged him with trying too hard to get media attention for himself and his ideas. Still others describe Schneider as an alarmist, as in their view, his statements are not always supported by evidence. Yet Schneider is aware that as a scientist he is ethically bound to tell the truth. At the same time, he is concerned that, to be effective, his ideas have to take hold of the public and its imagination. Schneider finds this a double ethical bind in which scientists frequently find themselves. This condition "cannot be solved by any formula. Each of us has to decide what the right balance is between being effective and being honest. I hope that means being both."

Schneider has served in many key positions as an academic and a policymaker. He has also received many honors for his research. He is the founder and editor of the journal *Climatic Change* and has authored or coauthored over 450 scientific papers, proceedings, legislative testimonies, edited books, and book chapters, as well as some 140 book reviews, editorials, published newspaper and magazine interviews and popularizations. He was a coordinating lead author in Working Group II IPCC TAR and is currently a coanchor of the Key Vulnerabilities Cross-Cutting Theme for the Fourth Assessment Report (AR4). For his ability to integrate and interpret the results of global climate research for both the academic community and the general public, Schneider was awarded the MacArthur Fellowship in 1992. For his furtherance of public understanding of environmental science and its implications for public policy, he also received, in 1991, the American Association for the Advancement of Science/Westinghouse Award for Public Understanding of Science and Technology. In 1998, he became a foreign member of the Academia Europaea, Earth and Cosmic Sciences Section. He was elected chair of the American Association for the Advancement of Science's Section on Atmospheric and Hydrospheric Sciences (1999–2001). Schneider was elected to membership in the U.S. National Academy of Sciences in April 2002.

SEE ALSO: Climate Change, Effects; Global Warming.

BIBLIOGRAPHY. Stephen H. Schneider, *Global Warming: Are We Entering the Greenhouse Century?* (Sierra Club Books, 1989); Stephen H. Schneider, *Laboratory Earth: The Planetary Gamble We Can't Afford to Lose* (HarperCollins, 1997); Stephen H. Schneider and Terry L. Root, eds., *Wildlife Responses to Climate Change: North American Case Studies* (Island Press, 2001); Stephen H. Schneider, Armin Rosencranz, John O. Niles, eds., *Climate Change Policy: A Survey* (Island Press, 2002).

Luca Prono
University of Nottingham

Scripps Institute of Oceanography

SCRIPPS INSTITUTE OF Oceanography is the world's preeminent center for ocean and earth research, teaching, and public education. Also known as SIO, Scripps Institute of Oceanography is located in La Jolla, California. The mission of the institute is to seek, teach, and communicate scientific understanding of the oceans, atmosphere, Earth, and other planets for the benefit of society and the environment. A graduate school of the University of California, San Diego, Scripps's leadership in many scientific fields reflects its continuing commitment to excellence in research, modern facilities and ships, distinguished faculty, and outstanding students—and its horizons continue to expand. The institute offers a number of undergraduate and graduate courses in a variety of marine and earth science disciplines.

Research at Scripps encompasses physical, chemical, biological, geological, and geophysical studies. Ongoing investigations include the topography and composition of the seafloor, waves and currents, and the interchanges between the oceans and atmosphere. Scripps's research ships are used in investigations throughout the world's oceans. Today, the Scripps staff of 1,300 includes approximately 100 faculty, 300 other scientists, and some 225 graduate students, with an annual budget of more than $140 million. Other observations and collections are made by ocean devices, airplanes, remotely operated aircraft, land stations, and satellites. Scripps's educational program has grown hand in hand with its research programs. In its most recent survey of graduate schools, the National Research Council ranked Scripps the number one oceanographic program in faculty quality, distinction, and scholarly publications.

Instruction is on the graduate level, and students are admitted as candidates for a Ph.D. degree in oceanography, earth sciences, or marine biology. Academic work is conducted through the graduate department and eight curricular groups: biological oceanography, physical oceanography, marine biology, geological sciences, marine chemistry and geochemistry, geophysics, climate sciences, and applied ocean sciences.

The Scripps Institute of Oceanography department offers over 45 undergraduate courses covering a wide breadth of earth and marine sciences on several different levels. There are several introductory classes for nonmajors, as well as upper-division courses intended for a wide range of students in natural science majors. For students interested in careers in earth sciences, the Scripps Institute of Oceanography offers a B.S. degree and a contiguous B.S./M.S. degree in earth sciences. In addition, students may follow a chemistry/earth sciences major, a physics major with a specialization in earth sciences, or an environmental systems/earth sciences major. The program also offers an academic minor in earth sciences.

Scripps is one of the oldest and largest centers for global science research and graduate training in the world. More than 300 research programs are now conducted at the institute, aimed at gaining comprehensive understanding of the oceans, atmosphere, and structure of the Earth.

Oceanography, by its very nature, is interdisciplinary. It spans many sciences including physics, chemistry, geology, biology, meteorology, climatology, and paleontology. Scripps scientists pioneered exploration of the world's marine environments. They are leaders in studies of climate change, plate tectonics, ocean circulation, marine biology and ecology, marine pharmaceuticals, seafloor mapping, seismology, coastal processes, the El Niño phenomenon, biodiversity and conservation, and atmospheric sciences.

Graduate students play an integral role in the Scripps missions of teaching and research. Scripps offers excellent graduate instruction, and students perform a significant part of Scripps research activities. The stature of the institution is manifested in the quality both of the students it attracts to the program and of the scientists it graduates.

Climate sciences concerns the study of Earth's climate system, with emphasis on the physical, dynamic, and chemical interactions of the atmosphere, ocean, land, ice, and the terrestrial and marine biospheres. One of the central challenges is developing the ability to predict future climate changes, whether they are the consequences of human activities or the result of natural climatic cycles. A related challenge is understanding how and why the climate of the Earth has changed in the past.

To understand Earth's climate system requires understanding the mechanistic links between physical and chemical changes in the atmosphere (e.g., changes in winds, clouds, rainfall, sunlight, greenhouse gas abundance, or stratospheric ozone) and changes in the oceans (e.g., shifts in the current systems, temperature structure, or ocean biota), in the ice sheets (e.g., advances and retreats), and in land biota (e.g., changes in length of growing seasons or habitat range).

The climate system includes powerful feedback mechanisms. The amount of moisture in the atmosphere, for example, increases with global temperature, but the moisture also contributes to additional warming through the greenhouse effect. Scientists studying the climate system need experience in many disciplines, including meteorology, oceanography, geography, ecology, geology, and paleontology.

Observing the climate system depends increasingly on new measurement technologies, such as satellite and in situ measurements of the atmosphere, oceans, and land surface. These developments are enabling Scripps scientists to develop a more precise and detailed understanding of various conditions that affect climate, such as winds, ocean currents, clouds, and amount of vegetation.

Charles David Keeling, a professor of oceanography at Scripps, received the Tyler Prize for Environmental Achievement, which is awarded for accomplishments in environmental science, energy, and medicine that confer great benefit on mankind. Keeling, a world leader in research on the carbon cycle and the increase of carbon dioxide (CO_2) in the atmosphere, known to influence the greenhouse effect, has been affiliated with Scripps since 1956. Keeling was the first to confirm the accumulation of atmospheric carbon dioxide by very precise measurements that produced a data set now known widely as the Keeling curve. Before Keeling's investigations, it was unknown whether the oceans and vegetated areas on land would absorb any significant excess carbon dioxide from the atmosphere produced by the burning of fossil fuels and other industrial activities. Keeling became the first to determine definitively the fraction of carbon dioxide from combustion that is accumulating in the atmosphere.

Keeling's major areas of interest include the geochemistry of carbon and oxygen and other aspects of atmospheric chemistry, with an emphasis on the carbon cycle in nature. He has been a world leader in the study of the complex relationships between the carbon cycle and changes in climate. The Keeling record of the increase in atmospheric carbon dioxide measured at Mauna Loa, Hawaii, represents what many believe to be the most important time series data set for the study of global change.

SEE ALSO: Geography; Oceanography; University of California; Weather.

BIBLIOGRAPHY. Climate Change and Carbon and Carbon Management, http://www-esd.lbl.gov/CLIMATE; Climate of Change, http://www.universityofcalifornia.edu; Scripps Institute of Oceanography, http://www.sio.ucsd.edu.

Fernando Herrera
University of California, San Diego

Sea Ice

SEA ICE IS frozen ocean water. It forms primarily in and near the polar regions, though it can grow closer to the equator as far as 40 degrees N latitude and 55 degrees S latitude. Sea ice has a strong seasonal variability. In the Northern Hemisphere, the annual maximum extent occurs in late winter (March), covering about 5,791,532 sq. mi. (15 million sq. km.) on average. It then melts during spring and summer to an annual minimum extent of about 2,702,715 sq. mi. (7 million sq. km.) in September. In the Antarctic, the annual maximum is about 7,335,941 sq. mi. (19 million sq. km.) during September, and the annual minimum is about 1,158,306 sq. mi. (3 million sq. km.) in February or March. Overall, roughly 10 percent of the world's ocean area is covered with sea ice at some point during the year.

The point at which the ocean begins to freeze is a function of the salt content of the water, which for typical ocean salinities is around 29 degrees F (minus 1.6 degrees C). The saline nature of the ocean makes the formation of sea ice distinct from freshwater ice growth in lakes and rivers. Freshwater becomes less dense as it approaches the freezing point, keeping the coldest water at the surface and allowing ice to form as soon as the surface cools to the freezing point. However, the presence of salt changes the character of near-freezing water such that as saline water nears the freezing point, it continues to increase in density. Thus, cooling surface waters will become denser and sink. This means that there is overturning, and subsurface waters must also cool before ice can begin to form.

Sea ice typically grows to an average level thickness of 3 to 6 ft. (1 to 2 m.) in the Antarctic and 10 to 13 ft. (3 to 4 m.) in the Arctic. The ice is thinner in the Antarctic because most ice melts during the austral summer, whereas in the Arctic a significant fraction (~40 percent) remains through the summer and can grow over several years. A larger ocean heat flux at the bottom of the ice in the Antarctic also keeps the ice thinner.

However, thicker ice is not uncommon because of the effect of ice motion. Most sea ice is almost constantly in motion mainly because of the force of winds and ocean currents (other factors include the Coriolis effect, the slope of the ocean surface, and the internal structure of the ice). The speed of sea ice motion varies considerably; it can move 31 mi. (50 km.) or more in a day, though 1.2 mi. (2 km.) per day is typical. The motion of the ice can result in convergence between different parts of the ice cover, causing the ice to pile up into features called ridges. Ridges may easily rise 16 to 33 ft. (5 to 10 m.) above the surrounding level ice (and many tens of m. below the surface).

Sea ice plays an important role in climate. It has a much higher albedo than the unfrozen ocean, meaning that 60 to 70 percent of the sun's energy is reflected by the sea ice surface, whereas the unfrozen ocean reflects less than 10 percent of the sun's energy, resulting in much less energy absorption where ice is present. Sea ice is also a physical barrier between the ocean and atmosphere. This prevents the transfer of heat and moisture between the two and during winter. Thus, sea ice keeps the polar regions cooler and drier than they would be without ice. Sea ice also reduces fetch and dampens waves, limiting coastal erosion.

Sea ice has important effects on wildlife and human activities. Polar bears, seals, and other creatures rely on the ice to traverse and hunt, and during summer, the sea ice edge is a fertile area for phytoplankton and other microorganisms. Native communities in the Arctic are intimately tied into the presence of sea ice, and it plays an important role in their traditional culture and ways of life, such as hunting and transporta-

tion. Navigation of surface ships is severely limited or curtailed altogether, and for ships that do sail in or near ice-infested waters, it represents a significant hazard. Finally, sea ice plays a role in military operations, providing a useful cover for submarine activities.

Because of its location near the poles, the thin nature of the ice cover, and its interaction with the ocean and the atmosphere, sea ice is a sensitive indicator of the climate state. Sea ice in the Arctic has been decreasing dramatically over the past several decades. Overall, the Arctic has lost approximately 20 percent of the average summer ice extent since the late 1970s. Reductions during winter are less but are still significant. On the basis of current trends and projects by climate models, the Arctic may be ice free during summer by 2050 or earlier. This reduction in sea ice has been linked to warming temperatures resulting from the anthropogenic emission of greenhouse gases, though other factors also play a role. Unlike in the Arctic, there has been no significant trend seen in Southern Hemisphere ice, likely because of its remoteness relative to other continental land areas, the seasonal nature of the ice, and a greater ocean influence.

Changes in Arctic sea ice cover will have profound effects on climate, human activities, and wildlife, some of which are already being felt. Polar bears and other animals may be endangered, as well as the traditions of native communities. Less ice may also have benefits by opening up shipping routes and facilitating extraction of natural resources. Nonetheless, most effects are expected to be negative, and their implications for future climate will extend to regions far beyond the Arctic.

SEE ALSO: Animals; Climate Change, Effects.

BIBLIOGRAPHY. Dan Lubin and Robert Massom, *Polar Remote Sensing, Volume I: Atmosphere and Oceans* (Springer Praxis Books, 2006); O.M. Johannessen, R.D. Muench, and J.E. Overland, eds., *The Polar Oceans and Their Role in Shaping the Global Environment* (American Geophysical Monograph 85, 1994); Norbert Untersteiner, ed., *The Geophysics of Sea Ice* (Plenum Press, 1986); Peter Wadhams, *Ice in the Ocean* (Gordon and Breach Science Publishers, 2000).

Walt Meier
Julienne Stroeve
University of Colorado

Sea Level, Rising

SEA LEVEL RISE is caused by thermal expansion of the oceans, melting of glaciers and ice caps, melting of the Greenland and Antarctic ice sheets, and changes in terrestrial storage. Changes in sea level will be felt through increases in the intensity and frequency of storm surges and coastal flooding; increased salinity of rivers, bays, and coastal aquifers resulting from saline intrusion; increased coastal erosion; loss of important mangroves and other wetlands (the exact response will depend on the balance between sedimentation and sea level change) and its effect on marine ecosystems (i.e., coral reefs).

Global sea level rose by about 394 ft. (120 m.) during the several millennia that followed the end of the last Ice Age (approximately 21,000 years ago) before stabilizing between 3,000 and 2,000 years ago. Sea level indicators suggest that global sea level did not change significantly from then until the late 19th century, when the instrumental record of modern sea level change shows evidence for onset of sea level rise. Estimates for the 20th century show that global average sea level rose at a rate of about 1.7 mm. per year.

Satellite observations available since the early 1990s provide more accurate sea level data with nearly global coverage. This decade-long satellite altimetry data set shows that since 1993, sea level has been rising at a rate of around 3 mm. per year—significantly higher than the average during the previous half century. Coastal tide gauge measurements confirm this observation and indicate that similar rates have occurred in some earlier decades.

Sea level rise is currently determined by the employment of two techniques: the use of tide gauges and satellite altimetry. Tide gauges provide sea level variations with respect to the land on which they lie. To extract the signal of sea level change resulting from ocean

The Lower Patuxent River in Maryland, showing the flooding of low-lying areas by extreme high tides. If climate change causes sea level to continue to rise, this type of flooding will become increasingly common.

water volume and other oceanographic change, land motions need to be removed from the tide gauge measurement. Sea-level change based on satellite altimetry is measured with respect to the Earth's center of mass and thus is not distorted by land motions, except for a small component resulting from large-scale deformation of ocean basins. The total 20th-century rise is estimated to be around 0.5 ft. (0.17 m.).

Sea-level rise is accelerating worldwide. Globally, 100 million people live within about 3 mi. (1 m.) of sea level. Eight to 10 million people live within 3 mi. (1 m.) of high tide in each of the unprotected river deltas of Bangladesh, Egypt, and Vietnam. Intergovernmental Panel on Climate Change (IPCC) reports estimate that the global average sea-level rose at an average rate of 1.8 (1.3–2.3) mm. per year between 1961 and 2003, and within that period, the rate of rise was faster between 1993 and 2003—about 3.1 (2.4–3.8) mm. per year. Overall, the IPCC concludes that there is high confidence that the rate of observed sea-level rise has risen from the 19th to the 20th century. The total 20th-century rise is estimated to be 0.17 (0.12–0.22) m. In 2001, IPCC projections were for a sea-level rise of between 9 and 88 cm. between 1990 and 2100 and a global average surface temperature rise of between 2.5–10.4 degrees F (1.4–5.8 degrees C). In 2007, IPCC projections based on different scenarios predict seal level rise from 0.18 to up to 0.59 mm. by 2099.

Toward the end of the 21st century, projected sea-level rise will affect low-lying coastal areas with large populations. The cost of adaptation could amount to at least 5 to 10 percent of gross domestic product. Mangroves and coral reefs are projected to be further degraded, with additional consequences for fisheries and tourism. Snowmelt runoff as a result of sea-level rise will have major consequences. For example, one change will be a change from spring peak flows to late winter peaks in snowmelt–dominated regions. Many species, both aquatic and riparian (i.e., riverine), have evolved to take opportunity of the spring flows as a result of snowmelt. For example, some fish time their reproduction strategies specifically to avoid the stress of springtime flows. Changes in springtime flow regimes, or high winter flows associated with rain or snow events, can scour streambeds and destroy eggs. Trees that provide riparian habitat along rivers may find it harder to reproduce, as they are dependent on high spring flows. Many species, such as salmon, that

are already under pressure from other environmental effects will be significantly affected by climate change. For example, higher temperatures and a reduced stream flow in the Columbia River Basin may be increasing the mortality of juvenile coho salmon, or in some cases, increased temperatures may be creating thermal barriers for the migration of adult salmon.

There are a number of associated events that are the result of climate change and that will also have affects on sea-level rise. For example, the Kangerdlugssuaq Glacier in Greenland is moving much faster, now melting at a rate of 8.7 mi. (14 km.) a year in comparison to just 3 m. (5 km.) a year in 1988. This loss will also have serious implications for sea-level rise, with some scientists predicting that within the next 100 years, ice cover in this region will completely disappear over summer, and hence species living within it, such as polar bears, will be threatened. The complete melting of the Greenland Ice Sheet and the West Antarctic Ice Sheet would lead to a contribution to sea-level rise of up to 23 ft. and about 16 ft. (7m. and 5 m.), respectively.

The potential socioeconomic impacts of sea-level rise are as follows: direct loss of economic, ecological, cultural, and subsistence values through loss of land, infrastructure, and coastal habitats. For example, it is estimated that the total of people at risk of sea-level rise in Bangladesh could be 26 million, in Egypt 12 million, in China 73 million, in India 20 million, and elsewhere 31 million. This makes an aggregate total of 162 million people affected by sea-level rise. There will be an increased flood risk for people, land, and infrastructure. There will be other affects related to changes in water management, salinity, and biological activities. A rise in sea level would inundate wetlands and lowlands, accelerate coastal erosion, exacerbate coastal flooding, threaten coastal structures, raise water tables, and increase the salinity of rivers, bays, and aquifers. Similarly, the areas vulnerable to erosion and flooding are also predominantly located in the southeast, whereas potential salinity problems are spread more evenly throughout the coast. Such a loss would reduce available habitat for birds and juvenile fish and would reduce the production of organic materials on which estuarine fish rely.

Some of the most important vulnerable areas are recreational barrier islands and spits such as found within the Atlantic and Gulf coasts. Coastal barriers are generally long narrow islands and spits (peninsulas) with the ocean on one side and a bay on the other.

Typically, the ocean-front block of an island ranges from 6.5 ft. to 13 ft. (2 to 4 m.) above high tide, whereas the bay side is less than a meter above high water. Thus, even a 1-m. rise in sea level would threaten much of this valuable land with inundation. Erosion, moreover, threatens the high parts of these islands and is generally viewed as a more immediate problem than the inundation of their bay sides. Although inundation alone is determined by the slope of the land just above the water, coastal engineer Per Bruun showed that the total shoreline retreat from a rise in sea level depends on the average slope of the entire beach profile. For example, most U.S. recreational beaches are less than 30 m. (100 ft.) wide at high tide, thus even a 30-cm. (1-foot) rise in sea level would require a response.

Finally, a rise in sea level would enable saltwater to penetrate farther inland and upstream in rivers, bays, wetlands, and aquifers, which would be harmful to some aquatic plants and animals and would threaten human uses of water. In Delaware in the United States, for example, salinity is seen as a factor resulting in reduced oyster harvests.

Coastal areas worldwide will become more vulnerable to flooding as a result of sea-level rise because higher sea levels provide higher bases for storm surges to build on. In this context, a 1-meter rise in sea level would mean that a 15-year storm will flood many areas that today are only flooded by a 100-year storm. Beach erosion will make land more vulnerable to storm waves, and higher water levels will increase the effects of flooding caused by rainstorms by reducing coastal drainage. Sea level will also raise water tables in various systems.

There are many examples of nations vulnerable to sea level rise. Japan, for instance, is particularly vulnerable to the effects of sea-level rise; a 1-meter rise in sea level would increase the area situated below mean high water from 332 sq. mi. to 903 sq. mi. (861 sq. km. to 2340 sq. km.). Future estimates show this would affect up to 4.1 million people and cost 109 trillion Yen ($1,300 billion). Over 57 and 90 percent of the existing sandy beaches would be eroded by sea-level rises of 0.3 and 1.0 m., respectively. In response, Japan has initiated a new coastal policy that combines disaster prevention, human resource use, and nature conservation. This policy includes an increase in monitoring of changes in mean sea level and the frequency of extreme events, the consideration of climate-change scenarios when developing plans for ports and landfills, and the preparation of a set of technological countermeasures to prevent effects on port facilities and maintain coastal protection.

Egypt's Nile Delta is one of the world's areas most vulnerable to sea-level rise. It is estimated that about 30 percent of the area will be lost because of inundation, almost 2 million people will lose their homes, and approximately 195,000 jobs will be lost, with a predicted economic impact of over US$3.5 billion over the next century.

In the chapter on coasts, the United Nations Environment Programme Handbook on Adaptation and Mitigation Methodologies specifically outlines a suite of strategic responses to sea-level rise. It cautions, however, that before applying these strategies, policymakers must decide whether or not their adaptation is to be autonomous or planned, reactive or proactive.

There are three management responses to sea-level rise: retreat, accommodation, and protection. Retreat involves no effort to protect the land from the sea. The coastal zone is abandoned, and ecosystems shift landward. This choice can be motivated by excessive economic or environmental effects of protection. In the extreme case, an entire area may be abandoned.

Accommodation implies that people continue to use the land at risk but do not attempt to prevent the land from being flooded. This option includes erecting emergency flood shelters, elevating buildings on piles, converting agriculture to fish farming, or growing flood- or salt-tolerant crops.

Protection involves hard structures such as sea walls and dikes, as well as soft solutions such as dunes and vegetation, to protect the land from the sea so that existing land uses can continue.

SEE ALSO: Climate Change, Effects; Floods; Refugees, Environmental.

BIBLIOGRAPHY. Diane Bates, "Environmental Refugees? Classifying Human Migrations Caused by Environmental Change," *Population and Environment,* (v.23/5, May 2002); Kevin Baumert, Jonathan Pershing, Timothy Herzog, and Matthew Markoff, *Climate Data: Insight and Observations* (Pew Center on Global Climate Change, 2004); N. Bindoff, V. Willebrand, A. Artale, J. Cazenave, S. Gregory, K. Gulev, C. Hanawa, S. Le Quéré, Y. Levitus, C. Nojiri, L. Shum, A. Talley, and A. Unnikrishnan, "Observations: Oceanic

Climate Change and Sea Level," in *Climate Change 2007: The Physical Science Basis. Contribution of Working Group I to the Fourth Assessment Report of the Intergovernmental Panel on Climate Change* (Cambridge University Press, 2007); M. El-Raey, K. Dewidar, and E. El-Hattab, "Adaptation to the Impacts of Sea Level Rise in Egypt," *Mitigation and Adaptation Strategies for Global Change* (v.4, 1999); J. Hay and N. Mimura, "Sea Level Rise: Implications for Water Resources Management," *Mitigation and Adaptation Strategies for Global Change* (v.10, 2005); Intergovernmental Panel on Climate Change, "A Common Methodology for Assessing Vulnerability to Sea Level Rise," in *Change and the Rising Challenge of the Sea. Report of the Coastal Zone Management Subgroup* (IPCC, 1992); R. Leafe, J. Pethick, and I. Townsend, "Realising the Benefits of Shoreline Management," *Geographical Journal*, (v.164/3, 1998); W. Mitchell, J. Chittleborough, B. Ronai, and G. Lennon, *Sea Level Rise in Australia and the Pacific.* Pacific Islands Conference on Climate Change, Climate Variability and Sea Level Change, Rarotonga, Cook Islands, April 3–7, 2000; James Neumann, Gary Yohe, Robert Nicholls, and Michelle Manion, *Sea-Level Rise and Global Climate Change: A Review of Impacts to U.S. Coasts,* (Pew Center on Global Climate Change, 2000); Nicholas Stern, *Stern Review on the Economics of Climate Change* (HM Treasury, 2007).

MELISSA NURSEY-BRAY
AUSTRALIAN MARITIME COLLEGE
ROB PALMER
RESEARCH STRATEGY TRAINING

Seasonal Cycle

THERE ARE SEVERAL versions of seasons. The classical concept of season is of the four seasons that divide the year—spring, summer, fall, and winter. Some regions of the globe have weather-based seasons, such as rainy or dry seasons. Certain natural occurrences are more frequent during particular times of year; therefore, we have hurricane season as well as tornado season.

Seasonal cycles for weather patterns occur because of the atmosphere. Although the atmosphere is hundreds of miles thick, weather occurs in the base 7 mi. (11 km.), called the troposphere. Wind patterns in the troposphere are affected by the earth's rotation. This pattern is stereotypical and is the cause for seasonal cycles such as monsoon season in the northern Indian Ocean.

Other seasonal cycles include patterned flooding of rivers, such as the Nile River in Egypt. Ancient Egyptians utilized flood patterns for agriculture; today many cultures try to alter flood patterns. For example, the Colorado River floods created the Grand Canyon; today those floods no longer occur, due to dams and rerouting of the river.

A well-known example of a seasonal cycle is El Niño. It usually occurs in late December, around Christmas; hence it was named with the word for "child" in Spanish. El Niño is a warm ocean current, typically based off of Australia, that once every three to five years travels northeast towards Ecuador and Peru. Its opposite cycle, La Niña, contains cooler winds. Additionally, El Niño weakens trade winds, while La Niña brings stronger trade winds.

Tornados typically follow a seasonal cycle as well. They usually occur in the spring, yet they can happen at any time. Tornados occur when warm air at the surface of the earth rises quickly, and for a long distance. Air that is actively warmed at the surface, as occurs during the spring, is more buoyant and more likely to form a tornado.

Although hurricanes can form year-round, June 1 through November 30 is officially the Hurricane Season on the Atlantic coast of the United States. Hurricanes generally arise in the Caribbean and travel westward to the coast. Typically during these months, the most frequent type of storm is a tropical storm. Hurricanes are more frequent in the months of August, September, and October; however, severe hurricanes are usually limited to September, the midpoint of hurricane season.

SEE ALSO: Atlantic Ocean; Australia; Ecuador; Egypt; El Niño and La Niña; Floods; Hot Air; Hurricanes and Typhoons; Indian Ocean; Monsoons; Peru; Somali Current; Trade Winds; Trophosphere; Weather.

BIBLIOGRAPHY: W.J. Burroughs, *Weather Cycles: Real or Imaginary?* (Cambridge University Press, 2004); K. Emanuel, *Divine Wind: The History and Science of Hurricanes* (Oxford University Press USA, 2005); C.W. Landsea, "A Climatology of Intense (or Major) Atlantic Hurricanes",

Monthly Weather Review (1993); J.A. Miron, *The Economics of Seasonal Cycles* (The MIT Press, 1996).

Claudia Winograd
University of Illinois at Urbana-Champaign

Seawater, Composition of

SEAWATER IS A solution of salts of nearly constant composition, dissolved in variable amounts of water. It is denser than fresh water. It is risky to drink seawater because of its high salt content. More water is required to eliminate the salt through excretion than the amount of water that is gained from drinking the seawater. Seawater can be turned into potable water by desalination processes or by diluting it with freshwater. The origin of sea salt is traced to Sir Edmond Halley, who in 1715 proposed that salt and other minerals were carried into the sea by rivers, having been leached out of the ground by rainfall runoff. On reaching the ocean, these salts would be retained and concentrated as the process of evaporation removed the water. There are more than 70 elements dissolved in seawater as ions, but only six make up more than 99 percent of all the dissolved salts; namely, chloride (55.04 weight percent [wt%]), sodium (30.61 wt%), sulphate (7.68 wt%), magnesium (3.69 wt%), calcium (1.16 wt%), and potassium (1.10 wt%). Trace elements in seawater include manganese, lead, gold, and iodine. Biologically important elements such as oxygen, nitrogen, and iron occur in variable concentrations depending on utilization by organisms. Most of the elements occur in parts per million or parts per billion concentrations and are important to some positive and negative biochemical reactions. Properties such as salinity, density, and pH could be used to highlight the composition of seawater.

Salinity is the amount of total dissolved salts present in 1 L. of water and is used to express the salt content of seawater. Normal seawater has a salinity of 35 g./L. of water; that is, 3.5 percent. The salinity of seawater is made up by the dissolved salts. Seawater is more enriched in dissolved ions of all types than freshwater. Salts dissolved in seawater come from three main sources: volcanic eruptions, chemical reactions between seawater and hot newly formed volcanic rocks of spreading zones, and chemical weathering of rocks. Because of some chemical reactions between seawater and hot newly formed volcanic rocks, the composition of seawater has been nearly constant over time. Salinity affects marine organisms because the process of osmosis transports water toward a higher concentration through cell walls. Marine plants and many lower organisms have no mechanism to control osmosis, which makes them very sensitive to the salinity of the water in which they live. The density of surface seawater ranges from 1,020 kg. per cu. m. to 1,029 kg. per cu. m., depending on the temperature and salinity: the saltier the water, the higher its density. Seawater pH is limited to the range from 7.5 to 8.4 and increases with phytoplankton production. The speed of sound in seawater is about about 4,921 ft. (or 1,500 m.) per second and varies with water temperature and pressure.

Carbon (IV) oxide in the sea exists in equilibrium with that of exposed rock containing limestone ($CaCO_3$). Seawater also contains small amounts of dissolved gases such as nitrogen, oxygen, carbon (IV) oxide, hydrogen, and trace gases. Water at a given temperature and salinity is saturated with gas when the amount of gas entering the water equals the amount leaving during the same time. Surface seawater is normally saturated with atmospheric gases such as oxygen and nitrogen. The concentrations of oxygen and carbon (IV) oxide vary with depth. The surface layers are rich in oxygen, which reduces quickly with depth to reach a minimum between 656 and 2,625 ft. (200 and 800 m.) in depth. The amount of gas that can dissolve in seawater is determined by the water's temperature and salinity. Increasing the temperature or salinity reduces the amount of gas that can be dissolved. As water temperature increases, the increased mobility of gas molecules makes them escape from the water, thereby reducing the amount of gas dissolved. The gases dissolved in seawater are in constant equilibrium with the atmosphere, but their relative concentrations depend on each gas' solubility. As salinity increases, the amount of gas dissolved decreases because more water molecules are immobilized by the salt ion. Inert gases like nitrogen and argon do not take part in the processes of life and are thus not affected by plant and animal life, but gases like oxygen and carbon (IV) oxide are influenced by sea life. Plants reduce the concentration of carbon (IV) oxide in the presence of sunlight, whereas animals do the opposite in either light or darkness.

The world under water is different from that above in the availability of important gases such as oxygen and carbon (IV) oxide. Whereas in air about one in five molecules is oxygen, in seawater this is only about four in every thousand million water molecules. Whereas air contains about one carbon (IV) oxide molecule in 3,000 air molecules, in seawater this ratio becomes four in every 100 million water molecules. Thus, carbon (IV) oxide is much more available in seawater than is oxygen. All gases are less soluble as temperature increases, and particularly nitrogen, oxygen, and carbon (IV) oxide, which become about 40 percent to 50 percent less soluble with an increase of 45 degrees F (25 degrees C). When water is warmed, it becomes more saturated, resulting in bubbles leaving the liquid.

SEE ALSO: Chemistry; Climate Change, Effects.

BIBLIOGRAPHY. *Marine Biology and Oceanography* (Florida International University, 2002); *Ocean Acidification Due to Increasing Atmospheric Carbon Dioxide* (Royal Society, 2005).

AKAN BASSEY WILLIAMS
COVENANT UNIVERSITY

Senegal

THIS WEST AFRICAN country, formerly a French colony, has a land area of 75,955 sq. mi. (196,723 sq. km.), with a population of 12,379,000 (2006 est.) and a population density of 153 people per sq. mi. (59 people per sq. km.). Some 31 percent of the country is forested, with 12 percent being arable and a further 16 percent being used for meadows or pasture, mainly for cattle and sheep.

Senegal's entire electricity production comes from fossil fuels. The country has maintained a relatively stable level of carbon dioxide emissions from 0.4 metric tons per capita in 1990, rising to 0.44 metric tons by 2003. About 85 percent of this carbon dioxide comes from liquid fuels, with the remainder coming from cement manufacturing. In terms of sectors, that covering electricity and heat production accounts for 38 percent of carbon dioxide emissions, with 33 percent from transportation, 18 percent from manufacturing

and construction, and 10 percent for private residences. The high level of emissions from transportation comes from the lack of adequate public transport systems and the heavy use of old minibuses, which have inefficient fuel usage rates. Although there is a railway in Senegal, connecting Dakar with Bamako, the capital of Mali, few people use it for travel within the country.

The effects of climate change and global warming will result in potential flooding on the Atlantic coast of the country. This has already led to coastal erosion in some parts of the country, with illegal mining adding to this problem, especially around Malika and Rufisque to the north of the Cap Vert peninsula. A rise in temperature will lead to more arable land becoming marginal, and probably to increased desertification. The Senegal government of Abdou Diouf took part in the United Nations Framework Convention on Climate Change, signed in Rio de Janeiro in May 1992, and ratified the Vienna Convention in the following year. On July 20, 2001, Senegal signed the Kyoto Protocol to the

Net fishing in Nianing, near Dakar, Senegal. Senegal has experienced coastal erosion in some areas.

UN Framework Convention on Climate Change, with it entering into force on February 16, 2005.

SEE ALSO: Climate Change, Effects; Deserts.

BIBLIOGRAPHY. Sharon Gellar, *Senegal: An African Nation Between Islam and the West* (Westview Press, 1982); Katharina Kane, *The Gambia & Senegal* (Lonely Planet, 2006); "Senegal—Climate and Atmosphere," http://earthtrends.wri.org (cited October 2007).

ROBIN S. CORFIELD
INDEPENDENT SCHOLAR

Serbia and Montenegro

SERBIA AND MONTENEGRO are two of six independent countries that formed after the disintegration of the former Yugoslavia in the early 1990s. These two countries neighbor each other in eastern Europe and cover geographic regions with diverse climates. Serbia is landlocked and geographically positioned in the region of mild continental climate, whereas Montenegro also has a coastal region with a Mediterranean climate. Both countries have significant local climate variations resulting from local atmospheric circulations, as well as mountainous and hilly terrain away from Montenegro's Adriatic coast. In general, the subalpine climate with short cold winters and relatively hot summers is typical for elevations of from 2,000 ft. (600 m.) to 4,000 ft. (1,200 m.), and the Alpine climate with long and snowy winters and short warm summers is characteristic for elevations above 4,000 ft. (1,200 m.).

The capital of Serbia is Belgrade. Belgrade has a moderate continental climate that has hot summers, cold winters, and moderate precipitation. The yearly average temperature in Belgrade is around 53.1 degrees F (11.7 degrees C). The capital of Montenegro is Podgorica. Podgorica has a Mediterranean climate that has dry and hot summers, as well as rainy and mild winters, resulting in yearly average temperature of 61.5 degrees F (16.4 degrees C). In the summer, the temperatures in both capital cities go over 104 degrees F (40 degrees C). Interestingly, a town in the costal region of Montenegro has the highest yearly average rainfall in Europe of 183 in. (465 cm.). Nevertheless, the inland regions of these counties have been in "on" and "off" drought conditions resulting from extremely hot summers and unusually low rainfalls and snowfalls since 2000. These drought problems are characteristic for the entire region of the eastern part of Europe, with the drought in 2007 being particularly severe.

SEE ALSO: Climate Change, Effects; Drought.

BIBLIOGRAPHY. David C. King, *Serbia: Cultures of the World*, 1st ed. (Marshall Cavendish, 2005); Michael A. Schuman. *Serbia and Montenegro* (Facts on File, 2004).

JELENA SREBRIC
PENNSYLVANIA STATE UNIVERSITY

Seychelles

THIS COUNTRY CONSISTS of 115 islands spread over 250,966 sq. mi. (650,000 sq. km.) of ocean and has a land area of 176 sq. mi. (451 sq. km.), with a population of 87,000 (2006 est.) and a population density of 458 people per sq. mi. (178 people per sq. km.). With 18 percent of the land forested, 2 percent is arable, and 13 percent is used for meadows and pasture.

The entire electricity production of the country comes from fossil fuels, and the country, largely because of this and a burgeoning luxury tourist industry, has a relatively high level of per capita carbon dioxide emissions. Although it was 1.6 metric tons per person in 1990, it has risen steadily, reaching 2.6 metric tons in 1996, and then, with a massive rise in tourist numbers, rose to 5.5 metric tons per person in 1997, rising again to 6.9 metric tons in 2003. All the carbon dioxide emissions of the country come from liquid fuels, with most of this resulting from little availability of public transport and heavy use of gasoline-operated generators.

The rising water levels and temperatures in the Indian Ocean are likely to have dramatic effects on Seychelles. The former might lead to the loss of some of the smaller islands in the Seychelles, and the latter has led to some coral bleaching, affecting some of the coral reefs around the Seychelles, along with its marine life. As a result, many of them are heavily protected, with the government keen on promoting ecotourism and with money from tourists being invested in environmental protection. This has been particularly true of Cousine Island in the Seychelles, where there have been serious attempts to protect the turtles breeding there and prevent soil erosion.

In 1990, the Seychelles drew up a 10-year environmental management plan, being the first African country to do this. The Seychelles government of France-Albert René took part in the United Nations Framework Convention on Climate Change, signed in Rio de Janeiro in May 1992, and in the following year ratified the Vienna Convention. The government signed Kyoto Protocol to the UN Framework Convention on Climate Change on March 20, 1998, which was then ratified on July 22, 2002, with it entering into force on February 16, 2005.

SEE ALSO: Climate Change, Effects; Floods; Tourism.

BIBLIOGRAPHY. Jan Dodd, *Mauritius, Réunion & Seychelles* (Lonely Planet, 2004); Jacques Hodoul, *Seychelles Industrial Development Policy 1989–1993* (Ministry for National Development, Victoria, 1988); "Seychelles—Climate and Atmosphere," http://earthtrends.wri.org (cited October 2007); Geordie Torr, "Green Island," *Geographical* (v.777/6, June 2005).

ROBIN S. CORFIELD
INDEPENDENT SCHOLAR

Sierra Leone

SIERRA LEONE IS on the West African coast, covering 27,699 sq. mi. (71,740 sq. km.), which is equivalent to an area the size of South Carolina. Sierra Leone has a population of 6,017,643 people (est. in July 2005), with nearly one million people living in the capital, Freetown. Sierra Leone was the lowest ranking, among 177 countries surveyed, in the United Nations Human Development Indicators (2006), in part because of an 11-year civil war (1991–2002) that resulted in 50,000 deaths and the displacement of 2 million people. The civil war also accelerated the depletion of natural resources including diamonds, tropical timber, and wild game. Sierra Leone has a per capita gross domestic product of $600 (2004 est.), with 68 percent of the population living in poverty, most of whom are subsistence farmers. Products of economic significance to the country include cocoa, coffee, rice, palm oil, and fish. Alluvial diamond mining is the major source of hard currency earnings, whereas other extracted minerals include titanium ore, bauxite, iron ore, gold, and chromite. The geography is characterized by coastal mangrove swamps and wooded uplands inland, and the climate is tropical. Four environmental issues facing the country are overharvesting of tropical timber, clearing of forests for cattle grazing, deforestation and related soil erosion, and overfishing.

The contributions that Sierra Leone makes to human-induced climate change are minimal compared with the rest of sub-Saharan Africa. Per capita CO_2 emissions in 1998 were only 100 metric tons, compared with an average of 800 tons for the subcontinent and a global average of 4,100 metric tons. The burning of liquid fuels (petroleum products) represented 90 percent of the country's CO_2 emissions. Other non-CO_2 air pollution in Sierra Leone is low compared with the rest of the continent and the world. Nitrogen oxide and carbon monoxide emissions (in 1995) were 64,000 and 1,380,000 metric tons, respectively, making up just 0.007 and 0.008 percent of the totals for sub-Saharan Africa.

Climate change could have significant consequences on the people and the environment in Sierra Leone. With 248.5 mi. (400 km.) of coastline, significant areas, including the capital, could become more prone to flooding. A rising sea level would also destroy the extensive network of mangrove forests that covers much of the coastline. Climatic change leading to a shortened rainy season, especially inland, could facilitate the conversion of tropical forests to grazing land for livestock. With these land-use changes could come new diseases, as well as the elimination of others. For example, Lassa fever, which is endemic to the rainforest along the eastern border, could disappear with changes in climate and landscapes. However, the intensification of rainfall could accelerate soil erosion and instigate flash flooding (which is already a problem in Freetown), threatening people and their livelihoods.

SEE ALSO: Carbon Dioxide; Climate Change, Effects; Floods; Global Warming.

BIBLIOGRAPHY. Central Intelligence Agency, *The World Factbook* (Central Intelligence Agency, 2006); Economist Intelligence Unit, *Country Report: Sierra Leone* (*The Economist*, 2006); John F. McCoy, *Geo-Data: The World Geographical Encyclopedia* (Thomson-Gale, 2003); United Nations Human Development Indicators (United Nations, 2006); World Resources Institute, *Earthtrends*, http://earthtrends.wri.org/.

MICHAEL JOSEPH SIMSIK
U.S. PEACE CORPS

Simulation and Predictability of Seasonal and Interannual Variations

MEASUREMENTS OF CHANGES in atmospheric molecular oxygen using a new itechnique shows that the oxygen content of air varies seasonally in both the Northern and Southern Hemispheres, and is decreasing from year to year. The seasonal variations provide a new basis for estimating global rates of biological organic carbon production in the ocean, and the interannual decrease constrains estimates of the rate of anthropogenic CO_2 uptake by the oceans.

One example of research into variations are the interannual and interdecadal zooplankton population changes that have been observed in parallel with temperature (SST) changes at Helgoland, in the North Sea, over a period of 32 years.

Temperature determines the phenological timing of populations for each species in a unique way as to be seen in multiannual regressions. Sign, inclination of the regressions of phenophases with temperature and determination coefficient vary from species to species. Besides the limited predictability of annual temperature dynamics the species specificity limits the predictability of future phenophase timing. However, the strong correlation of phenophase timing with preceding SST permits the prediction of the annual seasonality based on statistical models separately for each population.

The regressions determined in the correlation analyses are the first approach to the phenological prognoses which the the Senckenberg Research Institute calculates daily for 192 phenophases of zooplankton including ichthyoplankton on a daily basis and which it publishes since April 2004 on the home page of the institute at www.senckenberg.de/dzmb/plankton. The calculations are based on more than 30 years of weekly and more frequent sampling at Helgoland Roads (54°11'18" N 7°54' E), the only proper offshore island of the North Sea. Temperatures were provided by the Biologische Anstalt Helgoland and the German Weather Service. Beyond the historic data used, current temperature measurements for the operative daily calculations are obtained from the websites. They are corrected according to the historic deviations stemming from current temperature measurements for the position at Helgoland Roads, published on the internet and weather report measurements and are then used for the calculation of the minimum current error for phenophase prediction. The prognoses is exclusively restricted to temporal prognoses. Abundance predictions are not included.

SEE ALSO: Oceanic Changes; Oceanography.

BIBLIOGRAPHY. Helgoland Marine Research, *The Warming Trend at Helgoland Roads, North Sea: Phytoplankton Response* (Springer Berlin/Heidelberg, 2004); Research Institute Senckenberg, German Centre for Marine Biodiversity Research, www.senckenberg.de/dzmb/plankton.

WULF GREVE
INDEPENDENT SCHOLAR

Singapore

THE REPUBLIC OF Singapore covers the main island of Singapore and 57 outlying islands, most of which are uninhabited. It covers a land area of 270 sq. mi. (704 sq. km.) and has a total population of 4,680,600 (July 2003 est.). This means that Singapore has one of the highest population densities in the world—more than 17,335 people per sq. mi. (6425 people per sq. km.). Combined with a high standard of living, Singapore has a high per capita emission level of carbon dioxide, with 15 metric tons of carbon dioxide per capita in 1990, rising to 19.1 metric tons in 1994 and then falling to 16.8 metric tons in 1997, after which date it has fallen significantly, reaching 11.3 metric

The Tanjong container port in Singapore. Singapore is the busiest port in the world in terms of tonnage shipped, and has a population of over 4.5 million people. It is noted for business, finance, manufacturing, exports, refining, and imports.

tons in 2003. Some 98 percent of the carbon dioxide emissions come from liquid fuel, with the remainder being from the manufacture of cement.

Many of the problems associated with global warming are prominent in Singapore, which has a large urban area and increasingly smaller wooded areas. The heavy usage of electricity has come from widespread use of air conditioners, as Singapore is located in the tropics. Not only found in homes and offices, the air conditioning of shops and shopping centers has resulted in significant levels of carbon dioxide emissions, as has heavy economic reliance on industrial development and oil refineries. Another very important contributing factor has been the tourist industry, which involves well over 7 million tourists visiting the country each year—the vast majority of them arriving by airplane, with the consequent effects on the ozone layer; Singapore's Changi airport is one of the busiest airports in the world.

As shown by its reduction in carbon emissions in recent years, Singapore has long been aware of the problems faced by the government. As early as 1963, Singapore's prime minister, Lee Kuan Yew, launched a "Garden City" program that encouraged the planting of trees and also the incorporation of grass verges, trees, and parks into city developments. Emphasis was made on local flora, and the project was extremely successful. The ministry of the environment was formed in 1972, with Lim Kim San as the minister, himself being succeeded by Edmund William "Eddie" Barker, who remained minister until 1979. Recent ministers have included Mah Bow Tan; Yeo Cheow Tong, who was joint minister of health and environment from 1997 to 1999; Lee Yok Suan; Lim Swee Say, acting minister and then minister; and Dr. Ibrahim Yaacob, who presided over the new ministry of the environment and water resources from 2004, reflecting the importance of water in the long-term development of Singapore.

On the international scene, Singapore has been a member of the United Nations Framework Convention on Climate Change, signed in Rio de Janeiro in 1992, and also ratified the Kyoto Protocol to the UN Framework Convention on Climate Change on April 12, 2006, coming into force on July 11, 2006.

The smog found in Singapore in 1999 and some succeeding years following the burn-off of forests in Sumatra has led to Singapore doing much to help combat global warming. There are limitations on the use of private automobiles, with the cost of running a car being high, and extra charges to alleviate city congestion. Combined with this there has been one of the best integrated transport systems in the world, with heavy public use of buses and trains (mass rapid transport), including an efficient bus and train service to neighboring Malaysia. In the center of Singapore island, and over to the northwest, there have been reforestation programs, especially on Bukit Batok and around the water reservoirs.

SEE ALSO: Climate Change, Effects; Pollution, Air; Tourism; Transportation.

BIBLIOGRAPHY. Gavin Chua Hearn Yuit, *Key Environmental Challenges Facing Singapore and the Region* (Singapore Institute of International Affairs, 2007); Justin Corfield and Robin Corfield, *Encyclopedia of Singapore* (Scarecrow Press, 2006); *The Singapore Green Plan 2012* (Ministry of the Environment and Water Resources, 2006).

ROBIN S. CORFIELD
INDEPENDENT SCHOLAR

Singer, S. Fred (1924–)

CONTROVERSIAL ATMOSPHERIC PHYSICIST, distinguished research professor at George Mason University, emeritus professor of environmental science at the University of Virginia, and founder of the Science and Environmental Policy Project—a policy institution on climate change and environmental issues, S. Fred Singer has been a leading skeptic of the scientific consensus on global warming. He points out that the scenarios pictured by most scientists are alarmist, that computer models reflect real gaps in climate knowledge, and says that future warming will be inconsequential or modest at most. He has also challenged the connection between ultraviolet-B radiation and melanoma and between secondhand smoking and lung cancer. Singer's critics have pointed out that the financial ties of his nonprofit organizations to tobacco and oil companies make Singer a case of clear conflict of interest.

Dr. Singer was born in Vienna on September 27, 1924. He did his undergraduate work in electrical engineering at Ohio State University and holds a Ph.D. in physics from Princeton University. He has served in numerous government and academic positions such as acting as director of the Center for Atmospheric and Space Physics at the University of Maryland (1953–62); as special adviser to President Eisenhower on space developments (1960); as first director of the National Weather Satellite Service (1962–64); as founding dean of the School of Environmental and Planetary Sciences at the University of Miami (1964–67); as deputy assistant secretary for water quality and research, U.S. Department of the Interior (1967–70); as deputy assistant administrator for policy, U.S. Environmental Protection Agency (1970–71); as professor of environmental sciences, University of Virginia (1971–94), and as chief scientist, U.S. Department of Transportation (1987–89).

To Singer, climate change is not something humans should fear. He argues that the climate has changed constantly throughout this and previous centuries and that people have always successfully adapted to it. In addition, he believes that humans can affect climate at a local level. Yet, whether they can cause global weather changes has still to be proved. Singer has repeatedly claimed that the atmosphere has not warmed up in recent decades. In fact, he has claimed that since 1979, it has slightly cooled down. Surface records that show increases in temperature are not, according to Singer, reliable sources of information, as thermometers tend to be placed in or very near to urban areas, which are traditionally warmer than other locations. Singer claims that models and observations about global warming do not agree. Although climatic models show that there should be an increase of about 1 degree F per decade in the middle troposphere, observations contradict these models. Singer is critical of arguments based on laboratory experiments, as the atmosphere is much more complicated and does not function under controlled circumstances. He recognizes that the increase in atmospheric CO_2 might lead to a slight warming, yet he says that this phenomenon is counterbalanced by increased evaporation of the oceans. The production of aerosols also causes cooling, which may counterbalance the effects of carbon dioxide. Yet, although Singer admits that aerosols last for a maximum of few weeks but CO_2 stays for decades, he is critical of models emphasizing the role of aerosols in connection to carbon dioxide.

Singer argues that because aerosols are mostly emitted in the Northern Hemisphere, where industrial activities are rampant, we would expect the Northern Hemisphere to be warming less quickly than the Southern Hemisphere. Actually, according to such models, the Northern Hemisphere should be cooling. To him, however, the data show the opposite, as both the surface data and the satellite data agree that, in the last 20 years, the Northern Hemisphere has warmed more quickly than the Southern Hemisphere. This fact contradicts the whole idea that aerosols make an important difference and proves that aerosols cannot be invoked as an explanation for the discrepancies between models and observations.

Singer does not have much faith in computer models, which he describes as having been "tweaked" to produce the present climate and the present short-term variation. He also points out that the two dozen models presently used are not entirely consistent with each other. These models also fail to depict all types of clouds, which to Singer is a fundamental flaw. He compares the current concern over global warming and the urgent calls for action to buying insurance with a high premium against a risk that is small. The Kyoto Protocol, to Singer, is part of the high insurance premium. The reduction of energy use by about 35 percent within 10 years implies, according to Singer's estimate, giving up one-third of all energy use, using one-third less electricity, and demolishing one-third of all cars. In spite of accusations aimed at other scientists that they use an apocalyptic tone when describing global warming, Singer too uses apocalyptic overtones to describe the Kyoto scenario: "It would be a huge dislocation of our economy, and it would hit people very hard, particularly people who can least afford it."

To Singer, global warming is a big business, with governments pumping about $2 billion into climate research. Thus people have to justify this expenditure, which supports jobs and research. Yet George Monbiot has emphasized that Singer has strong ties with oil and tobacco company—a fact that constitutes a conflict of interest given his stance on CO_2 emissions and secondhand smoking: "In March 1993, APCO sent a memo to Ellen Merlo, the vice president of Philip Morris, who had just commissioned it to fight the Environmental Protection Agency: 'As you know, we have been working with Dr. Fred Singer and Dr. Dwight Lee, who have authored articles on junk science and indoor air quality (IAQ) respectively.'" Singer's Science and Environmental Policy Project also received multiple grants from ExxonMobil.

SEE ALSO: Climate Change, Effects; Climate Models.

BIBLIOGRAPHY. George Monbiot, "The Denial Industry," *The Guardian*, September 19, 2006, www.guardian.co.uk/environment/2006/sep/19/ethicalliving.g2; PBS Interview with Dr. S. Fred Singer, "What's Up With the Weather?" www.pbs.org/wgbh/warming/debate/singer.html; Fred Singer, *Global Climate Change: Human and Natural Influences* (Paragon House, 1989); S. Fred Singer, *The Greenhouse Debate Continued* (Institute for Contemporary Studies Press, 1992); S. Fred Singer, *Hot Talk, Cold Science: Global Warming's Unfinished Debate* (The Independent Institute, 1997); S. Fred Singer, *The Scientific Case Against the Global Climate Treaty* (Science and Environmental Policy Project, 1997).

LUCA PRONO
UNIVERSITY OF NOTTINGHAM

Slovakia

LOCATED IN CENTRAL Europe, formerly the eastern part of Czechoslovakia, Slovakia has a land area of 18,932 sq. mi. (49,035 sq. km.), with a population of 5,390,000 (2006 est.) and a population density of 287 people per sq. mi. (111 people per sq. km.). A strongly agricultural country, some 31 percent of Slovakia's land is used for arable purposes, with an additional 17 percent used for meadows and pasture.

In terms of its per capita carbon dioxide emissions, Slovakia ranks 56th in the world, with 8.1 metric tons of emissions per person in 1992, falling to 7 metric tons in 2003. Slovakia sources 35.3 percent of its electricity production from fossil fuels, with 47.6 percent nuclear power and 17.1 percent from hydro power. Because of the heavy use of coal, solid fuels account for 45 percent of the carbon dioxide emissions in the country, with 34 percent from gaseous fuels, 17 percent from liquid fuels (reflecting the fact that transportation accounts for 12 percent of emissions), and the remaining 4 percent from cement manufacturing. The high amount of fossil fuels used in electricity also shows itself in the emissions by sector, with 32 percent of emissions from electricity and heat production and 43 percent from manufacturing and construction. The fossil fuels also account for the nation's relatively high sulfur dioxide and carbon monoxide emissions.

The Slovakian government took part in the United Nations Framework Convention on Climate Change, signed in Rio de Janeiro in May 1992, and ratified the Vienna Convention in 1993. The government signed the Kyoto Protocol to the UN Framework Convention on Climate Change on February 26, 1999, ratified it on May 341, 2002, and entered into force on February 16, 2005.

SEE ALSO: Climate Change, Effects; Transportation.

BIBLIOGRAPHY. Catherine Albrecht, "Environmental Policy in the Czech Republic and Slovakia," in Uday Desai, ed., *Ecological Policy and Politics in Developing Countries: Economic Growth, Democracy, and Environment* (State University of New York Press, 1998); *Environmental Performance Reviews: Slovak Republic* (Organisation for Economic Cooperation and Development, 2002); "Slovakia—Climate and Atmosphere," http://earthtrends.wri.org (cited October 2007).

Robin S. Corfield
Independent Scholar

Slovenia

LOCATED IN THE Balkans, and formerly one of the constituent parts of Yugoslavia, Slovenia has a land area of 7,827 sq. mi. (20,251 sq. km.), with a population of 2,030,000 (2006 est.) and a population density of 251 people per sq. mi. (99 people per sq. km.). About 12 percent of the country is arable, with a further 24 percent covered in meadows and pasture and 54 percent in forests.

In spite of its large amount of forestry, because of its high standard of living, Slovenia ranks 50th in the world in terms of its per capita carbon monoxide emissions, with 6.3 metric tons per person in 1992, falling to 5.5 metric tons in 1994 and then rising dramatically to 7.1 metric tons in 1995 before reaching a peak of 8.1 metric tons in 1997, falling slightly to 7.8 metric tons in 2003. Heavily reliant on nuclear power from the nuclear power plant at Krsko in Dolenjska, which accounts for 35.4 percent of the country's electricity production, 34.9 percent of the electricity comes from fossil fuels and 29.4 percent from hydropower. There is heavy use of fossil fuels, especially by the thermal electric power stations using coal to provide electricity for Ljubljana, the capital, and also Sostanj and Trbovlje. This results in electricity and heat production accounting for 42 percent of the country's carbon dioxide emissions.

Apart from electricity generation, there is also heavy use of private automobiles, with 365 vehicles per 1,000 people, similar to the proportions in Germany and the United Kingdom. This leads to regular traffic congestion, especially on roads around Ljubljana and Celje, Koper, and Maribor. Combined with relatively cheap petroleum, compared with prices in Western Europe, this has resulted in transportation contributing 28 percent of all carbon dioxide emissions by sector and 48 percent of emissions by source. This is in spite of the country's small size and an excellent public transport network, with an electrified train network run by Slovenske Zeleznice; the five steam trains are maintained solely for occasional journeys for tourists. In addition, there is an extensive bus network.

The effects of global warming and climate change on the country have been seen first with higher temperatures resulting in the melting of the snow in the mountains, limiting the periods available for skiing and beginning to affect arable crop production. To combat this, the government has succeeded in reducing sulphur dioxide emissions by half between 1985 and 1995 and also in reducing nitrogen oxide emissions by 20 percent. The Slovenian government took part in the United Nations Framework Convention on Climate Change, signed in Rio de Janeiro in May

1992, ratifying the Vienna Convention in the same year. In 1999, the government drew up a National Environmental Protection Program, and although it signed the Kyoto Protocol to the UN Framework Convention on Climate Change on October 21, 1998, it was not ratified until August 2, 2002, and entered into force on February 16, 2005.

SEE ALSO: Automobiles; Climate Change, Effects; Transportation.

BIBLIOGRAPHY. "Slovenia—Climate and Atmosphere," http://earthtrends.wri.org (cited October 2007); Neil Wilson and Steve Fallon, *Slovenia* (Lonely Planet, 2001).

ROBIN S. CORFIELD
INDEPENDENT SCHOLAR

Smagorinsky, Joseph (1924–2005)

AMERICAN METEOROLOGIST AND the first director of the National Oceanic and Atmospheric Administration's Geophysical Fluid Dynamics Laboratory (GFDL), Joseph Smagorinsky developed influential methods for predicting weather and climate conditions and lectured at Princeton for many years. With his decision to move the GFDL to Princeton, Smagorinsky made the university a leading center for the study of global warming.

Joseph Smagorinsky was born to Nathan Smagorinsky and Dina Azaroff. His parents were from Gomel, Belarus, but fled during the pogroms. Smagorinsky's father was the first to immigrate to the United States in 1913, settling in Manhattan's Lower East Side, where he opened a paint store. Three years later, he was joined by his wife and their children. Joseph was born on January 29, 1924, when the family was already living in the United States. Similar to his other three brothers, he worked in his father's paint store. He attended Stuyvesant High School for Math and Science in Manhattan. After high school, he expressed his wish not to stay in the family business and to go to college instead. As his intellectual skills had already become apparent, the whole family decided to support him in his decision. Smagorinsky

earned his B.S. (1947), M.S. (1948), and Ph.D. (1953) at New York University. During his sophomore year there, he joined the Air Force and became a member of an elite group of recruits who had been selected for their talents in mathematics and physics. Because of his scientific interests, Smagorinsky was included in the Air Force meteorology program. As a part of the scheme, he was sent to Brown University to specialize in mathematics and physics for six months. Smagorinsky was then sent to the Massachusetts Institute of Technology to learn dynamical meteorology, under Ed Lorenz, the author of chaos theory. During World War II, Smagorinsky worked as a weather observer for the Air Force. In May 1948, Smagorinsky married Margaret Frances Elizabeth Knoepfel—one of the first female weather statisticians.

After the war, Smagorinsky concluded his studies. Although he had planned a career as a naval architect, the rejection of the Webb Institute led him to choose meteorology as a field. After a question-and-answer session with prominent Princeton meteorologist Jule Charney, Smagorinsky was invited to carry out the research for his doctoral dissertation at the Princeton Institute for Advanced Studies. In 1950, Smagorinsky was part of Charney's team of scientists who successfully solved Charney's equations on the Electronic Numerical Integrator and Computer, also known as ENIAC. This was a milestone event in modern meteorology, as it pioneered the use of computers for weather forecasting. At the Institute for Advanced Studies, Smagorinsky and Charney developed the technique of the so-called numerical weather prediction. This technique relied on data collected by weather balloon, which were then elaborated by computers according to the laws of physics. This enabled researchers to forecast the interaction of turbulence, water, heat, and other factors in the production of weather patterns.

After completing his doctorate, in 1953 Smagorinsky accepted a position at the U.S. Weather Bureau and was among the founders of the Joint Numerical Weather Prediction Unit. Two years later, at the suggestion of eminent meteorologist John von Neumann, the U.S. Weather Bureau created a General Circulation Research Section and appointed Smagorinsky to direct it. Smagorinsky conceived his task as the completion of the von Neumann/Charney computer modeling program. He wanted to obtain a three-dimensional, global, primitive-equation general circulation model of the atmosphere. The section was initially located in Suitland, Maryland, but was moved to Washington, D.C., where it was renamed the General Circulation Research Laboratory in 1959. In 1963, it became the GFDL before moving to Princeton University, where it is still located, in 1968. Smagorinsky continued to serve as director of the lab until his retirement in January 1983.

Under Smagorinsky's directorship, the GFDL expanded, and Smagorinsky was able to attract respected international scientists such as Syukuro Manabe and Kirk Bryan to work there. The laboratory's work profoundly influenced the practice of numerical weather prediction around the world. Thanks to the GFDL's climate models, scientists have been able to assess more precisely humans' capabilities to affect climate change. In his years at Princeton, Smagorinsky was also appointed as visiting professor in geological and geophysical sciences at the university. As a member of the teaching staff, he helped to develop the Program in Atmospheric and Oceanic Sciences, a doctoral program in the Department of Geosciences that collaborates closely with the GFDL. After his retirement as director of the GFDL in 1983, Smagorinsky became a visiting senior fellow in atmospheric and oceanic sciences at Princeton until 1998.

It is thanks to the connection established by Smagorinsky between Princeton and the GFDL that the university became a major center for the study of global warming. From the 1970s onward, scientists working under Smagorinsky created the first models illustrating how climate could change in the face of increasing levels of carbon dioxide in the atmosphere. These models provided the first modern estimates of climate sensitivity and stressed the importance of water vapor feedback and stratospheric cooling. Research at the laboratory also allowed the development of the first models coupling atmosphere–ocean climate for studies of global warming, establishing the important differences between "equilibrium" and "transient" responses to the growing levels of carbon dioxide.

SEE ALSO: Climate Models; History of Meteorology.

BIBLIOGRAPHY. S. Manabe, J. Smagorinsky, and R.F. Strickler, "Simulated Climatology of General Circulation With a Hydrologic Cycle," *Monthly Weather Review* (v.93, December, 1965); J. Smagorinsky, "The Beginnings of Numerical Weather Prediction and General Circulation Modeling: Early Recollections," *Advances in Geophysics* (v.25, 1983); J. Smagorinsky, "General Circulation Experiments With the Primitive Equations," *Monthly Weather Review* (v.91/3, 1963); J. Smagorinsky, "On the Numerical Integration of the Primitive Equations of Motion for Baroclinic Flow in a Closed Region," *Monthly Weather Review* (v.86/12, 1958); J. Smagorinsky, S. Manabe, and J.L. Holloway, "Numerical Results From a Nine-Level General Circulation Model of the Atmosphere," *Monthly Weather Review* (v.93, 1965).

Luca Prono
University of Nottingham

Snowball Earth

IN THE EARLY 1960s, Brian Harland, a geologist at Cambridge University, observed that rocks on several continents, dating from the Neoproterozoic era (approximately 800–680 million years ago), contain glacial debris. Some of the glacial debris included carbonate rocks, which are known to form in the tropics (e.g., in the present-day Bahama Banks). This conclusion later gained additional support from paleomagnetic data. One potential explanation is that the

The Snowball Earth hypothesis proposes that the Earth was entirely covered by ice in part of the Cryogenian period.

entire Earth was covered by ice and snow during the Neoproterozoic. This has come to be known as the "Snowball Earth" hypothesis.

One early problem was understanding how a global ice age could have commenced. During the 1960s, the Russian climate scientist Mikhail Budyko used a computer simulation to establish that a runaway ice–albedo feedback effect could lead to global glaciation. The term albedo refers to the amount of the sun's energy that is reflected by the Earth's surface. As glaciers grow in extent, they reflect more of the sun's energy, which causes the atmosphere to cool. This in turn causes the glaciers to grow. Budyko showed that if the glaciers extended beyond a certain critical point, this ice–albedo feedback could lead to a global ice age.

A second obstacle was understanding how a global ice age could ever end once it began. In the early 1990s, Joseph Kirschvink of the California Institute of Technology observed that during a global ice age, the carbon cycle would shut down. Volcanoes sticking up through the ice cover would continue to add carbon dioxide to the atmosphere. Having nowhere else to go, the carbon dioxide would then accumulate over millions of years until a runaway greenhouse effect caused the ice to melt.

One important rival to the Snowball Earth hypothesis is the high obliquity hypothesis. If the tilt of the earth's axis had been much different during the Neoproterozoic, the poles could have received more solar energy than the tropics. If so, it would be possible to explain the evidence for glaciers in the tropics without supposing that the entire planet had frozen over.

In his widely cited 1992 paper, Kirschvink also proposed an explanation for banded iron deposits observed in Neoproterozoic glacial debris. Iron is not soluble in seawater in the presence of oxygen. During a true Snowball Earth episode, the oceans would have become deoxygenated over time. Iron from thermal vents would build up in the seawater. Then, when the ice finally melted, and oxygen was once again exchanged between the oceans and atmosphere, oxidized iron would have been left along with the debris from the retreating glaciers.

During the 1990s, two Harvard scientists, Paul Hoffman and Daniel Shrag, gathered additional, highly suggestive evidence that seemed to favor the Snowball Earth theory. They found that in many places, the Neoproterozoic glacial debris occurs right below thick layers of carbonate rock (which are known as "cap carbonates"), and they showed how Kirschvink's proposal could account for this. During a Snowball Earth episode, very large amounts of carbon dioxide would have built up in the atmosphere. As the ice receded and the carbon cycle resumed, large amounts of carbon would have been washed out of the atmosphere during storms and ended up in the form of carbonate rock on the ocean floor. More controversially, Hoffman and Shrag also studied the ratio of carbon-12 to carbon-13 isotopes in the cap carbonates. They argued that an unusual dip in the carbon isotope ratio signified a temporary shutdown of photosynthetic activity in the earth's oceans.

CHALLENGES TO THE SNOWBALL EARTH THEORY

One potentially serious challenge to the Snowball Earth theory comes from paleontology. Today, most geologists agree that there were at least two major ice ages during the Neoproterozoic: the Sturtian, around 750 million years ago, and the Varanger, around 590 million years ago. The second of these episodes occurred shortly before the Cambrian Explosion of metazoan life. However, a true Snowball Earth episode would have killed off nearly all eukaryotic life, and it is not clear that there was enough evolutionary time for life to recover from a global ice age. Some scientists have used computer models to show that softer versions of the Snowball Earth episode might have been possible—for example, a mostly ice-covered planet with massive continental ice sheets in the tropics but largely ice-free tropical oceans.

Although scientists generally agree that there was low-latitude glaciation during the Neoproterozoic, they continue to use a combination of fieldwork and numerical modeling techniques to work out the details. The Snowball Earth scenario remains an intriguing live hypothesis.

SEE ALSO: Abrupt Climate Changes; Albedo; Carbon Cycle; Climate Feedbacks; Climate Models; Climate Thresholds; Computer Models; Earth's Climate History; Glaciology; Greenhouse Effect; Historical Development Of Climate Models; Ice Ages; Ice–Albedo Feedback; Ice Component of Models; Modeling of Ice Ages; Modeling of Paleoclimates; Ocean Component of Models; Paleoclimates.

BIBLIOGRAPHY. P.F. Hoffman, A.J. Kaufman, G.P. Halverson, and D.P. Schrag, "A Neoproterozoic Snowball Earth," (*Sci-*

ence (v.281/5381, 1998); P. F. Hoffman and D. T. Schrag, "Snowball Earth," *Scientific American* (v.282/1, 2000); W. T. Hyde, T.J. Crowley, S.K. Baum, and W.R. Peltier, "Neoproterozoic 'Snowball Earth' Simulations With a Coupled Climate/Ice Sheet Model," *Nature* (v.405/6785, 2000); J.L. Kirschvink, "Late Proterozoic Low-Latitude Global Glaciation: The Snowball Earth," in J.W. Schopf and C. Klein, eds., *The Proterozoic Biosphere: A Multidisciplinary Approach* (Cambridge University Press, 1992).

Derek Turner
Connecticut College

Social Ecology

SOCIAL ECOLOGY IS an ecological vision for the future developed by the anarchist thinker Murray Bookchin. This theory is part of a left-wing tradition that rejects notions of hierarchy, domination, power, and place to advocate political reformism, or restructuring that will resolve basic issues of societal, gender, and environmental imbalance. Social ecology is based on the understanding that all our present ecological problems are a result of deep-seated social problems. As Bookchin states, "economic, ethnic, cultural, and gender conflicts, among many others, lie at the core of the most serious ecological dislocations we face today." Specifically, social ecologists argue that the chief source of ecological destruction is the capitalist system and its products, such as overconsumption, consumerism, and concomitant economic growth.

Trade for profit, industrial expansion, and the association of progress with corporate self-interest are among others. Bookchin argues, therefore, that to separate ecological from social problems underplays not only the sources of the environmental crisis but also the interplay among all of these factors. Human beings must not downplay the importance of how they deal with each other as social beings. This, social ecologists argue, is the key to addressing the environmental crisis. The social ecological vision is to see a society that is based along social ecological lines. In this context, there are a number of principles that characterize social ecology.

First, a society based on social ecology would be one in which ecological regeneration would be insep-

arable from social regeneration. For example, social regenerative strategies might include the formation of ecocommunities and the adoption of ecotechnologies that establish a creative intersection between humanity and human nature. Spirituality, or what Bookchin calls regeneration of the spirit, is another principle, signifying the growth and development of a whole society. Such a society would be diverse and holistic in nature. Spirituality is defined as a natural phenomenon—one that focuses on the ability of humans to act as moral agents and actively promote the end to needless suffering, undertake ecological restoration, and foster aesthetic appreciation of all living things. Building on this spirituality will ensure the presence of liberty (in the sense of encouraging and nurturing individual and collective creativity, imagination, and personality) as a "continuum of natural evolution," resulting in a healthy society.

SOCIAL ECOLOGY AND THE ENVIRONMENT

Social ecology goes further, however, addressing the deep structural failures within society, and seeks to redress the ecological effects humans are having on the environment. Social ecology here seeks to change the definition of the very idea of a society based on hierarchy, class, and domination to one based on equity and the ethics of complementarity. In this case, social ecologists argue that humans must play a supportive role in maintaining the integrity of the planet. As such, they promote the establishment of community institutions that can embrace community-based ethical systems that in turn encourage the qualities of wholeness so integral to the social ecology vision. Societal structures that are supported by social ecology include confederal municipalism, in which municipalities conjointly gain rights to self-governance through the networks of confederal councils; empowerment of people; ecocommunities that are linked into the confederations of economy, fostering a healthy interdependence; and shared property.

Bookchin notes that "Social ecology calls upon us to see that nature and society are interlinked by evolution into one nature that consists of two differentiations: first or biotic nature, and second or human nature." By first nature, social ecologists refer to the way in which human beings are ultimately connected to their biological and evolutionary history. Second nature refers to the way in which humans produce or have a distinct social nature, as opposed to animals. As reflexive reflective beings, humans have the responsibility of being the voice of first nature. To understand and work within society, social ecologists argue we must understand and embrace both natures. Consistent with anarchist theory, social ecologists see social hierarchy to be the enemy of natural order.

In his early work, *The Ecology of Freedom* (1971), Bookchin in fact highlights a model of evolutionary human social development that suggests that social hierarchy first emerged in the Neolithic period with simple forms of governance within and between different social groups. As such, social ecologists perceive that humans are always rooted within their own biological evolutionary history. The separation of the current society of the human from the biological is a failure to think organically and to recognize wholeness. In this context, social ecologists argue that the human and the nonhuman must be seen as being part of an evolutionary continuum and, as such, we are in a state of continual becoming. Unlike deep ecology, which is underpinned by the belief that all organisms have intrinsic rights, proponents of social ecology argue that the environment has rights when and if they are conferred by humans.

Social ecologists also believe, however, that human beings have, by virtue of their innate creativity and powers of reason, intrinsic value. As such, Bookchin maintains that the most ethical standpoint is for humans to understand different forms of hierarchy in nature, that is, different ecosystems, patterns, and orders, without implying that humans thus have the right to dominate those hierarchies. Social ecology attacks capitalism at a fundamental level, believing that modern capitalism is "structurally amoral and hence impervious to any moral appeals." It believes that the driver for capitalism is to grow or die, and as such, that the system is inherently ecologically destructive. They advocate instead a society based on complementarity and mutual aid. In this way, social ecologists advance the need to redress the ecological effect humans have had on society by calling for social reconstruction based along ecological lines. The ethics of complementarity are based on a system in which human beings play a supporting role in upholding the integrity of the biosphere and the planet. Social ecologists argue we have a moral responsibility to do this, as well as to enshrine the ethics of complementarity within social institutions in ways that will enable active participation of all in the process of reconstruction. In such a society, property would be shared, which would ultimately give rise to individuals for whom there is no separation between the individual and collective interest, the private and the personal, the political interest and the social.

Social ecologists stress the social causes and consequences of the degradation of the environment above all else. In this context, social ecologists seek answers to societal problems that are organic. They argue that as human forms of hierarchy have evolved, so has the human capacity to impose forms of domination on nature. Hence, it is the responsibility of the human race to redress the inequalities and problems attendant on the imposition of our own social order on nature. If social change in line with the organic elements ascribed to does not occur, social ecologists believe that the biosphere as we know it is heading toward complete destruction.

Overall, the position of social ecologists has attracted critique from those who argue that their position is going too far in its interrogation of the link between society and the environment. Nonetheless, Bookchin's analysis of the relationship between society and the environment, and how this relationship constitutes and causes forms of domination—and hence ecological destruction—has played an important role in highlighting the nexus between social and natural dimensions of environmental decision making.

SEE ALSO: Climate Change, Effects; Ecological Footprint.

BIBLIOGRAPHY. Murray Bookchin, *The Ecology of Freedom* (Cheshire Books, 1982); Murray Bookchin, *Post-Scarcity Anarchism* (Ramparts Press, 1972); Murray Bookchin, *The Rise of Urbanization and the Decline of Citizenship* (Sierra Club Books, 1987); Erica Cudworth, *Environment and Society A Reader* (Routledge, 2003).

MELISSA NURSEY-BRAY
AUSTRALIAN MARITIME COLLEGE
ROB PALMER
RESEARCH STRATEGY TRAINING

Soil Organic Carbon

THE EARTH'S TERRESTRIAL ecosystems store over 2000 gross tons (Gt; 1 Gt = 1,015 g) of soil organic carbon (SOC), which is about four times more carbon than is stored in the atmosphere. Annually, soils release over 60 Gt Carbon to the atmosphere, which is about 10 times that amount released by fossil fuel combustion. Warming can increase the rate at which fresh organic matter (e.g., recently senesced leaves, fruits) decomposes, with the highest rates found where it is wet and warm. Less is known about how more thoroughly decomposed material (e.g., SOC) responds to changes in temperature, but it has been assumed that global warming will increase SOC decomposition rates, with the sensitivity of SOC decomposition to warming determining the extent to which SOC storage will be altered by climate change. Because the sensitivity of SOC decomposition to temperature remains poorly quantified, it cannot be accurately predicted whether global soils will change from a net sink to a net source of CO_2 as the planet warms.

SOC serves many important ecosystem roles. Because the supply of organic carbon exerts a dominant control on the activity of soil heterotrophic organisms—from bacteria to insects—SOC is critical to regulating the structure and functioning of soil communities. Further, during SOC decomposition, large quantities of nutrients are released from organic to mineral forms, and so SOC provides a critical source of nutrients to growing vegetation. The amount of SOC can also affect the water-holding capacity of a soil, as well as water movement through soils. Despite these important roles, the tremendous complexity of SOC in natural and agricultural systems presents important challenges to quantifying SOC formation and decomposition, including the sensitivity of these processes to climate change. This complexity results from the fact that very large quantities of organic matter are cycled through soils annually, but only a very small fraction remains in soils, typically in a highly transformed state that can persist in soils for millennia as a result of chemical recalcitrance or protection by clay minerals.

Given the important effect that climate may have on SOC decomposition, it is critical that tools be developed to accurately predict how SOC formation, decomposition, and storage respond to climate change. Of particular importance is quantifying interactions among driving variables, as these interactions will influence responses in difficult-to-predict ways. For example, warming in cold and wet climates may result in the loss of SOC, as warming can dry out often anaerobic soils in which oxygen supply limits decomposition rates. In contrast, warming in temperate or tropical climates may have little effect on or even slow SOC decomposition rates, especially if moisture is limiting for the soil microbes responsible for decomposing SOC. Reducing uncertainty is important for accurately predicting how the terrestrial carbon cycle, and hence the climate, will respond to global warming.

SEE ALSO: Chemistry; Climate Change, Effects; Soils.

BIBLIOGRAPHY. E.P. Paul and F.E. Clark, *Soil Microbiology and Biochemistry* (Academic Press, 1996); Richard G. Zepp, *The Role of Nonliving Organic Matter in the Earth's Carbon Cycle* (Wiley & Sons, 2001).

CHRISTIAN GIARDINA
INSTITUTE OF PACIFIC ISLANDS FORESTRY

Soils

SOIL CONTAINS PULVERIZED rock, organic matter, and microorganisms, which convert the organic matter to humus. Soils range from loosely-packed sandy soil, to finely packed, sticky clay soil. Between these two extremes is loam, whose high content of organic matter makes it easy to cultivate. Soils are thin in the Mediterranean basin, and thick in northern Europe, Russia, and the American Midwest and Mississippi delta. As the basis of agriculture and as a factor in the formation of climate, soils have shaped the destiny of humans.

Soils warm when absorbing sunlight, and cool when reflecting sunlight. Soil temperature coincides with an area's climate. In tundra, for example, the air warms in spring and summer much faster than the soil does. Only the outermost layer of the soil warms enough to thaw, and then for only a few months, before winter returns. Not all soils absorb and reflect sunlight at the same rate. Wet soil darkens, and absorbs the most sunlight and gains the most heat. Dark soil absorbs as much as 86 percent of sunlight, gray soil absorbs 80 percent, and light soil absorbs only 20 percent of sunlight, reflecting the rest into the atmosphere. When water saturates soil, however, its color lightens. Being light, saturated soil reflects sunlight, yet the large amount of water that soil can hold increases the capacity of soil to absorb heat. Saturated soil absorbs and radiates heat slowly and in large quantities. Water evaporating from soil liberates heat, warming the surrounding air. Dry soil reflects more sunlight and heat than wet soil. Yet dry soil, because it contains little moisture, warms quickly on a cloudless day and becomes hot. At night, dry soil quickly cools, radiating heat back into the atmosphere.

The term *albedo* is the ratio of the amount of sunlight an object reflects to the amount it absorbs. A solid black object absorbs all sunlight and reflects none, and so has an albedo of zero. A white object absorbs no sunlight, reflects all of it, and has an albedo of one. The mean albedo of earth is 0.36. Dark soil has an albedo between 0.1 and 0.2. Clay soil has an albedo between 0.15 and 0.35. Sandy soil has an albedo between 0.25 and 0.45, and light soil has an albedo between 0.4 and 0.5.

At most sunlight penetrates only a few feet of soil. Infrared light, for example, penetrates to a depth of 3 ft (.9 m.). Because the amount of sunlight that soil receives varies during the day, soil temperatures vary. Soil temperatures also vary throughout the year. In Bridgewater, Massachusetts, for example, soil at a depth of one inch varied from a low of 35.2 degrees F (1.7 degrees C) in February to a high of 63.5 degrees F (17 degrees C) in August. Soil temperature therefore varies less than the temperature of air. Sunlight penetrates more deeply in rocky and wet sandy soil than in wet clay. Sunlight penetrates least in dry sandy soil.

Vegetation affects the capacity of soils to absorb sunlight. Plant cover decreases the albedo of light soil and increases the albedo of dark soil. Dense foliage blocks sunlight from reaching the soil, and so lessen the soil's absorption of sunlight and heat. Before the evolution of plants, soils must have been, along with the oceans, the major reservoirs of heat by absorbing sunlight. Like water, soil absorbs the most heat when the sun is overhead. Soil reflects increasing amounts of light as dusk approaches. Along with the oceans, the soil shapes the climate and influences the course of life.

SEE ALSO: Albedo; Climate; Plants.

BIBLIOGRAPHY. Endre Dobos, "Albedo," http:kyclim.wku.edu/bolder/Lawrence&Chase.ppt; D.F. Post, A. Fimbres, A.D. Matthias, E.E. Sano, L. Accioly, A.K. Batchily and L.G. Ferreira, "Predicting Soil Albedo from Soil Color and Spectral Reflectance Data," http://soil.scijournals.org/cgi/content/full/64/3/1027.

CHRISTOPHER CUMO
INDEPENDENT SCHOLAR

Solar Energy Industries Association (SEIA)

THE SOLAR ENERGY Industry Association (SEIA) is an American trade association for the solar industry, working to expand markets, strengthen and develop research, and improve education for the employment of solar energy. SEIA is affiliated with the PVNow coalition of photovoltaic companies, which aims to expand the North American–distributed, grid-connected photovoltaic market opportunities and eliminate market barriers. They are pursuing this goal through lobbying key state legislatures, utility rate–making authorities, and other state energy policymaking agencies. SEIA represents over 700 companies and 20,000 employees in the U.S. energy sector.

SEIA headquarters are located in Washington, D.C. The organization is divided into 14 state SEIA organizations. Members of the state organizations monitor and advocate in the state governments and also supply grassroots support for solar consumers

SEIA's mission is to reduce regulatory barriers to photovoltaic installations, increase photovoltaic markets across the nation.

and small businesses. SEIA chapters have up-to-date information on retailers and distributors in their area and frequently host workshops and discussion groups. SEIA's mission is to reduce regulatory barriers to photovoltaic installations, increase photovoltaic markets across the nation through meaningful and appropriate incentive programs at the state and federal levels, guarantee continuing federal research into high standards of photovoltaic devices, and supporting technologies such as inverters and balance-of-system equipment. SEIA campaigns for legislation in favor of renewable energy and technologies that could provide emission-free energy sources. It also advocates for a reduction of CO_2 emissions.

Rhone Resch is the president of SEIA. Together with his organization, he applauded the U.S. Congress and President George W. Bush for producing the strongest national policy for solar power since the 1980s. The solar tax benefits in the 2005 Energy Bill were acclaimed by SEIA as an important victory and as the measure necessary to allow the solar industry to meet the challenge of playing a significant role in supplying energy for the United States. To Resch, the United States has the best solar resources in the industrialized world, and by choosing solar energy, Americans can make a real contribution to their country's energy independence: "Installing solar energy on your roof is one of the most meaningful steps an individual can take to reduce our reliance on foreign sources of energy and help declare energy independence. Now solar comes with a more affordable price tag, and more consumers will take a step towards energy independence by choosing solar power. That means cleaner air, more jobs, and greater energy security for all." SEIA also supports the Solar Energy Research and Advancement Act of 2007—legislation that Resch described as "helping solar energy to make major strides in contributing to a clean, domestic, renewable supply of electricity."

SEE ALSO: Alternative Energy, Solar; Carbon Dioxide; Sunlight.

BIBLIOGRAPHY. Solar Energy Industry Association (SEIA), http://www.seia.org/index.php; PVNow, www.pvnow.com/.

LUCA PRONO
UNIVERSITY OF NOTTINGHAM

Solar Wind

TWO COMPETING THEORIES for global warming and their effect on Earth's changing climate persist today. The first theory suggests that the driver for global warming is the increasing amount of greenhouse gases dumped into the atmosphere as a result of humanity's burning of fossil fuels. The second theory posits that the solar wind and its associated magnetic field alters the Earth's cloud cover and adjusts the atmosphere's water vapor content, which leads to the steady temperature rise known as global warming.

The latter theory involves a stream of plasma, or high-energy charged particles, propelled from the sun's upper atmosphere. This stream of electrons and protons escapes the gravitational pull of the sun and creates the solar wind. It varies in speed from 190 to 500 mi. (306 to 805 km.) per second and passes by Earth as the sun rotates in space. The solar wind affects Earth's magnetic field and, in turn, is believed to have a major effect on climate change.

Opponents to the greenhouse gas theory of global warming argue that increasing radiation activity from the sun over the past 300 years has been the primary culprit—not an increase of atmospheric carbon dioxide. Researchers believe that because the doubling of the sun's magnetic flux recorded in the 20th century had led to increased sunspot activity as it follows its periodic 11-year cycle, the ferocity of the solar wind and the overall brightness of the sun also increased.

Proponents of global warming who subscribe to increasing carbon dioxide emissions as the cause of the problem avoid citing work done by NASA's Goddard Institute for Space Studies (GISS) or other scientific evidence offering credence to the solar wind theory. Just like the greenhouse gas theory, the GISS climate model is used to show that changes in the solar wind throughout the ages have varied surface warming. Climate researchers determined that the sun has played a role in modulating the atmosphere's moisture content and its circulatory patterns, causing droughts in ancient times. Backing up these computer-generated data are a number of natural records that correlate with the model's projections. Lake sediment analysis, fire records, and tree-ring measurements from the Yucatan Peninsula, Mexico, and Peru, to name a few locations, illustrate that periods of drought occurred during times of heightened solar output.

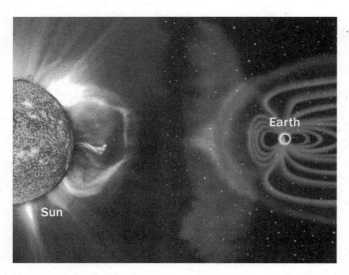

The Earth's magnetosphere, or magnetic field, protects us from most effects of the solar wind and from solar storms.

Increasing solar wind produces more ozone in Earth's upper atmosphere by breaking up oxygen molecules and heating the atmosphere. As a consequence, the circulation of the atmosphere is affected right down to the surface, which, in general, warms and reinforces existing rainfall patterns. Wet regions receive more rain, and dry regions become more susceptible to drought as the warmer air temperature pulls more moisture out of the soil. Droughts become more intense.

Although such facts are rarely disputed, the scope of the influence of solar wind is hotly debated. Some researchers state that the current period of global warming cannot be caused by the changes in solar output alone. Other researchers suggest that a double effect is in play, in which a more vigorous solar wind increases the global temperature, which in turn causes the oceans to warm, as made evident by melting sea ice. Because warm water absorbs less carbon dioxide, more of that greenhouse gas remains in the atmosphere. The debate boils down to whether all, some, or none of the burning of fossil fuels leads to global warming. As of this writing, most scientists and the popular media believe that human activity adds so much greenhouse gas to the atmosphere that this round of global warming could be catastrophic to life on Earth.

Solar-focused satellites have been monitoring the sun since the 1970s. More recently, the Solar and Heliospheric Observatory and the Wind and Advanced Composition Explorer have kept their instruments trained on the sun to measure the sun's temperature, capture the ion content of the solar wind, determine how the solar wind is accelerated, and more. Recently, the twin STEREO spacecraft were launched to expand on and augment existing satellite measurements by tracking the sun together and reporting on its solar behavior. The Solar Terrestrial Relations Observatory became operational in 2007 and will provide researchers the first three-dimensional space forecasts associated with solar activity.

Whatever the conclusion as to the ultimate cause of global warming, it is clear that the "third rock from the sun" will play a big role. The question for humanity is to determine which theory best describes global warming—greenhouse gas emissions or solar wind influences—and to develop policies to mitigate the effects on human civilizations.

SEE ALSO: Climatic Data, Historical Records; Computer Models; Goddard Institute for Space Studies; Sunlight.

BIBLIOGRAPHY: Jane S. Shaw, *Global Warming* (Farmington Hills, 2002); Fred S. Singer and Dennis T. Avery, *Unstoppable Global Warming: Every 1500 Years* (Lanham, 2007).

ROBERT KOSLOWSKY
INDEPENDENT SCHOLAR

Solomon Islands

LOCATED IN THE Pacific, east of Papua New Guinea, the Solomon Islands have a land area of 11,157 sq. mi. (28,896 sq. km.), covering 249,000 sq. nautical mi., with a population of 496,000 (2006 est.) and a population density of 43 people per sq. mi. (17 people per sq. km.). Only 1 percent of the land is arable, with a further 1 percent being used for meadows and pasture and 91 percent of the country forested, although a massive timber industry is resulting in heavy deforestation.

The entire electricity production in the country is from fossil fuels, with liquid fuels making up the entire carbon dioxide emissions from the country. Because the country is largely undeveloped, the per capita carbon dioxide emissions rate is low, being 0.5

metric tons per person in 1990 and falling to 0.39 metric tons in 2003. The increasing deforestation of the islands, however, will start to lead to a heavy increase in the country's carbon dioxide emissions in subsequent years.

The possible effects of global warming and climate change on the Solomon Islands are significant. There is the strong probability of increased flooding, which in turn could lead to a rise in the prevalence of insect-borne diseases such as malaria and dengue fever. The flooding might also result in the inundation of some of the several hundred islands that make up the country.

The Solomon Islands government took part in the United Nations Framework Convention on Climate Change, signed in Rio de Janeiro in May 1992, and ratified the Vienna Convention in the following year. The government signed the Kyoto Protocol to the UN Framework Convention on Climate Change on September 29, 1998, and ratified it on March 13, 2003, with it entering into force on February 16, 2005.

SEE ALSO: Climate Change, Effects; Deforestation; Floods.

BIBLIOGRAPHY. Ross McDonald, *Money Makes You Crazy: Custom and Change in the Solomon Islands* (University of Otago Press, 2003); "Solomon Islands—Climate and Atmosphere," http://earthtrends.wri.org (cited October 2007); *Solomon Islands: Rebuilding an Island Economy* (Department of Foreign Affairs and Trade, Australia, 2004).

ROBIN S. CORFIELD
INDEPENDENT SCHOLAR

Somalia

LOCATED IN NORTHEAST Africa—the "Horn" of Africa—Somalia has a land area of 246,201 sq. mi. (637,657 sq. km.), with a population of 8,699,000 (2006 est.) and a population density of 34 people per sq. mi. (13 people per sq. km.). About 80 percent of the population is dependent on agriculture, though only 2 percent of the land is arable, with a further 69 percent used for meadows or pasture, mainly low-intensity grazing of cattle, goats, and pigs. Some 14 percent of the land is forested.

Because the country is underdeveloped, it has a very low level of electricity usage, with the carbon dioxide emissions per capita being the lowest of any country in the world, even though accurate statistics have not been available for the last 10 years. Official statistics from 2001 give the entire electricity production for the country at 245 million kWh, with consumption levels at 228 kWh, with 100 percent of all electricity coming from fossil fuels.

The effects of global warming and climate change on Somalia are expected to be extensive. The rising temperature is expected to make more of the arable land unusable for the growing of crops and to render the pasture land even less productive than it is at the moment. There is also the possibility of flooding in some low-lying parts of the country.

The Somali government sent an observer to the United Nations Framework Convention on Climate Change, signed in Rio de Janeiro in May 1992, and ratified the Vienna Convention on 2001. Because of the instability in the country, there have been few measures introduced to combat some of the effects of climate change, and the current government has so far not expressed any opinion on the Kyoto Protocol to the UN Framework Convention on Climate Change.

SEE ALSO: Climate Change, Effects; Deserts; Floods.

BIBLIOGRAPHY. A. S. Abbas, *The Health and Nutrition Aspect of the Drought in Somalia* (Ministry of Health, Community Health Department, Nutrition Unit, 1978); Robert Caputo, "Tragedy Stalks the Horn of Africa," *National Geographic* (v.184/2, August 1993); "Somalia—Climate and Atmosphere," http://earthtrends.wri.org (cited October 2007).

JUSTIN CORFIELD
GEELONG GRAMMAR SCHOOL, AUSTRALIA

Somali Current

THE SOMALI CURRENT can be found on the surface of the northern Indian Ocean, serving as a western boundary of this ocean. It is a movement of waters around the Indian Ocean, dispersing heat. Atmospheric circulation and ocean circulation together are the major mechanisms for global heat distribution. As atmospheric circulation defines large-scale air movements around the globe, ocean circulation refers to the patterned movement of particular waters.

In summer, a southwest monsoon blows upward from the east coast of the Horn of Africa. Carried along with the monsoon are the waters of the western Indian Ocean, moving in a northeast direction underneath, and powered by winds. These waters may reach speeds of 9 mi. per hour (14 km. per hour). As the current reaches Somalia, the waters turn eastward. Some stay on near the Arabian Peninsula to form the East Arabian Current. Those that continue eastward eventually become the northeast monsoon during the autumn and winter, flowing southwest back to their origins. During the months of December and March, the Somali Current typically hovers between 5 degrees and 1 degree of latitude North of the equator, with this reach extending to span between 10 and 4 degrees latitude North during the central months of January and February.

The Somali Current is of interest because it creates an upwelling of cold water that is the only other region of such low surface temperatures within 10 degrees of the equator outside of Peru, and perhaps even colder. The cold surface temperatures around Peru are caused by the Peruvian or Humboldt Current which is related to El Niño.

The waters of the Somali Current swirl into what is known as the Great Whirl, an eddy with a diameter that can reach 500 km. (approximately 311 mi.), spinning in an anticyclonic direction. Anticyclonic direction is opposite to the earth's rotation; in the Northern Hemisphere the eddy therefore spins clockwise. The upwelling occurs during the months of May through September, and can lower the surface temperature in the western Indian Ocean by up to 9 degrees F (5 degrees C). Ocean surface temperatures are an important data source for monitoring global warming; therefore it is important to record the temperatures found during the northern (summer) swing of the Somali Current.

The Somali Current and other phenomena in the Indian Ocean were investigated at length in the year 1995 in a study that began in late 1994 and concluded in early 1996. It was an ambitious project that attempted to record all data related to the Indian Ocean during that year, and was undertaken by the World Ocean Circulation Experiment (WOCE) Indian Ocean Expedition.

SEE ALSO: El Niño and La Niña; Hadley Circulation; India; Indian Ocean; Modeling of Ocean Circulation; Monsoons; Oceanic Changes; Oceanography; Peru; Peruvian Current; Somalia; Thermocline.

BIBLIOGRAPHY: J.R.E. Lutjeharms, *The Agulhas Current* (Springer, 2006); J. Madeleine Nash, *El Niño: Unlocking the Secrets of the Master Weather-Maker* (Grand Central Publishing, 2003); S. George Philander, *Our Affair with El Niño: How We Transformed an Enchanting Peruvian Current into a Global Climate Hazard* (Princeton University Press, 2006).

CLAUDIA WINOGRAD
UNIVERSITY OF ILLINOIS AT URBANA-CHAMPAIGN

South Africa

THE REPUBLIC OF South Africa has a land area of 471,443 sq. mi. (1,221,037 sq. km.), with a population of 48,577,000 (2006 est.) and a population density of 101 people per sq. mi. (39 people per sq. km.). Some 10 percent of the country is devoted to arable purposes, with a further 67 percent used as meadows and pasture, much of it for low-intensity grazing of cattle and goats. Only 7 percent of the country is forested.

As the nation is a major producer of coal, some 92.6 percent of South Africa's electricity production comes from fossil fuels, with 6.7 percent from nuclear power and 0.7 percent from hydropower. In spite of its location, South Africa has made little use of solar power. In 1990, South Africa produced 7.8 metric tons per capita of carbon dioxide, and this level of emissions remained relatively stable until 2003, rising to 9.2 metric tons per capita in the following year. Coal and other solid fuels contribute to 80 percent of the country's carbon dioxide emissions, with 18 percent being from liquid fuels and 1 percent from gaseous fuels. By sector, 63 percent of South Africa's carbon dioxide emissions come from the generating of electricity, with 21 percent from manufacturing and 13 percent from transport.

Recently observed climates in South Africa, possibly attributable to global warming, include the following: The temperature of the Benguela Current has recently been increasing, resulting in a decrease in the fishing catch off the coast of the country. This, in turn, has led to a significant reduction in the diversity of species of fish in the region. In January 2000, following one of the driest Decembers on record, with many days having temperatures above 104 degrees F (40 degrees C), there were many bush fires along the coast of the Western Cape, leading to the destruction of woodland and other vegetation.

The South African government of Frederik W. de Klerk took part in the United Nations Framework Convention on Climate Change, signed in Rio de Janeiro in May 1992. They accepted the Kyoto Protocol to the UN Framework Convention on Climate Change on July 31, 2002, with it entering into force on February 16, 2005. In August 2005, Marthinus van Schalkwyk, the environmental affairs minister for South Africa, attended a week-long ministerial meeting in Greenland to discuss what became dubbed the Greenland Dialogue to examine further ways of making the Kyoto Protocol more effective.

A windmill near Johannesburg: Wind resources could be used to reduce carbon dioxide emissions from generating electricity.

SEE ALSO: Carbon Dioxide; Climate Change, Effects.

BIBLIOGRAPHY. Douglas H. Chadwick, "A Place for Parks," *National Geographic* (v.190/1, July 1996); Clive Grylls, "Return of the World," *Geographical* (v.78/12, December 2006); Michael E. Meadows and Timm M. Hoffman, "Land Degradation and Climate Change in South Africa" *The Geographical Journal* (v.169/2, 2003); "SA Faces up to Global Warming," http://www.southafrica.info/ess_info/sa_glance/sustainable/update/climate-190805.htm (cited October 2007); Kennedy Warne, "Oceans of Plenty," *National Geographic* (v.202/2, August 2002); World Resources Institute, "South Africa—Climate and Atmosphere," http://earthtrends.wri.org (cited October 2007).

JUSTIN CORFIELD
GEELONG GRAMMAR SCHOOL, AUSTRALIA

South Carolina

SOUTH CAROLINA IS 32,020 sq. mi. (82,931 sq. km.), with inland water making up 1008 sq. mi. (2,610 sq. km.), coastal water making up 72 sq. mi. (186.5 sq. km.), and territorial water making up 831 sq. mi. (2,152 sq. km.) South Carolina's average elevation is 350 ft. (106 m.) above sea level, with a range in elevation from sea level on the Atlantic Ocean to 3,560 ft. (1,085 m.) at Sassafras Mountain. The major natural regions include what is referred to as the low country—the Coastal Plain with offshore islands (inland is rolling, and along the coast is flat)—and what is referred to as the up-country—piedmont (higher elevation, with forests and pastureland) and Blue Ridge (mountainous and forested). South Carolina has river systems; the large lakes are artificially created.

The state's hot summers and warm winters come from the combination of the state's relatively low latitude, low elevation, proximity of the warm Gulf Stream in the Atlantic, and the Appalachian Mountains. During the winter, the mountains limit cold air entering the interior of the continent. Low country summers are hot and humid, though a sea breeze brings some relief. The up-country is usually cooler than the low country, even in the middle of the summer. The highest temperature recorded in the state was 111 degrees F (44 degrees C) on June 28, 1954, and the lowest temperature recorded was minus 19 degrees F (minus 28 degrees C) on January 21, 1986. Winters in South Carolina are mild, with brief periods of cold (lakes and rivers rarely freeze over), throughout the state. Precipitation is abundant all year; snow falls in the mountains but is rare in other parts of the state. Most of the state receives about 48 in. of rain a year, though the mountains areas receive more.

Forests cover much of the state, and wood is commercially harvested for lumber, wood pulp, paper, and furniture. Agriculture products include tobacco, cotton, corn, soybeans, and wheat. South Carolina's electricity is generated by nuclear power plants and coal-fueled power plants, and a small portion is hydroelectric.

South Carolina is already experiencing the effects of higher temperatures and rising sea levels (nine in. in the last century), and hurricanes and other major storms have increased in intensity and duration by about 50 percent since the 1970s; they are linked to increases in average sea surface temperatures and eroding coastlines. The coastal plains have already experienced problems with water supplies, and the increased use of groundwater for irrigation has lowered groundwater levels.

Although climate models suggest an increase in temperature of 5.4 degrees F (3 degrees C) by the end of the 21st century, potential risks anticipated include sea levels rising an additional 19 in. or 48 cm. (causing beach erosion and saltwater incursion); decreased water supplies; population (both human and animal) displacement; changes in food production, with agriculture improving in cooler climates and decreasing in warmer climates; forest loss, with persistent drought and loss of trees unsuited to higher temperatures; change in rain pattern to downpours, with the potential for flash flooding (causing sediments, agricultural chemicals, and other substances to leach into water sources); and increased health risks of certain infectious diseases stemming from water contamination or disease-carrying vectors such as mosquitoes, ticks, and rodents and of heat-related illnesses.

In populated areas at higher elevations, as in northwestern South Carolina, water quality and amount may both become issues. Should the levels of streams, groundwater, and lakes decrease, shortages of water for industrial and municipal markets would occur. Shallow wells would be affected, with rural communities losing water supplies. Higher rainfall would increase flooding and erosion, as well as contaminating the remaining water supplies with runoff.

South Carolina's economy relies on tourism, and beach erosion resulting from climate change would damage the coastline.

On the basis of energy consumption data from the Energy Information Administration's State Energy Consumption, Price, and Expenditure Estimates (SEDS), released June 1, 2007, South Carolina's total CO_2 emissions from fossil fuel combustion in million metric tons for 2004 was determined to be 88.56, made up of contributions from commercial (1.53), industrial (14.59), residential (2.36), transportation (32.10), and electric power (37.98) sources.

South Carolina established the Governor's Climate, Energy, and Commerce Advisory Committee (CECAC) in February 2007 to research and evaluate, for presentation to the governor, recommended policy options to mitigate the effects of global warming. Appointed by the governor, the CECAC comprises a diverse group of stakeholders who bring broad perspective and expertise to the topic of climate change in South Carolina. Members represent the following sectors: energy, manufacturing, agriculture, forestry, tourism and recreation, heath care, nongovernmental organizations, academia, and state and local government. The advisory committee has held regular meetings considering options including renewable energy and recycling options.

Over the past three years, the University of South Carolina has purchased 70 flex-fuel vehicles that run on E-85, a mixture that is 85 percent ethanol and 15 percent gasoline. Much of South Carolina's renewable energy potential comes from biomass—organic matter such as plant fibers and animal waste that can be converted into electricity and fuel.

A number of programs have been undertaken for the conservation and improvement of South Carolina's soils, forests, and wildlife resources. Conservation efforts, including the protection of critical land areas, are carried on by a number of federal and state agencies as well as private organizations.

Watershed management considers economic interests as well as the protection of natural resources. Management plans are used to guide watershed management for water quality, recreation, wildlife management, and agricultural and forestry practices, balanced against community economic development.

The principal soil conservation effort is directed toward covering the badly eroded lands with pasture grasses or trees to prevent further soil removal. Reforestation, supervised cutting and replanting, and fire protection are practiced to provide adequate timber supplies in the present and future.

SEE ALSO: Climate Change, Effects; Floods; Tourism.

BIBLIOGRAPHY. Department of Natural Resources, "Watershed Planning," http://www.dnr.sc.gov; Natalie Goldstein, *Earth Almanac* (Onyx, 2002); National Wildlife Federation, "Global Warming and South Carolina" (May 7, 2007).

Lyn Michaud
Independent Scholar

South Dakota

SOUTH DAKOTA'S GEOGRAPHICAL location near the center of the North American continent provides it with a high degree of sensitivity to climatic change, as any variations from mean weather conditions will be readily apparent. The state is situated in a region that experiences all of the conditions of the continental climate classification, which usually means pronounced seasonality with long, cold winters; hot summers; midlatitude cyclonic storms; and variable precipitation. A border, running north-south, between the semihumid and semiarid precipitation regimes of North America, essentially divides the state (generally corresponding to the 100th meridian) and responds to changes, oscillating from east to west, in relation to overall global weather patterns. Thus, if there were significant changes in the global atmospheric system, such as warming resulting from anthropogenic or natural factors, climatic conditions would manifest themselves across the state.

This brief essay presents a summary of historical climate research and recent climate modeling in an effort to illuminate possible consequences of climate change on South Dakota's people and landscape. The state's climate parameters are summarized as follows: mean normal precipitation varies from 25 in. (635 mm.) in the southeastern part of the state to less than 14 in. (350 mm.) in the northwestern sector; temperatures range from summer highs in the lower 100 degrees F (40 degrees C) to minus 40 degrees F/C in winter, with an average annual temperature of less than 39 degrees F (4 degrees C). Record high and low temperatures are 120 degrees F (49 degrees C) and –58 degrees F (–50 degrees C.) Strong surface winds patterns, principally blowing from the north and northwest during the colder part of the year, also persist across the northern Great Plains, of which the state is a part. The region also experiences severe weather episodes such as tornadoes, hail storms, and blizzards in their respective seasons. The state's Black Hills subregion represents an anomaly to the general southeast-northwest precipitation and north-south temperature gradients, which vary temporally, with average wetter and warmer conditions prevailing at the higher elevations that support coniferous forest vegetation cover.

South Dakota's state climatologist presented research in 2001 revealing that the state had experienced a "definite climate shift" during the 1990s, recording an average increase of two or three in. (51–76 mm.) of precipitation annually over the previous three decades. The agricultural effect of increased moisture allowed advancement of the production of corn and soybeans westward toward the Missouri River—a feature that essentially divides the state into eastern and western sectors. Semihumid climatic characteristics associated with western Minnesota shifted as much as 200 mi. (320 km.) to the west, according to the report.

Contrary-wise, in the current decade, the semiarid extreme southwestern part of the state has been in the grips of a seven-year-long drought with severe consequences in terms of agricultural losses and forest and range wildfires, plus tendencies toward desertification conditions. The drought-affected area is the northeast extension of a multistate region in the western United States that has been experiencing prolonged, abnormally dry weather patterns.

In 2005, research was published indicating that future global warming may adversely affect wetlands within the eastern half of the state (Johnson et al., 2005) and, by implication, the general area as well. Eastern South Dakota's geologic landscape was produced mainly by Wisconsin age glacial deposits. The retreating ice left a terrain of ground moraines, outwash plains, and end moraines and a swath of tens of thousands of prairie potholes and larger kettles (shallow lake basins), which can be very productive waterfowl breeding areas when moisture conditions are favorable. The area is a part of the Prairie Pothole Region (PPR), which covers parts of five other states and two Canadian provinces. With large-scale Anglo-American settlement and development, beginning in the second half of the 19th century, much of the area has been transformed through drainage projects and intensive cultivation into a productive agricultural region, although a considerable acreage of natural wetlands remains. Western, or West River, South Dakota was unglaciated during the Pleistocene epoch, and the landscape is heavily influenced by the presence of the semiarid climatic regime. Contemporary agriculture is focused predominantly on cattle production, with some row crops, across vast areas of grasslands and rough terrain cut by a network of valleys occupied by mostly intermittent stream courses.

Johnson's research, based on climate modeling methodology, suggested that the most productive habitat for ducks in the PPR would shift under a drier climate from eastern South Dakota to areas to the east and much farther north—locations currently less productive or where most wetlands have been drained. The conclusion that was drawn indicates negative effects on both waterfowl production and commercial agriculture. This work concurs with other climate change studies that predict adverse consequences with a warming global climate in various areas of the world, especially in heavily populated or agriculturally productive regions of the Northern Hemisphere.

In sum, the climate region in which South Dakota is located has experienced a variable regime for millennia because of its geographical continentality. Temporally, the eastern half of the state was wetter than most of the early 20th century. However, if the current global climate change forecasts become reality, the outlook for South Dakota's environment for the next several decades, especially the eastern portion, becomes problematic, with anticipation of less precipitation and its associated effects on land and life. Some degree of mitigation may be possible in the shorter run, according to experts, if initiated in time, but the issue may be ultimately beyond any human control or even influence.

SEE ALSO: Climate Change, Effects; Global Warming.

BIBLIOGRAPHY. A. Bender, http://climate.sdstate.edu/archives/feb/2001/ovr_fe01.htm; Intergovernmental Panel on Climate Change, http://ipcc-wg1.ucar.edu/wg1/Report/AR4WG1_Print_TS.pdf; W. Carter Johnson, et al., "Vulnerability of Northern Prairie Wetlands to Climate Change," *BioScience* (v.55/10, 2005); John G. Lockwood, *World Climatology: An Environmental Approach* (St. Martin's Press, 1974); National Assessment Synthesis Team, *Climate Change Impacts on the United States: The Potential Consequences of Climate Variability and Change* (Cambridge University Press, 2001); Research Directorate of the National Defense University, *Climate Change to the Year 2000: A Survey of Expert Opinion* (National Defense University, 1978).

DONALD J. BERG
SOUTH DAKOTA STATE UNIVERSITY

Southern Ocean

THE GLOBAL OCEAN influences the Earth's climate by storing and transporting vast amounts of heat, moisture, and carbon dioxide. Huge quantities of carbon are cycled annually among the biosphere (forests, grasslands, and marine plankton), the atmosphere, and the oceans. The oceans are the largest active reservoir of carbon, containing 50 times more carbon than the atmosphere. Of the 6 to 7 billion tons of carbon currently released into the atmosphere by human activities, approximately 3 billion tons remain in the atmosphere, 1 to 3 billion are absorbed by the oceans, and up to 2 billion appear to be absorbed by the terrestrial biosphere.

Oceanographers commonly refer to the oceanic region that surrounds the continent of Antarctica as the Southern Ocean. The northern boundary of the Southern Ocean is not well defined, but it coincides approximately with a broad zone of transition between the warm, saline surface waters of the subtropical regime and colder, fresher subantarctic waters, called the Subtropical Front, which occur between 40 degrees S and 45 degrees S. Using this definition, the surface area encompassed by the Southern Ocean represents approximately 29.7 million sq. mi. (77 million sq. km.), or 22 percent of the global surface ocean. Its unique geography makes it a key player in global climate.

The Southern Ocean is the only ocean that encircles the globe unimpeded by a land mass. It is home to the largest of the world's ocean currents: the Antarctic Circumpolar Current (ACC). The ACC carries between 135 and 145 million cu. m. of water per second from west to east along a 12,427 mile- (20,000-km.) path around Antarctica, thus transporting 150 times more water around the globe than the total flow of all the world's rivers. By connecting the Atlantic, Pacific, and Indian Oceans, the ACC redistributes heat around the Earth and so exerts a powerful influence on global climate.

Near the Antarctic continent, the Southern Ocean is a source of cold, dense water that is an essential driving force in the large-scale circulation of the world's oceans. The cooling of the ocean and the formation of sea ice during winter increases the density of the water, which sinks from the sea surface into the deep sea. This cold, high-salinity water includes Antarctic Bottom Water and Antarctic Intermediate Water. Antarctic Bottom Water originates on the continental shelf close to Antarctica, spills off the continental shelf, and travels slowly northward, hugging the seafloor beneath other water masses, moving as far as the North Atlantic and North Pacific. Antarctic Intermediate Water is less saline and forms farther north, when cold surface waters sink beneath warmer sub-Antarctic waters at the Antarctic Convergence at about 55 degrees S. Together, these motions form a complex, three-dimensional pattern of ocean currents that extends around the globe, known as the thermohaline circulation, or "great ocean conveyor." The thermohaline circulation has a critical influence on climate by transporting heat efficiently around the globe and by controlling how much dissolved inorganic carbon is stored in the ocean.

At the sea surface, seawater exchanges gases such as oxygen and carbon dioxide with the atmosphere at the same time that it is being cooled. As a result, sinking water efficiently transfers changes in temperature, fresh water, and dissolved gases into the deep ocean 2.5 to 3 mi. (4 to 5 km.) beneath the sea surface; in terms of carbon sequestration, this is called the solubility pump. Biological processes also play a role in the surface layer, where photosynthesis by single-celled marine phytoplankton can sequester carbon dioxide in the surface water and, through the process of sedimentation, transfer this organic carbon to deeper waters—the so-called biological pump.

The Southern Ocean is distinguished as a region of high levels of dissolved nutrients, but with modest rates of annual net primary production, so that the biological pump appears to be operating well below its maximum capacity. An interesting idea of recent years is that it may be possible to sequester much more atmospheric carbon if iron, an essential micronutrient, is added to the ocean to encourage the growth of marine phytoplankton, and thus stimulate the biological pump. The overall effect would be to lower the concentration of dissolved carbon dioxide in surface waters, allowing more atmospheric carbon dioxide to dissolve into the sea.

Understanding the global circulation and conditions under which surface waters penetrate into the deep ocean is critical for scientists estimating the timing and magnitude of climate change. At this time, the Southern Ocean is considered to be a net sink for atmospheric carbon dioxide; however, the magnitude of this sink has a high uncertainty, with mean annual estimates ranging between 0.5 and 2.5 billion tons. The degree of interannular variability in the Southern Ocean carbon sink, and its possible future response to climate change, is still poorly understood. However, climate model projections indicate that the Southern Ocean overturning circulation will slow down as the Earth warms. A decrease in the rate of overturning circulation will result in a decrease in the rate of carbon dioxide absorbed by the Southern Ocean, which represents a positive feedback and tends to increase the rate of climate change.

The presence of sea ice in the Southern Ocean is another factor that contributes to the Southern Ocean's important role in climate. Sea ice formation during the winter months is the largest single seasonal phenomenon on Earth, with approximately 7.7 million sq. mi. (20 million square km.) of ice formed annually, effectively doubling the size of Antarctica. This has a profound effect on both regional and global climate processes. Because of its high albedo, sea ice reflects the sun's heat back into space, intensifying the cold. However, it can also act as a blanket, insulating against heat loss from the ocean to the atmosphere. Its yearly formation injects salt into the upper ocean, making the water denser and causing it to sink downward as part of the deep circulation. As ocean temperatures increase in response to the global warming, the amount of sea ice is expected to decrease; this has already been observed in the Arctic Ocean. The resulting decrease in the planetary albedo would act as a positive feedback, increasing the amount of energy from the sun absorbed by the Earth and tending to further increase the rate of climate change.

SEE ALSO: Albedo; Antarctic Circumpolar Current; Arctic Ocean; Carbon Cycle; Climate Models; Phytoplankton; Sea Ice; Thermohaline Circulation.

BIBLIOGRAPHY. Sayed Z. El-Sayed, *Southern Ocean Ecology: The Biomass Perspective* (Cambridge University Press, 1994); Kate A. Furlong and Kate A. Conley, *Southern Ocean* (ABDO Publishing Company, 2003); George A. Knox, *Biology of the Southern Ocean*, 2nd ed. (CRC, 2006); A.T. Ramsey, J.G. Baldauf, eds., *Reassessment of the Southern Ocean Biochronology* (Geological Society Publishing House, 1999).

AL GABRIC
GRIFFITH UNIVERSITY

Southern Oscillation

FIRST DESCRIBED EXTENSIVELY by British meteorologist Sir Gilbert T. Walker in the 1920s, the Southern Oscillation refers to the periodic exchange of mass across the equatorial Pacific that is recorded in sea level pressure fluctuations between the eastern and western Pacific. Under normal conditions in the tropical Pacific, surface high (low) pressure prevails in the eastern (western) Pacific, with the easterly trade winds dominating surface wind and ocean flow.

This pressure pattern, also known as the Walker circulation, tends to support rising air motions and convectional precipitation near eastern Australia, as well as sinking air motions and dry conditions near coastal northern Peru. Every two to seven years, this generalized atmospheric surface pressure pattern weakens as equatorial Pacific air pressure rises in the west and lowers in the east. This shift in the pressure field considerably weakens the trade winds and promotes the eastward movement of warm surface water across the tropical Pacific. The associated abnormal warming in the eastern Pacific is known as El Niño. Because the reversals in pressure and associated ocean temperature fluctuations are often simultaneous, this coupled climate variability between the tropical Pacific Ocean and atmosphere is often collectively referred to as the El Niño/Southern Oscillation (ENSO).

MEASURING SOUTHERN OSCILLATION

The mode and relative strength of the Southern Oscillation during a given time period is determined using one of several indices that signifies changes in the Walker circulation. A relatively simplistic and common method employed to gauge this change is the Southern Oscillation Index (SOI), which measures the monthly or seasonal sea level pressure differences between two stations, one located in the central Pacific at Tahiti and the other in the western Pacific at Darwin, Australia. Negative SOI values result from abnormally low pressure occurring in Tahiti and high pressure occurring at Darwin, which tends to indicate an El Niño episode; positive SOI values indicate the cold phase of ENSO, or La Niña. The sea level pressures at these two stations thus are negatively correlated and are associated with significant, yet contrasting shifts in regional temperature and precipitation patterns. Some of the most severe Australian summer droughts and heat waves (e.g., in 1983) have been associated with a strongly positive SOI.

ENSO events often affect the temperature and precipitation regimes in tropical regions. The magnitude of these effects differs with the intensity of individual ENSO events. Climatic anomalies associated with ENSO's warm phase in other tropical regions include dry summers and autumns for northern South America, Central America, and southeastern Africa (including Madagascar), as well as less rainfall during the Indian monsoon. Drier-than-normal conditions negatively affect crops—a particularly serious concern in developing regions. Such atmospheric conditions are also conducive to the threat of wildfires. Wetter conditions pervade the Chilean coast, as well as parts of east-central Africa.

EFFECTS OF ENSO

Despite being primarily a tropically located phenomenon, ENSO also has extensive effects on extratropical global precipitation and temperature variability. This is achieved, in part, by shifts in storm

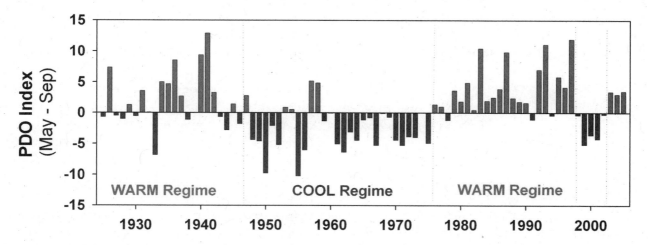

The term PDO Index refers to Pacific Decadal Oscillation. The horizontal scale is marked in units of decades from 1925 to 2006, in the months of May through September. The vertical lines show positive (warm) years and negative (cool) years.

tracks. Changes in large-scale atmospheric circulation include deviations from the normal jet stream paths and persistent pressure systems, which in turn steer storms in new directions. During the warm ENSO phase in winter, a deepened Aleutian Low moves southeast of its average position. This is coupled with a strong subtropical jet stream and a weak polar jet stream over eastern Canada, setting up the circulation pattern that redirects storms into the southern United States. The winter cold ENSO phase is characterized by a blocking high forming in the Gulf of Alaska and a split polar jet. The main branch flows from Alaska and northern Canada south toward the western and northern United States; the jet's southern branch moves from the Pacific Ocean toward the Pacific Northwest.

Winter tends to bring the strongest North American precipitation and temperature responses to ENSO, though effects are noted in other seasons. Warm ENSO events typically result in wetter-than-normal conditions for much of the southern United States, with California often experiencing flooding as a result of the position of the stronger-than-normal subtropical jet stream, which directs storms into the region. Conversely, dry conditions tend to occur in the midwestern United States. Warm ENSO events also bring mild, warm winters into western Canada and Alaska, as well to Canada's Maritime Provinces. Cold ENSO events bring considerably drier, warmer winters to the American Southeast and below-average temperatures to western and central Canada and the northern tier of the United States.

Global climate models have been used in recent decades to predict how changes in our current climate will affect the frequency, strength, and position of cycles of interannular variability such as the Southern Oscillation. Although the predictability of the Southern Oscillation has been subject to differences between observed and predicted timescales, researchers have shown that the current climate models do tend to place warm ENSO events within the observed two- to seven-year timescale. Ensemble forecasts, in which average forecasts are generated by running models with slight variations in initial conditions, produce good results, with multimodel ensembles generally outperforming single-model ensembles. These models serve to enhance our understanding of atmosphere-ocean interactions, such as ENSO, for improving long-range weather and climate forecasts.

SEE ALSO: El Niño and La Niña; Walker Circulation.

BIBLIOGRAPHY. C. Donald Ahrens, *Meteorology Today* (Thomson Brooks/Cole, 2007); S. George Philander, ed., *El Niño, La Niña, and the Southern Oscillation* (Academic Press, 2006); Cynthia Rosenzweig and Daniel Hillel, *Climate Variability and the Global Harvest: Impacts of El Niño and Other Oscillations on Agro-Ecosystems* (Oxford University Press, U.S., 2007); John M. Wallace and Peter V. Hobbs, *Atmospheric Science: An Introductory Survey*

(Academic Press, 2006); Warren M. Washington and Claire L. Parkinson, *Introduction to Three-Dimensional Climate Modeling* (University Science Books, 2005).

JILL S. M. COLEMAN
PETRA A. ZIMMERMANN
BALL STATE UNIVERSITY

Spain

THE COUNTRY OF Spain lies on the Iberian Peninsula on the Mediterranean and Atlantic and is separated from Europe by the Pyrenees Mountains. The central plateau is drained by three rivers. Special geographic features include two sets of archipelagos, with the Balearic Islands in the Mediterranean and the Canary Islands in the Atlantic off the coast of Africa. Spain has a variety of ecosystems: snowcapped peaks in the Pyrenees, green meadows in Galicia, orange groves in Valencia, and desert in Almeria. The Partido Socialista Obrero Espanol government set aside more than 400 protected areas covering 15,444 sq. mi. (40,000 sq. km.) and protecting a broad range of ecosystems including mountains, wetlands, islands, wood and forest, and volcanic landscapes.

With extremes in temperature and low rainfall, water is a valuable resource. Droughts in the 1950s, 1960s, and 1990s were the driving force behind water policy. The Tajo-Segura water diversion system can divert 600 cu. m. of water per year from Tajo region in central Spain to the Valencia and Murcia regions. In 2000, a National Water Plan (Plan Hidrologico Nacional) intended to double the amount of water diverted from Rio Ebro basin was heavily protested, and in 2004 the plan was cancelled. Although climate models vary on what the temperature increase will be for Spain, from a conservative 3 degrees –4 degrees C to a pessimistic 6 degrees–7 degrees C by the end of the century, Spain is already experiencing the effects of global warming, with glacier melt in the Pyrenees (continued warmer winters would reduce snow cover and tourism to the area for winter sports); Europe's first hurricane (Hurricane Vince), which made landfall on the southwestern coast of Spain in October 2005; and an 11 percent decrease in average rainfall in 2005–06, which followed an extreme drought the year before.

Precipitation is expected to decrease, though not in all areas. Spain's northeast could see an increase in autumn and winter rainfall, and there may be a decrease in rainfall in the arid southwest. Spring and summer precipitation would be expected to decrease, except in the Canary Islands. Higher storm activity under climate change over the adjacent Atlantic is likely to lead to an increase in the intensity of winds over some parts of the country by the end of the century. Maximum wind speeds could increase by 2–4 percent in northwestern Spain by the end of the century, whereas in Galicia, the number of days with high winds could increase by up to 10 percent. Northern Spain could expect increased yield for most crops, and southern Spain could expect decreased yields.

In May 2002, Spain ratified the Kyoto Protocol, an international and legally binding agreement to reduce greenhouse gas emissions worldwide, entered into force on February 16, 2005, with Spain's entry into force on the same date.

In 2005, Spain became the second-highest wind capacity electric generator by adding 1,764 megawatts of wind power for a total of 10,027 megawatts, surpassing their 2010 goal.

In July 2007, the Spanish government approved numerous legislative measures related to environmental preservation and mitigating climate change. Spain has made fighting climate change a priority in the government's working agenda. By approving the Spanish Strategy for Climate Change and Clean Energy, Horizon 2007–2012–2020, issued by the Minister of environment, the government reinforced a commitment to the Kyoto Protocol. In addition to a plan for reducing government energy consumption and greenhouse gas emissions from government buildings and for the increasing use of blended biofuel in official vehicles, increasing energy efficiency with equipment and appliances as well as heating and cooling is a priority. Vehicle registration taxes will be directly related to emissions and responsibility will be put on the consumer (low-emission vehicles will be exempt from the tax, and higher-emission vehicles will be taxed at the highest rate).

SEE ALSO: Climate Change, Effects; Kyoto Protocol.

BIBLIOGRAPHY. "Climate Change in Spain", http://www.iberianature.com; "The Government Approves More Than 80

Urgent Measures to Tackle Climate Change," http://www.la-moncloa.es/idiomas/en-gb/actualidadhome/200707-consejo.htm; *Worldwatch Institute, Vital Signs 2006–2007* (W. W. Norton, 2006).

LYN MICHAUD
INDEPENDENT SCHOLAR

Species Extinction

CHANGING CLIMATES INCREASE the uncertainties of life for all organisms. A long-term warming trend would alter the distribution of life on the planet as colder habitats shrink and warmer ones expand. Some species would become more common, and others would become rarer. We cannot predict with any precision which species will become extinct—or when. Plants and animals that are highly adapted to already extreme (hot, cold, or dry) climates are most likely to be the first and most drastically affected.

We consider a species to be extinct once all known individuals of that type have died. Many interacting factors affect the survival of individual organisms, and therefore the persistence of their species. In general, extinction results when a species' requirements and abilities no longer match the resources and hazards in its environment. For animals, these factors include food, water, and shelter from predators and weather extremes. For plants, they include water and nutrients and the action of herbivores and pollinators. Sometimes a factor is critically important, like rainfall in a desert. It determines whether enough individuals will survive that a species can persist. If that "limiting factor" changes in some way, survival rates may rise or fall. If they fall far enough, extinction results. Climate is a major limiting factor for life on Earth. When it changes, life on Earth also changes. A continuing trend of global warming, cooling, or drying will lead to extinctions that might otherwise not occur as soon. We still know little about the precise climate limits or thresholds of most species. Because it is also difficult to predict precisely what the climatic conditions will be like in any given place at any given time in the future, it is even harder to predict which species will become extinct as a direct result of climate change, and when it will happen. In addition, because species interact and rely on each other in many ways, climate change produces many sometimes indirect or complex effects among them. This adds further layers of uncertainty to predictions about extinctions. For the most part, we can make only very general predictions about climate change and extinctions. This uncertainty leads many climate scientists and ecologists to conclude that humans would be wise to avoid or resist contributing to the uncertain risks of climate change whenever it lies within our power to do so.

CLIMATE, BIOGEOGRAPHY, AND EXTINCTION

Climate change is complex. Tracking the local effects of regional or global change requires a great deal of data. Much of this information is now collected via remote sensing devices like radar and satellite-mounted cameras. So much data is collected that they can only be compiled into a usable form with very high-speed computers. However, those technologies are very recent. Scientists began collecting accurate and extensive climate measurements in the 18th century, recording data by hand. Naturalists like Alexander von Humboldt and H.C. Watson first correlated climates and species distributions in the early 19th century. Thus began the study of biogeography.

Long before there were biogeographers, it was evident that different kinds of plants and animals occupied different kinds of places. Biogeography added mathematical precision to the folk knowledge that temperatures were lower at higher elevations and higher latitudes and that mountain ranges received more precipitation to windward than to leeward and were warmer on the sunnier slopes facing the equator. More climate and biogeographical data became available at the same time that cartography and species inventories were improving. All were necessary for accurately describing what lived where and for predicting what sorts of species would live in various places. Repeated inventories, measurements, and mapping were needed to show whether and how biogeography was changing.

Among the first patterns understood by biogeographers was that average temperatures on the earth's surface changed with latitude. Temperatures tended to be low in polar regions and higher near the equator. At the same time, they saw that temperatures near sea level tended to be warmer than temperatures at higher elevations. They found that even near the equator, the tops of very tall mountains (such as the Andes) were as cold as the poles and discovered that the plants and animals of polar and alpine locations were very similar. As observations accumulated, biogeographers were able to begin mapping the ranges of different species. Climate measurements helped biologists determine the limits of heat, cold, and precipitation that various species could tolerate.

It was long debated whether species actually could become extinct. It was not until large, easily observed birds like the dodo and the great auk could no longer be found and the fossilized remains of large, otherwise unknown animals were being discovered, that the fact of extinction was established. Not until the third quarter of the 19th century did it become clear that extinctions might regularly follow as the unintended consequences of intensive human activities. Climate changes traceable to human activities were hardly recognized for over another century, during which our population doubled twice over. During the roughly same period, our major technologies changed from being mostly animal, wind, and water powered to being combustion powered, using wood and fossil fuels.

CLIMATE DYNAMICS AND LIFE ON EARTH

Conditions on the Earth's surface and in its atmosphere have undergone many changes over time. Some of these changes were quite drastic and had proportionally drastic effects on living things. We can say with some confidence that the Earth's climate has sometimes been much warmer than it is today, and we know that at other times it has been much colder. This much can be inferred partially from recorded history but more reliably from fossils and other geological evidence. Scientists have proposed many plausible explanations for these climate changes, but since the events cannot really be modeled in detail or rep-

With global warming, newly ice-free Arctic lands would potentially become available for mining, oil and gas development, and manufacturing activities. Greater human density could affect survival rates of large mammals such as the Caribou, above.

licated for study, they can only agree about general effects, rather than local specifics.

Paleontologists and others who study the evolutionary history of life on earth have concluded that most of the species that ever inhabited the planet are now extinct. They have also estimated, as a sort of rule of thumb, that any given species, on average, persists for about a million years. Estimates of the total number of species that have existed on the planet range from tens of millions to hundreds of millions. Some of these species are known from the fossil record to have persisted much longer than a million years, and others for much shorter periods. Estimates and averages are only as good as the actual data and methods used to make them. Even if the data we have to work with are reliable, the fossil record is far from complete, and different methods of analysis continue to yield different estimates.

Because our planet is changeable, or dynamic, extinction seems to be normal and inevitable for species, much as death is inevitable for individuals. At the same time, however, evolution also generates new species from some of the old ones, as the average characteristics of a population change and "adapt" to emerging conditions. Overall, there is still life on Earth because the rate at which species evolve has exceeded the rate at which they become extinct.

No one is credibly predicting that all life on earth would end, and all species would go extinct, as a result of human-caused global warming, but many scientists are concerned that any continuing trend in climate change would increase the rate of extinctions, changing life as we know it and perhaps making life more difficult or less interesting for humans as a result. Many people want to preserve life as we know it and to prevent extinctions of other species caused by human activities. Global warming is one of many environmental changes human activities may bring about. The combined effects of human activities, along with those of geological and even cosmic events, are complex. Among the extinctions that occur during the foreseeable future, we will probably be able to blame very few solely, or even mostly, on human-caused climate change. However, if apparent trends continue, climate change will probably contribute in some way—large or small—to almost any extinction that occurs.

ECOLOGY AND CLIMATE CHANGE

It is difficult to distinguish extinctions caused mostly or mainly by climate change from those caused by other factors such as directly converting habitats to human uses. Conservation biologists have long considered habitat destruction to be the most likely cause of extinctions. Habitat has been described many ways, but it generally means an environment in which enough individuals of a particular species can survive and reproduce to keep their population from decreasing to zero. In other words, each species has a habitat, and each needs a persisting habitat to continue as a species.

Some habitats are more complicated than others, but all habitats can be thought of as having two general kinds of components. Biotic components are living things: all the other organisms that somehow affect the life of a plant or animal. Abiotic components are factors like terrain, minerals, water, sunlight, and temperature. Climate change can directly affect some of the abiotic components of a habitat. When particular places become warmer or cooler, or drier or wetter, the ability of any particular species to persist in that place also changes.

Some abiotic habitat components, such as temperature and humidity, will vary daily or seasonally. Organisms have to be able to tolerate the extremes of night and day, summer and winter, and wet and dry seasons. When the climate of a place changes to the point that one of an organism's tolerances is exceeded, a habitat literally ceases to exist.

Because of the shape of the Earth, less sunlight reaches the poles than the tropics. Habitats are limited by the climatic effects of latitude. If we could look down from space at the North Pole and see all the way to the equator, but still recognize all the land plants and animals, we would see that similar kinds of organisms are roughly arranged in a series of bands or zones centered on the pole, like a target. Working out from the center, each zone is slightly warmer than the one immediately inside it. When the average global temperature falls, the polar center of the target expands and the hottest equatorial zones shrink or even disappear. This happened during the Ice Ages, when glaciers covered much of the Northern Hemisphere. Animals and plants had to change, migrate, or become extinct. When the average global temperature rises, the polar center shrinks, and each climate zone moves toward it. The icy center may even disappear, and the next zone takes its place. Meanwhile, entirely new, hotter zones may appear at the equatorial edge.

Because the atmosphere is less dense in the mountains than at sea level, all habitats are also limited by the climatic effects of elevation. Higher elevations are colder. Seen from directly above, a tall mountain has bands of similar plants and animals, just as the whole planet does. A general trend in climate change means that these bands move down and up the mountain just as the latitude bands move toward and away from the poles.

CLIMATE-RELATED CAUSES OF EXTINCTIONS

Extinctions can occur gradually or suddenly. Large numbers of extinctions have sometimes occurred during relatively short periods of time. These "mass extinctions" resulted when a catastrophic event such as an asteroid impact suddenly made large areas of the earth's surface, or its oceans, uninhabitable. The effects produced by such catastrophes probably included sudden, drastic climate changes, but not enough evidence has been found to say with certainty how great these changes were or exactly how long they lasted.

Changing climates affect the survival prospects of individual organisms. As a result, changing climates affect the survival and reproduction rates of whole populations and species. Populations may rise or fall as climates change. Some increases or decreases will be dramatic and obvious. Others will be almost unnoticeable to us. Almost all such population changes will result from combinations of many small changes, rather than a few catastrophic ones.

As average global temperatures increase (or decrease), populations will migrate to follow shifts in local conditions. Some organisms can do this quickly and easily. Many animals already migrate to follow seasonal changes in food and water supplies. There are rare exceptions, but many individuals, such as rooted plants, cannot move at all. Their populations can migrate only as seeds are dispersed and new individuals germinate and survive in newly suitable locations. Meanwhile, the old individuals, trapped in increasingly unsuitable locations, gradually die out. When populations cannot shift to new locations quickly enough, species may become extinct. Extinctions also follow when no new locations become available or when potentially suitable locations exist but cannot be reached in time. We can easily imagine scenarios that include the extinction of plant species unable to disperse to new habitats. Because the phenomenon is so complex, scientists have been reluctant or unable to publish firm, reliable estimates of the numbers of species that could become extinct as a result of climate change, or to predict when such extinctions will occur.

Climate change is most likely to directly produce species extinctions in already extreme, barely survivable environments. These are the very cold, hot, wet, dry, or chemically unusual places in which only relatively few types of highly specialized organisms can exist. Where such extreme conditions are climate induced, even small temperature changes can be highly significant. Organisms in extreme environments are likely to be living near the limits of physiological possibility. When extreme environments become more extreme, some organisms die. When extreme environments become too extreme, nothing can survive in them—but that is only part of the story.

When extreme environments become more moderate, more species can move into them, leading to increased competition for living space and other resources. They may become "too moderate" for specialists that have lost, or perhaps never evolved, ways to compete or escape in highly diverse and densely populated environments. In other words, given a trend of global warming, hot, dry environments may become hotter and drier, crossing some survival threshold of survival for desert-adapted species. Individuals of those species will have to emigrate or die. However, some hot, dry environments might become wetter, or cooler, or both, even as average global temperatures are rising. This will allow species adapted to the new, more moderate conditions to immigrate and to compete with, prey on, or infect the existing populations in unprecedented ways.

DIRECT EFFECTS OF CLIMATE CHANGE ON SPECIES EXTINCTIONS

The direct effects of climate change are most likely to affect organisms of polar regions, mountaintops, and equatorial areas. Under a general trend of increasing temperatures, the very coldest climates—the Arctic and alpine tundras—could disappear, and along with them would likely disappear at least some of the species adapted to tolerate them. As the warmer habitats move toward the poles and up the mountains, their species will follow. Those that cannot migrate or disperse as fast as their potential habitats are shifting will either have to evolve new climatic tolerances or die out.

When abiotic factors change, some habitats may contract, even to the point of disappearing altogether. Others may expand, and new ones may appear. Overall, they can be imagined as flowing slowly across the landscape, expanding in some directions while retreating from others, sometimes forming and seemingly evaporating like puddles. The most obvious response for organisms that can move is to follow the changes in habitat or to find and occupy the new habitat. As long as enough individuals of a species can somehow keep up with these movements, their species may persist.

An expected direct effect of climate change with the potential for causing species extinction is a rise in sea level caused by the melting of polar and alpine glaciers. Large areas of low-lying coastal lands would be inundated by rising sea levels. In effect, some areas of terrestrial habitats would be converted to areas of aquatic habitats. Some low-lying oceanic islands would disappear, and along with them any land plants and animals that might be found nowhere else. Whether as a result of habitat inundation or other effects, species with very restricted ranges, called endemics, are likely to be more significantly affected by climate change than those with larger ranges.

Every species has different abiotic tolerances, so the edges of their potential habitats, based on moisture or temperature, rarely correspond exactly. Instead, these habitat edges usually overlap. Not only do they overlap, but climate change will affect each one differently, so different species habitats will move, grow, or contract at different rates. As we have seen, not all species are equally mobile. This means that two species may experience different direct, abiotic affects in the same place. These differences create the possibility of many indirect effects of climate change as species interact in new ways and places.

INDIRECT EFFECTS OF PERSISTENT CLIMATE CHANGE

Most effects on species resulting from any continuing climate change trend will be indirect. All animals rely on other species as sources of food. Many plants rely on insects and other animals to pollinate them or disperse their seeds. When different species come to depend on each other in predictable ways, their relationship is called a symbiosis. Symbioses range from pure exploitation, where only one species benefits, to cooperation or mutualism, where both species benefit. In many cases, such as those of internal parasites or intestinal bacteria, one organism actually becomes the entire habitat of another. Far more often, individuals of different species have no obvious interactions at all but do influence each other in much more subtle ways, such as by preying on another species' competitors, or its predators, or its pollinators, or by spreading its disease organisms.

The possibilities for changing species interactions are seemingly endless, but we can describe only a few examples here. Individuals of predatory species might find themselves able to range farther north, or higher into the mountains, where they will encounter potential prey species that have never seen them before. These prey animals may lack defensive or escape behaviors, and their populations may be significantly reduced. This does not mean that tropical cats like jaguars will be decimating caribou herds. Most of the land animals in the world are insects, as are most of the predators. We are hardly aware of predation at the insect level, but it is cumulatively enormous and enormously influential.

Most insects are unable to regulate their body temperatures except by seeking shelter. Flying insects have to meet minimum temperature requirements before their muscles work efficiently, allowing them to lift off. However, flying insects are highly mobile. Once aloft, they are often carried great distances by winds, sometimes to places where they normally cannot survive. However, if climates warm and their habitats move and expand, insects are likely to arrive in any newly suitable locations pretty quickly. If these pioneering insects are herbivores, they may find plants that have evolved no defenses against them. This could hasten the demise of individual plants and reduce populations that were physiologically capable of tolerating the direct effects of warmer temperatures.

Many plants rely on insects for pollination. Some plant populations could be reduced if predatory insects begin to survive in areas formerly unavailable to them because of climate factors and begin preying on the local pollinators. If pollinators become too scarce, plant reproduction could be reduced to levels that cannot maintain a population. Both the plants and the pollinators could be affected.

Polar ice caps and alpine glaciers are composed of accumulating snow. If they melt, the resulting water is fresh, not salty. There is not enough fresh water in these sources to significantly dilute the world's oceans and change the fundamental chemistry of seawater. However, fresh water is less dense than salt water, and until the two mix, fresh water entering the oceans actually floats as a surface layer. The addition of massive amounts of cold, fresh water to the Arctic, North Atlantic, and north Pacific oceans, and to the south polar regions of the Atlantic, Pacific, and Indian oceans, would affect the way currents flow and nutrients circulate in these areas. This would affect the types and distribution of plankton, and thus all the many levels of oceanic food webs in those areas. Reduced plankton production would ultimately mean less prey for aquatic predators of polar seas such as polar bears, penguins, and some toothed whales, seals, and sea lions. Added to the direct effects of reduced pack ice, such changes could lead to the extinction of animals highly specialized for life under cold polar conditions that would no longer exist.

Hot deserts have fewer rivers, lakes, and ponds, but many of them have springs and small water courses that support endemic aquatic species including fishes, amphibians, reptiles, and many invertebrates and microorganisms. If these hot deserts become even hotter or drier because of climate change, these "oases" could literally dry up. In the process, numerous rare aquatic species that cannot move to other habitats (even if they existed) would become extinct in the process.

Aquatic species endemic to small tributary streams in any watershed face various new conditions when a region becomes drier or wetter. Neither trend is automatically beneficial. If it becomes drier, the smaller tributaries become ephemeral or intermittent, forcing fully aquatic species downstream into larger, more permanent waters, where they may encounter more (and larger) predators, at least for a time. If the region becomes wetter, the small tributaries will become larger, and the larger predators may move upstream. In high, steep terrain, the physical characteristics of the newest small tributaries may make them unsuitable for colonization.

In wet tropical areas, the effect of climate change will most easily be seen if it results in changes to the flow of atmospheric moisture to the region and, as a result, to the seasonality and overall amount of precipitation. At its simplest, a rainforest with less rain will gradually become another kind of forest, having fewer species requiring high moisture or seasonal inundation by floodwaters and more that tolerate drier conditions. As in all cases, if suitable habitat disappears or appears only at an unreachable distance, some species could become extinct. The complexity and diversity of tropical forests is such that not only some tree species, but also their dependent animals (and, in turn, their own dependent animals), could become extinct in the process. Our knowledge of the flora and fauna of these regions is insufficient to support any precise estimate of the number of species present, much less the number that could be affected by any particular degree of climate alteration.

CLIMATE-RELATED EXTINCTIONS INVOLVING OTHER HUMAN ACTIVITIES

In anticipation of a continued warming trend in polar regions, various countries are already positioning themselves to take advantage of ice-free Arctic seas and increasingly temperate high latitudes. Others are bracing for possible desertification in tropical grass and scrublands. Areas likely to experience intensified human use may have higher likelihood of species extinctions.

Increasing commercial ship traffic in Arctic waters would produce the same sorts of side effects that shipping has elsewhere. Leaks and spills of fuel and cargo oil would affect the biota of littoral zones. Ballast water exchange would further redistribute aquatic species, leading to new predation and competition among aquatic species without prior experience of each other.

Newly ice-free Arctic lands would potentially become available for mining, oil and gas development, and allied manufacturing activities. This will require an influx of workers and equipment, along with creation of the physical, economic, and cultural infrastructure needed to support them. Each activity entails a direct conversion of some existing habitat to human use. This could fragment the habitats of migratory birds such as snow geese and affect survival rates for large mammals such as caribou and musk oxen.

More ship traffic, mining, and oil exploration would encourage more permanent human settlements to service, and be serviced by, these industries, leading to a greater likelihood of chemical pollution and of sewage and solid waste management issues. Human population centers would encourage the establishment of human commensals and inquilines, such as dogs and cats, rats and mice, cockroaches and houseflies. Each potentially adds a new challenge to the persistence of Arctic species.

Under a continued warming trend, farmers in Europe, Asia, and North America would experience the same northward and upward shift in habitat bands affecting uncultivated plants and wild animals. For example, grain production will likely be possible farther north, in the Canadian "Prairie Provinces." This would require "sodbusting" of existing grasslands or logging of forests to convert them into farms, reducing or eliminating their habitat value to most wildlife. All the world's major crops—corn, soybeans, wheat, rice, and cotton, along with most every other valued plant—would become economically viable in new areas while becoming impractical in others where they have been traditionally grown. The net effects on agricultural production are hard to estimate, as are the potential effects on other species.

SEE ALSO: Agriculture; Animals; Antarctic Circumpolar Current; Arctic Ocean; Atlantic Ocean; Biology; Botany; Cetaceans; Climate Zones; Conservation; Desertification; Deserts; Ecosystems; Geography; Glaciers, Retreating; History of Climatology; Ice Ages; Indian Ocean; Land Use; Marine Mammals; Modeling of Ocean Circulation; Modelling of Paleoclimates; Oceanic Changes; Pacific Ocean; Penguins; Phytoplankton; Plants; Polar Bears; Rainfall Patterns; Sea Level, Rising; Upwelling, Coastal; Upwelling, Equatorial.

BIBLIOGRAPHY. Miguel B. Araújo, et al., "Reducing Uncertainty in Projections of Extinction Risk from Climate Change," *Global Ecology & Biogeography* (v.14/6, November 2005); Thomas J. Crowley and Gerald R. North, "Abrupt Climate Change and Extinction Events in Earth History," *Science* (v.240/4855, May 20, 1988); Malte C. Ebach and Raymond S. Tangney, eds., *Biogeography in a Changing World* (CRC Press, 2007); Susan Joy Hassol, *Impacts of a Warming Arctic: Arctic Climate Impact Assessment* (Cambridge University Press, 2004); Robert L. Peters and Thomas E. Lovejoy, *Global Warming and Biological Diversity* (Yale University Press, 1992); Charles L. Redman, ed., *The Archaeology of Global Change: The Impact of Humans on Their Environment* (Smithsonian Books, 2004); Chris D. Thomas, et al., "Extinction Risk From Climate Change," *Nature* (v. 427, January 2004); Richard L. Wyman, ed. *Global Climate Change and Life on Earth* (Routledge, Chapman and Hall, 1991).

MATTHEW K. CHEW
ARIZONA STATE UNIVERSITY

Sri Lanka

THE DEMOCRATIC SOCIALIST Republic of Sri Lanka, formerly known as Ceylon, is a small, pear-shaped island of 25,332 sq. mi. (65,610 sq. km.) lying off the southeastern coast of India. Sri Lanka has an extraordinary diversity of wildlife and vegetation because of its location near the equator and its remarkable range of terrain and climate. Although only a minute contributor to greenhouse gas emissions, the island is vulnerable to the effects of climate change.

Sri Lanka's carbon emissions per capita are 161st out of 211 countries measured worldwide but more than doubled in the 10 years from 1990, largely in response to the country's increasing population. Like much of the rest of South Asia, Sri Lanka relies heavily on carbon-neutral biomass such as collected wood and animal waste for its domestic energy needs, particularly in rural areas. Biomass accounted for 80 percent of total residential energy consumption in 2005 and is expected to remain as high as 70 percent through 2020.

Oil consumption more than doubled between 1990 and 2005 in response to a growing demand for transport fuels. Sri Lanka imports all of its daily crude oil consumption of 87,000 barrels, and in recent years, it has further increased oil imports to avoid overreliance on hydroelectricity for industrial power. Hydropower currently provides the majority of Sri Lanka's electricity, making the country vulnerable to changing rainfall patterns. In an effort to diversify, the Sri Lankan government is developing fossil-fuel-fired power plants.

Sri Lanka's rich biodiversity includes an unusually large number of endemic species living in cloud forests, grasslands, and wetlands, as well as freshwater, coastal, and marine ecosystems. The island is both ecologically and economically vulnerable to climate change. Its famous tea plantations remain an important source of economic activity, and the island's rich cultural heritage, together with its tropical forests, beaches, and wildlife, make it a world-famous tourist destination. Climatic conditions and rich biodiversity are therefore key to maintaining Sri Lanka's economy. However, logging and population pressures continue to lead to deforestation and habitat loss. Large tracts of forest have been cut down for fuel wood or for timber export and have been replaced by rice, coconut, rubber, and coffee farms. Many species are in danger of extinction, including cheetahs, leopards, several species of monkeys, and wild elephants. Sri Lanka's coral reefs, already damaged by bleaching and the 2005 Asian tsunami, are being destroyed by human refuse and sewage and by dynamite fishing. Climate change–induced ecological stress will compound these eco-economic concerns.

Overall, the importance of environmentally sustainable behavior is underappreciated in the general population, as concerns such as poverty and the ongoing conflict between Tamil separatists and the Sinhalese government are dominant. Even so, the government of Sri Lanka has taken action to conserve wildlife. Over 13 percent of the land is protected, and the Sinharaja Forest Reserve, which protects the largest remaining stand of primary rainforest on the island, was declared a World Heritage Site in 1988. Indeed, Sri Lanka has a long history of conservation, being the first country in the world to establish a wildlife reserve.

SEE ALSO: Deforestation; Developing Countries; India; Rainfall Patterns; Tourism.

BIBLIOGRAPHY. Energy Information Administration, Country Analysis Briefs, www.eia.doe.gov/emeu/cabs/bhutan.html; Review of Progress Made Since the United Nations Conference on Environment and Development 1992, www.un.org/esa/earthsummit/lanka-cp.htm.

HARRIET ENNIS
UNIVERSITY OF YORK

Stanford University

STANFORD UNIVERSITY IS a private university located approximately 37 mi. (59 km.) southeast of San Francisco and approximately 20 mi. (32 km.) northwest of San Jose in Stanford, California. Stanford is situated adjacent to the city of Palo Alto, near Silicon Valley. The university enrolls approximately 6,700 undergraduates and 8,000 graduate students. The university has approximately 1,700 faculty members. Forty percent of the faculty is affiliated with the medical school, and a third serves in the School of Humanities and Sciences. Graduate studies in the Department of Geological and Environmental Sciences involve academic course work and independent research. Students are prepared for careers as professional scientists in research or for the application of the earth sciences to mineral, energy, and water resources. Programs lead to M.S., Engineer, and Ph.D. degrees. Course programs in the areas of faculty interest are tailored to the student's needs and interests with the aid of his or her research adviser. Students are encouraged to include in their program courses that are offered in other departments in the School of Earth Sciences, as well as courses offered in other departments in the university.

Launched in December 2002, the Global Climate and Energy Project (GCEP) at Stanford University seeks to find new solutions to one of the grand challenges of this century: supplying energy to meet the changing needs of a growing world population in a way that protects the environment. The primary goal of the project is to conduct fundamental research on technologies that will permit the development of global energy systems with significantly lower greenhouse gas emissions. With the support and participation of four international companies (ExxonMobil, General Electric, Schlumberger, and Toyota), the GCEP is a unique collaboration of the world's energy experts from research institutions and private industry. The project's sponsors will invest a total of $225 million over a decade or more as GCEP explores energy technologies that are efficient, environmentally benign, and cost effective.

The Stanford Climate Change Campaign, a project of Students for a Sustainable Stanford, in conjunction with People for the American Way Foundation, Global Exchange, and other student groups, has made a public commitment to lead the fight against global warming and to reduce its own carbon footprint.

The Center for Environmental Science and Policy (CESP) began as a specialized research center within the Freeman Spogli Institute for International Studies (FSI) in September 1998. It evolved as an outgrowth of the more informal Global Environmental Forum, which had existed within FSI for nearly a decade. CESP is also an affiliated center of the Woods Institute for the Environment. Formed in 2004, Woods is Stanford's principal initiative for assessing environmental science, technology, and policy on local, national, and global scales.

The center has grown considerably under the experienced leadership of a series of codirectors—Walter Falcon, Donald Kennedy, Pamela Matson, and Stephen Schneider. Leading scholars from the natural and social sciences, these individuals reflect CESP's integrative approach to research, which balances the analyses of environmental problems from both scientific and policy perspectives.

The CESP plays a crucial role in mobilizing a multidisciplinary network of scholars, students, policymakers, and leaders in understanding and helping to solve international environmental problems through science and policy research. The work of the center engages scholars from disciplines as varied as the biological and geological sciences, civil engineering, economics, and law to develop new methods for environmental assessment, negotiation, remediation, and protection.

Workshops, policy briefings, and publications link CESP with other public policy and scholarly institutions within and outside Stanford. The center houses the Program on Energy and Sustainable Development, a multiyear, interdisciplinary program that draws on the fields of engineering, political science, law, and economics to investigate how the production and consumption of energy affect sustainable development. CESP does not award degrees, but it is heavily engaged in graduate and undergraduate education. The center holds a close affiliation with Stanford's Interdisciplinary Graduate Program on Environment and Resources and also directs the Goldman Honors Program, an interschool honors program in environmental science, technology, and policy.

SEE ALSO: California; Impacts of Global Warming; Sustainability.

BIBLIOGRAPHY. Global Climate and Energy Project, www.gcep.stanford.edu; Stanford Solar Center, http://solar-center.stanford.edu; Students for a Sustainable Stanford, www.sustainability.stanford.edu.

Fernando Herrera
University of California, San Diego

Stockholm Environment Institute (SEI)

THE STOCKHOLM ENVIRONMENT Institute (SEI) is a nonprofit, independent research institute specializing in sustainable development and environmental issues. SEI headquarters are located in Stockholm, Sweden, with centers in the United States and the United Kingdom. They work at the local, national, regional, and global policy levels. Their mission is to support decision making and induce change toward sustainable development around the world by providing integrative knowledge that bridges science and policy in the field of environment and development. The institute was established in 1989 following an initiative by the Swedish government to develop an international environment/development research organization. Each center has its own personality and foci of interests, and each operates with significant autonomy while participating in the five SEI research programs.

SEI's Climate and Energy Program addresses these challenges in collaboration with a global network of partners, enabling them to perform work in locally defined interests and resources. For the last two decades, projects in Africa, Asia, Europe, and Latin America have spurred innovative energy strategies that support the goals of social equity, environmental sustainability, and efficient economic development. Emissions of air pollutants to the atmosphere have had, and continue to have, significant effects on human health and well-being, crops and food security, ecosystems and biodiversity, and materials and cultural monuments.

The Atmospheric Environment (AE) Program within SEI contributes to the goal of reducing the local, regional, and global effects of the emission of pollutants to the atmosphere.

SEI's U.S. center conducts a diverse program focusing on the social, technological, and institutional requirements for a transition to sustainability. Funding is received from the United Nations, the World Bank, and numerous foundations and national governments such as the United States, Sweden, Denmark, Germany, the Netherlands, and the United Kingdom.

In addition to providing policy-relevant analyses, the center works to build capacity in developing countries for integrated sustainability planning through training and collaboration on projects. Its decision support tools are widely used: LEAP for energy planning and climate change mitigation, WEAP for water resources planning, and PoleStar for evaluating sustainable development strategies.

The center is organized into three programs: the Climate and Energy Program, which conducts energy system analyses, examines environmental consequences of energy use such as global warming, and develops policies for a transition to efficient and renewable energy technology; the Water Resources Program, which brings an integrated perspective to freshwater assessment—one that seeks sustainable water solutions by balancing the needs for basic water services, development, and the environment; and the Sustainable Development Studies Program, which takes a holistic perspective in assessing sustainability at the global, regional, and national levels.

SEE ALSO: Climate Change, Effects; United Nations.

BIBLIOGRAPHY. Encyclopedia of Earth, www.eoearth.org; Stockholm Environment Institute, www.sei.se; Stockholm Environment Institute–US Center, www.seib.org.

FERNANDO HERRERA
UNIVERSITY OF CALIFORNIA, SAN DIEGO

Stommel, Henry (1920–92)

HENRY STOMMEL IS an American oceanographer and meteorologist whose theories on general circulation patterns in the Atlantic Ocean made him the creator of the modern field of dynamical oceanography. Stommel carried out a series of research studies and first suggested that the Earth's rotation is responsible for the Gulf Stream along the coast of North America. He theorized that its northward thrust must be balanced by a stream of cold water moving in the opposite direction beneath it. Carl Wunsch has described Stommel as "a transitional figure, being probably the last of the creative physical oceanographers with no advanced degree, uncomfortable with the way the science had changed, and deeply nostalgic for his early scientific days." Stommel has been praised for being both a creative theorist and an acute observer who was willing to spend months at sea.

Stommel was born in Wilmington, Delaware, on September 27, 1920, into a family of extremely mixed background. His ancestors came from such different places as the Rhine Valley, Poland, Ireland, the Netherlands, England, and France, and they also had a trace of Micmac Indian. Henry's father, Walter, was a chemist born in northern Germany and trained in Darmstadt and Paris. During World War I, Walter Stommel

Henry Stommel theorized the circulation patterns of oceans and cold water moving in the opposite direction beneath it.

emigrated to Wilmington, where he was employed by Dupont Chemical. While in the United States, he married Marian Melson. Their son Henry was born shortly after the marriage. Although the reason is not completely clear, perhaps because of anti-German sentiment following World War I, the family then moved to Sweden. Henry's mother, however, soon left Sweden with Henry and returned to Wilmington. Because of his mother's decision not to see her husband again, Henry and his sister Anne grew up in a single-parent family. When Henry was 5 years old, his mother moved with the two children to Brooklyn, New York, to live with her parents and other relatives. Marian supported the entire household thanks to her job as a fund-raiser and public relations officer at a hospital. Henry and his grandfather, Levin Franklin Melson, developed a meaningful relationship in a household dominated by women.

Stommel attended New York City's public schools. He spent one year at Townsend Harris High School but finished high school at Freeport, Long Island, because his family had moved there. Thanks to his receiving a full scholarship, he was able to enroll at Yale University, from where he graduated in 1942. He remained at Yale for two years following graduation, teaching analytic geometry and celestial navigation in the Navy's V-12 program. He also spent six months at the Yale Divinity School, but his lifelong ambivalence toward religion made the ministry an unsuitable vocation for him. In 1944, renowned astrophysicist Lyman Spitzer suggested that Stommel apply for work at the Woods Hole Oceanographic Institution in Woods Hole, Massachusetts—an organization that was fast becoming a decisive part of the U.S. war effort. Stommel was recruited to work in acoustics and antisubmarine warfare but disliked his assignment and tried to be employed in other areas.

In 1948, Stommel wrote "The Westward Intensification of Wind-Driven Ocean Currents," a paper that is unanimously regarded as constituting the starting point of dynamical oceanography. In it, he explained the Gulf Stream deductively by fluid dynamics. In particular, he discovered the mechanism (the latitudinal change of the Coriolis force on the rotating Earth) that produced the westward intensification of oceanic currents. Stommel proposed a global circulation model similar to a conveyor belt: surface water sinks in the far north to supply the deep, south-flowing current,

and water rises in the Antarctic region to contribute a northward flow along the eastern coasts of North and South America. His important book *The Gulf Stream* was probably the first true dynamical discussion of the ocean circulation. He put the Gulf Stream in the wider context of the general circulation and paved the way for the development of the so-called thermocline theories. Stommel also concluded that changes in density caused by cooling and evaporation at the sea surface can be responsible for deep flows in the ocean. He was thus responsible for establishing the basic factors that helped to establish theories about global circulation. His thermocline theories stressed the role of oceans and sea currents in the definition of global climate and thus anticipated debates on global warming.

In December 1950, Stommel married Elizabeth Brown. The couple had three children. Although Stommel liked working at the Woods Hole Oceanographic Institution, he did not get on well with his director Paul Fye. Therefore, he accepted an invitation to become a professor at Harvard University in 1959, lured by the prestige of the institution. He spent four unhappy years there, where his democratic ideals clashed with a rigid sense of hierarchies. After Harvard, Stommel went to work at the Department of Meteorology at the Massachusetts Institute of Technology (MIT). There he worked with the most famous meteorologists of the day, such as Jule Charney, Norman Phillips, Edward Lorenz, and Victor Starr. Stommel worked enthusiastically with these scientists to improve theories of general circulation. He also worked on other important topics such as the classification of estuaries and the effect of volcanoes on climate.

Stommel worked at MIT for 16 years as a professor of physical oceanography. He returned to the Woods Hole Oceanographic Institution when Fye retired and continued to work there until his death on January 17, 1992. Stommel established several stations for the study of ocean currents, including the PANULIRUS station (begun in 1954) in Bermuda. He was elected to the National Academy of Sciences in 1962 and received the National Medal of Science in 1989.

SEE ALSO: Oceanography; Thermocline.

BIBLIOGRAPHY. N.G. Hogg and R.X. Huang, eds., *Collected Works of Henry Stommel* (American Meteorological Society, 1996); Henry Stommel, *The Gulf Stream: A Physical* *and Dynamical Description* (University of California Press, 1958); Carl Wunsch, "Henry Stommel," *Biographical Memoirs*, www.nap.edu/readingroom/books/biomems/hstommel.html#FOOT1.

LUCA PRONO
UNIVERSITY OF NOTTINGHAM

Stratopause

THE STRATOPAUSE IS one of the layers into which the atmosphere is divided. It is the buffer region of the atmosphere that lies between the stratosphere and the mesosphere, from a height of about 31 to 34 mi. (50 to 55 km.) above the Earth's soil. The atmospheric pressure is about 1/1000th of the pressure at sea level. In the stratopause, the temperature reaches a peak because of the heating generated by the absorption of ultraviolet radiation by ozone molecules in the stratospheric ozone layer. In this region, the catalytic cycles, which are less efficient at colder temperatures because of reduced O density, produce a significant ozone increase (~15 percent). Because of the considerable ozone presence in the stratopause, the understanding of this region is considered crucial to understanding the changes in climate and in the composition of the ozone layer. Above the stratopause, the temperature starts again to decrease with height as a result of the reduced solar heating of ozone.

The depletion of the ozone layer resulting from the emission of halogen atoms and the photodissociation of chlorofluorocarbon compounds is of particular concern to scientists, as the layer prevents the most harmful ultraviolet-B wavelengths from passing through the Earth's atmosphere. Near the stratopause, the ozone reduction is slightly smaller in the drier stratosphere because of the stronger temperature dependence of the drier atmosphere.

Studies of the temperature in the stratopause have also been important to assess the validity of global circulation models. For example, a study published in 2002 by the University of Illinois at Urbana-Champaign, and the High Altitude Observatory of the National Center for Atmospheric Research in Boulder, Colorado, showed that wintertime warming caused by sinking air masses was not as strong as the

models had assumed. The study employed lidar laser measurements and balloon observations made at the Amundsen-Scott South Pole Station from December 1999 to October 2001. These measurements and observations were then used to calculate the monthly mean winter temperature profiles from the surface to about 63 mi. (110 km.). The measured temperatures during midwinter in both the stratopause and mesopause regions were 20–30 degrees Kelvin colder than current model predictions. These differences were caused by weaker than expected compressional heating associated with subsidence over the polar cap. The study showed that the greatest difference occurred in the month of July, when the measured stratopause temperature was about 0 degrees F (minus 18 degrees C) compared with the about 40 degrees F (4.4 degrees C) predicted by the models.

SEE ALSO: Atmospheric Composition; Atmospheric Vertical Structure.

BIBLIOGRAPHY. John M. Wallace and Peter V. Hobbs, *Atmospheric Science, Volume 92, Second Edition: An Introductory Survey (International Geophysics)* (Academic Press, 2006); Pan Weilin, Chester S. Gardner, and Raymond G. Roble, "The Temperature Structure of the Winter Atmosphere at South Pole," *Geophysical Research Letters* (v.29/16, 2002).

LUCA PRONO
UNIVERSITY OF NOTTINGHAM

Stratosphere

THE STRATOSPHERE IS a layer in the atmosphere that extends between about 9 to 31 mi. (15 and 50 km.) in altitude. It is characterized by a vertical temperature structure that is nearly isothermal (no temperature change with altitude) in the lowermost stratosphere and a pronounced inversion (increase of temperature with altitude) above. The stratosphere owes its name to the strong stratification, which is a consequence of this thermal structure.

The stratosphere plays an important role in the climate system. It contains the ozone layer, which shields the Earth's surface from harmful ultraviolet radiation and is responsible for the temperature of the stratosphere. Radiative processes in the infrared part of the electromagnetic spectrum also play an important role. Because, in the stratosphere, chemistry, dynamics, and radiative processes operate under very different conditions than in the troposphere, the stratosphere is susceptible to climatic forcings in a different way than the troposphere. As a consequence, stratospheric processes play an important role for climate variability and change.

The stratosphere was discovered independently by Teisserence de Bort and by Richard Assmann around 1900. It has been explored by balloon-borne observations since around the 1930s and by satellite observations since the 1970s.

The lower boundary of the stratosphere is the tropopause, the altitude of which varies with latitude (higher over the tropics than over the poles), season, and on a day-to-day scale related to weather systems. The upper boundary of the stratosphere is the stratopause. Below and above the stratosphere are the troposphere and mesosphere, respectively.

Because of its thermal structure (strong static stability), the circulation of the stratosphere is quasi-horizontal. The most important features of the zonal circulation are the vortices in the polar regions and the Quasi-Biennial Oscillation (QBO) in the tropics.

The polar vortices form over both poles during the corresponding winter season and vertically extend through the entire stratosphere. The Arctic vortex is subject to strong variability on short timescales (during so-called sudden stratospheric warmings, the vortex can break down completely within days) and on interannular timescales. The Antarctic vortex varies much less. The QBO is an oscillation of the zonal wind in the equatorial stratosphere, with changes from westerlies to easterlies and back to westerlies within approximately 28 months.

Compared with the zonal flow, the meridional circulation and associated vertical motion in the stratosphere are very weak but are, nevertheless, important. The meridional circulation is caused by planetary waves originating from the troposphere, which break and dissipate in the stratosphere and thereby deposit momentum, decelerating the zonal flow and inducing a meridional flow component. The meridional flow is compensated for by vertical motion in the tropics and in the polar areas, forming a single meridional circulation cell, which in the context of trace gas transport is often referred to as Brewer-Dobson circulation. Air enters the stratosphere in tropical areas. On passing the tropopause, the air loses almost all of its moisture; hence, the stratosphere is very dry and mostly cloud free. In the stratosphere, the air slowly moves upward and poleward toward the winter hemisphere (the summer hemisphere is dynamically quiet). In the subpolar and polar region, the air descends and can eventually enter the troposphere in conjunction with midlatitude weather systems. The stratospheric meridional circulation has a turnover time of one to three years.

Chemically, the stratosphere is characterized by a layer of ozone (O_3) formed from atomic (O) and molecular (O_2) oxygen in the presence of ultraviolet radiation. Ozone can be destroyed by catalytic processes that involve radicals of chlorine, bromine, nitrogen oxides, or hydrogen oxides. The most important source of chlorine radicals are manmade chlorofluorocarbons (CFCs), which have caused a reduction of the ozone layer since the 1970s, as well as, since the 1980s, the Antarctic ozone hole (a substantial reduction of the total stratospheric ozone amount over Antarctica). The Montreal Protocol of 1987 and its amendments have led to a strong reduction in CFC emissions worldwide. However, because of the long

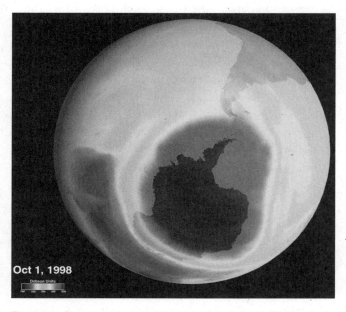

Oct 1, 1998
Dobson Units

The large ozone opening over the poles (dark area). Stratospheric ozone blocks harmful ultraviolet radiation produced by the sun.

lifetime of CFCs, a full recovery of the ozone layer is only expected for the mid-21st century.

The anthropogenic greenhouse effect, as well as ozone depletion, causes a cooling of the stratosphere, whereas volcanic eruptions lead to warming. The stratosphere plays an important role for climate at the Earth's surface. Perturbations of the stratospheric circulation can propagate downward and affect weather at the ground. This provides a pathway through which some of the forcings can affect climate. For instance, it is now believed that part of the climate effect of strong volcanic eruptions operates via the change in stratospheric circulation induced by the heating effect of volcanic aerosols. Similarly, changes in solar irradiance could affect climate via stratospheric ozone chemistry and their subsequent effects on circulation.

SEE ALSO: Atmospheric Vertical Structure; Climatic Data: Atmospheric Observations; Mesosphere; Stratopause; Tropopause; Waves, Rossby.

BIBLIOGRAPHY. Mark P. Baldwin and Timothy J. Dunkerton, "Stratospheric Harbingers of Anomalous Weather Regimes," *Science* (v. 244, 2001); Guy P. Brasseur and Susan Solomon, *Aeronomy of the Middle Atmosphere* (Springer, 2005); Karin G. Labitzke and Harry van Loon, *The Stratosphere. Phenomena, History, and Relevance* (Springer, 1999); World Meteorological Organization, Scientific Assessment of Ozone Depletion: 2006 (World Meteorological Organization, 2007).

STEFAN BRÖNNIMANN
ETH ZURICH, SWTIZERLAND

Sudan

SUDAN IS A sub-Saharan African (SSA) country that has been combating global warming for many decades. Similar to other SSA countries, Sudan can expect an increase in temperature of 0.4–0.9 degrees F (0.2–0.5 degrees C) per decade, which may lead to the climate becoming drier. In Sudan, rainfall is predicted to decrease by 25 percent over 30 years, leading to desertification in a country that is already 50 percent desert.

Since the 1930s, the desert in Sudan expanded between 31 and 124 mi. (50 and 200 km.), which has led to severe water shortages. It has been predicted that 350 to 600 million people will face water shortages in Sudan by the middle of the 21st century. Water shortages, especially in western Sudan, have led many herders from northern Sudan to migrate to southern areas onto farmers' lands in search of water. This conflict over water, which is a precious resource in Sudan, has led to violence and to 2.4 million people being displaced in 2003. Increases in livestock, which has degraded the land, and deforestation (loss of 12 percent of forests over the last 15 years) have also contributed to continuing desertification. The drying climate has threatened the food security of 1.7 million people, as 90 percent of the people in Sudan depend exclusively on rain-fed agriculture.

Positive adaptive measures have, however, been taking place since 1992 in drought-prone western areas of Sudan, such as the Bara Province. In this area, rangeland rehabilitation measures are being taken through community participation to manage land and water resources and promote agroforestry and sand dune fixation. This has helped prevent overexploitation of resources and restore the productivity of rangelands.

Although Sudan is facing many challenges because of climate change, it hardly contributes to worldwide greenhouse gas emissions. For instance, in 1998, Sudan's total carbon dioxide emission was 3,597 thousand metric tons (tmt) compared with 515,001 tmt for SSA and 24,215,376 tmt for the world. Non–carbon dioxide emissions for Sudan in 1995 were 132 tmt compared with 5,345 tmt for SSA and 141,875 tmt for the world. Even though Sudan hardly contributes to global warming, the government of Sudan intends to adapt to the consequences of the problem. It has prepared the National Adaptation Plans for Action, which highlight several policy measures to adapt to climate change. These include increasing irrigation and low-water crops, water management and conservation technology, sustainable forest resource consumption, and reduction of livestock. The government of Sudan has also ratified the United Nations Framework Convention on Climate Change.

SEE ALSO: Carbon Dioxide, Climate Change, Effects; Drought.

BIBLIOGRAPHY. Commission for Africa, "Our Common Future," www.commissionforafrica.org (cited July 2007); International Institute for Sustainable Development, "Sustainable Drylands Management: A Strategy for Securing Water Resources and Adapting to Climate Change," www.iisd.org (cited 2003); V.A. Orindi and L.A. Murray, "Adapting to Climate Change in East Africa: A Strategic Approach," IIED *Gatekeeper Series 117* (2005); Michael Renner, "Desertification as a Source of Conflict in Darfur," www.scidev.org (cited July 2007); United Nations Environment Program, "Synthesis Report Sudan Post-Conflict Environment Assessment," www.unep.org (cited July 2007); World Resources Institute, "Climate and Atmosphere—Sudan," earthtrends.wri.org (cited 2007).

MOUSHUMI CHAUDHURY
UNIVERSITY OF SUSSEX

Sulphur Dioxide

SULPHUR DIOXIDE (SO$_2$) is an important component of the atmosphere, present as the result of both natural and human activity. Although it is a primary pollutant in its own right, causing respiratory irritation and damage to plants, it is the secondary pollutants produced from SO$_2$ that are particularly important in connection with global climate change. Sulphur dioxide is notorious as the cause of acid rain, but it is also a precursor to the formation of clouds. Hence, its release to the atmosphere is a major contributor to global dimming, a process that is thought to offset some of the effects of global warming.

There are, therefore, important implications of SO$_2$ release for the global climate change agenda. The reduction in SO$_2$ pollution in recent decades, stimulated by health concerns and by the effects of acid rain, is removing an unexpected and previously unidentified protection against increasing global temperatures. This illustrates the complexity of climate science that compounds the social and political responses to the threat of climate change.

Once in the atmosphere, SO$_2$ is rapidly oxidized, ultimately producing sulphuric acid. Although this transformation is well known in the formation of acid rain, it also has broader climatological significance. The liquid sulphuric acid forms as an aerosol (tiny droplets suspended in the air), and this sulphuric acid aerosol attracts water vapor, which dissolves in the acid. In this way, the gas-to-liquid conversion of SO$_2$ to sulphuric acid brings about the nucleation of clouds: sulphuric acid aerosol is a cloud condensation nucleus (CCN).

Clouds play important roles in the atmosphere and in the climate, principally acting to transport water (and energy) between regions and to affect the Earth's radiation balance. Clouds have a very strong tendency to reflect sunlight (they have a high albedo) and also absorb energy from the sun, so that the amount of cloud present in the atmosphere affects the amount of sunlight reaching the surface: more clouds result in a dimmer planet. This dimming effect of clouds is well documented.

SULPHUR DIOXIDE AND CLIMATE CHANGE

Furthermore, the influence of SO$_2$, and subsequent aerosol formation, has been observed directly during volcanic eruptions. For example, the 1991 eruption of Mount Pinatubo in the Philippines released an estimated 20 megatons (20,000,000 tons) of SO$_2$ into the atmosphere. The force of the explosion injected a large fraction of this material, along with dust particles, directly into the stratosphere, from which removal via rainout is very slow. The aerosol and clouds that formed as a consequence of this lasted for many years, with measurable effects on global temperatures. In the year following the eruption, the global average temperature reduced by 0.9 degrees F (0.5 degrees C), and even in 1993 the temperature was depressed by as much as 0.45 degrees F (0.25 degrees C).

In the lower atmosphere, the rate of SO$_2$ gas-to-liquid conversion is increased in the presence of other materials, notably particles such as soot, and this has an important effect on cloud condensation. The typically hydrophobic—water-repelling—surfaces of soot particles catalyze the chemical reactions that convert SO$_2$ to sulphuric acid, so that the soot ends up coated with a water-loving, hydrophilic layer. The simultaneous emission of SO$_2$ and soot (e.g., from burning coal or diesel fuel) therefore increases the concentration of cloud condensation nuclei in the air, affecting both the amount and nature of cloud formation.

Because there are many more cloud condensation nuclei under these conditions than in the clean atmosphere, clouds form with smaller, more numerous droplets. More numerous particles means that the clouds reflect more light, and smaller droplets take longer to form raindrops. Hence, the clouds formed on sulphate/soot aerosol CCN are longer lived and have a higher albedo than ordinary clouds. In this way, SO_2 emissions in combination with soot increase the amount of incoming sunlight reflected away from the Earth, effectively dimming the planet's surface.

Sulphur dioxide pollution–related global dimming is thought to explain the slight global cooling trend in the period from 1950 to the late 1970s. With the recent, legislation-driven decrease in emissions of SO_2 and soot particles from industry and transport in industrialized nations, there has been a steady rise in the amount of sunlight reaching the earth. It is suspected that reducing this form of pollution is removing an effect that has been offsetting the full force of anthropogenic global climate change.

SEE ALSO: Aerosols; Cloud Feedback; Volcanism.

BIBLIOGRAPHY. Robert Henson, *The Rough Guide to Climate Change* (Rough Guides, 2006); Richard P. Wayne, *Chemistry of Atmospheres* (Oxford University Press, 2000).

CHRISTOPHER J. ENNIS
UNIVERSITY OF TEESSIDE

Sulphur Hexafluoride

SULPHUR HEXAFLUORIDE (SF_6) is the most powerful of all the greenhouse gases recognized by the Kyoto Protocol and evaluated by the Intergovernmental Panel on Climate Change. Although its concentration in the atmosphere is low, the combination of a high global warming potential and a very long lifetime make emissions of SF_6 a considerable concern. It is primarily used as an electrical insulator in the high-voltage distribution network, and major industrial users are beginning to restrict the use and emission of SF_6.

The Kyoto Protocol requires developed nations to cut their emission of six greenhouse gases. The Intergovernmental Panel on Climate Change periodically assesses these gases and estimates a global warming potential (GWP) for each. The gasses, with their respective GWPs, are as follows: CO_2 (1), CH_4 (21), N_2O (310), PFC (9200), HFC (11,700), and SF_6 (23,900). In this list, HFC and PFC are groups of chemicals, and the value quoted is for the member of the group with the highest GWP. Despite the low atmospheric concentration of SF_6 (5.6 parts per trillion), its extremely high global warming potential and long lifetime (probably in excess of 1,000 years) mean that present emissions will have an effect on climate for a long time to come.

As SF_6 gas is denser (heavier) and more electrically insulating than either dry air or dry nitrogen, it is an ideal electrical insulating material. The gas is used extensively in electrical applications, and its principal use is in the electrical generation and high-voltage distribution industry. There are two specific advantages of SF_6 in these applications. First, its highly insulating character means that less space is needed between high-voltage components, so that equipment can be made significantly smaller than is possible when air or nitrogen are use as insulators. Second, gas-insulated switch gear using SF_6 rather than air demands a controlled environment, and the equipment is consequently more robust with regard to environmental pollutants and weathering than would be the case with simpler air-insulated equipment. In addition to these advantages, the gas is unreactive, nontoxic and nonflammable. In the United States, the electric power distribution industry works on a voluntary basis with the SF_6 Emissions Reduction Partnership for Electric Power Systems to identify and implement technologies for reducing SF_6 emissions.

Another large-scale use of SF_6 is in magnesium metal manufacturing and casting. Magnesium metal is extremely reactive in air, particularly when hot or molten. The high density and low chemical reactivity of SF_6 make it a suitable choice as a protective gas layer preventing contact of the molten, highly reactive metal with oxygen and water in the air. A voluntary SF_6 Emission Reduction Partnership for the Magnesium Industry exists in the United States in association with the U.S. Environmental Protection Agency, which, together with the International Magnesium Association, is committed to eliminating SF_6 emissions from the industry by 2011.

SF_6 is also used in certain medical applications, including eye surgery and ultrasound scanning. Once again, it is the gas's high density and low toxicity that are used. In eye surgery, the gas is commonly used to form a plug to seal the retina during surgery. Its high density means that the gas stays in place and does not enter the blood at an appreciable rate. The density of the gas also makes it an excellent contrast agent in medical ultrasound scanning.

Similar to the perfluorocarbons, SF_6 is also used in the semiconductor industry, and there is concern about the growth of this industry leading to uncontrolled increases in the amount of SF_6 released to the atmosphere.

Paradoxically, because of its high chemical stability, low toxicity, and low natural abundance, SF_6 has been extensively used by atmospheric scientists as a tracer gas to understand the movements and mixing of air. The gas has, for instance, been injected into the exhaust plumes from power stations in an attempt to understand the origins of acid rain. In the United Kingdom, SF_6 tracer experiments have demonstrated that power stations are capable of delivering acid rain pollution to Scandinavia. For similar reasons, SF_6 is used to trace the movements of air within ventilation and air conditioning system tests. Recently, the gas was released on the London Underground in an attempt to understand the way toxic gases would spread throughout the system in the event of a terrorist attack.

SEE ALSO: Global Warming; Intergovernmental Panel on Climate Change; Kyoto Mechanisms; Kyoto Protocol; Perfluorocarbons.

BIBLIOGRAPHY. John Houghton, *Global Warming—The Complete Briefing* (Cambridge University Press, 2004); Intergovernmental Panel on Climate Change, Fourth Assessment Report, Working Group 1 Report, The Physical Science Basis, http://www.ipcc.ch/; Richard P. Wayne, *Chemistry of Atmospheres* (Oxford University Press, 2000).

CHRISTOPHER J. ENNIS
UNIVERSITY OF TEESSIDE

Sunlight

SUNLIGHT IS THE electromagnetic radiation given off by the sun. It is passed through the atmosphere to the Earth, where the solar radiation is reflected as daylight. Sunshine results when the solar radiation is not

blocked. Sunlight is the primary source of energy to the Earth. It provides infrared, visible, and ultraviolet (UV) electromagnetic radiation with different wavelengths. Small sections of the wavelengths that are visible to the human eye are reflected as rainbow colors. Sunlight may be recorded using a sunshine recorder. Electromagnetic waves are waves that are capable of transporting energy through the vacuum of outer space and that exist with an enormous continuous range of frequencies known as the electromagnetic spectrum. The spectrum is divided into smaller spectra on the basis of interactions of electromagnetic waves with matter.

The longer-wavelength, lower-frequency regions are located on the far left of the spectrum, and the shorter-wavelength, higher-frequency regions are on the far right. Two very narrow regions within the spectrum are the visible light region and the X-ray region. The visible light region is a very narrow band of wavelengths located to the right of the infrared region and to the left of the UV region. Though electromagnetic

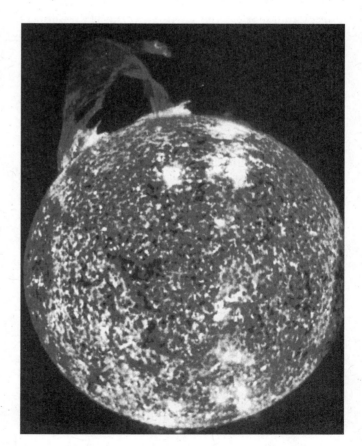

Sunlight is Earth's primary source of energy, providing infrared, visible, and ultraviolet electromagnetic radiation.

waves exist in a vast range of wavelengths, human eyes are only sensitive to the visible light spectrum. The visible portion of the solar spectrum lies between 400 and 700 nm. and separates the UV region of shorter wavelengths from the infrared region of longer wavelengths. A combination of waves results in white light. Red has the longest visible wavelength, whereas violet has the shortest. Waves longer than red are known as infrared, and waves shorter than violet are called UV.

The sun is the closest star to the Earth and the most closely studied. It is at the center of the solar system and accounts for about 99.8 percent of the mass of the solar system. The planets revolve around the sun. The sun is composed of hydrogen, helium, and other trace elements and goes around the center of the Milky Way galaxy at a distance of about 26,000 light years from the center of the galaxy. The amount of solar energy incident on the Earth's atmosphere is about 342 Watts per sq. m.r, based on the surface area of the Earth. Although the Earth's surface continuously radiates energy outward to space, only part of the surface area receives solar radiation at a time. Most of the solar energy incident on the Earth is in the UV region of shorter wavelengths. The sun is the source of heat that sustains life on Earth and controls the climate and weather. Only the sun's outer layers, which consist of the photosphere, the chromosphere, and the corona, can be observed directly. These three regions have different properties from one another, with regions of gradual transition between them. The sun has basically the same chemical elements as are present on the Earth. However, the sun is so hot that all of these elements exist in the gaseous state. Energy generated in the sun's core takes a million years to reach its surface. Solar energy is created deep within the core of the sun, where nuclear reactions take place.

SUNLIGHT AND THE DANGER OF SKIN CANCER

Every living thing exists because of the light from the sun. Sunlight is important in photosynthesis. For humans, UV light in small amounts is beneficial because it helps the body produce vitamin D from the UV region of sunlight. However, excessive exposure to sunlight is dangerous, as it can cause sunburns, skin cancer, and aging. UV light wavelengths are short enough to break the chemical bonds in skin tissue, and when the skin is exposed to sunlight, most skin will either burn or tan. The skin undergoes certain changes when exposed to UV light to protect itself against damage. The epidermis thickens, blocking UV light, and the melanocytes make increased amounts of melanin, which darkens the skin, resulting in a tan. Melanin absorbs the energy of UV light and prevents the light from penetrating deeper into the tissues. Sensitivity to sunlight varies according to the amount of melanin in the skin. Darker-skinned people have more melanin and therefore have greater protection against the sun's harmful effects. The amount of melanin present in a person's skin depends on heredity as well as on the amount of recent sun exposure. Albinos have little or no melanin. The more sun exposure a person has, the higher the risk of skin cancers, including squamous cell carcinoma, basal cell carcinoma, and malignant melanoma. Actinic keratoses (solar keratoses) are precancerous growths also caused by long-term sun exposure.

UV light, although invisible to the human eye, is the component of sunlight that has the greatest effect on human skin. Sunlight deficiency could increase blood cholesterol by allowing squalene metabolism to progress to cholesterol synthesis rather than to vitamin D synthesis, as would occur with greater amounts of sunlight exposure. Larger amounts of UV light damage the body's DNA and alter the amounts and kinds of chemicals that the skin cells make. UV light may also break down folic acid, sometimes resulting in a deficiency of that vitamin in fair-skinned people. UV light is classified into three types, UVA, UVB, and UVC, depending on its wavelength. Although UVA penetrates deeper into the skin, UVB is responsible for at least three quarters of the damaging effects of UV light, including tanning, burning, premature skin aging, wrinkling, and skin cancer. The amount of UV light reaching the Earth's surface is increasing, especially in the northern latitudes. This increase is attributable to chemical reactions between ozone and chlorofluorocarbons that are depleting the protective ozone layer, creating a thinner atmosphere with some holes. The key to minimizing the damaging effects of the sun is avoiding further sun exposure. It should be noted that damage that is already done is difficult to reverse.

SEE ALSO: Chemistry; Climate Change, Effects.

BIBLIOGRAPHY. T. Hartmann, *The Last Hours of Ancient Sunlight* (Hodder and Stoughton, 1998); S. Ring and Joseph

Moran, *The Sun* (Coughlan Publishing, 2003); Seymour Simon, *The Sun* (HarperCollins, 1989).

AKAN WILLIAMS
COVENANT UNIVERSITY

Suriname

LOCATED IN THE northeast of the South American mainland, Suriname, formerly a Dutch colony, has a land area of 63,251 sq. mi. (163,270 sq. km.), with a population of 458,000 (2006 est.) and a population density of 7 people per sq. mi. (2.7 people per sq. km.). With 97 percent of the land covered in forests, and a relatively small timber industry, there is little arable land available.

For electricity production, in 2001, 64.2 percent of the country's electricity came from hydropower, with the remaining 35.8 percent coming from fossil fuels. Much of the hydropower comes from a number of hydroelectric plants in the country. Although most provide electricity for the government, some were constructed to provide electricity for specific businesses, such as the Brokopondo Reservoir for the nearby Alcoa aluminum plant. Most of the fossil fuels used in Suriname come from petroleum, with liquid fuels being responsible for 95 percent of the carbon dioxide emissions from the entire country. On account of this reliance on liquid fuels, in spite of being largely undeveloped, Suriname had a per capita rate of carbon dioxide emissions of 4.5 metric tons per person in 1990, rising steadily to 5.1 metric tons in 2003.

The effects of global warming on Suriname include a greater possibility of flooding, with the increased risk of insect-borne diseases such as malaria and dengue fever. There is also the risk of alienating some of the arable land in the country, possibly making it dependent on imported food. Furthermore, the rising water temperature is already having an effect on the leatherback turtles at the Galibi Nature Reserve. The Suriname government took part in the United Nations Framework Convention on Climate Change, signed in Rio de Janeiro in May 1992, ratifying it four years later, with the government ratifying the Vienna Convention in 1997. On September 25, 2006, Suriname became the 163rd country to ratify the Kyoto Protocol to the UN Framework Convention on Climate Change.

SEE ALSO: Climate Change, Effects; Floods.

BIBLIOGRAPHY. Baijah Hunderson and Philip Mhango, *Revenue-Income Elasticity of Resource-Rich Developing Countries: The Case of Suriname* (Institute of Economic and Social Research, University of Suriname, 1982); John McCarry, "Suriname" *National Geographic* (v.197/6, June 2000); Switi Sranan, Surinam (K.I.T., 2003); World Resources Institute, "Suriname—Climate and Atmosphere," http://earthtrends.wri.org (cited October 2007).

ROBIN S. CORFIELD
INDEPENDENT SCHOLAR

Sustainability

THE LANGUAGE OF sustainability emerged during the 1970s, though the concept was introduced as sustainable development in 1980 in the World Conservation Strategy and was popularized in 1987 by the World Commission on Environment and Development (also known as the Brundtland Commission after its Norwegian chairperson Gro Harlem Brundtland). Today there are numerous definitions of sustainability, but the important question to ask is, What is to be sustained? Is it the planet, particular environments, individual species, current lifestyles, certain rates of economic growth, a specific level of profit?

Sustainability, especially as constructed in mainstream definitions of sustainable development, is very similar to the concept of conservation espoused by the American forester, Gifford Pinchot, in the late 19th century. Conservation emphasized using natural resources wisely, not depleting nonrenewable resources, ensuring that all American men received a fair share of the distribution of benefits, and that consideration be given to the needs of their descendents. Sustainable development globalizes the discourse. The World Commission on Environment and Development report in 1987 defined sustainable development as "development that meets the needs of the present without compromising the ability of future generations to meet their own needs." This definition is used today in many parts of the world by governments, businesses, environmental groups, and educators. The history of the concept, and the specific term, mean that it may be interpreted as a repackaging of environmental management. The managerial focus and faith in technological progress evident in this definition of sustainability mean that it is critiqued by more radical sustainability advocates.

In Australia, the term ecologically sustainable development emerged as a unique approach as a result of the power of major environmental groups in Australia in the early 1990s. In 1992, ecologically sustainable development was defined as "using, conserving and enhancing the community's resources so that ecological processes, on which life depends, are maintained and the total quality of life, now and in the future, can be increased." This terminology and definition, which arose as a result of the political power of environmental groups in the early 1990s in Australia, highlights the dependence of all life on ecological processes (thermodynamics, hydrological cycles, nutrient cycles, and so on).

The Australian definition leans toward what has been termed strong sustainability, meaning that humans should not be substituting human-made capital for natural capital. In contrast, weak sustainability advocates substitution provided the total store of capital is not diminished. Critics of the weak sustainability approach point out that this is what has been happening for thousands of years, leading to the destruction of the environment. Other critics reject the notion of turning nature into "natural capital" and therefore do not engage in the strong versus weak sustainability debates.

CONFERENCES ON SUSTAINABILITY

The concept of sustainable development was the basis for a massive conference in Rio de Janeiro in 1992 that was chaired by Maurice Strong and attended by 178 governments, including 118 heads of state. The United Nations Conference on Environment and Development (UNCED, otherwise known as the Earth Summit) was the five-year follow-up to the release of the Brundtland Report. The conference attempted to move from debates about the notion of sustainability and sustainable development to working out how to implement this idea. The idea of expanding the global economy, although contro-

versial, was accepted within sustainable development discourses because development was seen as being necessary to overcome poverty. Sustainable development was intended to allow economic growth to continue but to make this growth greener. Growth was seen as essential for developing countries and also for developed countries, so as to facilitate trade and help the poorer countries of the world. This concept of sustainability was compatible with that of the newly founded business organization, the World Business Council for Sustainable Development, which was influential in shaping the idea of sustainable development and how it would be implemented.

Implementation has been the focus of subsequent conferences in New York (1997) and Johannesburg (2002) and in the ongoing work of the United Nations Commission on Sustainable Development. Many countries, states/provinces, and local governments, as well as some businesses, have also introduced departments focused on implementing sustainability within their organization. Implementation is challenging because there are many barriers to implementing sustainable development. These include corporate cultures, countervailing market signals, and jurisdictional issues. Another issue is that although the temporal emphasis within the concept of sustainability is apparent, the spatial or geographical scale is unclear. This has led to various scales of analysis and implementation, including concepts such as sustainable lifestyles, sustainable cities, and sustainable regions. The UNCED Conference in 1992 produced five important documents including Agenda 21, which was a 40-chapter document outlining the actions needed to implement sustainable development. Importantly, chapter 28 highlighted the important role of local government in implementing the concepts introduced at the global level. This led to the development of Local Agenda 21 (LA21). At the Johannesburg Conference in 2002, LA21 was relaunched as Local Action 21, which is the second decade of this program containing a focus on action and implementation.

FUTURE PLANNING

There are also different ways of conceptualizing sustainable development vis-à-vis sustainability. Some authors present sustainability and sustainable futures as being the goals to be reached by a process called sustainable development. Other authors maintain a distinction between sustainable development and ecological sustainability on the basis of their approach to existing structures and institutions. Sustainable development is seen as more of a reformist approach by advocates who primarily support the existing institutions but want them to be greener, whereas those activists and authors who emphasize sustainability or ecological sustainability often question the structures that perpetuate unsustainable practices.

Today it is impossible, given the adoption of legislation related to sustainability, not to be planning for sustainability. However, many of the differences in various concepts of sustainability can be attributed to the relative weight given to the economic, social, cultural, and environmental components of sustainability. The differences are also caused by the perception of how these components fit together.

There are two main approaches to conceptualizing sustainability, with numerous variations on these approaches. The dominant, mainstream representation of sustainable development that emerged from the Brundtland Commission, and that has been adopted by many governments and business groups throughout the world, is the balanced approach. Although the notion of balance has been largely discredited in scientific ecology, it is still a powerful metaphor within the environmental literature. In many models of sustainable development, balance is achieved by the construction of three circles of equal size to represent the economy, society, and environment. At the intersection of these three equal-sized circles is sustainable development. In contrast, more radical advocates of sustainability may posit a hierarchical approach, in which the hierarchy may vary between models. It often includes ecological considerations at its base, followed by society—because there would be no society without an environment, and then a smaller economy—because there would be no economy without society. Variations may include the use of thermodynamic processes to support biochemical cycles that allow ecosystems to flourish, eventually reaching human social and individual scales.

Some environmental groups avoid using the term sustainable development, partly because of its perceived cooption. Other groups use the term sustainability, whereas some groups attempt to avoid this language altogether. The challenge for sustainability advocates is to be able to implement something that moves humankind and the rest of the planet away from a state of being unsustainable at a rate that is needed to avoid catastrophe.

SEE ALSO: Australia; Conservation; Culture; Norway; Resources; World Business Council for Sustainable Development (WBCSD).

BIBLIOGRAPHY. Ecologically Sustainable Development Steering Committee, *National Strategy for Ecologically Sustainable Development* (Australian Government Publishing Service, 1992); Martin Purvis and Alan Grainger, eds., *Exploring Sustainable Development: Geographical Perspectives* (Earthscan, 2004); World Commission on Environment and Development, *Our Common Future* (Oxford University Press, 1987).

PHIL McMANUS
UNIVERSITY OF SYDNEY

Sverdrup, Harald Ulrik (1888–1957)

HARALD ULRIK SVERDRUP is a Norwegian meteorologist and oceanographer known for his studies of the physics, chemistry, and biology of the oceans and considered as the founding father of modern physical oceanography. Sverdrup explained the equatorial countercurrents and helped develop the method of predicting surf and breakers. A unit of water flow in the oceans was named after him by oceanographic researchers: 1 sverdrup (Sv) is equal to the transport of 1 million cubic meters of water per second. The American Meteorological Society honored him with the Sverdrup Gold Medal, which recognizes researchers for outstanding contributions to the scientific knowledge of interactions between the oceans and the atmosphere.

Sverdrup was born on November 15, 1888, in Sogndal, Sogn, Norway, into an ancient and respected family of university lecturers, lawyers, politicians, and Lutheran ministers. His father Johan was a teacher and, following the family tradition, became a Lutheran minister of the State Church of Norway. In 1894, his father became minister in the island district of Solund, about 40 mi. (64 km.) north of Bergen, and then moved to Rennsö near Stavanger. In 1908, he became professor of church history in Oslo. Because of his father's different jobs, Sverdrup spent much of his boyhood in various sites in western Norway and was taught by governesses until he was 14 years old. At that age, he went to school in Stavanger. During his adolescence, Sverdrup experienced conflicts between his interest in natural science and the religious background of his family. It was particularly difficult for him to reconcile the concept of evolution with his religious upbringing.

SVERDRUP'S FORMAL TRAINING

As he was not aware of the possibility to study science at university, he first opted for the classical curriculum in 1903. Within this field, his major interest became astronomy. Sverdrup left the gymnasium with honors and spent a year in Oslo preparing for university preliminary examinations. Military service was compulsory at the time, so he decided to combine it with his scientific education, enrolling at the Norwegian Academy of War. This training was combined with the study of physics and mathematics. The physical training that he received while at the academy was extremely useful for his survival during his later long arctic expeditions.

When Sverdrup entered university, he decided to major in astronomy. In 1911, he was offered an assistantship with Professor Vilhelm Bjerknes, the preeminent Norwegian meteorologist and founder of the Bergen School, which allowed him to enter one of the brightest scientific circles in the country. The Bergen School was supported by an annual grant that Bjerknes received from the Carnegie Institution of Washington almost from its founding. Sverdrup initially planned to continue his research in astronomy, but he became increasingly interested in meteorology and oceanography and thus changed his major. When, in 1912, Bjerknes went to Germany to work at the University of Leipzig as professor and director of the new Geophysical Institute, Sverdrup followed him, remaining in Germany from January 1913 to August 1917. He also continued his thesis for the University of Oslo and received his doctorate in June 1917 on a published paper on the North Atlantic trade winds.

CAREER HIGHLIGHTS

In July 1918, Sverdrup joined Roal Amundsen's expedition in the Arctic, on the *Maud*, as a chief scientist. Although the planned duration of the expedition was from three to four years, it lasted for seven and a half years. Sverdrup did not return to Norway until December 22, 1925. He was enthusiastic about the experience and defined the most interesting period as the eight months between 1919 and 1920 spent in Siberia, living with nomadic reindeer herders, the Chukchi. Sverdrup's arctic expedition was interrupted for six months between 1921 and 1922, when the meteorologist had the chance of spending a profitable period of time at the Carnegie Institution. From the arctic expedition, Sverdrup gained better understanding of the basic physical oceanography of currents. He argued that the effect of the Earth's rotation, a fundamental aspect of the dynamics of the oceans, is best observed in the polar regions, because it reaches its greatest level there. On his return to Norway, Sverdrup married Gudrun Bronn Vaumund.

By 1926, Sverdrup was a well-established scientific researcher and was offered the chair of meteorology at Bergen, which had been previously held by Bjerknes. The Carnegie Institution had also offered Sverdrup a permanent position twice, but the Norwegian scientist refused the American offer and took up the position at Bergen. In this capacity, Sverdrup edited the scientific report of the *Maud*. He also continued to collaborate with the Carnegie Institution. In 1931, he led the scientific group in the Wilkins-Ellsworth North Polar Submarine Expedition, during which valuable information was gathered despite its failure to achieve the chief goal of the expedition, the submarine exploration of the Arctic in the *Nautilus*.

In 1936, Sverdrup accepted the position of director of the Scripps Institute of Oceanography in La Jolla, at the University of California, remaining there for almost 12 years. During his tenure as director, Sverdrup expanded the Scripps Institute, making it an institute with a research program, and developing closer ties between Scripps and the University of California, Los Angeles. During World War II, Sverdrup was involved in the U.S. war effort, although he did not directly work for the University of California Division of War Research. He worked on problems related to forecasting surf conditions for military beachhead assaults. His current and wave forecasting methods were applied by military weathermen to predict landing conditions for Allied invasions.

Sverdrup was a central figure in the postwar development of oceanography and allied sciences. He served on many scientific committees after the war, and his contributions to science were increasingly recognized. He was elected to the National Academy of Sciences in 1945. He joined the Executive Committee of the American Geophysical Union in 1945 and presided over the American Geophysical Union Oceanography Section. In 1946, he became president of the International Association of Physical Oceanography. He chaired the Division of Oceanography and Meteorology at the 1946 Pacific Science Conference. Sverdrup returned to Norway in 1948, where he worked as a professor of geophysics at the University of Oslo until his death on August 21, 1957.

SEE ALSO: History of Climatology; Oceanography.

BIBLIOGRAPHY. William A. Nierenberg, "Harald Ulrik Sverdrup," in *Biographical Memoirs* (National Academy of Science), http://www.nap.edu/readingroom/books/biomems/hsverdrup.html.

LUCA PRONO
UNIVERSITY OF NOTTINGHAM

Swaziland

THE KINGDOM OF Swaziland is landlocked, with its neighbors being South Africa and Mozambique. It has a land area of 6,704 sq. mi. (17,363 sq. km.), with a population of 1,141,000 (2006 est.) and a population density of 153 people per sq. mi. (59 people per sq. km.). Some 11 percent of the land is arable, with much of it used for subsistence farming, and also for growing maize, cotton, rice, sugar cane, and citrus fruits. In addition, 62 percent of the country is used as meadows or pasture for low-intensity grazing of cattle, sheep, and goats. About 6 percent of the country is forested, with a significant logging industry.

Because the country is largely undeveloped, there is a relatively low use of electricity, with a significant component being used for heating in winter. Electricity production in Swaziland comes from fossil fuels (55.8 percent) and hydropower (44.2 percent), with most of it imported from South Africa. In terms of its carbon dioxide emissions, Swaziland ranks 144th in the world, with emissions of 0.5 metric tons per person in 1990, falling to 0.1 metric tons per person in 1993 but rising steadily to 0.92 metric tons per person by 2003. All the carbon dioxide emissions in Swaziland are attributed to the use of solid fuels, with most of the electricity generated and also the residential and business heating done by coal or wood. This has resulted in a significant per capita emission of carbon monoxide.

As a result of global warming and climate change, Swaziland has seen the effects of water shortages for some of its crops, such as rice, and desertification of some areas previously used for farming. The Swaziland government took part in the United Nations Framework Convention on Climate Change, signed in Rio de Janeiro in May 1992, signing the Vienna Convention in the same year. On January 13, 2006, the country accepted the Kyoto Protocol to the UN Framework Convention on Climate Change, being the 155th country in the world to do so, with it coming into force on April 13, 2006.

SEE ALSO: Climate Change, Effects; Drought

BIBLIOGRAPHY. Michael E. Meadows and Timm M. Hoffman, "Land Degradation and Climate Change in South Africa," *The Geographical Journal* (v.169/2, 2003); Percy Selwyn, *Industrial Development in Peripheral Countries* (Institute for Development Studies, Sussex University, 1973); "World Resources Institute, Swaziland—Climate and Atmosphere," http://earthtrends.wri.org (cited October 2007).

ROBIN S. CORFIELD
INDEPENDENT SCHOLAR

Sweden

SWEDEN IS ONE of the few nations expected to meet or exceed its target according to the Kyoto Protocol and is lauded as one of the countries that has adopted the most stringent measures to address global warming on the basis of emission levels and trends and climate policy. Since 2000, greenhouse gas emissions in Sweden have been an average of 3.7 percent below levels in 1990 and are expected to be 4 percent lower in 2010. This meets Sweden's national target of reducing emissions by 4 percent between 1990 and 2012.

A traditional windmill in southern Sweden. Sweden has invested heavily in the development of wind and water power plants.

Sweden has committed to becoming the world's first oil-free nation by 2020. The Swedish Government's Commission on Oil Independence proposed measures necessary to eliminate Sweden's dependence on fossil fuels for transport and heating and promotes the use of renewable alternatives. In 1970, 77 percent of Sweden's energy came from oil, but this amount decreased to 32 percent by 2003. Nuclear power will be phased out, and Sweden has invested heavily in the development of wind and water power plants. Innovative programs including the use of boilers that use wood-based pellets have dramatically decreased the use of oil for home heating, and tax exemptions enabling drivers to use ethanol-based fuel have increased compliance among residents. Sweden's climate strategy involves partnerships between the business community, scientists, and politicians.

Recent modelling scenarios indicate an increase in annual mean temperature in Sweden of between 4.5–8.1 degrees F (2.5–4.5 degrees C) by 2100, with a greater increase in temperature and rate of precipitation during the winter than the summer. Specific effects include flooding resulting from increased precipitation and heavier rainfall. Summer drought and water shortages are expected in southern Sweden because of changes in precipitation rates and increased evaporation. Although the flora and fauna of Sweden may be enriched by a number of southern species, northern species and those indigenous to the Baltic Sea region will be displaced.

Agricultural yields are likely to benefit from warmer temperatures, with extended growing periods and better conditions for cultivation leading to increased harvest yields of around 20 percent and an increased number of commercial crops. A warmer climate would result in elevated levels of pests and disease, leading to the more frequent usage of pesticides. Anticipated changes in temperature and salinity in the Baltic Sea are expected to have an adverse effect on species of importance to the fisheries sector such as Baltic herring, cod, salmon, turbot, and plaice. Species composition is expected to shift as new fish and shellfish species are introduced from the south. Reduced sea-ice cover may also have an adverse effect on reproduction for flatfish, whereas warm-water species such as pike, perch, and carp may benefit from higher water temperatures.

SEE ALSO: Kyoto Protocol; Renewable Energy Policy Project (REPP); Stockholm Environmental Institute (SEI).

BIBLIOGRAPHY. Jon Moen, Karin Aune, Lars Edenius, and Anders Angerbjörn, "Potential Effects of Climate Change on Treeline Position in the Swedish Mountains," *Ecology and Society* (v.9/1, 2004); Markku Rummukainen, Sten Bergström, Gunn Persson, Johan Rodhe, and Michael Tjernström, "The Swedish Regional Climate Modelling Programme, SWECLIM: A Review," *Ambio* (v.33/4–5, 2004).

JOANNA KAFAROWSKI
UNIVERSITY OF NORTHERN BRITISH COLUMBIA

Switzerland

SWITZERLAND HAS A long-standing tradition of environmental awareness and protection and has been active in bringing the debate on climatic change to the forefront of international environmental affairs. Switzerland was a strong supporter of the Intergovernmental Panel on Climate Change from its inception in 1988 and an important negotiator for the UN Framework Convention on Climate Change, leading *inter alia* to the Kyoto Protocol. Climate research is high on the agenda of Swiss academia, with recognized expertise in paleoclimate reconstructions, climate modeling, and impacts studies. In 1987, the Swiss Academy of Science set up a unique scientific platform at the interface of science and policy (ProClim, the Swiss Forum on Climate and Global Change) to facilitate the transfer of knowledge to decision makers and to the media.

The keen awareness of Switzerland to climatic change is the result of many climate-driven changes in the Alpine environment already being perceptible, such as the retreat of mountain glaciers. The climate of Switzerland is rendered complex by the interactions between the Alpine topography and atmospheric flows and the competing influences of a number of contrasting climate regimes that converge into the region (the Mediterranean, continental, Atlantic, and polar systems). Temperatures have risen by up to 3.6 degrees F (2 degrees C) in many parts of Switzerland since 1900—well above the global average 20th-century warming of about 1.3 degrees F (0.7 degrees C).

Future climatic change in the Alps will be a complex aggregate of decadal- to century-scale forcing factors related to the North Atlantic Oscillation, the Atlantic Multidecadal Oscillation, and the anthropogenic greenhouse effect. Regional climate models suggest that by 2100, Swiss winters will warm by 5.5–9 degrees F (3–5 degrees C) and summers by 11–12.5 degrees F (6–7 degrees C); in parallel, precipitation is projected to increase in winter and to sharply decrease in summer. Strong heat waves similar to the 2003 European event are likely to become the norm by 2100, and both drought and intense precipitation are projected to increasingly affect the country.

The effects of climatic change on Switzerland will change the natural environment and economic activities. Alpine glaciers may lose between 50–90 percent of their current volume, and the average snow line will rise by 492 ft. (150 m.) for each degree of warming. Hydrological systems will respond in quantity and seasonality to changing precipitation patterns and to the timing of snowmelt in the Alps, with a greater flood potential in spring and drought potential in summer and fall. More extreme events will trigger frequent slope instabilities, and at high elevations, melting permafrost will compound these problems. The distribution of natural vegetation will change as plants seek new habitats with similar climatic conditions to those of today. A rapidly warming climate will result in a loss of mountain biodiversity, as not all species are capable of adapting to change. The direct and indirect effects of a warming climate will affect important economic sectors such as winter tourism, hydropower, agriculture, and the insurance industry, which will be confronted with more frequent natural disasters. Climate-related health risks (allergies, pollution) are expected to increase, with consequent economic effects resulting from prolonged morbidity and absenteeism.

SEE ALSO: Climate Change, Effects; Drought; Floods; Kyoto Protocol.

BIBLIOGRAPHY. Martin Beniston, *Climatic Change and Impacts: A Focus on Switzerland* (Kluwer Academic/Springer, 2004); Carla Riccarda S. Soliva, et. al., *Greenhouse Gases and Animal Agriculture: An Update: Proceedings of the 2nd International Conference on Greenhouse Gases and Animal Agriculture, Held in Zurich, Switzerland between 20 and 24 September 2005* (Elsevier, 2006).

MARTIN BENISTON
UNIVERSITY OF GENEVA, SWITZERLAND

Syria

LOCATED IN THE Middle East, the Syrian Arab Republic has a land area of 71,479 sq. mi. (185,180 sq. km.), with a population of 19,929,000 (2006 est.) and a population density of 267 people per sq. mi. (103 people per sq. km.). Some 28 percent of Syria is arable land, with a further 43 percent used as meadows or pasture, much of it for low-intensity grazing of sheep. Only a very small part of the country is woodland.

In terms of its per capita carbon dioxide emissions, Syria ranks 93rd in the world, with emissions of 2.8 metric tons per person in 1990, gradually falling to 2.7 metric tons per person by 2003, after which emissions experienced a significant rise to 3.72 metric tons per person in 2004. Fossil fuels make up 64.5 percent of electricity generation in the country, and hydropower contributes to the remainder, with dams located on the Euphrates River. In 1973, the Syrian government built the Tabaqah Dam on the Euphrates to create a new reservoir called Lake Assad to help with the irrigation of the region, and there have been other, smaller projects in recent years.

About 70 percent of the country's carbon dioxide emissions come from liquid fuels, with 19 percent from gaseous fuels, 7 percent from gas flaring, and 4 percent from the manufacture of cement. Solid fuels are not used in the country. By sector, 42 percent of the carbon dioxide emissions come from the generating of electricity, with 32 percent from manufacturing and construction, and 12 percent from transportation. In terms of the effects of global warming and climate change, Syria has experienced a higher average temperature, which has resulted in some level of desertification and the widespread alienation of marginal arable land as the country draws heavily on its water reserves. One positive benefit, although short term, has been that in the Jabal and Nusariyah mountains, parallel to the coastal plain, there has been a rise in temperature, which has led to the melting of the snows, helping with the irrigation of the heavily populated eastern slopes of the mountain range.

The Syrian government of Hafez al-Assad took part in the United Nations Framework Convention on Climate Change, signed in Rio de Janeiro in May 1992. The government of his son Bashar al-Assad accepted the Kyoto Protocol to the UN Framework Convention on Climate Change on January 27, 2006, with it entering into force on April 27, 2006.

SEE ALSO: Carbon Dioxide; Climate Change, Effects; Deserts; Kyoto Protocol.

BIBLIOGRAPHY. Hussein A. Amery, "Water Wars in the Middle East: A Looming Threat," *The Geographical Journal* (v.168/4, 2002); A. M. Lapshin, "Cascade of Hydroelectric Units on the Euphrates River in Syria," *Power Technology and Engineering* (v.24/8–9, August 2000); Peter Theroux, "Syria Behind the Mask," *National Geographic* (v.190/1, July 1996); World Resources Institute, "Syria—Climate and Atmosphere," http://earthtrends.wri.org (cited October 2007).

JUSTIN CORFIELD
GEELONG GRAMMAR SCHOOL, AUSTRALIA

Tajikistan

TAJIKISTAN IS LOCATED in Central Asia, east of China. It is a landlocked country dominated by the Pamir and Alay mountain ranges, with the highest peak being over 7,200 m. (23,600 ft.) high. The valley floors are considered a continental climate zone, whereas the mountains range from semiarid to polar. Climate change has begun to affect the lives of many of the country's seven million residents, particularly those who live in the mountain zones.

According to modeling scenarios published by the government, Tajikistan should see an average temperature increase of 3.2–5.2 degrees F (1.8–2.9 degrees C) by the year 2050. Mean monthly temperature increases varied within the models, but at least one showed a sharp increase in February and March temperatures of 8.5–8.8 degrees F (4.7–4.9 degrees C) over historical averages. Precipitation should increase by 3–26 percent by 2050 in most regions, with several models showing an average increase of 14 percent in the mountains and 18 percent in the valleys. More frequent rainfall will increase soil erosion in the main agricultural sectors.

Six percent of Tajikistan is covered with glaciers, and they are receding at an increasingly worrisome pace. Several thousand small glaciers will vanish entirely by 2050, and the major glacier fields will shrink by 15–20 percent. Over the course of the 20th century, the massive Garmo glacier retreated by 4.3 mi. (7 km.) and shrunk in area by 2.3 mi. (6 sq. km.); the 43.5 mi.-long (70-km.) Fedchenko glacier has retreated 0.6 mi. (1 km.) in length and lost 0.85 mi. (2 sq. km.) of thickness in recent years.

Water flow in the major river basins is expected to decrease by an average of 7 percent by 2050. With increased snowfall in the mountains and the melting of glaciers, the spring floods are anticipated to increase in duration. At the same time, reduced water flow will have a severe effect on irrigation and hydroelectric power production. Mountain villages are already feeling the effects of the changing climate. Over the last few years, mountain communities have seen increased snowfall, which leads to the closure of mountain roads for longer and longer periods and causes severe flooding and landslides when the snows melt. New precipitation patterns, characterized by unusually heavy downpours, have led to flash flooding and crop loss as fields are washed away.

Tajikistan is not a major contributor to global emissions, expelling just 5.1 million metric tons of CO_2 in 1998. Of this, 67 percent came from liquid fuels, 29 percent from gaseous sources, and 4 percent from solid fuels. The government of Tajikistan has developed a national mitigation plan as part of their participation in the UN Framework Convention on Climate Change, pledging to support sustainable agricultural practices, the development of renewable energy sources, the reduction of greenhouse gas emissions, and the protection and development of carbon sinks for mitigation.

SEE ALSO: Climate Change, Effects; Floods; Rain.

BIBLIOGRAPHY. "Climate Change in Tajikistan 2002," Central and Eastern Europe, Caucasus, and Central Asia Environmental Information Programme, www.enrin.grida.no/htmls/tadjik/vitalgraphics/eng/index.htm (cited November 1, 2007); "National Action Plan for Climate Change Mitigation," Government of Tajikistan, www.unfccc.int/resource/docs/nap/tainap01e.pdf (cited November 1, 2007); "TAJIKISTAN: Climate Change Threatens Livelihoods of Mountain Villagers IRIN: Humanitarian News and Analysis From Africa, Asia and the Middle East," www.irinnews.org/report.aspx?ReportId=72573 (cited November 1, 2007).

HEATHER K. MICHON
INDEPENDENT SCHOLAR

Tanzania

TANZANIA, SITUATED IN east Africa just south of the equator, is a country characterized by high environmental, ecosystem, and cultural diversity. Much of Tanzania is also characterized by low-lying plains extending in from the coastal area in addition to the low-lying island archipelagos dominated by the islands of Mafia, Pemba, and Zanzibar. Inland there is the Eastern Arc Mountain chain, which is formed from heavily metamorphosed pre-Cambrian basement rocks, periodically uplifted by faulting and weathering over millions of years.

The mountains rise to 8,350 ft. (2,600 m.) in altitude, although maximum altitudes of 2,200 to 2,500 m. are more typical and cover an area of 2,085 sq. mi. (5,400 sq. km.). Farther west there is a high plateau of gently undulating terrain commonly between 4,920 and 6,561 ft. (1,500 and 2,000 m.). To the north of Tanzania, the landscape is dominated by the quite recent (2 million years old) volcanic chain that includes Kilimanjaro and Mount Meru.

Rainfall patterns in Tanzania are associated with the passage of the Intertropical Convergence Zone (ITCZ), which migrates from approximately 10 degrees S during January to 10 degrees N during July. The southeast trade winds are driven by annual oscillation of the ITCZ, bringing monsoonal rainfall to the east of Tanzania. Wet and dry seasons are clearly defined: Northern Tanzania experiences a rainy season from March to May and from October to December, whereas southern areas have one long rainy season from November to May. The elevational gradient on the eastern slopes of the Eastern Arc Mountains is relatively steep, whereas the western sides are relatively gently sloping. It is estimated that forests and woodlands cover 45 percent of Tanzania. The widespread flora on mountain islands led to the view of a continent-wide, archipelago-like center termed Afromontane; however, the distinct flora of the Eastern Arc Mountains suggests that they are floristically different from other African mountains. The altitudinal distribution of forest types comprises three major vegetation categories: upper montane forest, upper montane herb and shrub, and montane forest. Much of the coastal area is characterized by a coastal forest ecosystem that has been fairly heavily degraded, and the central part of the country is dominated by savanna bordered to the west by the Eastern Arc Mountains; these make up one of the world's hot spots of biodiversity because of their great variety of plant and animal species and their unusually high number of endemic species.

Relatively few palaeoecological records have been generated from Tanzania, with those records that have been produced being largely associated with the Rift Valley lakes located in the west of the country—the Empaki Crater Lake and an interesting ice core from the permanent ice on Kilimanjaro. One record from the Eastern Arc Mountains indicates relatively little ecosystem change over the past 40,000 years; this is in marked contrast with those records from the lowland lakes, which show expansion of montane forest into present savannah environments under the cold dry climate of the last glacial period. The Holocene, similar to other places in East Africa, is marked by human effects from around 4,000 years ago, and particularly after around 2,000 years ago as the agricultural transformation took place. Superimposed on this era are numerous climate change events, such

Lions in Tanzania stay close to their water source. Lion attacks have increased in rural Tanzania, the increase mirroring the dramatic rise in population, which grew by nearly 50 percent between 1988 and 2002 and encroached on the lions' habitat.

as the shift to more arid condition centered around 4,000 years ago and detected as a rapid increase in dust from Mount Kilimanjaro. The complicated picture of human-induced effects within a background of changing climate is something to be explored by generating new records on environmental history in a much underresearched part of the world.

SEE ALSO: Climate Change, Effects; Deforestation.

BIBLIOGRAPHY. C. Mumbi, R. Marchant, H. Hooghiemstra, and M. Wooller, "Late Quaternary Ecosystem Stability From an East African Biodiversity Hot Spot: The Eastern Arc Mountains of Tanzania," *Quaternary Research* (in press); L.G. Thompson, E. Mosley-Thompson, M.E. Davis, K.A. Henderson, H.H. Brecher, V.S. Zagorodnov, T.A. Mashiotta, P.N. Lin, V.N. Mikhalenko, D.R. Hardy, and J. Beer, "Kilimanjaro Ice Core Records: Evidence of Holocene Climate Change in Tropical Africa," *Science* (298, 2002); A. Vincens, Y. Garcin, and G. Buchet, "Influence of Rainfall Seasonality on African Lowland Vegetation During the Late Quaternary: Pollen Evidence From Lake Masoko, Tanzania," *Journal of Biogeography* (2007).

ROB MARCHANT
UNIVERSITY OF YORK

Tata Energy Research Institute (TERI)

THE ENERGY AND Resources Institute (TERI) began in 1974 as the Tata Energy Research Institute. Motivated by concerns about finite, nonrenewable energy resources and pollution, Darbari Seth, a chemical engineer working for Tata Chemicals, proposed a research institute dedicated to the collection and dissemination of information regarding energy production and utilization. R.J.D. Tata, then chairman of the Tata Group, actively supported the institute, and TERI was formally established in Delhi in 1974. By 1982, TERI had expanded to include research activities in the fields of energy, environment, and sustainable development. As the scope of activities continued to widen, TERI maintained its acronym, although it was renamed the Energy and Resources Institute in 2003. Now, TERI's staff of over 700 conducts research and provides professional support to governments, institutions, and corporations worldwide. TERI's global leadership in efforts to mitigate the threat of climate change has been further endorsed by the election of its director-general, Dr. Rajendra K. Pachauri, as chairman of the Intergovernmental Panel on Climate Change in April 2002.

Although spawned by the Tata Group, the largest conglomerate in India accounting for 96 companies operating in over 40 countries and exporting to 140 countries, TERI operates as a not-for-profit, nongovernmental organization. TERI's work is sponsored by over 900 organizations (Tata Group included), and more than 200 organizations from 43 countries serve as partners in TERI projects.

The Energy and Resources Institute currently operates through divisions that include energy–environment technology, environmental and industrial biotechnology, biotechnology and management of bioresources, regulatory studies and governance, resources and global security, action programs, information technology and services, sustainable development outreach, and policy analysis. Examples of TERI-led research projects vary widely and include microbial bioremediation of oil spills and oil sludge deposits; design and dissemination of biomass gasifiers; wasteland reclamation and biodiesel production through Jatropha Curcas, a nonedible, oil-bearing crop; e-waste recycling; green buildings; and ecovillages.

The Asian Development Bank declared TERI a clean energy knowledge hub in 2006. As evidence of its dedication to clean energy, TERI established the Green Rating for Integrated Habitat Assessment, the first of its kind in India. Other work in this area includes research and training initiatives to advance large-scale use of renewable and clean energy, energy efficiency, and response to climate change in Asia and the Pacific region. Although large-scale projects with international cooperation remain key to many of TERI's goals, TERI also promotes empowering disadvantaged populations and generating employment through small-scale entrepreneurial endeavors.

The institute established the TERI University in 1998. It became a deemed university in India in 1999. The TERI University offers degree programs only at the master's and Ph.D. levels, and faculty and students participate in research conducted by the institute.

TERI continues to collect, generate, and make available a wide range of publications on issues related to its areas of research and training. Given its long and successful history of publication, the institute established TERI Press.

During its relatively brief existence, TERI has expanded to include research, training, and support efforts throughout India, and it claims to be the only institution in a developing country to have established a significant presence in North America, Europe, the larger Asian continent, Japan, Malaysia, and the Middle East.

SEE ALSO: Biomass; Climate Change, Effects; Intergovernmental Panel on Climate Change (IPCC); Japan; Malaysia; Nongovernmental Organizations (NGOs).

BIBLIOGRAPHY. P. K. Bhattacharya and S. Ganguly, "The TERI ENVIS Centre: An Indian Information Centre on Renewable Energy and the Environment," *Information Development* (v.22/3, 2006); The Energy and Resources Institute, www.teriin.org.

JENNIFER ELLEN COFFMAN
JAMES MADISON UNIVERSITY

Technology

TECHNOLOGY IS DEFINED as applying science to manipulate or change the human environment. Although it is usually thought to involve some form of machinery or physical equipment, technology can just as effectively be intangible in form, such as with management technology, which provides different ways of understanding how resources, including people, may be organized for more efficient production or operation. Historically, technology changed and developed very slowly around the world. However in more recent years, the improvements in infrastructure—and particularly in communications—have meant that technological advances have increased at an ever-quicker rate. Meanwhile, the dissemination of that technology has spread around the world, although there are still many hundreds of millions of people too poor to benefit from it. Nevertheless, for most people, especially in the Western world, life and society have been transformed completely by the technologies that have emerged over the last two decades.

Because the rate of change of new technological innovation continues to accelerate, it seems likely that life and society in the future will be at least as difficult to predict now as it would have been a few decades in the past. Because of its prevalence in society, its ability to reduce the time needed for generally undesirable domestic tasks, and its ability to improve leisure opportunities, among other attributes, most people welcome technology and believe it to be beneficial to their lives. However there are still individuals and groups of people who may, perhaps for ideological or religious reasons, reject the use of technology. Because global climate change has come to be associated with the use of technology and the energy required to power so much of it, technology as a whole has come to be regarded by some people as an enemy that must be resisted and eradicated. In the extreme case, there are people who believe that only by returning to a form of society in which all forms of technology are rejected can humanity survive the forthcoming environmental crisis.

Philosophers such as Michael Foucault, meanwhile, consider technology to be a tool most commonly used by the powered elites of society to suppress the masses. They would point out that the introduction of technology is customarily followed by the imposition of restrictions that prevent the majority of people from accessing the benefits of that technology. For example, internet technology in China is regularly used to spy on the activities of ordinary people and keep their discussions heavily monitored. This is an instance of technology being used to suppress people and to maintain the existing architecture of power. In contrast, it is possible to argue that the very same technology actually represents a liberation of people because of the many new ways it enables people to communicate with each other and to share information.

Most people tend toward a more moderate position, recognizing the better lifestyle that some aspects of technology provide and being unwilling to abandon these forms, while accepting the need for greater efficiency in the use of resources. They would be unwilling to voluntarily choose to forego the use of that technology to cause some future effects to abate. In other words, people will not vote for significant reductions in the use of personal technologies to reduce future damage caused by climate change. To change their minds, some activists believe that it is necessary to startle or scare people into realizing what sort of changes are likely in the future. Those more skeptical of those future changes, meanwhile, accuse such activists of regularly committing this act and concluding, as a result, that all calls for changes in behavior are overstated and, possibly, politically motivated. This argument has been successfully deployed, in that it has muddied the waters of debate and, hence, reduced the likelihood of future changes in behavior.

THE ROLE OF TECHNOLOGY IN ABATING CLIMATE CHANGE

Technology can be employed to abate current and future climate change in a number of ways. At a large or macrolevel, there are plans to place enormous mirrors into orbit around the earth so that they reflect light energy from the sun back out into space, reducing atmospheric temperature. At a medium or mesolevel is the attempt to develop new and cleaner technologies in terms of energy production, which would reduce carbon emissions. At a microlevel, there are the efforts to reduce resource use inefficiencies by such means as recycling waste products, turning off unattended electronic devices, and generally developing technologies to mitigate future climate change. The extent to which it is possible to abate future climate change by action at the microlevel is not clear, and many estimates vary widely.

Clearly, turning off unused computers or televisions that are idle will save some energy and reduce carbon emissions, but many people believe that this amount of saved energy will be dwarfed by the increased amount of emissions resulting from the rapid and rather dirty industrialization occurring in India and

Concurrent with the rise of technology is the growth of cities, such as Hong Kong, China, above. One of the principal causes of increased atmospheric heat is the presence of cities, especially large cities, across most of the inhabited world.

China, in particular. Indeed, there is an argument that because the effects elsewhere are so great, there is no point trying to reduce emissions on a personal level. This argument does not bear rational examination: in the first place, the reduction of the rate of acceleration of global warming must of itself be a necessary and important thing; second, the people and governments of India and China (and most of the rest of the world) are also aware of the problems of global climate change and are willing to do what they can to bring about changes in their own lives.

Other arguments suggest that changes at the microlevel can have significant changes in the extent of future climate change. The noted skeptical environmentalist Björn Lomborg, for example, has argued that one of the principal causes of increased atmospheric heat is the presence of cities, especially large cities, across most of the inhabited world. Cities are built or have organically grown, in general terms, to maximize population density and, as such, are dry areas without much greenery or standing water. Further, many of the buildings or infrastructure within cities are dark in nature and, as a result, absorb a great deal of energy in a needless fashion. The result is that cities are several degrees F/C or more hotter than surrounding areas. This problem fuels increased use of air conditioning systems and other energy use (e.g., for refrigerators and other cooling devices), which leads to a vicious cycle. Lomborg argues that low-technology solutions can reduce the urban effect: paint buildings white, introduce more water areas, replace some tarmac with grass, and so forth. Taking these steps may reduce the temperature in the local city areas by several degrees. This would have a knock-on effect, too, as in many cases political priorities are determined by urban electorates. As a consequence, demonstrating that technology—even comparatively low-technology solutions—can lead to a measurable improvement in quality of life, which might then lead to more positive sentiment toward the use of technology to improve future lifestyles.

THE SEARCH FOR ALTERNATIVE ENERGY SOURCES

Because it is the use of hydrocarbon fuels that leads to carbon dioxide emissions and is the largest contributor to global climate change, it follows that finding alternative forms of energy that do not emit carbon to the same extent would represent the best means of abating future climate change. Further, the world's reserves of hydrocarbon fuels are finite, and there is a need to develop alternatives if current and projected lifestyles are to be maintained in the future. It is not clear exactly when the reserves of oil will be depleted under current trends of usage, as it is possible (although increasingly less likely) that areas of significant previously unexploited reserves will be found, and probably more importantly, improvements in technology have made it possible to extract profitably existing reserves that have to date been too difficult or expensive to obtain. In addition, as the price of oil increases in general—and continues to increase as demand increases—with respect to supply, more and more known but problematic reserves will become commercially viable. Already, using oil-soaked earth in parts of Canada that were previously prohibitively expensive to process has become viable because of the effect of supply and demand. Other difficult-to-access reserves will, likewise, become more viable.

Nevertheless, although the figures are controversial and contested, it seems likely that all oil reserves will be exhausted within about 120 years, based on current rates of consumption. It is possible, although far from certain, that peak oil production has already been reached. Although declining supply relative to demand will stimulate some increased efficiency of use of oil, it is nevertheless clear that new forms of energy will need to be developed within the next few decades. Clean technologies such as wind and wave power are expected to make up an increasing part of a portfolio of alternative sources of energy. They already contribute significantly to the energy production of some European countries, although there are problems with nimbyism—or not in my back

yard-ism—as people complain that wind turbines are unsightly and noisy. Solar panels have been used with moderate success in many parts of the world, particularly those areas with high levels of sunshine. However improvements in this technology are still needed. Photovoltaic cells have been used to collect power from the sun, and water pipes have been heated by placing them in the sun, but there is a need for more integrated solutions to ensure that a higher level of efficiency is achieved and also to reduce initial start-up costs, which may be high. This is likely to come about through regulation rather than market influences.

Other technologies that might be developed include the tapping of geothermal energy, which has been used for thousands of years in a nonsystematic manner, in the form of hot springs. There remains considerable scope for the further development of the use of geothermal energy on a more systematic basis. Hydroelectricity is a further form of alternative energy and it has been used effectively on many rivers. Countries such as China plan a massive increase in the number of hydroelectric stations, with attendant dams, on the rivers passing through its territory. This includes plans for as many as 12 dams on the River Mekong, for example. However this form of energy is problematic because of the effect on people living downstream of reduced flow of water and because of the effect of building large dams on the populations living in the vicinity, many of whom must be resettled—some of them forcibly. The ownership of a river passing through the territory of more than one country is also a problematic issue.

The one means of producing energy that is available without the consideration of geography is that of nuclear power. The technology involved is to employ certain heavy elements such as plutonium and uranium, which undergo atomic decay on a largely predictable manner, releasing considerable amounts of energy at the same time. A more advanced approach is to employ nuclear fission, which involves causing atoms to collide with each other so as to release more subatomic particles such as neutrons and, hence, more power. The amount of power, which is converted to electricity, that may be released through these processes is limited only by the availability of the appropriate heavy elements. These elements are scarce, and the existing amounts are controlled, although not always effectively. There is a need for control because of the radioactive nature of the substances involved, which makes them very dangerous to life, as well as the possibility that they may be used to create highly destructive nuclear bombs.

Many governments are planning to increase, perhaps quite significantly, their reliance on nuclear power. This causes problems because of the threat of an accident releasing nuclear material, as happened at Chernobyl in the Ukraine, causing thousands of deaths. The use of power plants in known earthquake zones is of particular concern. In addition, depleted uranium or other material, which is no longer productive, remains dangerously radioactive for many thousands of years, and it is not clear where that waste may be safely stored over the long term. There is also the potential problem that nations that develop nuclear power plants might also employ this knowledge and technology to develop nuclear weapons. One consequence of this is that there is widespread public concern about the use of nuclear energy and opposition to it in democratic countries. Even so, improvements in the technology of safety suggest that more governments will wish to augment their nuclear power production capacity and begin building new plants, knowing that it takes several years or more between deciding to construct such a plant and when power from it is ready to enter the grid.

MACROLEVEL TECHNOLOGY APPROACHES TO CLIMATE CHANGE ABATEMENT

Various macrolevel technologies have been suggested as a means of reducing climate change. These are generally very expensive and time-consuming to create and maintain and, hence, tend to be regarded as something of a last-ditch attempt. These technologies include placing large mirrors or reflective dust in space to absorb or reflect away the sun's energy, or to use huge series of tubes leading to the ocean floor, through which excess heat energy may be circulated. Many of these ideas derive from the United States, which has a long tradition of optimism in terms of technology and also large firms and organizations possessing the kind of capital necessary to develop and, in time, implement such solutions. Because none of these solutions has, to date, been operationalized, it is not yet clear whether all or any of them will in fact

be viable. Nevertheless, it is becoming clear that companies are starting to realize the market opportunities emerging for green or clean technologies. At the individual level, technology will be consumed by households to meet their own mandated requirements; for example, in terms of recycling or energy use reduction. At the state level, in contrast, public-sector support of megaprojects can provide sustained funding for a number of years sufficient to underwrite very large research and development operations. Because these technologies are not yet proven to be feasible and may be extremely expensive in practice, it would be much more cost-effective to make extensive use of micro- and mesolevel technologies immediately rather than waiting for a last-ditch attempt to maintain the planet as a place for human life. This will require some sacrifices in the short and medium terms.

TECHNOLOGY AND REGULATION

The Montreal Protocol of 1987, which helped to resolve the problem of atmospheric ozone depletion, demonstrated the ability of states to work together to solve transboundary environmental issues effectively. Market-based attempts to achieve similar goals, for example, through creating markets in tradable carbon emission permits, have foundered without a strong institutional shaping of the rules of the market and supervision of its activities. Technologists around the world have created numerous efficiencies that would help abate future climate change but that will only be implemented when state regulation requires it to be introduced. Just as the numbers of deaths from road traffic accidents was reduced (proportionate to the amount of traffic) after governments introduced legislation requiring safety belts to be worn, so too have buildings become more resistant to earthquakes after stricter building codes were introduced. In countries such as Germany, new regulations about energy production enable households under certain conditions to produce and sell their own power, such as that generated from solar power, online, for example. Good regulations provide appropriate incentives to encourage people to behave in the desired way and disincentives to dissuade people from behaving in undesired ways. The extent to which this can affect behavior and make measurable changes to energy use may be seen in California, where state-level legislation, transparently introduced over a sustained period, has ensured that energy use per capita has not increased in a number of years, despite the significant amount of electronic consumer goods owned per household.

SEE ALSO: Alternative Energy, Overview. Alternative Energy, Solar.

BIBLIOGRAPHY. Michel Foucault, *Politics, Philosophy, Culture: Interviews and Other Writings, 1977–84* (Routledge, 1990); Gerhard Hausladen, Michael de Saldana, Petra Liedl, and Christina Sager, *Climate Design: Solutions for Buildings That Can Do More With Less Technology* (Birkhauser, 2005); Debra Justus and Cédric Philibert, *International Energy Technology Collaboration and Climate Change Mitigation: Synthesis Report* (OECD, 2005); Neil Leary, James Adejuwon, Vicente Barros, Ian Burton. Jvoti Kulkarni, and Rodel Lasco, eds., *Climate Change and Adaptation* (Earthscan, 2007); Björn Lomborg, "Paint It White," *The Guardian* (November 18, 2007), http://commentisfree.guardian.co.uk/bjrn_lomborg/2007/11/paint_it_white.html.

JOHN WALSH
SHINAWATRA UNIVERSITY

Tennessee

TENNESSEE IS 42,143 sq. mi. (109,150 km.) in size with inland water making up 926 sq. mi. (2,398 sq. km.) Tennessee's average elevation is 900 ft. (274 m.) above sea level, with a range in elevation from 178 ft. above sea level on the Mississippi River to 6,643 ft. (2,025 m.) at Clingman's Dome. The variety of topographic features includes mountains, forested ridges, cultivated valleys, rugged plateau with valleys cut by streams, and the Highland Rim (an upland plain surrounds the Nashville Basin, and the Tennessee River separates the Highland Rim from the relatively flat coastal plain in the western part of Tennessee that extends almost to the low-lying area on the Mississippi River, which forms Tennessee's western border). The major rivers are the Mississippi, the Tennessee River, and the Cumberland River, along with many tributaries. Natural lakes and reservoirs store water.

Tennessee has hot summers, mild winters, and abundant precipitation, with variations based on region. West Tennessee (the Gulf Coastal plain), Middle Tennessee (Highland Rim and Nashville Basin is made up of the Highland Rim [mountains] and East Tennessee [mountainous high region]). Average July temperatures range from less than 70 degrees F (21 degrees C) in the Blue Ridge region to 80 degrees F (27 degrees C) in Nashville and Memphis, and even summer nights can be warm and muggy in central and western Tennessee, with cooler temperatures in the eastern mountains. Average January temperatures range from freezing in the eastern mountains to 42 degrees F (22 degrees C) in the southwestern region. The highest temperature recorded in the state is 113 degrees F (45 degrees C) on August 9, 1930, and the lowest temperature recorded in the state is 32 degrees F (0 degrees C) on December 30, 1917. The average annual precipitation is 52 in. or 132 cm. (ranging from 60 in. or 152 cm. in mountain areas to 45 in. or 114 cm. in protected ridges and valleys). Severe blizzards rarely hit Tennessee, but some snow falls every year. West Tennessee receives about 5 in. (13 cm.), and the east can expect twice as much. The precipitation is made up of a combination of rain, snow, and sleet. Heavy rain falls in March and April, causing rivers to overflow their banks.

Tobacco is grown in much of the state. Western Tennessee is the largest agriculture region, with crops of cotton, corn, soybeans, tobacco, and others. Livestock (cattle, sheep, hogs, and poultry) predominates in the Nashville Basin, as well as dairy farming. Crops grown in the region are used to feed the livestock, except tobacco, which is a cash crop. Tennessee's electricity is generated by coal-fired, nuclear power, and hydroelectric plants.

Although climate models vary on predicted temperature increase for Tennessee, estimates range from 1–5 degrees F (1.8–9 degrees C) in all seasons. Precipitation is estimated to increase only slightly in winter, perhaps 10–30 percent in spring and autumn and by 10–50 percent in summer. This increased rainfall could increase flooding (already a concern in the mountains in eastern Tennessee, in unregulated streams, and in growing urban areas near Chattanooga, Nashville, and Memphis).

With changes in climate, the extent of forested areas in Tennessee could change little or decline slightly, though the types of trees would be likely to change. Pine and scrub oaks would replace eastern hardwoods. The success of tree planting in environmental restoration areas (as in around mines) might decrease. Increased temperatures could pose a risk of wildfires. The agriculture may change little, however, with cotton yields unaffected and corn and hay yields possibly increasing.

If rainfall and runoff increase in the Tennessee region, then higher stream flows and lake levels could benefit hydropower production, enhance recreational opportunities, and improve water availability for water supplies. Increased water flow would dilute pollutants, though increased runoff including pesticides and fertilizers may shift levels of contamination higher, with the river basins in western Tennessee being especially susceptible.

Flooding increases the possibility of contamination of water supplies by sediment erosion, increased levels of pesticides and fertilizers, and runoff from grazing, mining, and urban areas.

Human health risks include, but are not limited to, contracting certain infectious diseases from water contamination or disease-carrying vectors such as mosquitoes, ticks, and rodents. Warmer temperatures would increase the incidence of heat-related illnesses and lead to higher concentrations of ground-level ozone pollution causing respiratory illnesses (diminished lung function, asthma, and respiratory inflammation).

On the basis of energy consumption data from the Energy Information Administration, State Energy Consumption, Price, and Expenditure Estimates, released June 1, 2007, indicate Tennessee's total CO_2 emissions from fossil fuel combustion for 2004 were 125.38 million metric tons CO_2, made up of contributions from: commercial sources (3.72 million metric tons CO_2), industrial sources (20.41 million metric tons CO_2), residential sources (4.41 million metric tons CO_2), transportation sources (44.93 million metric tons CO_2), and electric power sources (51.90 million metric tons CO_2).

Tennessee joined the Climate Registry, a voluntary national initiative to track, verify, and report greenhouse gas emissions, with acceptance of data from state agencies, corporations, and educational institutions beginning in January 2008. The Tennessee Environmental Council is including climate change strategy in their meetings.

Tennessee participates in a program called Rebuild America—an organization committed to assisting state and local government and school systems to implement energy-saving improvements. The Tennessee Valley Authority (TVA) has developed a program called Green Power Switch that enables customers to purchase 150-kilowatt-hour blocks of renewable energy (making up about 12 percent of a typical household's monthly energy consumption). In addition, TVA must meet federal and other environmental statutes and regulations for air and water quality as well as managing the disposal of wastes (including hazardous materials). These regulations are becoming more stringent with clean air requirements and reducing greenhouse gas emissions.

Federal and state agencies are responsible for conservation inducing forest management and environmental protection including protecting against soil erosion and water and air quality.

SEE ALSO: Carbon Dioxide; Carbon Emissions; Climate Change, Effects; Greenhouse Gases.

BIBLIOGRAPHY. "Climate Change in Tennessee," www.yosemite.epa.gov/OAR/globalwarming.nsf; National Wildlife Federation, "Global Warming and Tennessee" (June 25, 2007); "Tennessee Greenhouse Gas Emissions Mitigation Strategies," www.tennessee.gov/ecd/pdf/greenhouse/entiredocument.pdf; Tennessee Valley Authority, "Environmental Performance Update," www.tva.gov/environment/reports/envreports/06update/.

LYN MICHAUD
INDEPENDENT SCHOLAR

Tertiary Climate

THE TERTIARY PERIOD (ca. 66.4 to 1.8 million years ago [Ma]) was an interval of enormous geologic, climatic, oceanographic, and biologic change. It spans the transition from a globally warm world of relatively high sea levels to a world of lower sea levels, polar glaciation, and sharply differentiated climate zones. Over the past decade, however, it has become increasingly clear that Tertiary climatic history was not a simple unidirectional cooling driven by a single cause but a much more complicated pattern of change controlled by a complex and dynamic linkage between changes in atmospheric CO_2 levels and ocean circulation, both probably ultimately driven by tectonic evolution of ocean-continent geometry. Although satisfactory explanations for many aspects of Tertiary climate history are available, many areas remain incompletely understood.

The early Tertiary (Paleocene and most of the Eocene epochs, ca. 66–50 Ma) was characterized by a continuation of Cretaceous warm equable climates extending from pole to pole. Global temperatures may have been as much as 18–22 degrees F (10–12 degrees C) higher than present, and pole-to-equator temperature gradients were about 9 degrees F (5 degrees C) during the Paleocene, as compared with about 45 degrees F (25 degrees C) today.

The Paleocene-Eocene boundary (about 54 Ma) was marked by a geologically brief episode of global warming known as the Paleocene-Eocene thermal maximum (PETM), characterized by an increase in sea surface temperatures of 9–11 degrees F (5–6 degrees C), in conjunction with ocean acidification, a decline in productivity, and a large and abrupt decrease in the proportion of isotopically heavy terrestrial sedimentary carbon in the oceans. The PETM is thought to have lasted only about 170,000 to 220,000 years, with most of the temperature and isotopic change occurring in the first 10,000 to 20,000 years. Its causes remain unclear, but it was probably associated with dissolution of methane hydrates on the ocean floor, which would then have caused greenhouse warming. Possible triggers for this hydrate release include an increase in volcanism, leading to an increase in atmospheric CO_2 and consequent sudden initiation of greenhouse warming; a change in ocean circulation; or massive regional submarine slope collapse.

Global temperatures warmed still further during the early Eocene, reaching their highest levels of the past 65 million years during an interval sometimes called the early Eocene climatic optimum (52–50 Ma). Global cooling began during the early middle Eocene (ca. 50 Ma) and accelerated rapidly across the Eocene-Oligocene boundary (ca. 34 Ma), at which time Antarctic continental glaciation began. This shift is frequently referred to as a change from a greenhouse to an icehouse climate regime, and it was one of the most fundamental reorganizations of global climate known in the geological record.

Initiation of Antarctic glaciation has long been attributed to the tectonic opening of Southern Ocean gateways, especially the Drake Passage between South America and the Antarctic Peninsula, which allowed establishment of the Antarctic Circumpolar Current and the consequent isolation of the southern continent from warmer low-latitude waters. This has been questioned recently, however, as a result of the redating of the formation of these gateways, as well as modeling results that point to a greater role for reduced atmospheric CO_2.

Most estimates of early Cenozoic atmospheric pCO_2 range between two and five times the present values in the middle to late Eocene and then decline rapidly during the Oligocene to reach approximately present levels in the latest Oligocene. This decline in CO_2 may, in turn, have been at least partly a result of the tectonic uplift of the Tibetan plateau, beginning around 40 Ma, leading to increased rates of chemical weathering. Levels of CO_2 remained relatively constant throughout the Miocene, suggesting that the substantial climate changes during this time were driven by other factors, including changes in weathering or ocean circulation.

Global temperatures warmed again in the late Oligocene, followed by a brief (ca. 200,000 years) but deep glacial interval at the Oligocene-Miocene boundary (ca. 24 Ma). Temperatures then stabilized or slightly increased (punctuated by several more brief glacials), leading to what is sometimes referred to as the mid-Miocene climatic optimum around 17 to 15 Ma, during which time deep water and high-latitude sea surface temperatures were 11–18 degrees F (6–10 degrees C) warmer than at present. The causes of this warming are not clear, but they may have been related to increased northward oceanic heat transport in the North Pacific brought via intensified currents primarily triggered by narrowing of the Indonesian Seaway in the western equatorial Pacific.

Another major cooling occurred between 14.2 and 13.7 Ma and is associated with increased production of cold Antarctic deep waters and a growth spurt of the East Antarctic Ice Sheet, leading to an increased latitudinal temperature gradient and drying in midlatitudes. A further episode of aridity occurred between 8 and 4 Ma. This cooling trend continued into the Quaternary period, with a short warming interval in the early to mid-Pliocene (ca. 5–3.2 Ma), characterized by warmer sea and air temperatures across at least much of the North Atlantic region.

Northern Hemisphere ice sheets first expanded about 3.5 Ma, with a major pulse of growth occurring 2.5 Ma, at which time the Earth is usually said to have passed over a thermal threshold initiating the latest so-called Ice Age, in which mode the planet is still today. The initiation of Northern Hemisphere glaciation has been attributed to completion of the formation of the Central American Isthmus at around 3.5 Ma, which deflected warm low-latitude currents flowing westward from Africa northward into the Gulf of Mexico and through the Florida Straits to join the Gulf Stream. This strengthened Gulf Stream then transported more moisture to high latitudes, where it supplied an increase in snowfall, leading to increased albedo and temperature decline.

SEE ALSO: Ice Ages; Paleoclimates.

BIBLIOGRAPHY. J. E. Francis, J. A. Crame, and D. Pirrie, eds., *Cretaceous-Tertiary High-Latitude Palaeoenvironments: Special Publication No. 258* (Geological Society Publishing House, 2006).

WARREN D. ALLMON
PALEONTOLOGICAL RESEARCH INSTITUTION

Texas

IN A POST-WORLD War II climate of mass consumption, urban disinvestment, and the emerging dominance of the automobile as the preferred mode of transportation, Texas and its economy grew dramatically. Fleeing postindustrial urban decay and the loss of manufacturing economies, millions of Americans and immigrants flocked to the wide-open and nonunionized spaces of the southwest United States. Home to almost 25 million residents, the State of Texas ranks second only to California in population and is experiencing the largest net population growth of any state in the nation.

Driven by this continuing growth in population, diversifying economic development, persistent low-density suburban development, almost exclusive reliance on the automobile for transportation, and a warming climate, the demand for cheap and plentiful energy—and attendant emissions of greenhouse gases—is growing rapidly. Texas leads the United States in greenhouse gas emissions—40 percent of the national total—largely because of its reliance on existing coal-burning power plants.

In a recent study analyzing urban sprawl, Fort Worth/Arlington and Dallas metro areas rated 10th and 13th on the list of the nation's 83 most sprawling urban areas. Having increased by 30 percent over the past 10 years, the 2000 Environmental Protection Agency ozone National Ambient Air Quality Standard report estimated that Texas vehicle miles traveled would increase between 2007 and 2030 by over 44 percent, and Dallas/Fort Worth, Houston, and San Antonio metropolitan regions would remain in nonattainment for ground-level ozone. These urban areas are experiencing significant population growth, continuing sprawl development, and increasingly severe highway congestion that contribute significantly to their climate change effect.

The consequences for Texas of impending climate change are serious, especially given the already extreme nature of much of its regional weather.

CONSEQUENCES OF CLIMATE CHANGE

Texas exhibits a wide variety of climates within its boundaries, from subtropical in the southeast to high desert in the north and west. Although all regions are likely to experience an increase in mean annual temperature (both daily maximums and minimums) and increasing shortages of freshwater, other challenges faced by the state from climate change differ as a function of geography. Texas's 370 mi. (595 km.) of coastline will experience higher sea levels and resulting beach erosion, saltwater infiltration, and subsidence. Increased water temperatures in the Gulf of Mexico may result in more frequent and widespread algal blooms toxic to indigenous fish and plant species. Warmer ocean and Gulf waters will also contribute to the intensity, if not the frequency, of coastal storms and hurricanes.

Climate change also will result in a redistribution of rainfall across the state, significantly affecting agricultural economies and freshwater supplies available to increasingly urban populations. As it continues to drain its aquifers, Texas increasingly relies on freshwater captured in surface reservoirs. A future that is markedly warmer and drier in many regions of the state will jeopardize these supplies (as demonstrated by the drought that gripped the state in recent years). Increasing temperatures will contribute to already pronounced urban heat islands, resulting in increased frequency of heat-related illnesses and deaths, as well as in increased severity of isolated weather events, especially thunderstorms bearing isolated, flooding rains.

Agricultural and forestation patterns and productivity, staples of regional and state economies, will be disrupted not only by the changes in rainfall, increased uncertainty in available irrigation water, and higher mean temperatures but also by a changing variety of natural weeds and pests that will migrate northward as the climate warms. Infestations of insects new to Texas crops will result in reduced crop productivity and, as farmers try to respond, a likely increase in the number and environmental toxicity of herbicides and pesticides.

RENEWABLE ENERGY SOURCES

Although it confronts serious climate challenges across the state, Texas is also blessed with sources of renewable energy that have only begun to be exploited. It is famous for its scorching summers and sunshine that will "peel the chrome off a trailer hitch." The largest portion of its electricity needs are generated by coal-fired and nuclear power plants. Recently, the state backed away from approving the construction of as many as 11 new coal-fired power plants and is in the process of redefining its energy policies. As technological advances reduce the price and increase the efficiency of solar energy-generation devices, Texas, especially in its western reaches, will be able to capitalize on the abundant radiant energy provided by the sun. In addition, wind-generated electricity is being produced in increasingly economical quantities by west Texas wind farms. Although a transmission infrastructure is evolving to supply the state's urban demand, Texas ranks first in wind power generation among U.S. states. It is also the country's leading producer of biodiesel transportation fuel. Its leadership role was highlighted by the success of country music icon Willie Nelson in promoting locally produced BioWillie biodiesel fuel, primarily marketed for the long-haul trucking industry.

LOCAL ACTION

In the absence of a meaningful climate protection policy at the federal or state levels, many Texas cities have joined municipalities in other states in a variety of nongovernmental organization–led initiatives to reduce their carbon footprints and to lobby for action on climate change and related environmental issues at both the state and federal levels. Among other actions taken by Texas municipalities, 17 Texas cities—including Austin, Dallas, and San Antonio—have signed the U.S. Conference of Mayors Climate Protection Agreement

The Dallas metro area is rated 13th on the list of the nation's 83 most sprawling urban areas. Texas is home to almost 25 million residents, and is experiencing the largest net population growth of any state in the United States.

committing their respective cities to carbon dioxide reductions similar to those contained in the Kyoto Protocol. Both Austin and San Antonio are also members of the Cities for Climate Protection, a global campaign of local and regional entities led by the International Council of Local Environmental Initiatives Local Governments for Sustainability. Through process implementation and performance monitoring, these cities are committed to a rigorous accounting and reduction of their greenhouse gas emissions.

SEE ALSO: Alternative Energy, Wind; Land Use.

BIBLIOGRAPHY. Reid Ewing, Rolf Pendall, and Don Chen, *Measuring Sprawl and Its Impact* (Smart Growth America, 2002); ICLEI Local Governments for Sustainability, *Cities for Climate Protection Campaign*, www.iclei.org (cited September 7, 2007); Texas Department of Transportation, *TxDOT has a Plan: Strategic Plan for 2007–2011*, www.dot.state.tx.us (cited September 7, 2007); U.S. Conference of Mayors, *Climate Protection Agreement*, www.usmayors.org/climateprotection/agreement.htm; U.S. Environmental Protection Agency, *Climate Change and Texas* (1997), www.epa.gov (cited September 7, 2007).

KENT HURST
UNIVERSITY OF TEXAS AT ARLINGTON

Thailand

LOCATED IN SOUTHEAST Asia, Thailand has a land area of 198,115 sq. mi. (513,115 sq. km.), with a population of 62,828,700 (December 2006) and a population density of 317 people per sq. mi. (122 people per sq. km.). Bangkok, the capital and the largest city, has a population of 6,593,000 and a population density of 9,418 per sq. mi. (3,630 per sq. km.). Traditionally, the economy of Thailand has been agricultural, and 34 percent of the land is arable, with an additional 2 percent used as meadows and pasture. With the increase in the tourist industry, as well as manufacturing, the importance of agriculture has declined, but it remains the country's major employer. Some 30 percent of the country is forested, with the timber industry being heavily regulated, although there are regular allegations of illegal logging.

Thailand has a relatively low per capita rate of carbon dioxide emissions—1.8 metric tons in 1990, rising steadily to 4.28 metric tons per person by 2004. Some 92.3 percent of the electricity in the country comes from fossil fuels, with most of the remainder drawn from hydropower. The heavy use of automobiles, and also private generators, has led to 56 percent of the carbon dioxide emissions from the country being from liquid fuels, with 15 percent from gaseous fuels and 21 percent from solid fuels. Some 8 percent of Thailand's carbon dioxide emissions come from the manufacture of cement.

In terms of the sector causing the carbon dioxide emissions, 38 percent comes from the generation of electricity, with air conditioning—especially for the tourist sector—being a very important part of the demand. Some 33 percent comes from transportation, with the traffic problems in Bangkok often leading to a pall of smog in the city. Some 25 percent of carbon dioxide emissions come from manufacturing and construction, with the remaining 3 percent from residential use.

Thailand can best be described as tropical and humid for the majority of the country during most of the year. The area of Thailand north of Bangkok has a climate determined by three seasons while the southern peninsular region of Thailand has only two. In northern Thailand the seasons are clearly defined. Between November and May the weather is mostly dry; however, this is broken up into the periods November to February and March to May. The later of these two periods has the higher relative temperatures, although the northeast monsoon does not directly affect the northern area of Thailand, it does cause cooling breezes from November to February. The other northern season is from May to November and is dominated by the southwest monsoon, during which time rainfall in the north is at its heaviest.

The southern region of Thailand really has only two seasons— the wet and the dry. These seasons do not run at the same time on both the east and west side of the peninsula. On the west coast the southwest monsoon brings rain and often heavy storms from April through October, while on the east coast the most rain falls between September and December.

Overall the southern parts of Thailand get by far the most rain with around 2,400 millimeters

Driving during a monsoon in Thailand. The Boxing Day tsunami in Thailand in 2004, which resulted in the deaths of an estimated 8,200 people, including many hundreds of foreign tourists, has been partially blamed by some experts on climate change.

every year, compared with the central and northern regions of Thailand, both of which get around 55 in. (1,400 mm.).

Thailand has been heavily affected by global warming and climate change. An increase in flooding in southern Thailand has resulted in a rise in the prevalence of insect-borne diseases such as malaria and dengue fever. The Boxing Day tsunami in 2004, which devastated the Phi Phi Islands and other islands along Thailand's Indian Ocean coastline, such as Ko Tapu, and resulted in the deaths of an estimated 8,200 people, including many hundreds of foreign tourists, has been partially blamed by some experts on climate change. The gradual bleaching of coral reefs in that region, and also in the Gulf of Thailand, is certainly attributable to global warming.

The Thai government took part in the United Nations Framework Convention on Climate Change, signed in Rio de Janeiro in May 1992. It signed the Kyoto Protocol to the UN Framework Convention on Climate Change on February 2, 1999, and it was ratified on August 28, 2002, entering into force on February 16, 2005.

SEE ALSO: Climate Change, Effects; Floods; Tsunamis.

BIBLIOGRAPHY. Noel Grove, "Thailand," *National Geographic* (v.189/2, February 1996); Jonathan Rigg and Philip Stott, "Forest Tales: Politics, Policy Making and the Environment in Thailand," in Uday Desai, ed., *Ecological Policy and Politics in Developing Countries: Economic Growth, Democracy, and Environment* (State University of New York Press, 1998);

World Resources Institute, "Thailand—Climate and Atmosphere," www.earthtrends.wri.org (cited October 2007).

Justin Corfield
Geelong Grammar School, Australia

Thermocline

THE THERMOCLINE IS the region of the ocean where temperature decreases most rapidly with increasing depth. It separates the warm, well-mixed upper layer from the colder, deep water below. A thermocline is present throughout the year in the tropics and middle latitudes. It is more difficult to discern in high latitudes, where temperature is more uniform with depth. The presence of a very shallow thermocline in the eastern equatorial Pacific Ocean has important implications for global climate.

The thermocline exists because the ocean absorbs most of the sun's heat in a shallow layer near the surface. The heat absorbed from the sun increases the temperature of the surface relative to that of the deep ocean, maintaining the thermocline. This is in contrast to the atmosphere, where a much larger portion of incident solar radiation passes through to the Earth's surface.

Two important properties of the thermocline are its depth and its strength, or how rapidly temperature decreases with increasing depth. The thermocline's depth is influenced by the winds at the surface of the ocean. In the Atlantic and Pacific oceans, surface winds push warm surface water away from the equator toward the poles, bringing the thermocline close to the surface at the equator.

Water that diverges at the equator accumulates in the subtropics, increasing the depth of the thermocline there. The thermocline is generally (82 to 656 ft. (25 to 200 m.) deep in the equatorial regions and up to 3,281 ft. (1,000 m.) deep in the subtropics.

The thermocline is strongest in the tropics and weakest in high latitudes. This reflects the fact that the surface temperature of the ocean generally decreases from the tropics to the poles, whereas the temperature of the deep ocean is nearly the same at all latitudes. As a result, the temperature contrast between the upper ocean and the deep ocean is greatest in the tropics. The temperature can drop by as much as 18 degrees F (10 degrees C) in less than 164 ft. (50 m.) in the tropical thermocline.

In the extratropical oceans, the strength and depth of the thermocline vary from season to season. There is a main thermocline throughout the year between 656–3,281 ft. (200–1,000 m.). During summer, the sun heats the ocean's surface more strongly than in winter. Most of the additional heat is absorbed in a very shallow surface layer, generating a sharper "seasonal" thermocline above the main thermocline. The seasonal thermocline is similar to the tropical thermocline in terms of its strength and depth. It erodes in the winter as the surface cools relative to the temperature in the main thermocline.

TROPICAL OCEANS

The existence of a strong and shallow thermocline in the tropical oceans has important implications for climate. In the equatorial Pacific Ocean, westward surface winds lead to an accumulation of warm surface water in the west, depressing the thermocline there and raising it to near the surface in the east. The shallow thermocline in the east enables cold, nutrient-rich water to be mixed upward into the surface layer. Every few years the thermocline in the eastern equatorial Pacific deepens in association with an El Niño event. The mixing of cold, nutrient-rich thermocline water into the surface layer is reduced, the surface temperature of the eastern equatorial Pacific Ocean increases, and biological productivity decreases. The warmer surface temperatures associated with El Niño affect atmospheric circulation in the tropics and alter weather patterns throughout the world.

The depth of the eastern equatorial Pacific thermocline has varied significantly in association with changes in global climate. For example, during the early Pliocene period (between 4.5 and 3 million years ago; the most recent period with global temperatures significantly higher than today), the eastern Pacific thermocline was much deeper than it is today, much like it is during a modern El Niño event.

SEE ALSO: El Niño and La Niña; Mixed Layer; Wind-Driven Circulation.

BIBLIOGRAPHY. S. George Philander, James R. Holton, and Renata Dmowska, *El Niño, La Niña, and the Southern Oscillation* (Academic Press, 1989); George Pickard and

William Emery, *Descriptive Physical Oceanography* (Butterworth-Heinemann, 1990).

GREGORY R. FOLTZ
NOAA/PACIFIC MARINE ENVIRONMENTAL LABORATORY

Thermodynamics

THE SCIENCE OF thermodynamics, a branch of physics, aims to describe transformations in energy. Thermodynamics comprises three laws. The first holds that energy can neither be created nor destroyed. Energy in various forms may be transformed into heat (thermal energy) and heat may be transformed into another form of energy so long as the total energy in the system remains constant. The second law states that entropy, a measure of the amount of energy dissipated as heat, increases over time in a closed system. The conversion of energy into heat increases the entropy of a system and the dissipation of heat likewise increases the entropy of a system. The third law states that as temperature approaches absolute zero, the theoretical minimum temperature in the universe, entropy approaches a maximum.

The first law of thermodynamics accounts for the relative constancy of the climate, averaged over long durations. Were Earth simply a reservoir nergy in the form of sunlight, it would heat up to a very high but finite temperature. Earth does not heat up to this magnitude because it radiates heat back into space. The dissipation of energy as heat, according to the second law of thermodynamics, describes the Earth's shedding of radiant energy received from the sun as heat. This law, functioning as a heat accountant, is at the heart of understanding the role of heat in determining the climate. The third law of thermodynamics does not operate as long as the Sun generates energy. Rather, the third law anticipates the end of the universe. The Sun will one day burn out. Bereft of its heat, Earth's climate will be eternally cold, as its temperature approaches absolute zero. Not only will the Sun be extinguished, but all stars in the universe will one day burn out. The heat from these stars will dissipate in all directions in the universe, bringing the temperature, uniform throughout the universe, near absolute zero.

The science of thermodynamics traces the origin of energy in the solar system to the Sun. Energy from the Sun is the basis of Earth's climate, but not all sunlight reaches Earth. The thermosphere lies 190 mi. (306 km.) above Earth's surface, and is the outermost layer of the atmosphere. It absorbs ultraviolet light so efficiently that its temperature rises as high as 570 degrees F (299 degrees C). This conversion of the sun's radiant energy into thermal energy obeys the second law of thermodynamics. The next layer of the atmosphere, the mesosphere, is 50 mi. (80 km.) above earth. Its temperature, cooler than the thermosphere, is 200 degrees F (93 degrees C). Carbon dioxide (CO_2) in the mesosphere absorbs infrared light as heat, and that light radiates from Earth back into space. CO_2 molecules absorb a portion of this light before it reaches space. The larger the number of CO_2 molecules, the more heat they will absorb. The heating of the atmosphere by the absorption of infrared light causes the Greenhouse Effect, the warming of Earth's climate. Beneath the mesosphere is the ozone rich stratosphere, roughly 15 mi. (24 km.) above Earth. The ozone in the stratosphere blocks some 90 percent of sunlight from reaching Earth. Ozone, like the thermosphere, absorbs ultraviolet light. Beneath the ozone layer is the troposphere, a variable layer 5 mi. (8 km.) thick at the poles and 20 mi. (32 km.) thick at the equator. The troposphere holds water vapor, which absorbs both infrared and ultraviolet light, heating the atmosphere. These layers of the atmosphere both absorb and radiate heat. The heat that they radiate either scatters into space or reaches Earth.

Earth absorbs sunlight, chiefly at the equator. This sunlight, in the form of heat, moves to the poles through the currents of the oceans and air. This distribution of heat from an area of greater concentration (the equator) to a region of lesser concentration (the poles) obeys the second law of thermodynamics. Heat supplies the energy for the movement of the oceanic and air currents, which in turn transform the potential energy of stasis into the kinetic energy of motion. On an idealized Earth on which the oceans and air distributed heat evenly throughout the planet, heat would reach thermodynamic equilibrium, the point at which entropy would be at a maximum. Earth is much less efficient than this idealized model. For all the motion of the oceanic and air currents, heat nevertheless concentrates at the equator, which is always warmer than the poles. The waters at the equator hold enormous amounts of heat. Because the oceans liberate their heat slowly, heat accumulates at the equator and is slowly transferred toward the poles.

In accord with the second law of thermodynamics, entropy would increase as heat moves from equator to poles, but the Sun continuously adds heat to Earth, keeping the equator warmer than the poles. Entropy does not increase because the equator remains warmer than the poles. Without the oceanic and air currents, heat would accumulate at the equator and would not circulate to cooler regions of Earth. The currents therefore perform an important function in carrying heat from the equator to temperate and cold latitudes.

Earth and the atmosphere reflect roughly one-third of the sunlight they receive and radiate the other two-thirds into space. Earth sheds the same amount of heat as it receives, keeping earth on average at 60 degrees F (16 degrees C). By contrast outer space, which has no atmosphere to absorb heat, is much colder at minus 454 degrees F (minus 270 degrees C). Earth absorbs sunlight as ultraviolet and visible light and continually radiates it back into space as infrared light.

Earth also reflects light back into space. The oceans reflect half the sunlight they receive, whereas ice and fresh snow reflect 90 percent. In accord with the second law of thermodynamics, entropy decreases when Earth absorbs heat, and increases when the oceanic and air currents diffuse heat to other regions of the planet. Similarly entropy increases when Earth reflects light back into space, thereby dissipating heat.

Entropy is least in equatorial waters because they retain heat and slowly liberate it to other regions of Earth. Heat is not evenly-distributed in equatorial waters, as thermodynamic equilibrium would suggest. In holding heat, the oceans at the equator moderate the climate, keeping lands near them warmer than inland stretches of territory.

The land warms four times faster than the oceans; the air warms faster still. Land and air also radiate heat faster than the oceans. The climate of a desert underscores the rapidity of heating and cooling on land. Temperatures in a desert rise rapidly during the day, often surpassing 100 degrees F (38 degrees C). At night a desert cools with equal speed, dipping as low as freezing. In accord with the second law of thermodynamics, entropy decreases as a desert absorbs heart and increases as it dissipates heat.

Warm climates hold heat not only in water and land, but also in air. Warm air holds more moisture than cool air in the form of water vapor, a greenhouse gas. Water vapor holds more heat than CO_2, methane, other greenhouse gases. Water in all three phases absorb and emit heat. Ice absorbs the least heat and reflects the most sunlight back into space. Liquid water and water vapor are efficient reservoirs of heat.

The laws of thermodynamics work because Earth and its atmosphere absorb and radiate heat. One might argue that the absorption and radiation of heat give Earth its distinctive characteristics and its ability to sustain life.

SEE ALSO: Climate; Greenhouse Effect.

BIBLIOGRAPHY. Tim Li, Timothy F. Hogan, and C.P. Chang, "Dynamic and Thermodynamic Regulation of Ocean Warming," *Journal of Atmospheric Sciences* (v.57, 2000): 3353-3365; Hisashi Ozawa, Atsumu Ohmura, Ralph D. Lorenz and Toni Pujol, "The Second Law of Thermodynamics and the Global Climate System: A Review of the Maximum Entropy Production Principle," *Review of Geophysics* (v.41, 2003).

CHRISTOPHER CUMO
INDEPENDENT SCHOLAR

Thermohaline Circulation

THERMOHALINE CIRCULATION IS global oceanic circulation generated by buoyancy fluxes resulting from heat and freshwater exchange between the ocean, atmosphere, cryosphere, and land. External forcing leading to an increase in water density (i.e., cooling or salinity rise) causes the sinking of more dense water (so-called thermohaline convection) and compensating transport of more light shallower waters of the upper mixed layer and thermocline. This process forms thermohaline overturning, which is one of the principal mechanisms of meridional heat transport in the world's ocean and global coupled ocean-atmosphere system.

Thermohaline circulation is characterized by two regimes, as was first pointed out by Henry Stommel. They are caused by thermal and haline effects and, in turn, account for large-scale temperature and salinity distribution in the World Ocean and, hence, influence the global climate. In general, in recent climate conditions, just thermal overturning circulation prevails in the world oceans, because global thermohaline circulation is formed mostly by the sinking of cold high-latitude waters and the compensating transport of warm shallower water. In general, there are two principal sources of deep and bottom waters. They are in the North Atlantic and Antarctic regions, respectively. These sources produce North Atlantic deep water (NADW), the core of which is at 1.2 to 1.5 mi. (2 to 2.5 km.), and Antarctic bottom water (its core deepens below 2.5 mi. [4 km.]). Haline circulation prevails in some specific regions of the world oceans, such as in the semiclosed Black and Red seas. The effect of salinity changes on the density field is also enhanced in subtropical oceanic regions, especially in the subtropical Atlantic, which is close to the Sahara desert. There, the upper mixed layer depth is mostly controlled by thermohaline convection as a result of salinity effects.

Thermohaline and superimposed wind-driven forcing has caused recent large-scale general oceanic circulation. The relative importance of these two sources for integral volume transport of principal large-scale oceanic currents varies from one region to another. In the North Atlantic, for instance, thermohaline and wind-driven shares in the general circulation of the upper 1.5 mi. (2 km.) layer are discussed in a recent study by Alexander Polonsky.

Global warming may cause, in principle, a change in circulation regime as a result of ice/glacier melting and increased freshwater input into the polar zone of the North Atlantic. This may lead to surface water lightening and blocking of thermohaline overturning. The Gulf Stream should dramatically weaken as a result of that. Such a regime has been called thermohaline catastrophe because it should be accompanied by strong climate shift in the North America and Europe. It is expected that a new climate will be much more severe and will be accompanied by much more frequent and strong North Atlantic cyclones because even just eddy meridional heat transport prevails in the midlatitude atmosphere, and it must now compensate for the reduced meridional thermohaline heat transport in the ocean after thermohaline catastrophe. However, as follows from recent multimodel simulations published by Ronald Stouffer and coauthors, the likelihood of thermohaline catastrophe happening in the next 100 years is quite small, taking into account recent tendencies of ice/glacier melting.

As follows from the simulation results of Stefan Rahmstorf, during the Last Glacial Maximum (about 21,000 years ago), thermohaline circulation was characterized by more shallow meridional cell and reduced meridional heat transport in the North Atlantic. The core of NADW was at about 1 km. In general, this was the result of severe ice conditions in the North Atlantic, where NADW has being produced. Different paleodata analyzed recently by Jean Lynch-Stieglitz and coauthors (2007) confirm in part such a scenario.

There is some evidence that blocking thermohaline circulation in the North Atlantic occurred about 8,200 years ago, just after the last glacier period. Most likely, it was a result of a plume of freshwater from juvenile lakes that rose in the end of a glacier period. Another possibility is the relatively fast melting of an armada of icebergs spreading from a Greenland glacier. However, it has not been proven by the analysis of deep ocean sediments provided by Christopher Ellison and coauthors (2006).

SEE ALSO: Abrupt Climate Changes; Modeling of Ocean Circulation; Modeling of Paleoclimates; Mixed Layer; Thermocline.

BIBLIOGRAPHY. Christopher Ellison, et al., "Surface and Deep Ocean Interactions During the Cold Climate Event 8200 Years Ago," *Science* (v.312/5782, 2006); Jean Lynch-Stieglitz, et al., "Atlantic Meridional Overturning Circulation During the Last Glacial Maximum," *Science* (v.316/5821); Alexander Polonsky, *Role of the Ocean in the Climate Change* (Naukova Dumka, 2007); Stefan Rahmstorf, "Rapid Climate Transitions in a Coupled Ocean-Atmosphere Model," *Nature* (v.372/6501); Henry Stommel, "Thermohaline Convection With Two Stable Regimes of Flow," *Tellus* (v.13, 1961); Ronald Stouffer, et al., "Investigating the Causes of the Response of the Thermohaline Circulation to Past and Future Climate Changes," *Journal of Climate* (v.19/8, 2006).

ALEXANDER BORIS POLONSKY
MARINE HYDROPHYSICAL INSTITUTE, SEBASTOPOL

Thermosphere

THE EARTH IS surrounded by a blanket of air, called the atmosphere. The atmosphere is a thin layer of gases that envelope the Earth. The gases are held close to the earth by gravity and the thermal movement of air molecules. Life on Earth is supported by the atmosphere, solar energy, and the magnetic fields. Five layers have been identified in the atmosphere, using thermal characteristics, chemical composition, movement, and density. The atmosphere is divided into the troposphere, the stratosphere, the mesosphere, the thermosphere, and the exosphere. The thermosphere, from the Greek word (thermos) for heat, is the fourth atmospheric layer from Earth, separated from the mesosphere by the mesopause. It begins about 50 mi. (80 km.) above the Earth and is the layer of the atmosphere directly above the mesosphere and below the exosphere. The lower part of the thermosphere, from 50 to 342 mi. (80 to 550 km.) above the Earth's surface, contains the ionosphere, which is the region of the atmosphere that is filled with charged particles. Beyond the ionosphere, extending out to perhaps 6,214 mi. (10,000 km.), is the exosphere.

The Earth's thermosphere is the layer of the atmosphere that is first exposed to the sun's radiation and so is first heated by the sun; it is the hottest layer of the atmosphere. Within the thermosphere, temperatures rise continually to well beyond 1,832 degrees F (1,000 degrees C). In the thermosphere, ultraviolet radiation causes ionization. At these high altitudes, the residual atmospheric gases sort into strata according to their molecular mass. Thermospheric temperatures increase with altitude as a result of the absorption of highly energetic solar radiation by the small amount of residual oxygen present. Temperatures in the thermosphere are highly dependent on solar activity. Radiation causes the air particles in this layer to become electrically charged, enabling radio waves to bounce off and be received beyond the horizon.

The few molecules that are present in the thermosphere receive extraordinary amounts of energy from the sun, causing the layer to warm to high temperatures. Air temperature, however, is a measure of the kinetic energy of air molecules—not of the total energy stored by the air. The air is so thin that a small increase in energy can cause a large increase in temperature. Because the air is so thin within the thermosphere, such temperature values are not comparable to those of the troposphere or stratosphere. Again, because of the thin air in the thermosphere,

The Northern Lights occur in the thermosphere, which is the fourth atmospheric layer from Earth. It is the first layer of the atmosphere that is exposed to the sun's radiation, so it is the hottest layer of the atmosphere.

scientists cannot measure the temperature directly. Instead, they measure the density of the air by how much drag it puts on satellites and then use the density to determine the temperature.

Although the measured temperature is very hot, the thermosphere would actually feel very cold to humans because the total energy of the few air molecules residing there would not be enough to transfer any appreciable heat to our skin. In addition, it is so near vacuum that there is not enough contact with the few atoms of gas to transfer much heat. A normal thermometer would read significantly below 32 degrees F (0 degrees C). The dynamics of the lower thermosphere are dominated by the atmospheric tide, which is driven, in part, by the very significant diurnal heating.

The atmospheric tide dissipates above this level because molecular concentrations do not support the coherent motion needed for fluid flow. The International Space Station has a stable orbit within the upper part of the thermosphere, between 199–236 mi. (320–380 km.). The Northern Lights also occur in the thermosphere.

SEE ALSO: Atmospheric Composition; Atmospheric Vertical Structure; Climate Change, Effects.

BIBLIOGRAPHY. Lecture Notes in Plant Biology, University of Maryland, 2000; Windows to the Universe, 2000, www.windows.ucar.edu/; University Corporation for Atmospheric Research, www.wikipedia.org.

AKAN WILLIAMS
COVENANT UNIVERSITY

Worldwide, it is estimated that 50,000 thunderstorms occur every day, and over 18 million occur per year.

Thunderstorms

A THUNDERSTORM IS a localized storm that is produced by a cumulonimbus cloud and always contains thunder and lightning. Thunderstorms form in conditionally unstable environments, meaning there is cold, dry air aloft over warm, moist surface air. This causes the air to become buoyant and allows for rising air motion. A lifting mechanism is also needed to start the air moving. Such lifting mechanisms include surface heating, surface convergence, lifting caused by mountains, or lifting along frontal boundaries.

The heat and the humidity of the summertime can often produce what are called ordinary thunderstorms or air mass thunderstorms. These are the type of thunderstorms that seem to suddenly pop up, last less than an hour, and are rarely severe. A severe thunderstorm is defined by the National Weather Service as having three-quarter-inch diameter hail or surface winds exceeding 58 mi. (93 km.) per hour or producing a tornado. Ordinary thunderstorms also do not usually have excessive vertical wind shear, meaning that the wind speed or direction does not change greatly with height. They usually go through a series of stages from birth to decay. The first stage is known as the cumulus stage and is dominated by updrafts. The updrafts bring in warm, moist air, which then cools and condenses as it rises. When the clouds further develop and precipitation starts to fall, a downdraft is produced. This marks the beginning of the mature stage, which is the most intense stage. During this stage, the strong updraft is still present, supplying the warm, moist air, but the strong downdraft is also evident. The gust front is located at the boundary of the updraft and the downdraft. This is an area where the wind velocity changes rapidly. Eventually, the downdraft will cut off the supply of warm, moist air in the updraft. When this occurs, typically 15 to 30 minutes after the mature stage, the thunderstorm will start to weaken and enter the dissipating stage as a result of the deprivation of energy from the updraft.

If the vertical wind shear increases, this allows for the thunderstorm to tilt. Therefore, the downdraft is less likely to cut off the updraft, which allows the thunderstorm to persist longer. Sometimes the downdraft can slide underneath an updraft, which can produce multiple cell thunderstorms or simply multicell storms. If the vertical wind shear becomes extremely strong, the shear can produce a large rotational thunderstorm known as a supercell. Supercell thunderstorms are large storms that last longer than an hour and are often severe and can produce tornadoes. The strong wind shear creates horizontal spin, which can then rotate vertically when the updraft encounters the vortex.

Thunderstorms can occur as a line of multiple-cell thunderstorms known as squall-line thunderstorms. These usually form along or slightly ahead of a cold front. The line of thunderstorms can extend over 500 mi. (805 km.) and often exhibit severe characteristics. When thunderstorms occur in a large circular pattern, they are known as a mesoscale convective complex, or MCC. An MCC is a large, convectively driven system that usually lasts more than 12 hours and covers more than 38,610 sq. mi. (100,000 sq. km.) Many thunderstorms are embedded within the MCC and often form during the summer in the Great Plains. As warm, moist air is brought in from the Gulf of Mexico, the tops of the very high clouds cool rapidly by emitting radiation into space. This makes the atmosphere very unstable and allows for the MCCs to generate and persist. Because MCCs are usually located underneath weak upper-level winds, they tend to travel very slowly, which can cause locally heavy rains and flooding events.

All thunderstorms have lightning—the electrical discharge—and thunder—the resulting shockwave produced by the extreme heating. Lightning has a temperature of approximately 54,000 degrees F (29,982 degrees C), which is five times hotter than the surface of the sun. Lightning occurs during the mature stages of thunderstorms and can appear within a cloud and travel from one cloud to another or from cloud to ground. Most lightning strikes are within a cloud.

Worldwide, it is estimated that 50,000 thunderstorms occur every day, and over 18 million occur per year. Thunderstorm frequency is most common in the tropics, especially near the Intertropical Convergence Zone, which is an area of low pressure near the equator. Thunderstorms occur with lower frequency in drier regions near 30 degrees N/S, which is dominated by the subtropical high pressure, as well as in the polar regions. In the United States, thunderstorm activity is predominantly found in the southeast, with a maximum located over Florida. Florida has over 90 days of thunderstorms per year as a result of the convergence of wind from the Gulf of Mexico and the Atlantic.

Thunderstorms release a massive amount of latent heat energy through condensation. This is a major mechanism for the Earth to transfer heat from areas of energy surplus near the equator to areas of energy deficit toward the poles. It is generally agreed that increased greenhouse gas emissions are causing global temperatures to rise. With increased global surface temperatures, it is expected that more clouds will be produced with increased evaporation rates. However, the potential effects of increased cloud coverage and what it means for potential rainfall and thunderstorm activity is not fully understood.

The degree to which the increased aerosols and clouds will reflect solar energy back into space is the main discrepancy. Some scientists think that global warming will increase evaporation rates, thereby producing more precipitation. Others have speculated that with the increased clouds and aerosols in the atmosphere, this will greatly reduce the amount of radiation reaching the earth. As a result, this will make the lower and midlevels of the atmosphere warmer, therefore reducing evaporation rates. Because the thermal gradient, that is, the difference in temperature, will be reduced from the surface to the atmosphere, the reduced evaporation rates could then potentially make for drier conditions.

SEE ALSO: Climate Change, Effects; Rain; Rainfall Patterns.

BIBLIOGRAPHY. C. Donald Ahrens, *Meteorology Today* (Brooks/Cole, 2007); John Houghton, *Global Warming* (Cambridge University Press, 2004); Fred T. Mackenzie, *Our Changing Planet* (Prentice Hall 2003).

KEVIN LAW
MARSHALL UNIVERSITY

Togo

THIS WEST AFRICAN country, officially the Togolese Republic, is located between Ghana and Benin, with a northern border with Burkina Faso. It has a land area of 21,925 sq. mi. (56,790 sq. km.), with a population of 6,585,000 (2006 est.) and a population density of 280 people per sq. mi. (108 people per sq. km.). Some 38 percent of the country is arable, with a further 4 percent used for meadows and pasture and another 28 percent of the land forested.

For the electricity production in Togo, 97.9 percent comes from fossil fuels, with 2.1 percent coming from hydropower. These generate 102 million kilowatt-hours (kWh; in 2001), with 520 kWh, largely drawn from hydropower, more imported from Ghana. Although before World War I the Germans tried to establish Togoland (as it was then called, including part of what is now eastern Ghana) into a model colony, the country has remained relatively poor, with its carbon dioxide emissions being 0.2 metric tons per person in 1990, rising gradually to 0.38 metric tons per person in 2003. About 68 percent of the country's carbon emissions come from liquid fuels, with the remainder coming from the manufacture of cement. By sector, 37 percent comes from manufacturing and construction, with 35 percent from transportation and 14 percent from electricity and heat production. The high figure for transportation is because there is only one train line, going from the capital, Lomé, to Blitta, with very few other forms of public transport.

The coastal part of Togo, around Lomé, is low lying, and as such, it is at risk from global warming and climate change. The rising average temperatures are also likely to lead to increased desertifi-cation in the north of the country. The Togo government ratified the Vienna Convention in 1991 and took part in the United Nations Framework Convention on Climate Change, signed in Rio de Janeiro in May 1992. The government accepted the Kyoto Protocol to the UN Framework Convention on Climate Change on July 2, 2004, with it coming into force on February 16, 2005.

SEE ALSO: Climate Change, Effects; Deserts.

BIBLIOGRAPHY. David Bovet and Laurian Unnevehr, *Agricultural Pricing in Togo* (The World Bank, 1981); World Resources Institute, "Togo—Climate and Atmosphere," www.earthtrends.wri.org (cited October 2007).

ROBIN S. CORFIELD
INDEPENDENT SCHOLAR

Tonga

LOCATED IN THE South Pacific, the Kingdom of Tonga has a land area of 289 sq. mi. (748 sq. km.), with a population of 100,000 (2006 est.) and a population density of 396 people per sq. mi. (153 people per sq. km.). The country consists of 169 islands, but only 36 of these are permanently inhabited. With 24 percent of the country listed as arable, some 6 percent is used for meadows and pasture, with 12 percent of the country being forested; Tonga has a very restricted timber industry program.

Tonga also has a low per capita rate of carbon dioxide emissions, with 0.8 metric tons per person in 1990, rising to 1.12 metric tons by 2003, ranking Tonga 136th in the world in terms of emissions. Although the electricity production in the country is low, it is all generated from fossil fuels. With all the country's carbon dioxide emissions coming from liquid fuels, this accounts for not just electricity production but also the use of automobiles and small household or business generators.

Global warming and climate change are already having a major effect on Tonga, with the flooding of parts of the country, including a number of the uninhabited islands, and the very real risk of large parts of the country being lost as the water level rises. In addition, there is a problem over the alienation of arable land, deforestation leading to soil erosion, and off-shore coral reef bleaching and loss of marine life.

The Tonga government took part in the United Nations Framework Convention on Climate Change, signed in Rio de Janeiro in May 1992 and ratified in 1998, in the same year as the ratification of the Vienna Convention. The current Tonga government has not expressed a position on the Kyoto Protocol to the UN Framework Convention on Climate Change.

SEE ALSO: Climate Change, Effects; Floods.

BIBLIOGRAPHY. Paul Smitz, *Samoan Islands and Tonga* (Lonely Planet, 2006); "Climate Change and the Pacific Islands," United Nations Economic and Social Commission for Asia and the Pacific (UNESCAP), http://www.unescap.org/mced2000/pacific/background/climate.htm; World Resources Institute, "Tonga—Climate and Atmosphere," hwww.earthtrends.wri.org (cited October 2007).

JUSTIN CORFIELD
GEELONG GRAMMAR SCHOOL, AUSTRALIA

Toronto Conference

SCIENTISTS FROM VARIOUS international organizations, such as the World Meteorological Organization in Geneva, met with their peers in groups at various locations for three years. Following the signing of the United Nations Vienna Convention on the Protection of the Ozone Layer (1985) and the Villach Conference (1985), these meetings helped to develop the basis for further action. From the discussions at these meetings, a scientific accord on the main aspects of how much climate warming can be expected emerged. The confluence of this emerging consensus and other events led to the Toronto Conference in 1988.

The scientists' efforts gained the support of the United Nations, the World Meteorological Organization, the Canadian government, and other international organizations. The scientists then came together in Toronto, Canada, from June 27 to 30, 1988. In attracting national policymakers as well as 300 scientists from 46 countries and organizations, this conference became the first such international conference to combine science and policy.

Entitled the International Conference of the Changing Atmosphere: Implications for Global Security, the meeting highlighted atmospheric issues in a comprehensive way. The concern for the potential damage to the planet was compared with the consequences of nuclear war, and the scientific consensus at the conference astonished its chair, Stephen Lewis, who was then Canada's ambassador to the United Nations. Lewis also brokered the strongly worded final declaration. Identifying the existing situation as "an unintended, uncontrolled, globally pervasive experiment," the Conference Statement claimed that the consequences of this experiment would be second only those of a global nuclear war.

Recognizing that attempts to address issues affecting the atmosphere as a whole had been fragmentary to date, the Toronto Conference took a more global approach. The initiative was to integrate the existing Vienna Convention (1985) and the 1979 Geneva Convention on Long-Range Transboundary Air Pollution and to provide a basis for including issues that had not yet been addressed or recognized. Such an integrated approach to considering the atmosphere as a whole would conceivably permit a more complex approach to interrelated issues and solutions. As such, this initiative raised the possibility of a comprehensive Law of the Air.

RECOMMENDATIONS

The comprehensive approach and wide representation enabled attention to be paid to the scientific, economic, and social concerns. The attendees generated specific calls for action to governments, industry, and nongovernmental organizations. Working groups within the conference made specific recommendations to address a wide range of issues that were related and relevant to the health of the global atmosphere.

Issues that were recognized were those that arose both directly out of usage of the atmosphere and indirectly, through human effects on land and water. The atmospheric effects of the manner and form of human settlement—including the increasing urbanization of populations and acid rain—were directly relevant. Indirect atmospheric effects resulted from the full range of human activities including food production, industry, energy usage, trade, and investment.

Changing climate and human effects on coastal and marine resources were also pertinent. Human decision making involving forecasting, uncertainty, futures, and geopolitical issues—higher-order considerations resulting from the integration of programs and legal issues—were also addressed.

The precise form of a global pact was debated, with the Canadian government favoring the concept of an International Law of the Air. Canadian Prime Minister Brian Mulroney pointed out that the groundwork for such an approach exists in the Montreal Protocol to protect the ozone layer and in the impending international protocol on nitrogen oxide control. Norwegian Prime Minister Gro Brundtland recommended a global convention on protection of the climate.

The meeting recommended a global pact to protect the atmosphere and a world atmosphere fund to facilitate global solutions, which recognized differential issues in usage and effects. For instance, the different historical consumption of and contribution to the atmosphere of already industrialized nations and those in the process of industrializing would be balanced by having the fund financed in part by taxes on fossil fuels consumed in industrialized nations. The proposed atmosphere fund would then be used partly to provide economic assistance to developing countries pursuing environmentally friendly strategies such as reducing deforestation.

The delegates concluded that immediate action is imperative to address ozone depletion, global warming and sea-level rise, and acidification by atmospheric pollutants. The potential role of nuclear power as a clean energy source was debated, but no official recommendations emerged. Reduction of other greenhouse gases, substances that deplete the ozone layer, and acidifying emissions were recommended.

Specifically on the issue of global warming from greenhouse gases and climate change, the conference reached a consensus on the likelihood of a rise in the global mean temperature of between 2.7–8 degrees F (1.5–4.5 degrees C) by about 2050, but not on whether such warming has begun. The conference statement called for a 20 percent cut in present (1988) levels of global carbon dioxide emissions by the year 2005, about half of which could be achieved through conservation, leading to an eventual cut of 50 percent. This statement was possible as a result of the participation of governments that voluntarily committed to cutting carbon dioxide emissions by 20 percent by the year 2005. This became the so-called "Toronto target" for greenhouse gas emissions and went beyond the emissions targets recommended by most later international conferences, as well as the 1992 UN Framework Convention on Climate Change and the core goal of Kyoto.

The Toronto Conference was also influential in other developments. The Intergovernmental Panel on Climate Change (IPCC), an international grouping of over 300 of the world's best climate scientists, charged with peer reviewing and reporting on the latest international science, effects, and responses to climate change, had been formed just before the conference. The conference was instrumental in promoting the IPCC and in the eventual appointment of Swedish scientist Bert Bolin to head it.

PUBLIC AWARENESS

Discussions at the conference also led to the allocation of resources to the World Climate Programme and other global research institutions, to the support for technology transfer solutions, to the advocacy of reduction of deforestation, and to raising public awareness of issues related to the atmosphere.

As a follow-up to June's Toronto Conference, a smaller meeting of legal and policy experts was held in Ottawa, Canada, from February 20 to 22, 1989, to begin developing an international accord for the protection of the atmosphere. The 80 participants, acting in a personal capacity, constituted a broad spectrum of experts and officials from developed and developing countries, nongovernmental organizations, and academic institutions and discussed the legal and institutional framework for dealing with emerging atmospheric problems, agreed where possible on the basis of an umbrella convention framework, and identified areas of possible disagreement.

With the 1982 Law of the Sea as a precedent, the meeting recommended that one or more international conventions such as a Law of the Atmosphere and a narrower Climate Change Convention with appropriate protocols were urgently needed, especially to limit greenhouse warming. The statement from this meeting presented early drafts of the proposed documents.

The Law of the Atmosphere approach was criticized as being more unrealistic than the narrower Climate Change Convention and did not receive much attention from subsequent negotiators. In carrying the ideas from the Toronto Conference forward, the Ottawa meeting proposed broad terminology for atmosphere and atmospheric interference; discussed the obligation of nations to protect the atmosphere, recognizing the relationship between the atmosphere and other aspects of the environment; recognized the need to balancing of development internationally; and proposed an international notification process for harmful activities, liability, compensation, and dispute resolution mechanisms, as well as details of the Atmospheric Trust Fund.

Features that were eventually included in the 1992 Climate Convention included the approach of a framework treaty that deals with the central issues with protocols for particular details. This would be similar to the Vienna Convention with the Montreal Protocol. Two years after the Toronto Conference, the IPCC issued assessments that provided the basis for the United Nations Framework Convention on Climate Change in 1992, followed by the Kyoto Protocol in 1997.

SEE ALSO: Climate Change, Effects; Villach Conference; World Meteorological Organization.

BIBLIOGRAPHY. Environment Canada Library, www.msc-smc.ec.gc.ca/library/index_e.html?; D. Zaelke and J. Cameron, "Global Warming and Climate Change: An Overview of the International Legal Process," *American University Journal of International Law and Policy* (v.5/2, 1990).

LESTER DE SOUZA
INDEPENDENT SCHOLAR

Tourism

THE RELATIONSHIP BETWEEN tourism and global warming is a paradoxical one: global warming has become a threat to tourism, yet tourism remains a major cause of global warming. This vicious circle is well known to all stakeholders of the tourism industry, but implementing meaningful change has proven difficult because of three types of resistance: politico-economic resistance (from policymakers in regions and countries that rely heavily on tourism as a source of income), commercial resistance (from the tourism industry itself), and sociocultural resistance (from tourists who are not ready to change their behavior).

Several factors account for the considerable development of tourism since World War II: growing affluence, longer holidays, cheaper transportation, the availability of preorganized packaged tours, and the development of an industry catering both to mass tourists and to independent travellers. The subsequent increase in demand has resulted in an exponential rise in visitor numbers, both domestically (within countries) and internationally (especially from developed countries to developing countries). Although domestic tourism is statistically much more important (e.g., it accounts for 99 percent of all U.S. tourism and for 85 percent of all Australian tourism), international tourism is easier to measure (through a simple head count at borders); in addition, international tourism corresponds much more to the mainstream imagery of tourism: an island-hopping cruise in the Caribbean, a romantic holiday in Paris, a big game safari in Kenya. According to the World Tourism Organization, the number of international tourists increased from a mere 25 million in 1950 to 800 million in 2005. This number is predicted to double and to reach 1.8 billion by 2020, as more and more people want to travel. They may well know that they contribute to global warming and climate change, and some may feel a pang of guilt and remorse, but their desire to travel is stronger.

Climate is a key resource for tourism: favorable climatic conditions are key attractions for tourists, be it to ski in the mountains, to relax on a beach, or to experience nature. As soon as climatic conditions fluctuate and become less predictable, the tourism demand is affected and tourist flows move elsewhere: tourism, as a geographic phenomenon, is fickle and versatile. The mass media occasionally run stories about tourism hot spots that are victims of climate change and see their tourism appeal decrease; examples abound from all across the world, from less snowfall and shorter skiing seasons in Aspen, Colorado, or in Chamonix in the French Alps, to damage to coral reefs and rising ocean

Warmer temperatures at ski resorts: As soon as climatic conditions fluctuate and become less predictable, the tourism demand is affected and tourist flows move elsewhere; tourism, as a geographic phenomenon, is fickle and versatile.

water in Australasia, not to mention hurricanes that affect island resorts and the cruising industry.

These media stories are not just anecdotes or isolated incidents: they are part of a wider concern already well documented in the tourism literature, both in the academic literature (with seriously researched case studies, a nascent modelization of the relationship between climate change and tourism, and an increasing number of specialists, such as the Canadian Daniel Scott, the Dutch Bas Amelung, and the French Jean-Paul Ceron) and in the professional literature (industry publications such as professional bodies' reports and newsletters, as well as travel guides for tourists).

ENVIRONMENTAL AWARENESS AND TOURISM

International tourism organizations endeavor to raise awareness, to harness energies, and to articulate realistic plans of action. In 2003, the World Tourism Organization held its first Summit on Climate Change and Tourism. This resulted in the so-called Djerba Declaration on Tourism and Climate Change, whose signatories encouraged all governments to act to control climate change. The Djerba Declaration also asked the travel industry to adjust its activities to minimize climate change, and it invited consumer associations and the media to further raise public awareness, both at destinations and in generating markets. Taking place in 2007 in Davos, Switzerland, the Second International Conference reviewed and emphasized the key aims and intentions of the Djerba Declaration, strengthening its urgency and exploring concrete ways for tourism to respond to climate change, while still ensuring tourism development as a tool of economic growth and sociocultural well-being. At another level, in 2007 the World Travel and Tourism Council launched an international campaign on the same topic, calling for an open and mature dialogue on issues of tourism, climate change, and the environment; the campaign ran full pages in major publications such as *The Daily Telegraph*, *Newsweek*, and *The Wall Street Journal*, as well as travel trade media around the world.

Tourism professionals are aware of their responsibilities with regard to the environment, but they also know how much tourism contributes to the world economy. In 2007, the tourism industry globally represented 231.2 million jobs, which corresponded to 8.3 percent of total employment; that is, one in every 12 jobs worldwide. By 2017, this figure is expected to rise to 262.6 million jobs. Through their statements and declarations, tourism policymakers at all levels (local, regional, national, and supranational) and from all sectors (public, private, and voluntary) show that they understand the seriousness of the situation with regard to global warming and climate change, but one cannot ignore the fact that tourism is one of the world's largest economic sectors.

Tourism is not just about holidays and recreation, it is a powerful economic force that cannot be obliterated; rather, it has to be managed by implementing ways to maximize its benefits and minimize its costs. Through the concept of sustainable tourism, sustainable development has now become a key agenda in tourism, covering a range of social, cultural, and environmental issues, including references to climate change and global warming, yet solutions and ways forward are difficult to find. Simply controlling the number of international flights and limiting the amount of international tourists may not be a viable alternative at all: such a short-term measure could have devastating effects on many regions and countries whose economies are dependent, if not overreliant, on tourism income; for instance, small island states such as the Maldives, the Netherlands Antilles, and the Seychelles.

In 2007, global tourism generated over $7 trillion, and this number is likely to double within the next decade. The ongoing democratization of tourism means that more and more people can afford to travel and readily do so, even when they are aware of the effect on climate change and global warming; their arguments are usually twofold: first, that their own individual contribution is minimal, and second, that tourism is only one cause of climate change and global warming, among others. Many factors account for the ongoing increase in tourists' numbers: technological developments (epitomized by the superjumbo A380, with its capacity of 850 passengers), market deregulation (leading to

more competition, which keeps prices low, especially in the airline industry), and the multiplication of specialized niche markets (such as sports tourism, senior tourism, gay tourism, or industrial tourism, making the demand more fragmented but also easier to target and satisfy). Neither financial penalties (tourism ecotaxes, imposed on airlines or at the destination) nor ethical appeals (campaigns asking would-be tourists to reconsider because of their carbon footprint) are proving effective deterrents for what is increasingly regarded as a right and not a privilege. Even the most vocal critics of tourism like to travel to conferences (business tourism) and go away on holidays (leisure tourism), which weakens the arguments of the antitourism lobby.

THREATS TO TOURISM

Climate change poses several risks to tourism; not only direct risks (climate variability affecting immediate demand as well as tourists' comfort and safety in the short term) but also indirect risks (such as causing damage to ecosystems or reducing water supplies, which may jeopardize tourism in the long term). This is ironic, inasmuch as tourism as a whole is partly responsible: by definition, tourism relies on methods of travel that generate air pollution (greenhouse gas emissions from vehicles that transport tourists, especially aircraft), so by its very existence, tourism heavily contributes to climate change. Rather than attempting a difficult—if not impossible—balancing act, specialists have identified two strategies. The first strategy involves innovation (e.g., with regard to sources and production of energy) and disseminating best practice in terms of sustainable development (e.g., through benchmarks and rewards). This first approach takes tourism in its wider industrial context, applying to tourism some policies and measures from other sectors, such as building and manufacturing.

The second strategy involves analyzing how climate change affects tourism to proactively restructure the tourism industry itself, both in terms of tourism demand and tourism supply; for instance, extremely hot temperatures in summer in seaside destinations may lead to a decrease of the demand in summer but to higher rates in other warm times of the year, such as warmer winter periods. As seasonality has always been a plague of the tourism industry, this climate change may eventually prove beneficial, though it will require adapting and revisiting established patterns. This second approach considers tourism as a specific and idiosyncratic system, although some related sectors, such as agriculture, local transport, and the entertainment industry, are likely to be affected too (a phenomenon conceptualized as backward linkages). The two strategies are not mutually exclusive: they can be combined, as they are both underpinned by a mix of idealism and pragmatism (adaptation and mitigation are two concepts used by scholars and policymakers alike).

Because of the intrinsic diversity of the tourism industry (exemplified by differing tourists' needs and expectations, from a backpacking teenager touring Europe to jetsetters staying in exclusive resorts), there may not be a one-size-fits-all solution. Case studies of destinations ranging from the Fijian archipelago to Banff National Park in Canada show that each region needs to develop its own methodologies and planning scenarios to anticipate changes in tourism demand and distribution, while remembering that it is not only the complex tourism system that is affected by global warming and climate change but also the local communities. Collaboration between partners and agencies is always heralded as a necessary mechanism; a good illustration is the official cooperation between the World Tourism Organization and the World Meteorological Organization: in 2006, an Expert Team on Climate and Tourism was jointly set up by both agencies. Such meetings of experts can result in sharing intelligence to help with research projects, as there is a wide recognition that decisions need be evidence based. The tourism industry is aware of the problems posed by global warming and climate changes, and it wants to be part of the solution.

SEE ALSO: Economics, Cost of Affecting Climate Change; Transportation.

BIBLIOGRAPHY: Stefan Gossling and Michael C. Hall, eds., *Tourism and Global Environmental Change* (Routledge, 2005); Michael C. Hall and James Higham, eds., *Tourism, Recreation and Climate Change* (Channel View Publications, 2005); *Journal of Sustainable Tourism* special issue on climate change, 2006; Andreas Matzarakis and Chris de Freitas, eds., *Proceedings of the First International Workshop on Climate, Tourism and Recreation* (Porto Carras, 2001); Daniel Scott, et al., eds., *Climate, Tourism and Recreation: A Bibliography* (University of Waterloo, 2006).

LOYKIE L. LOMINE
UNIVERSITY OF WINCHESTER

Trade Winds

THE TRADE WINDS are a large-scale component of Earth circulation, occupying most of the tropics straddling the equator between approximately latitude 30 degrees N and latitude 30 degrees S, with a seasonal shift of the entire trade wind belt system about 5 degrees of latitude northward during summer (July) and southward during winter (December).

In the Northern Hemisphere, warm equatorial air rises and flows north toward the pole, the Coriolis Effect (caused by the Earth's rotation) deflects the current, and as the air cools, it descends, blowing southwestward from the northeast. In the Southern Hemisphere, warm equatorial air rises and flows south toward the pole, the Coriolis effect deflects the current, and as the air cools, it descends, blowing northwestward from the southeast. The rising air is associated with deep atmospheric convection, heavy precipitation, and weak wind speeds, with an influence on global weather patterns Air heated by the sun rises and releases moisture through rain and thunderstorms. Once the air cools, it descends as drier air. In the equatorial low, the air rises and travels aloft to the subtropical highs, where it then sinks.

Mariners called these reliable wind currents for sailing the trade winds, or westerlies. The name trade winds comes from an old sailing term meaning that the winds could be counted on to blow steadily from the same direction at a constant speed. The trade

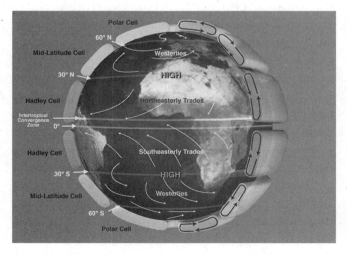

A NASA graphic of climatic zones and trade winds. In the tropics, the easterly trade winds dominate.

winds, or easterlies, carried air from east to west at low latitudes and are less regular over land areas than they are over the oceans. The trade winds meet at the Intertropical Convergence Zone. The doldrums (downward branch of the Hadley Cell, named for George Hadley, whose 1735 paper linked rising air and the Earth's rotation in causing the trade winds) are the calm winds at the Intertropical Convergence Zone in the area between latitude 5 degrees N and latitude 5 degrees S, where a sailing ship might not move because of the calm winds. In satellite imaging, the Intertropical Convergence Zone appears as a band of clouds. The strength and position of the Intertropical Convergence Zone influences tropical and global weather patterns.

Air temperature differences across the Earth's surface (both land and water) create winds, with warm air being lighter than cold air. Near the equator, the sun heats the sea surface, causing the warm air at the surface to rise and be replaced by the trade winds blowing from subtropical high pressure systems into equatorial low-pressure troughs. The trade winds blow steadily for days and are among the most consistent on earth. When trade winds move over warm tropical waters, they pick up moisture and bring heavy rainfall to the windward-facing slopes of mountainous areas, contrasting with the downward motion of dry air that creates desert areas on land. Because the area of Earth between the Tropic of Cancer and Tropic of Capricorn, lying at approximately 23 degrees latitude on either side of the equator, receives more solar heat than the rest of the earth, the warm air creates clouds and rain with thundershowers there almost every day.

The influence of the trade winds on weather and climate is seen with El Niño, La Niña, and the development of hurricanes/cyclones. The differences in pressure and temperature between the two sides of the Pacific are caused by the trade winds; air blowing from east to west pushes water, making the sea level higher in the western Pacific, and makes cold water rise toward the surface, making the eastern Pacific approximately 14 degrees F (7.7 degrees C) cooler than the western Pacific.

During El Niño years, the eastern Pacific sea surface is warmer, and the Intertropical Convergence Zone is closer to the equator, causing rainfall over the Pacific. The warm surface temperature is associated with reversed air pressure patterns and decreasing strength of trade winds, so more water stays in the eastern Pacific off the coast of South America. With the rain pattern shift eastward, the western Pacific can become drier over India and much of southeast Asia. A similar pattern sets up in the Atlantic, resulting in extreme drought in the eastern United States and reduced tropical storm development in the Atlantic Ocean.

During La Niña years, the trade winds are stronger than normal, causing more cold water to rise to the ocean surface. The cooler surface temperature is associated with a rain pattern shift westward. The eastern areas thus become drier, with an increased probability of flooding from monsoons in both India and much of southeast Asia.

Hurricanes (Atlantic) and cyclones (Indian Ocean) are tropical storms of low-pressure cells. Formation of hurricanes in the Atlantic comes from solar heating of water off the West African coast along the Intertropical Convergence Zone, with high cumulus cloud formation in the low-pressure area along the edge. These systems are pushed westward by the trade winds, and the rotation is set in motion by the Coriolis Effect. A similar pattern sets up in the Pacific, causing cyclones.

SEE ALSO: Coriolis Force; Doldrums; El Niño and La Niña; Winds, Westerlies.

BIBLIOGRAPHY. Tom Farrar and Robert Weller, "Where the Trade Winds Meet: Air-Sea Coupling in the Inter-Tropical Convergence Zone," www.research.noaa.gov; Jack Fishman and Robert Kalish, *The Weather Revolution Innovations and Imminent Breakthroughs in Accurate Forecasting* (Plenum Press, 1994); National Weather Service Southern Region Headquarters, "Effects of ENSO in the Pacific," www.srh.noaa.gov.

LYN MICHAUD
INDEPENDENT SCHOLAR

Transportation

TRANSPORTATION CAN BE simply defined as the movement of people, goods, and services from one place to another. Its system consists of the fixed facilities, flow entities, and control systems that permit the free flow and efficient movement of people and goods from place to place across geographical boundaries. Basically, there exist three forms of transportation: the road, water, and air transport.

The road transport systems are made up of the vehicular transport system and the rail transport system. The vehicular transport system comprises the different grades, sizes, and types of automobiles, and the rail transport system comprises the train systems of transportation. Water transportation also comprises the different shapes and sizes of water vehicles, known primarily as ships, boats, ferries, canoes, and so on. Air transport, in turn, comprises the different grades, shapes, and sizes of airplanes and helicopters.

FOSSIL FUELS

Each one of these means of transportation runs on fossil fuels of crude oil distillates and coal, apart from modern train systems in developed countries, which may run on automated power. These fuels are subjected to internal cycles of combustion, giving gaseous by-products of carbon dioxide (CO_2), carbon monoxide, and sulphur dioxide. Methane and nitrous oxide are also emitted by cars. World over, the greenhouse gas contribution of transportation is very high as a result of factors that include an increase in the number of vehicles, volume of passengers, and freight traffic. The percentage contribution of CO_2 from transportation alone varies from state to state and country to country. In 1990, Japan's contribution was put at about 19 percent, and that of the United States surprisingly doubled between 1960 and 2001. Specifically, it is reported that emissions of CO_2 in the United States jumped from 2 billion metric tons in 1960 to almost 5.7 billion metric tons in 2001, accounting for over a 100 percent increase, with over 20 percent of this emission linked to transportation. Transportation is also reported to account for 40 percent of volatile organic compounds, 77 percent of carbon monoxide, and 49 percent of nitrogen oxide emissions in the United States. In Canada, also, it is reported that transportation is the largest single anthropogenic source of outdoor air pollution. On average, each of the several million vehicles registered across the country emits approximately 5 tons of air pollutants and gases annually. This trend in record is the same for all industrialized nations and several developing nations, such as Nigeria in West Africa,

because of an increased population and a rapid rate of economic growth, bringing about increased use of automobiles and other means of transportation. This increase has brought with it increasing emissions of air pollutants and greenhouse gases because of the type of engines in use and the nature of the fuels in place.

In general, it is reported that combustion engines emit nitrogen oxides (NO_x), carbon monoxide, and unburned hydrocarbons capable of chemical transformation in the atmosphere, creating other gaseous matter such as ozone. Ozone is a triatomic molecule consisting of three atoms of oxygen; it is an allotrope of oxygen but is much less stable. Its instability makes it a strong oxidizing agent, having the ability to decompose to oxygen in the atmosphere within 30 minutes. In its physical undiluted state, at standard temperature and pressure, it is a pale blue, odorless gas. In the troposphere, ozone acts as a greenhouse gas and has a radiative forcing of about 25 percent that of CO_2. Around the Earth's surface, it poses a regional air pollution problem damaging human health and agricultural crops.

More so, residual fuel oils, particularly the heavy oil used aboard ships, contain sulphur, which reacts with atmospheric water and oxygen to produce sulfate particles and sulphuric acids, also known as acid rain. It lowers soil and freshwater pH, resulting in damage to our natural environment, and also causes chemical weathering. Its ability to increase the reflection of part of the sunlight that should come into the Earth's surface creates a cooling effect. Road traffic on its own is a major contributor to environmental degradation and global warming. It clearly provides the largest net contribution to warming through its large emissions of CO_2 and significant emissions of ozone and soot. Soot particles emitted by diesel engines have the ability to absorb sunlight, thus heating up the climate. Total warming from road traffic is reported to be about 0.19 Watts per sq. m. (W/m^2), forming about 7 percent of the total climate forcing, as a result of an increased concentration of ozone, soot, and greenhouse gases.

Air traffic, as a sector in the transportation industry, also shows a trend toward increased environmentally unfriendly emissions. Airplanes fly in the upper edges of the atmosphere, where the air is rarefied and the planes release large quantities of greenhouse gases. CO_2, the main constituent in the exhaust gases, slowly descends into the lower altitude. However, the large number of planes flying across the sky has caused the average amount of these greenhouse gases to increase. CO_2 and other greenhouse gases get cut up in the stratosphere, where they become much more potent than at the Earth's surface, blocking radiant energy from reaching the planet. Thus, the global warming effect of air traffic pollution in the stratosphere is very high. In addition, the NO_x emitted have an especially large effect on ozone formation. A more recent research report on air traffic suggests that the occurrence of ice clouds, called cirrus clouds, at flying altitudes is increasing in areas with heavy air traffic because the trails of vapor left by aircraft at high altitude under certain meteorological conditions can expand. These clouds, found at altitudes of between 5 and 7.5 mi. (8 and 12 km.), have a warming effect on the climate because their greenhouse effect is stronger than their cooling effect through the reflection of light. This is a result of the low temperatures at this height.

Moreover, the various gases emitted by engines of transportation units pose serious challenges to the environment. CO_2, the most popular of the greenhouse gases, is a colorless, odorless gas with a covalent bond between its atomic constituents. It is the most potent greenhouse gas, being highly atmospherically stable and having a life of over a hundred years, with a strong ability to absorb radiations below the visible light spectrum, thus trapping heat attempting to escape from the Earth's surface and causing an increase in the temperature of the planet's surface. It has a radiative forcing of 1.5 W/m^2 and is regarded as the most powerful greenhouse gas because of its long atmospheric stability period.

The presence of a ton of CO_2 put into the atmosphere thus has a deleterious environmental warming effect for over a hundred years. With the increasing anthropogenic emission of this gas from transportation and other sources, the global warming effect of CO_2 on the environment has never been ignored. It is heavily concentrated in the atmosphere.

Methane (NH_4), in contrast, is a covalent compound, colorless and odorless. It is not as stable as CO_2 but has a stronger effect as a greenhouse gas than CO_2. Its stability period in the Earth's atmosphere is 10 years. It absorbs infrared radiation and affects tropospheric ozone. Methane may not be as popular a greenhouse gas as carbon dioxide, but its effects on the climate are

stricter than it, and methane has been rated the second most potent greenhouse gas, after CO_2, with the exception of water vapor because of its short life and the quantity of it found in the Earth's atmosphere.

Another gas emitted from the burning of fossil fuels is nitrous oxide. It is a colorless, nonflammable, sweet-smelling gas having two nitrogen and one oxygen atoms covalently bonded together. When released into the atmosphere, nitrous oxide is the third largest greenhouse gas contributor to global warming, and has more effect than an equal amount of CO_2. It is reported that nitrous oxide is 296 times stronger a greenhouse gas than CO_2. It attacks ozone in the stratosphere, increasing the amount of ultraviolet light entering the Earth's surface. This ultraviolet light has deleterious effects on the human immune system, as well as the eye, and on the skin. It causes sunburn, inflammation, immunosuppression, tanning, and the accelerated aging of the skin.

GASEOUS POLLUTANTS

Other gaseous pollutants from transportation (mentioned earlier) are carbon monoxide and sulphur dioxide. Carbon monoxide is a colorless, odorless, and tasteless gas, formed by the thermal composition of excess carbon with oxygen. It is made up of a carbon atom covalently bonded to an oxygen atom. It is also released from the exhaust of motor vehicle engines, having gone through an incomplete internal combustion process of burning excess fossil fuels in the presence of oxygen. It is a toxic air pollutant. Its reaction in the atmosphere with some atmospheric constituents like the hydroxyl radical (OH^-) can increase the amounts of atmospheric methane and tropospheric ozone, thus causing an indirect forcing effect. It has the ability to combine with oxygen in the atmosphere to give CO_2, thereby contributing to greenhouse effects and global warming. Apart from this environmental degradation, CO has deleterious effects on human health.

Exposure to excess carbon monoxide can lead to heart and respiratory problems and has an effect on pregnancy, the central nervous system, and the heart—to mention a few adverse effects. Sulphur dioxide, in contrast, is a covalently bonded chemical compound made up of an atom of sulphur and two atoms of oxygen. It is anthropogenically produced from the combustion of coal and petroleum, which is commonly used by cars. It is a colorless gas with the smell of burning sulphur and is able to undergo serial combination to form sulphuric acid, which is an acid rain with corrosive tendencies found in the atmosphere. Sulphur dioxide is also toxic and has caused damage to humans in times past. However, its effects as a regulatory measure on the global warming effects of the greenhouse gases is limited by its lifespan on the Earth's surface, which is not more than a week.

With the present system of transportation still in use the world over, the rate of emissions of these harmful gases will continue to be on the increase, and their deleterious effects will become severe, as anthropogenic emissions of CO_2, methane, and nitrous oxide from industry and transportation are among the major causes of global warming. Gaseous pollutants of carbon monoxide and sulphur dioxide also contribute to dangerous effects on humans and the environment. However, various reports of the Intergovernmental Panel on Climate Change exist, each pointing to the damaging effects of such greenhouse gases as carbon dioxide, methane, and nitrous oxide. The contribution of transportation to the emission of these gases is high, and determining what must be done to reduce these emissions must be a global exercise that involves scientific contribution. Transportation effects on global warming and environmental pollution cannot be reduced without a redesign of the present system of engines vis-à-vis the available fuel systems. The emissions from modes of transportation can only increase as the numbers of vehicles, airplanes, ships, and other transport units migrating from one place to another increase.

SEE ALSO: Automobiles; Climate Change, Effects; Tourism.

BIBLIOGRAPHY. M. Myhre, E.J. Highwood, K.P. Shine, and F. Stordal, "New Estimates of Radiative Forcing due to Well Mixed Greenhouse Gases," *Geophysical Research Letters* (v.25/14, 1998); J.R. Partington, *A Short History of Chemistry*, 3rd ed. (Dover Publications, 1989); C.N. Sawyer, P.L. McCarty, and G.F. Parkin, *Chemistry for Environmental Engineering and Science*, 5th ed. (McGraw-Hill, 2003); M. Wang, *The Greenhouse Gases, Regulated Emissions and Energy Use in Transportation (GREET) Version 1.5* (Center for Transportation Research, Argonne National Laboratory, 1999).

AJAYI OLUSEYI OLANREWAJU
COVENANT UNIVERSITY

Trexler and Associates, Inc.

TREXLER CLIMATE AND Energy Services, Inc. (TC+ES) was founded as Trexler and Associates, Inc. (TAA) in the year 1991 by Dr. Mark C. Trexler, formerly of the World Resources Institute in Washington, D.C. TC+ES is based in Portland, Oregon. TC+ES was the company that wrote the first contracts for carbon offset, and designed the first methane carbon offset project for a coal mine.

Until 1997, the company was the only one serving the private sector in climate change mitigation services. That same year, TAA worked with Stonyfield Farm and its New Hampshire facility, helping the company to become the United States' first "greenhouse gas (GHG) neutral" facility. In 2000, TAA assisted Shaklee Corporation in its application to the Climate Neutral Network for the pilot certification of "Climate Neutral". The Climate Neutral Network awarded these certifications to companies who submitted worthy applications; it was to close down in 2007.

In 2002, TAA became partly run by Japan's Sumitomo Corporation. Sumitomo was established in 1919 as Osaka Hokko Kaisha Ltd., and adopted the English name Sumitomo Corporation in 1978.

Since its founding, TC+ES has worked with over 100 companies, large and small, in over 20 nations. Some of its better-known clients include the Chevron Research and Development Corporation, Fannie Mae, Nike Inc., Stonyfield Farm Inc., The Nature Conservancy, and the U.S. Department of Energy, and other institutions.

Services provided by TC+ES include achieving GHG neutrality, building ghg competitive advantage, customized price curve development, GHG inventory support, internal cost curve development, mitigation portfolio development, power plant siting and offset strategies, project design document (pdd) services, risk and opportunity assessment, and Sarbanes-Oxley compliance.

The company frequently publishes papers and reports regarding environmental strategizing and financial planning for companies interested in incorporating environmental responsibility.

SEE ALSO: Department of Energy, U.S.; Global Warming; Greenhouse Effect; Greenhouse Gases; Japan; Nongovernmental Organizations (NGOs); Oregon; World Resources Institute (WRI).

BIBLIOGRAPHY. Eberhard Jochem, Jayant A. Sathaye, and Daniel Bouille, *Society, Behaviour, and Climate Change Mitigation (Advances in Global Change Research)* (Springer, 2001); Intergovernmental Panel on Climate Change, *Climate Change 2007—Mitigation of Climate Change: Working Group III Contribution to the Fourth Assessment Report of the IPCC (Climate Change 2007)* (Cambridge University Press, 2007); Mohammad Yunus, Nandita Singh, and L. J. de Kok, *Environmental Stress: Indication, Mitigation, and Eco-conservation* (Springer, 2000).

CLAUDIA WINOGRAD
UNIVERSITY OF ILLINOIS AT URBANA-CHAMPAIGN

Triassic Period

THE TRIASSIC PERIOD is the geologic time period that extends from about 251 to 199 million years ago. This is the first period of the Mesozoic era, following the Permian and preceding the Jurassic period. Both the start and end of the Triassic are marked by major extinction events. During the Triassic period, both marine and continental life showed an adaptive radiation, beginning from the starkly impoverished biosphere that followed the Permian-Triassic extinction. The first flowering plants may have evolved during the Triassic, as did the first flying vertebrates, the pterosaurs. The Triassic period is further separated into Early, Middle, and Late Triassic epochs.

During the Triassic period, almost all the Earth's land mass was concentrated into a single supercontinent centered more or less on the equator, known as Pangaea. This supercontinent began to rift during the Triassic period but had not yet separated.

The Triassic climate was generally hot and dry, forming typical red bed sandstones and evaporites. There is no evidence of glaciation at or near either pole. The polar regions were moist and temperate—a climate suitable for reptile-like creatures. Pangaea's continental climate was highly seasonal, with very hot summers and cold winters. It probably had strong, cross-equatorial monsoons. The interior of Pangaea

was hot and dry during the Triassic period. This may have been one of the hottest times in Earth history. Rapid global warming at the very end of the Permian may have created a super hothouse world that caused the great Permo-Triassic extinction.

The Permian-Triassic extinction event, also known as the Great Dying, was an extinction event that occurred 251.4 mya (million years ago). This was the Earth's most severe extinction event, with up to 96 percent of all marine species and 70 percent of all terrestrial vertebrate species becoming extinct. There are several proposed mechanisms for the extinction event, including both catastrophic and gradualistic processes, similar to those theorized for the Cretaceous extinction event. The former include large or multiple impact events, increased volcanism, or sudden release of methane hydrates from the seafloor. The latter include sea-level change, anoxia, and increasing aridity. Evidence that an impact event caused the Cretaceous-Tertiary extinction event has led naturally to speculation that impact may have been the cause of other extinction events, including the Permian-Triassic extinction. Several possible impact craters have been proposed as possible causes of this extinction event, including the Bedout structure off the northwest coast of Australia and the so-called Wilkes Land crater of east Antarctica. In each of these cases, the idea that an impact was responsible has not been proven and has been widely criticized. If impact was a major cause of this extinction event, it is possible or even likely that the crater no longer exists. Seventy percent of the Earth"s surface is

Aetosaur bones from the Triassic period in the Petrified Forest in Arizona. The Triassic climate was generally hot and dry.

sea, so an asteroid or comet fragment is over twice as likely to hit sea as to hit land. There is evidence that the oceans became anoxic toward the end of the Permian. There was a noticeable and rapid onset of anoxic deposition in marine sediments around east Greenland near the end of the Permian. The most likely causes of the global warming that drove the anoxic event were a severe anoxic event at the end of the Permian, causing sulphate-reducing bacteria to dominate the oceanic ecosystems and causing massive emissions of hydrogen sulfide, which poisoned plant and animal life on both land and sea. These massive emissions of hydrogen would have severely weakened the ozone layer, exposing much of the life that remained to fatal levels of ultraviolet radiation.

Pangaea's formation would also have altered both oceanic circulation and atmospheric weather patterns, creating seasonal monsoons near the coasts and an arid climate in the vast continental interior. Marine life suffered very high but not catastrophic rates of extinction after the formation of Pangaea—rates almost as high as in some of the Big Five mass extinctions. The formation of Pangaea seems not to have caused a significant rise in extinction levels on land, and in fact, most of the advance of Therapsids and the increase in their diversity seems to have occurred in the late Permian, after Pangaea was almost complete. Thus it seems likely that Pangaea initiated a long period of severe marine extinctions but was not directly responsible for the Great Dying and the end of the Permian.

The possible causes, which are supported by strong evidence, appear to describe a sequence of catastrophes, each one worse than the previous. The resultant global warming may have caused perhaps the most severe anoxic event in the oceans' history. The oceans became so anoxic that anaerobic sulphur-reducing organisms dominated their chemistry.

SEE ALSO: Global Warming; Paleoclimates.

BIBLIOGRAPHY. D. Beerling, "CO_2 and the End-Triassic Mass Extinction," *Nature* (v.24, 2002); R.A. Kerr, "Paleontology: Biggest Extinction Hit Land and Sea," *Science* (v.289, 2000); Triassic Period, www.ucmp.berkeley.edu (cited May 2007).

Fernando Herrera
University of California, San Diego

Trinidad and Tobago

THE ISLANDS OF Trinidad and Tobago, located in the Caribbean close to the South American coast, have a land area of 1,979 sq. mi. (5,128 sq. km.), with the island of Trinidad accounting for 1,864 sq. mi. (4,769 sq. km.), and Tobago the remaining 116 sq. mi. (300 sq. km.). The country has a population of 1,333,000 (2006 est.), of whom 96.3 percent live on the island of Trinidad, which has a population density of 660 people per sq. mi. (254 people per sq. km.). Some 69 percent of the population live in urban areas. With 15 percent of the land being arable, 9 percent is under permanent cultivation, with 2 percent being used as meadow or pasture. Some 44 percent of the country is forested, with efforts being made by the government to conserve the forests.

Much of the economy of Trinidad and Tobago comes from the petroleum industry, with petroleum and petroleum products making up the vast majority of the country's exports. The country also has coal deposits. The consumption of these deposits has led to an increase in greenhouse gases, with the result that Trinidad and Tobago has one of the highest carbon dioxide emissions per capita in the world. Although data published by the U.S. Department of Energy's Carbon Dioxide Information Analysis Center puts it at 9th in the world, Trinidad is the fifth highest independent country. Its emissions were 13.9 metric tons of carbon dioxide per capita in 1990, and with the exception of a low figure for 1993, emissions levels have risen significantly since then, reaching 22.1 metric tons per capita in 2003. The emissions are largely from gaseous fuels (72 percent), with 16 percent from liquid fuels, 10 percent from gas flaring, and 2 percent from the manufacture of cement. Not only is the heavy use of petroleum a major contributing factor but there is also poor public transport on Trinidad, resulting in widespread use of automobiles.

There is also extensive use of air conditioning for private houses and businesses. The effect of global warming can be seen by rising water temperatures detrimentally affecting the population of leatherback turtles, as well as affecting other Caribbean nations far more significantly. The government of Trinidad and Tobago has also been worried about the effects of global warming on the tourism industry, as well as the problems that might be posed to yachting and the cruise liners operating in the Caribbean.

The government of Patrick Manning took part in the United Nations Framework Convention on Climate Change, signed in Rio de Janeiro in May 1992, and 2 years later, Trinidad and Tobago was represented at the Global Conference on the Sustainable Development of Small Island Developing States held in Barbados. On January 7, 1999, the Trinidad government of Basdeo Panday signed the Kyoto Protocol to the UN Framework Convention on Climate Change, with it being ratified exactly 3 weeks later, but only entering into force on February 16, 2005.

SEE ALSO: Climate Change, Effects; Floods; Tourism; Transportation.

BIBLIOGRAPHY. Julie Cohen and Stuart Conway, "Flying Colours," *Geographical* (v.74/5, May 2002); Michael J. Day and M. Sean Chenoweth, "The Karstlands of Trinidad and Tobago: The Land Use and Conservation," *Geographical Journal*, (v.170/3, September 2004); Jasmine Garraway, "Climate Change and Tourism," *Trinidad Guardian* (August 30, 2005); Leila Ramdeen, "Environmental Justice," *Trinidad Guardian* (June 6, 2005); A. R. Williams, "Trinidad and Tobago," *National Geographic* (v.185/3, March 1994).

ROBIN S. CORFIELD
INDEPENDENT SCHOLAR

Tropopause

THE TROPOPAUSE IS the boundary region dividing the troposphere, the lowest layer of the atmosphere, from the overlying stratosphere. Since the tropospheric and stratospheric air masses have rather distinct features, in correspondence to each surface location, the tropopause height is the level in the vertical where abrupt changes in the physical and chemical properties of the atmosphere are observed.

Three different definitions are typically adopted. The thermal tropopause is related to the change of the sign of the vertical derivative of the temperature (lapse rate), which is negative in the troposphere and positive in the stratosphere.

The World Meteorological Organization defines the tropopause as the lowest level where the absolute value of the temperature lapse rate decreases to 2K/km. or less, with the average lapse rate between this level and all higher levels within 1.2 mi. (2 km.) not exceeding 2K/km. The dynamical tropopause is defined in terms of sharp changes in the potential vorticity (much higher in the stratosphere), which measures stratification and rotation of the air masses. An abrupt increase (decrease) with height of the ozone (water vapor) mixing ratio indicates the presence of the chemical tropopause. In spite of the necessity of choosing phenomenological thresholds, the three definitions of the tropopause are quite consistent.

TROPICAL TROPOPAUSE

Typically, the tropopause height decreases with latitude, being around 3.7 mi. (6 km.) near the poles and 11 mi. (18 km.) near the equator. Whereas radiative and convective processes with time scale of the order of one week to one month basically determine the properties of the tropical tropopause, in the midlatitudes a relevant role is played also by baroclinic-fuelled extra-tropical cyclones, having a typical time scale of a few days, in such a way that the tropopause readjusts its height in such a way as to act effectively as a stabilizing mechanism limiting the growth of the weather perturbations. The tropopause is not a hard boundary: exchanges of tropospheric and stratospheric air occur through various mechanisms, including vigorous thunderstorms and midlatitude perturbations.

The globally averaged tropopause height tends to increase if the troposphere warms up and/or the stratosphere cools down, and the height change is approximately proportional to the difference between the tropospheric and stratospheric temperature changes. Therefore, the mean tropopause height can act as a robust indicator of climate change. Recent climate simulations have shown that the estimated increase after 1979 of about 492 ft. (150 m.) may be primarily explained by anthropogenic causes, namely the stratospheric cooling driven by ozone depletion and the tropospheric warming driven by increases in the greenhouse gases concentration. Considering natural processes, episodic, short-lived strong reductions of the globally averaged mean tropopause height are caused by large explosive volcanic eruptions, which warm the troposphere and cool the stratosphere.

SEE ALSO: Atmospheric General Circulation Models; Atmospheric Vertical Structure; Cyclones.

BIBLIOGRAPHY. World Meteorological Organization, "Meteorology: A Three Dimensional Science", *WMO Bulletin* (v.6, 1957); B.D. Santer "Contributions of Anthropogenic and Natural Forcing to Recent Tropopause Height Changes," *Science* (v.301/5632, 2003).

VALERIO LUCARINI
UNIVERSITY OF BOLOGNA

Troposphere

ON THE BASIS of thermal characteristics, the atmosphere is normally subdivided into four major vertical layers: the troposphere, stratosphere, mesosphere, and thermosphere. The troposphere makes up the lowest of these layers, extending from the surface to a global average height of 7.5 mi. (12 km.). Coined in 1908 by French scientist Leon Philippe Teisserenc de Bort, the name troposphere is derived from the Greek word tropos, meaning to turn, mix, or change. The term aptly describes the extensive vertical mixing and stability changes of this layer, which generates clouds, precipitation, and other meteorological events. For this reason, the troposphere is commonly referred to as the weather sphere.

The depth of the troposphere is relatively thin, yet it contains approximately 80 percent of the atmosphere's mass. Because the atmosphere is compressible, air molecules are more compact closer to the surface, thereby increasing the density and pressure of the air at lower altitudes. The relationship between density and pressure with altitude is nonlinear, decreasing at a decreasing rate with increasing altitude. In the lower troposphere, the rate of pressure decrease is about 10 mbars. for every 330 ft. (100-m.) increase in elevation.

Temperature in the troposphere generally decreases with height, contrasting considerably between its lower and upper boundaries. Temperature in this layer is largely affected by the radiant energy exchanges from the underlying surface and insolation intensity. The global average temperature at the surface is 59 degrees F (15 degrees C) but decreases to around minus 82 degrees F (minus 63 degrees C) at the top of the troposphere. On the basis of mean tropospheric depth, the average rate of temperature decrease is 3.6 degrees F per 1,000 ft. (6.5 degrees C per km.), a measurement known as the normal lapse rate. This rate represents average global conditions, deviating substantially depending on latitude, time, and local modifications. The actual temperature change with height is the environmental lapse rate, which is measured remotely, using satellites, or directly, using Radiosondes (a balloon-borne instrument package). Eventually, temperature ceases to decline with height, transitioning into a zero lapse rate region (or isothermal layer), where temperature is neither increasing nor decreasing. This shift demarcates the boundary between the troposphere and the stratosphere, known as the tropopause.

The mean height of the tropopause can have considerable spatial and temporal variability. In the tropics, the depth of the troposphere is around 16 km. (10 mi.), but near the poles, the depth dwindles to about 8 km. (5 mi.) or less. The tropopause also varies seasonally, with higher heights occurring during the summer than the winter. Warm surface temperatures occurring at low latitudes and high sun periods encourage vertical thermal mixing, thereby extending the depth of the troposphere. Accordingly, the environmental lapse rate in these regions continues to remain positive (i.e., temperature decreases with height), and tropopause temperatures are typically lower in the tropics than for high latitudes. Occasionally the tropopause is difficult to discern because of extensive mixing between the upper troposphere and the lower stratosphere.

This situation is common in portions of the midlatitudes, usually defining the location of jet streams (a narrow belt of high-velocity winds often in excess of 185 km. per hour (115 mi. per hour) that steer midlatitude cyclones. Because the height of the tropopause is dependent on the average temperature of the troposphere, temperature changes in this layer can influence the location of extratropical storm tracks and cloud depth.

Embedded frequently within the troposphere are thin sublayers in which the temperature actually increases with height, known as temperature inversions. Radiation inversions result from nocturnal surface cooling. Under certain ambient conditions (e.g., cloudless nights), terrestrial radiation loss to space is enhanced and the ground (and air above) cool rapidly, thereby establishing a shallow inversion layer. Conversely, subsidence inversions occur from mid-upper tropospheric processes that produce areas of sinking air that are being warmed by compression; hence, lower tropospheric temperatures are actually colder than those aloft. This setting tends to stabilize the air, inhibiting vertical mixing and cloud growth.

A semipermanent sublayer of the troposphere is the planetary boundary layer (PBL), a section directly influenced by surface daily conditions. Comprising typically the lowest 1 km. (3,300 ft.) of the troposphere, the PBL is characterized by turbulence generated by frictional drag from the surface beneath and rising thermals (heated air parcels). The depth of the PBL amplifies and diminishes with the daily solar cycle, such that the greatest thickness is during the day when the atmosphere is most turbulent.

Evidence suggests that the troposphere has undergone a significant rate of warming during the past century. The tropospheric temperature trend in the latter half of the 20th century is estimated at a 0.18 degree F (0.10 degrees C) increase per decade, similar to the surface temperature rate change. Higher temperatures mean increased surface evaporation and tropospheric water vapor content. As a consequence, cloud cover has also shown an increase, and extra-tropical precipitation in the Northern Hemisphere has increased 5 to 10 percent since 1900.

Other climate-forcing agents (e.g., anthropogenic-induced greenhouse gas emissions) can alter the Earth's radiation balance and may also explain the upward temperature trend. For instance, tropospheric ozone (O_3), a greenhouse gas and surface pollutant, has increased by nearly 35 percent since the preindustrial era.

SEE ALSO: Atmospheric Boundary Layer; Atmospheric Vertical Structure; Tropopause.

BIBLIOGRAPHY. Intergovernmental Panel on Climate Change, *Climate Change 2001: The Scientific Basis* (Cambridge University Press, 2001); Frederick K. Lutgens, Edward J. Tarbuck, and Dennis Tasa, *The Atmosphere: An Introduction to Meteorology,* 10th ed. (Prentice Hall, 2006); Timothy R. Oke, *Boundary Layer Climates,* 2nd ed. (Routledge, 1987).

Jill S. M. Coleman
Ball State University

Tsunamis

TSUNAMIS (SOMETIMES CALLED seismic sea waves) are large sea waves that are created by underwater earthquakes, volcanic eruptions, or even nonseismic events such as landslides and meteorite impacts. Tsunamis are also known as tidal waves, even though this is a misnomer because the waves have nothing to do with tides. The word *tsunami* is a Japanese word meaning "harbor wave." Tsunamis are not easily seen on the open water, as they have extremely long wavelengths on the order of tens of kilometers. The speed of the wave is directly related to the depth of the water; therefore, as the water depth decreases, the tsunami moves slower. As the waves propagate toward the coast, the speed will decrease, but the amplitude or the height of the waves can achieve extraordinary levels. Tsunamis lose energy as they approach the coast, but they still have incredible amounts of energy, as they often cause beach erosion and undermine trees and other types of coastal vegetation. The fast-moving water is capable of flooding several hundreds of meters inland, well above normal flood levels, and destroying buildings and other structures. Tsunamis can extend to heights well above sea level, in extreme cases sometimes as high as 30 m. or 100 ft.

Volcanic activity and earthquakes are the prime causes of tsunamis. When the seafloor starts to buckle, the overlying water will begin to displace. As the seafloor rises and sinks, the displaced water will form waves because of the effects of gravity. Most of the major earthquakes occur at plate boundaries.

There are three different types of plate boundaries. A divergent boundary takes place where two plates move away from each other. At this type of boundary, volcanoes will form, out of which molten material will flow. Also, weaker, shallow-focus earthquakes can occur along these boundaries. Divergent boundaries are very common in the midocean such as at the Mid-Atlantic Ridge, the East Pacific Ridge, the Mid-Indian Ridge, and the Southeast-Indian Ridge. Convergent boundaries occur where two plates moving in opposite directions collide. One plate will be denser and will subduct underneath the other. These subduction zones are a very common location for earthquake activity. There are three different types of convergent boundaries: oceanic-continental convergence, oceanic-oceanic convergence,

The third and largest wave of the tsunami of December 26, 2004, invades the promenade on Ao Nang Beach, Thailand. Waves were reported as high as 30 m. or 100 ft., and it was the deadliest in recorded history, with an approximate 230,000 lives lost,

and continental-continental convergence. At an oceanic-continental convergent boundary, the oceanic crust is denser and will subduct underneath the continental crust. Volcanoes will form along the continental boundary, whereas deep trenches will form off the coast. Shallow-focus earthquakes often form along these subduction zones, such as along the west coast of South America. At an oceanic-oceanic convergent boundary, two ocean plates collide, forming a volcanic island arc on the ocean floor. Examples of this type of boundary include the Aleutian Islands, the Mariana Islands, and Japan. At a continental-continental convergent boundary, two continental plates collide, typically forming huge mountain ranges such as the Himalayas or the Alps. Under this type of convergence, volcanic activity is rare, but earthquake activity is very common. The final type of boundary is called a transform boundary, where two plates slide past each other. Transform boundaries occur along vertical fractures called faults, which are noted for great magnitudes of earthquake activity. Most faults are found near midoceanic ridges, but they can also extend through continents, as evidenced by the San Andreas Fault in California.

Tsunamis can be formed by anything that displaces a large volume of water from its equilibrium state. When earthquakes or volcanoes generate tsunamis, water is displaced as a result of the uplift or subsidence of the seafloor and water column. Sometimes submarine landslides, which are common with large earthquakes as well as volcanic collapses, can displace great volumes of water. However, these types of disturbances disturb the water from above, rather than from below. Tsunamis derived from these types of mechanisms usually do not last long and have minimal impacts on the coastlines.

The most recent deadly tsunami was the Asian tsunami that occurred after the 2004 Indian Ocean earthquake on December 26 of that year. The epicenter, which is the location at the Earth's surface directly above the focus of the earthquake, took place off the coast of Sumatra, Indonesia. The magnitude of this earthquake has been estimated to be between 9.1 and 9.3 on the Richter scale (the scale devised to estimate the amount of energy released in an earthquake). This made it the fourth most powerful recorded earthquake since 1900. The earthquake lasted almost 10 minutes and was the second most powerful ever recorded on a seismograph (an instrument that measures the seismic waves from an earthquake). Waves were reported as high as 30 m. or 100 ft., and it was the deadliest in recorded history, with an approximate 230,000 lives lost, mostly in the countries of Indonesia, India, Sri Lanka, and Thailand. Before this, the 1782 Pacific Ocean tsunami was the deadliest in recorded history, with an estimated 40,000 casualties in the South China Sea. Other powerful tsunamis include the 1883 tsunami following the eruption of Krakatoa, a volcanic island in Indonesia, and the 1908 tsunami that occurred in the Mediterranean Sea near Messina, Italy.

There has been speculation about possible effects of global warming after the 2004 tsunami, with proponents suggesting that the increase in average temperature allows the atmosphere and the oceans to gather energy, which may cause more earthquake activity. Critics, however, claim that if this were the case, there would be more of a correlation between El Niño and tsunamis, as El Niño warms the ocean over an active region with many plate boundaries. Therefore, links to global warming and tsunamis are unsubstantiated because it is difficult to associate what is happening at the surface of the ocean with the depth at which the focus of the earthquakes takes place.

SEE ALSO: Climate Change, Effects; Oceanic Changes.

BIBLIOGRAPHY. Patrick L. Abbott, *Natural Disasters*, 6th ed. (McGraw Hill, 2007); Thomas A. Easton, *Taking Sides* (McGraw Hill/Dushkin, 2005); Tom McKnight and Darrel Hess, *Physical Geography* (Prentice Hall, 2005).

KEVIN LAW
MARSHALL UNIVERSITY

Tunisia

LOCATED IN NORTH Africa between Algeria and Libya, the Republic of Tunisia has a land area of 63,170 sq. mi. (163,610 sq. km.), with a population of 10,327,000 (2006 est.) and a population density of 161 people per sq. mi. (62 people per sq. km.). In spite of its having an arid climate, 19 percent of the land in Tunisia is arable, with 13 percent under permanent cultivation. There is also an additional 20 percent used as meadows or pasture, mainly for the raising of sheep, goats, and cattle. Only 4 percent of the country is forested.

Tunisia has its own supplies of crude oil and natural gas, and as a result, 99 percent of the country's electricity generation comes from the use of fossil fuels, with 1 percent coming from hydropower. In terms of its per capita rate of carbon dioxide emissions, they have risen from 1.6 metric tons per person in 1990 to 2.29 metric tons in 2004. Some 65 percent of Tunisia's carbon dioxide emissions come from liquid fuels, with 23 percent coming from gaseous fuels and 10 percent from the manufacture of cement. Heavy use of air conditioning, especially in the economically important tourist sector, results in electricity production making up 39 percent of the country's carbon dioxide emissions, with 25 percent coming from transportation, 24 percent from manufacturing and construction, and 11 percent from residential use.

The effect of global warming and climate change on Tunisia has seen the alienation of some arable land, with the rising temperature affecting the level of crop production. It has also led to gradual shortages of water in some inland parts of the country and in the increasing use of desalination, which in turn leads to increases in the amount of electricity that needs to be generated. The Tunisian government of Zine El Abidine Ben Ali took part in the United Nations Framework Convention on Climate Change, signed in Rio de Janeiro in May 1992. It accepted the Kyoto Protocol to the UN Framework Convention on Climate Change on January 22, 2003, with it entering into force on February 16, 2005.

SEE ALSO: Climate Change, Effects; Drought.

BIBLIOGRAPHY. Jamie Carstair, "A Room With no View" *Geographical* (v.73/3, March 2001); Jennifer Hill and

Wendy Woodland, "Contrasting Water Management Techniques in Tunisia: Towards Sustainable Agricultural Use," *Geographical Journal* (v.169/4, December 2003); World Resources Institute, "Tunisia—Climate and Atmosphere," www.earthtrends.wri.org (cited October 2007).

JUSTIN CORFIELD
GEELONG GRAMMAR SCHOOL, AUSTRALIA

Turkey

LOCATED MAINLY IN Asia, but with a small part of its land considered to be in Europe, the Republic of Turkey has a land area of 302,535 sq. mi. (779,452 sq. km.), with a population of 74,877,000 (2006 est.) and a population density of 240 people per sq. mi. (93 people per sq. km.). Istanbul, the former capital and the largest city, has a population density of 13,256 people per sq. mi. (5,137 per sq. km.). The present capital, Ankara, has a population of 3,641,900, with a population density of 13,328 per sq. mi. (1,424 per sq. km.). Some 32 percent of the land area of Turkey is arable, with 4 percent under permanent cultivation, and an additional 16 percent is used as meadows or pasture. Some 26 percent of the country remains forested, and there is only a small timber industry. In terms of its per capita rate of carbon dioxide emissions, it was 2.6 metric tons per person in 1990, rising to 3.14 metric tons per person by 2004.

Electricity generation in Turkey comes largely from fossil fuels (74.1 percent), with 25 percent being from hydropower. The use of hydropower has not been without controversy. Although it has generated much hydroelectricity, there have been environmental problems with the Illisu Dam in southeastern Turkey, as well as with some other projects. The production of electricity is responsible for 38 percent of Turkey's carbon dioxide emissions, with 8 percent coming from other energy industries, 25 percent being from manufacturing, 20 percent from transportation, and 13 percent from residential use. Heavy use of coal means that 44 percent of the emissions are the result of the use of solid fuels, with 37 percent being from liquid fuels, 10 percent from gaseous fuels, and 9 percent from the manufacture of cement.

Turkey has suffered some effects of global warming and climate change, with a rise in the average temperature in the country resulting in the snowmelt on some of the previously snow-covered peaks in the east of the country. More important, it has also led to the alienation of agricultural land throughout the country, with the need for water for irrigation and drinking leading to water shortages in some rural areas. There have also been problems in the agriculturally poor eastern part of Turkey, where there is already low productivity in farms and high levels of poverty.

The Turkish government took part in the United Nations Framework Convention on Climate Change, signed in Rio de Janeiro in May 1992. It has not expressed an opinion on the Kyoto Protocol to the UN Framework Convention on Climate Change.

SEE ALSO: Climate Change, Effects; Drought.

BIBLIOGRAPHY. Hussein A. Amery, "Water Wars in the Middle East: A Looming Threat," *Geographical Journal* (v.168/4, 2002); Aykut Coban, "Community-Based Ecological Resistance: The Bergama Movement in Turkey," *Environmental Politics* (v.13/2, Summer 2004); Gabriel Ignatow, *Transnational Identity Politics and the Environment* (Lexington Books, 2007); World Resources Institute, "Turkey—Climate and Atmosphere," http://earthtrends.wri.org (cited October 2007).

ROBIN S. CORFIELD
INDEPENDENT SCHOLAR

Turkmenistan

LOCATED IN CENTRAL Asia, Turkmenistan is a country primarily made of desert. It lies between 35 degrees 08 minutes and 42 degrees 48 minutes N and between 52 degrees 27 minutes and 66 degrees 41 minutes E, north of the Kopet-Dag mountains, between the Caspian Sea to the west and the Amu-Darya River to the east. Turkmenistan covers 188,455 sq. mi. (488,096 sq. km.) and had a population of 5,097,028 in 2007.

Despite its extensive oil and gas resources, Turkmenistan remains a poor, predominantly rural country, with the majority of the population relying on intensive agriculture in irrigated oases, and it is extremely vulnerable to climate change.

Turkmenistan has a distinctive continental climate, with an average annual air temperature ranging from 54–62 degrees F (12–17 degrees C) in the north to 59–64 degrees F (15–18 degrees C) in the southeast and to the absolute maximum temperature of 118–122 degrees F (48–50 degrees C) in the Central and South-East Kara-Kum. The highest amount of annual rainfall is observed in the mountains (up to 398 mm. in Koyne-Kesir), as is the least rainfall (less than 95 mm. above the Kara-Bogaz-Gol Bay). Up to 80 percent of Turkmenistan's land consists of desert. Meteorological data reveal an increase of annual and winter temperatures in Turkmenistan dating from the beginning of the past century. Since 1931, the mean annual temperature has increased by 1 degree F (0.6 degrees C) in the northern part of the country and by 0.7 degrees F (0.4 degrees C)in the south.

At the same time the number of days with temperatures higher than 104 degrees F (40 degrees C) has increased since 1983. Climate Models predict temperature increases in Turkmenistan of 5–7 degrees F (3–4 degrees C) by the middle of the 21st century. Precipitation projections are highly uncertain, but given the existing aridity and high interannular and interseasonal variability of Turkmenistan's climate, even a slight temperature increase is likely to exacerbate the existing water stress in the region.

Turkmenistan signed and ratified the UN Framework Convention on Climate Change in 1995. In January 1999, Turkmenistan ratified the Kyoto Protocol and published the Initial Communication on Climate Change. Because Turkmenistan is covered under the Kyoto Protocol's Clean Development Mechanism, it can trade carbon credits with other countries that fall under the Joint Implementation Mechanism.

The major sources of carbon emissions in Turkmenistan include oil and gas extraction, petroleum refining, and the chemical industry, as well as motor transport concentrated mainly in Ashgabat, Turkmenbashi, Balkanabat, Mary, Turkmenabat, and Dashoguz.

Turkmenistan has taken some steps to reduce carbon emissions, such as a massive tree planting project throughout the country (Green Belt Project), modernization of Turkmenbashi and Seyidi refineries to conform to modern ecological standards, and relocation of the cement factories away from inhabited areas.

However, the nation's widespread poverty, recent decline in the educational system, misuse of hydrocarbon revenues, and high economic dependence on cotton production and exports leave Turkmenistan particularly vulnerable to high climatic variability, desertification, and droughts.

SEE ALSO: Climate Models; Deserts; Kyoto Mechanisms.

BIBLIOGRAPHY. M. H. Glanz, "Water, Climate, and Development Issues in the Amu Darya Basin," *Mitigation and Adaptation Strategies for Climate Change* (v.10/1, January 2005); E. Lioubimtseva, R. Cole, J. M. Adams, and G. Kapustin, "Impacts of Climate and Land-Cover Changes in Arid Lands of Central Asia," *Journal of Arid Environments* (v.62/2, July 2005); *Turkmenistan: Initial National Communication on Climate Change* (UN Framework Convention on Climate Change, 1999).

ELENA LIOUBIMTSEVA
GRAND VALLEY STATE UNIVERSITY

Tuvalu

TUVALU HAS BEEN independent since 1978, joining the United Nations in 2000. The resident population has been estimated to be about 10,000 people in 2007, living on nine small, low-lying atoll islands. The total land area is only 9.4 sq. mi. (24.4 sq. km.); when the Exclusive Economic Zone is included, the total national area of Tuvalu is 289,576 sq. mi. (750,000 sq. km.). Tuvalu does not have substantial natural resources—the coral atoll soil is of poor quality, alkaline, shallow, and with low water-holding quality. Added to these environmental challenges are a lack of rainfall, lack of natural resources, the occurrence of cyclones, and the threat of sea-level rise. About 44 percent of the population lives on urban Funafuti, and the internal migration rate is expanding. The overcrowded urban situation has meant that having a sufficient freshwater supply is difficult.

As a small and fragmented Pacific Island nation, Tuvalu is increasingly vulnerable to cyclones, flooding, and sea-level rise, accompanied by environmental threats, such as land loss and coastal erosion, soil salinization, and intrusion of saltwater into the atolls' freshwater lenses and groundwater. Two incidences of flooding have been noted as being particularly severe for the urban island of Funafuti—one in 1977 and another one in 1993. Other climatic effects include coral bleaching in reef systems. Tuvalu's vulnerability, however, is also because of its impoverished economic situation, with large employment rates in the public sector, aid dependency, and a lack of natural resources for export; fragmentation and isolation of the islands; and a lack of capacity to support its population, which makes sea-level rise a problematic development issue. Current estimations of average sea-level rise in the South Pacific are 0.7 mm. per year. However, in relation to sea-level rise, the land of Funafuti island is sinking, which problem is accelerated by tectonic movement. As a consequence, and including indications of tide-gauge data, the sea-level rise in the Funafuti area has been estimated by scientists to be 2 + 1 mm. per year.

ENVIRONMENTAL REFUGEES

Tuvalu entered international negotiations on climate change in 2002, when Koloa Talake, the prime minister at the time, suggested pursuing legal actions against greenhouse gas emissions leading to global warming and the consequences of sea-level rise for Tuval, on the international level. The National Summit on Sustainable Development in Tuvalu in 2004 emphasized the need to promote awareness and strategies for adaptations on the national level. Migration as a final solution will be difficult for Tuvaluans and is also culturally contested. Existing migration schemes for Tuvalu are small.

Different from many other Pacific nations, Tuvalu, similarly to Kiribati, does not have the privilege of free access to the Pacific Rim countries. Current access allows 75 Tuvaluans to migrate each year to Australia and New Zealand. The strong relationship of Tuvaluans to their land, however, makes migration a last resort. Recent publications have suggested that Tuvalu and Tuvaluans are being publicly victimized when they are represented as potential places of disaster and as environmental refugees, respectively.

SEE ALSO: Climate Change, Effects; Floods.

BIBLIOGRAPHY. Michael A. Toman and Brent Sohngen, *Climate Change* (Ashgate Publishing, 2004); Official Website of the Government of Tuvalu, http://map.tuvalu.tv (cited May 2007).

MARIA BOROVNIK
MASSEY UNIVERSITY

Tyndall, John (1820–93)

JOHN TYNDALL WAS born in Leighlinbridge, County Carlow, in Ireland on August 2, 1820. After working as a surveyor and a mathematics teacher, he attended the University of Marburg in Germany, where he received his Ph.D. In 1854, he became a professor of natural philosophy at the Royal Institution in London (a scientific research center founded in 1799). In 1867, he was made superintendent of the institution, taking over from Michael Faraday.

Tyndall's most well-known scientific studies included the nature of sound, light, and radiant heat and observations on the structure and movement of glaciers. Glaciers had become a scientific area of interest during that time because in the 1830s, Louis Agassiz (considered the father of glaciology) had discovered that a large portion of Europe and North America had once been covered with ice.

Tyndall developed an interest in meteorology as a result of his love for mountain climbing. He studied alpine glaciers and took meteorological

John Tyndall was a 19th-century physicist. He is renowned for his studies in the absorption of heat by atmospheric gases.

measurements on Mont Blanc in the Alps. Using a spectrophotometer he designed, Tyndall studied the absorption of infrared light (at the time called radiant heat) by atmospheric gases. Infrared light is felt as heat and has wavelengths of approximately 0.7 to 1.0 μm. Visible light for comparison has wavelengths of approximately 0.4 to 0.7 μm.

Some of the invisible gases (oxygen, nitrogen, and hydrogen) were transparent to radiant heat, whereas water vapor and carbonic acid (now known as carbon dioxide) absorbed and reemitted infrared light, thereby warming the atmosphere close to the earth. From these experiments, Tyndall realized that water vapor, carbon dioxide, and ozone—even in small quantities—were the best absorbers of heat radiation, and he later speculated on how changes in these gases could correlate to climate change.

GREENHOUSE GASES

Among the various atmospheric gases in the troposphere, Tyndall discovered the importance of water vapor and carbon dioxide as greenhouse gases that trap heat on the surface of the earth. Without these atmospheric gases to trap heat, the heat would rapidly radiate back into space, and the Earth would be much colder. During cold nights, there is an enhanced chance of fog or dew in the mornings because the moisture in the air (water vapor) condenses into droplet as the air cools. In deserts—hot and dry climates—there is a lack of water vapor in the air, and the sand radiates heat easily into space. Changes in the atmospheric levels of gases produce changes in the climate as well.

In the 1860s, Tyndall began to suggest that slight changes in the atmospheric composition could bring about climatic variations. Tyndall was the first scientist to explain the Ice Ages as being caused by greenhouse effect. In particular, he noted that variations in water vapor resulted in a change in the climate and realized the importance of the greenhouse effect in maintaining ecosystems necessary for life. He thought changes in the composition of the atmosphere may have produced all changes in the climate.

Using the laws of thermodynamics, Tyndall proposed that changes in carbon dioxide cause an initial change in temperature, and when the humidity changes accordingly, the change in humidity leads to a second change in the temperature.

During the Victorian era, he was the contemporary and friend of other important scientists. In addition to his scientific research, he promoted scientific education of the public with scientific demonstrations and advocated using hands-on laboratory experimentation to teach science. In 1874, Tyndall used his address before the British Association for the Advancement of Science to proclaim rational thought and skepticism as the aim and superiority of science.

In the 1870s, he made a lecture tour of the United States and included his lab demonstrations. Over the course of his professional life, he was the recipient of numerous scientific awards and recognition. In addition to his dynamic public speeches using laboratory experimentation, he published a wide range of papers, treatises, and books. Digital versions of John Tyndall's books can be found at http://books.google.com and include *Lectures on Light Delivered in the United States in 1872–1873*, first published in 1873; *Hours of Exercise in the Alps*, first published in 1871; and *The Glaciers of the Alps*, first published in 1861.

John Tyndall died on December 4, 1893, from an overdose of chloral hydrate. At the inquest, his wife testified she mistakenly gave him chloral hydrate instead of his normal medication, leading to his death.

TYNDALL EFFECT

The Tyndall effect, a scientific principle on the dispersion of light beams through colloidal suspensions or emulsions, was named for him. The importance of his research continues, as his work continues to shape scientific research on climate change.

The Tyndall Centre, with headquarters at the School of Environmental Sciences at the University of East Anglia, founded in 2000, is named after John Tyndall, in honor of his being one of the first scientists to recognize the Earth's natural greenhouse effect and observing, identifying, and proposing the climate effects of the radiative properties of atmospheric gases. The mission of the Tyndall Centre is to research and educate policymakers on climate change, to develop and apply research methods for climate change, and to promote international dialogue on managing future climate change.

In support of this goal, research at the Tyndall Centre includes action to provide information through data collection and interpretation and modeling to assess possible scenarios of the effect of human and natural causes on climate change and to disseminate this information by encouraging international discussion and policymaking. By using empirical research, the Tyndall Centre proposes protection of coastline ecosystems, city-scale emissions testing, and accountability for contributions to climate change. The Tyndall Centre focuses on this objective, with a strategy to investigate and identify behavior modification and education opportunities to promote sustainable approaches to limiting human-induced climate change.

SEE ALSO: Carbon Dioxide; Greenhouse Effect; Greenhouse Gases.

BIBLIOGRAPHY. Mark Bowen, *Thin Ice: Unlocking the Secrets of Climate in the World's Highest Mountains* (Henry Holt, 2005); Earth Observatory, "John Tyndall," www.eobglossary.gsfc.nasa.gov; Tyndall Centre, "Research at the Tyndall Centre," www.tyndall.ac.uk (cited July 2007).

LYN MICHAUD
INDEPENDENT SCHOLAR

Uganda

UGANDA IS SITUATED in the interlacustrine high-lands of central Africa; these highlands are generally characterized by a high plateau that is associated with the Great Rift Valley that dominates Uganda. The plateau generally lies between 3,280 to 6,562 ft. (1,000–2,000 m.). One characteristic feature of the Great Rift Valley is the development of numerous lakes; these include Tanganyika and Kivu, while in the Albertine Rift region in the north, the lakes include George, Edward, and Albert. Uplift along the shoulders of the Western Rift Valley resulted in the reversal of the previously west-flowing rivers and the creation of the basin now occupied by Lake Victoria. The Nile River, which forms the eastern boundary of the interlacustrine highlands, is the major outlet of Lake Victoria, flowing northward from Lake Victoria. Farther to the west, Uganda is dominated by the Ruwenzori mountain range—the infamous mountains of the moon—that straddle the border with the Democratic Republic of Congo.

The climate of Uganda varies from humid through seasonally arid conditions. Precipitation is dependent largely on the Intertropical Convergence Zone (ITCZ); the movement of the ITCZ follows the position of maximum surface heating associated with the overhead position of the sun. Climate is further modified by topography and by the proximity of the large lakes. The distribution of vegetation is mainly dependent on altitude, levels of precipitation, and human activities. Because of human pressure, cultivated and grazed land occupies much of Uganda. Much of the interlacustrine highlands lie within a phytogeographical zone known as the Lake Victoria Regional Mosaic, where five distinct floras meet; namely, the Afromontane, Guineo-Congolian, Somalia-Masai, Sudanian, and Zambazian Regional Centers of Endemism. Above 6,562 ft. (2,000 m.) there are three distinct vegetation belts; namely, Afroalpine, Ericaceous, and Montane forest belts. The Afroalpine Belt occurs mainly above 11,810 ft. (3,600 m.) in the high mountains of the region, which include Mount Elgon, the Virunga volcanic mountains, and the Rwenzori.

Environmental histories of Uganda have been reconstructed via a long history of palaeoecological research from the numerous lakes and swamps. Recent interest in other proxy records, such as archaeology, historical linguistics, oral traditions, and meteorological data have provided a detailed understanding of the environmental history of Uganda and the associations between environmental and human histories in the region. Under the cool and dry climate conditions of the last glacial maximum, vegetation from higher altitudes extended to lower elevations. From around 15,000 years before present (yr. BP), as global climates warmed, montane forest types retreated back to mountainous areas, reaching the minimal extent around 6,000 yr. BP under the warm and wet conditions of the mid-Holocene.

The climate history from 5,000 yr. BP to the present is more complex, as ecosystems have been increasingly influenced by human activity. A period of relatively drier climate between 5,000 and 2,000 yr. BP is marked by relatively low lake water levels. Vegetation also reflected this climate change, experiencing a marked transition to drier forest types from about 3,500 yr. BP. Uganda experienced a return to relatively moist environmental conditions after about 2,000 yr. BP; during this same period, areas around Lake Victoria experienced a decrease in evergreen forest, with it being replaced by semideciduous forest. The introduction of agriculture to Uganda is commonly associated with Bantu-speaking people, who are reported to have spread throughout much of central Africa after about 3,000 yr. BP.

SEE ALSO: Climate Change, Effects; Food Production.

BIBLIOGRAPHY. A.C. Hamilton, *Environmental History of East Africa: A Study of the Quaternary* (Academic Press, 1982); B.J. Leiju, D. Taylor, and P. Robertshaw, "The Earliest Record of Banana in Africa," *Journal of Archaeological Science* (v.33, 2006); R.A. Marchant, D.M. Taylor, and A.C. Hamilton, "Late Pleistocene and Holocene History at Mubwindi Swamp, South-West Uganda," *Quaternary Research* (v.47, 1997); D.M. Taylor, "Late Quaternary Pollen Records From Two Ugandan Mires: Evidence for Environmental Change in the Rukiga Highlands of Southwest Uganda," *Palaeogeography Palaeoclimatology Palaeoecology* (v.80, 1990).

ROB MARCHANT
UNIVERSITY OF YORK

Ukraine

UKRAINE IS A republic in east Europe, part of the former Soviet Union. Its capital and largest city is Kiev (50 degrees 27 minutes N, 30 degrees 30 minutes E, population of 2.6 million in 2007, estimation by the Department of Ukrainian State Statistics). The total area of Ukraine is 233,090 sq. mi. (or 603,700 sq. km., the 44th country in size and second in Europe after European Russia). The total population is around 46.6 billion (in 2007, estimation by the Department of Ukrainian State Statistics). The word Ukraine originally meant "near the border" (of Russia and Poland).

The country has a mostly temperate continental climate, though a more Mediterranean climate is found on the southern Crimean coast. Precipitation is disproportionately distributed, being highest in the west and north (up to 31–39 in., or 800–1,000 mm. per year) and lesser in the east and southeast (around 16–18 in., or 400–450 mm. per year). Winters vary from cool along the Black Sea to cold farther inland. Summers are warm across the greater part of the country but are generally hot in the south. The average temperature in January is 18–19 degrees F (minus 7 to minus 8 degrees C) in the northern Ukraine, while it is 37–39 degrees F (3–4 degrees C) in the southern Crimea. The absolute January minimum/maximum temperature minus 43 degrees F to 67 degrees F (minus 42 degrees C to 19.5 degrees C) was observed in the northeast/Crimea. January precipitations vary from 1–2.5 in. (35–65 mm.) per month. The average temperature in July is around 64–75 degrees F (18–24 degrees C). The absolute July maximum/minimum temperature 106/36 degrees F (41.0/2.4 degrees C) was observed in the northeast. July precipitations vary from 1–7 in. (30–185 mm.) per month.

The linear temperature trend varies from around 1.8–3.6 degrees F (1–2 degrees C) per 100 years for the northern Ukraine in winter/spring to 30 degrees F (minus 1 degree C) per 100 years for the southern Ukraine in summer/fall. The linear trend of precipitation varies from around 4 in. (100 mm.) a year per 100 years for drier regions to minus 4 in. (100 mm.) a year per 100 years for wetter regions. So, the continental climate of Ukraine is becoming more mild. The interannular climate variability of Ukraine (especially its northern part) is under the strong control of the North Atlantic Oscillation.

The Ukrainian landscape consists mostly of fertile plains, or steppes, and plateaus, crossed by rivers such as the Dnieper, Seversky Donets, Dniester, and Southern Buh as they flow south into the Black Sea and the smaller Sea of Azov. To the southwest, the delta of the Danube forms the nation's border with Romania. The Danube is the largest European river, which rises in the Alpine region. Its average discharge from 1947–2001 was about 6,350 cu. m. per second. The maximum/minimum discharge (9,180/4,420 cu. m. per second) is observed in May/October. There is a positive linear trend of mean annual Danube discharge, which is about 200 cu. m. per second per 50 years. The country's only mountains are the Carpathian Mountains in the west, of which the highest is Mount Hoverla at 6,762 ft. (2,061 m.), and those in the Crimean peninsula in the extreme south, along the Black Sea coast.

The Black Sea is the southern border of Ukraine and is the largest natural reservoir of sulfate dioxide in any sea in the world. As a result of cyclonic general circulation in the Black Sea, the upper boundary of the sulfate dioxide zone is situated at a typical depth of about 262–328 ft. (80–100 m.) in the internal parts of the sea, and it deepens to up to 492–590 ft. (150–180 m.) along the shelf slope. The intensity of the principal element of the Black Sea general circulation (the so-called Rim current) varies from one decade to another as a result of quasi-periodic low-frequency changes of external forcing (such as wind stress, fresh/heat balance, and Mediterranean water inflow). The upper boundary of the sulfate dioxide zone moves up and down because of such quasi-periodical variability. There is also small negative trend of depth of sulfate dioxide zone (around minus 10 m. per 50 years), which points to its shoaling in the internal part of the sea in recent decades. This is

the result of Rim current intensification caused by stronger external forcing.

SEE ALSO: Climate Change, Effects; Oceanic Changes.

BIBLIOGRAPHY. Nikolai M. Dronin and E.G. Bellinger, *Climate Dependence and Food Problems in Russia, 1900-1990: The Interaction of Climate and Agricultural Policy and Their Effect on Food Problems* (Central European University Press, 2005); James Salinger, et al., *Increasing Climate Variability and Change: Reducing the Vulnerability of Agriculture and Forestry* (Springer, 2005); Valentina Yanko-Hombach, et al., *Black Sea Flood Question: Changes in Coastline, Climate and Human Settlement* (Springer-Verlag, 2006).

ALEXANDER BORIS POLONSKY
MARINE HYDROPHYSICAL INSTITUTE, SEBASTOPOL

UN Conference on Trade and Development/Earth Council Institute: Carbon Market Program

THE UNITED NATIONS Conference on Trade and Development (UNCTAD)/Earth Council Carbon Market Program (CMP) started in 1991 as the UNCTAD Emissions Trading Project and has opened the way for a collaboration between governments and the private sector in the development of a global carbon market. The program aims to reduce greenhouse gas emissions and is targeted primarily to economies in transition and third world countries. An emission trading system aims to control pollution by providing economic incentives for the reduction of polluting emissions. A central authority establishes a limit on the amount of a pollutant that can be emitted. Companies or other groups that release polluting emissions have to hold an equivalent number of credits or allowances, which represent the right to emit a specific amount. The total amount of credits cannot exceed the limit, so that total emissions cannot go beyond that level. Groups that need to increase their emissions must buy credits from those who pollute less. The transfer of allowances is described as a trade, although the buyer is actually

The Black Sea, bordering the south of the Ukraine, is the largest natural reservoir of sulfate dioxide.

being fined for polluting, while the seller is being compensated for their reduction of the emissions.

The starting point of the program is the recognition that future economies will be carbon constrained. UNCTAD and the Earth Council conceive their function as preparing developing countries for the likely changes in relative prices and relative production costs that the introduction of climate policies and measures will cause. They are also trying to facilitate a smoother transition to a postunregulated carbon economy.

The CMP aims to reduce the effect of climate change by contributing to the development of an integrated global emissions trading system in which all participant countries accept the principle of common but differentiated responsibilities. The United Nations Secretariat issued a major report on the subject in May 1992, as a contribution to the work of the Earth Summit held in Rio de Janeiro in June 1992. Since then, UNCTAD and the Earth Council have stimulated and encouraged research and capacity building in the area of greenhouse gas emissions trading. As one of their primary activities, UNCTAD and the Earth Council have lent their support to interested governments, corporations, and nongovernmental organizations for the development of a multilateral market for trading in greenhouse gas emission allowance and certified emission reduction credits, in accordance with the Kyoto Protocol and the decisions of the Conference of the Parties. The program also includes an annual Carbon Market Policy Forum, which brings together buyers, sellers, and market makers from the public and private sectors in both developed and developing countries.

A significant element of CMP is its work with economies in transition in developing their capacity to implement the Kyoto Protocol. CMP has devised a wide range of capacity-building activities and materials to assist accession countries, which form the Central Group 11 (CG11), in their development of national Kyoto strategies. CMP also has developed stocktaking reports of the current situation in relation to emissions trading and the Kyoto Protocol in accession countries. In addition, it has discussed a policy framework that identifies the various policy options and practical steps that CG11 countries may take, as well as the identification of specific capacity-building needs. In addition, UNCTAD and Earth Council have launched the Carbon Market E-learning Centre (CMEC), which contains modules related to emissions trading. The CMEC supplies learning opportunities for global audiences on the use of emissions trading as an economic resource to implement the UN Framework Conventions on Climate Change and the Kyoto Protocol. The center offers online courses and, more important, virtual workshops. These help other institutions to develop their own courses through the e-learning facilities of the CMEC.

UNCTAD and Earth Council Institute support the efforts of developing and transitional countries to participate effectively in the emerging trade and investments in allowances and credits for carbon emissions. As part of the CMP, UNCTAD and Earth Council Institute have also encouraged investments from the private sector for the Clean Development Mechanism (CDM) of the Kyoto Protocol, particularly in Least Developed Countries. The program has supported the creation of public-private bodies for the implementation of the CDM and has attempted to develop the CDM through capacity building, starting from the needs of local communities. The CMP has lent its support particularly to Brazil, as well as to African countries such as Tanzania, Uganda, Mozambique, Zambia, and Malawi.

SEE ALSO: Climate Change, Effects; Greenhouse Effect; Greenhouse Gases; Kyoto Protocol; United Nations; United Nations Environment Programme (UNEP).

BIBLIOGRAPHY. Earth Institute—Geneva, http://earthcouncil.com; United Nations Conference on Trade and Development, http://www.unctad.org/Templates/Page.asp?intItemID=4348&lang=1.

Luca Prono
University of Nottingham

United Arab Emirates

LOCATED IN THE Persian Gulf, the United Arab Emirates (U.A.E., formerly the Trucial States) has a land area of 32,278 sq. mi. (83,600 sq. km.), with a population of 4,380,000 (2006 est.) and a population density of 139 people per sq. mi. (64 people per sq. km.). Its economy is heavily dependent on petroleum and natural gas, with the country enjoying a very high standard of living. With little natural freshwater and regular sand and dust storms, the U.A.E. has little arable land, with most of the food in the country being imported.

With a dry desert climate and great wealth, widespread use of private automobiles and air conditioning has resulted in heavy use of electricity, all of which comes from fossil fuels. This has resulted in the country having the third highest rate of per capita carbon dioxide emissions in the world—29.3 metric tons in 1990, falling to 16.9 metric tons in 1996, and then rising steadily to 37.8 metric tons per person by 2004. Some 67 percent of the carbon dioxide emissions have come from gaseous fuels, with 27 percent from liquid fuels, and 3 percent from gas flaring. In terms of the sector generating the carbon dioxide emissions, 43 percent comes from electricity production, 45 percent from manufacturing and construction, and 8 percent from transportation.

An effect of global warming and climate change in the U.A.E. has been the raising of the average temperature in the country, making the establishment of arable land even harder. It has also forced the country to invest heavily in desalination plants to provide the country with enough freshwater. There has been large-scale use of water to transform Dubai, and some environmentalists have been critical of the use of water for the maintenance of gardens and golf courses in the country. The rise in the temperature of the water in the Persian Gulf has also led to the bleaching of coral reefs and further depletion of fish stock.

The U.A.E. government took part in the United Nations Framework Convention on Climate Change, signed in Rio de Janeiro in May 1992. The government accepted the Kyoto Protocol to the UN Framework Convention on Climate Change on January 26, 2005, with it entering into force on April 26, 2005.

SEE ALSO: Carbon Dioxide; Climate Change, Effects; Global Warming; Kyoto Protocol.

BIBLIOGRAPHY. Hussein A. Amery, "Water Wars in the Middle East: A Looming Threat," *Geographical Journal* (v.168/4, 2002); A.S. Goudie, A.G. Parker, and A. Al-Farraj, "Coastal Change in Ras Al Khaimah (UAE): A Carto-graphic Analysis," *Geographical Journal* (v.166/1, March 2000); Afshin Molavi, "Dubai: Sudden City," *National Geographic* (v.211/1, January 2007); World Resources Institute, "United Arab Emirates—Climate and Atmosphere," www.earthtrends.wri.org (cited October 2007).

ROBIN S. CORFIELD
INDEPENDENT SCHOLAR

United Kingdom

THE UNITED KINGDOM of Great Britain and Northern Ireland has a land area of 94,526 sq. mi. (244,100 sq. km.), with a population of 60,587,300 (July 2006) and a population density of 637 people per sq. mi. (246 people per sq. km.). London, the capital and the 16th largest city in the world, has a population density of 11,927 per sq. mi. (4,597 per sq. km.). Some 25 percent of the land of the United Kingdom is devoted to agriculture, with a further 46 percent used for meadow or pasture, and 10 percent of the land being forested.

Traditionally, most of the electricity generation in the United Kingdom has come from coal, which has been mined in parts of Scotland, Yorkshire, Nottinghamshire, and South Wales. The continued use of coal, and also of oil, Britain having made use of the North Sea oil fields since the 1970s, has meant that 73.2 percent of Britain's electricity generation was, in 2001, still coming from fossil fuels—coal, oil, and natural gas—with 23 percent coming from nuclear fuel and only 1.5 percent from hydropower. Although recent governments have tried to use nuclear power more extensively, this move has been widely opposed by many people, who are concerned about the safety of nuclear power, with political pressure over the location of the various nuclear power stations.

The United Kingdom ranks 37th in terms of its carbon dioxide emissions per capita, with 10.0 metric tons in 1990, falling steadily to 9.2 metric tons by 1998, and then rising to 9.79 metric tons by 2004. A third of all carbon dioxide emissions in the country are from the generation of electricity, with 27 percent from transportation, through heavy use of private automobiles, and large traffic jams and tailbacks in London and many other major cities, some 17 percent generated for residential use, and 15 percent from manufacturing and construction. In terms of the source of these emissions, 27 percent is from solid fuels, with 36 percent from liquid fuels and 35 percent from gaseous fuels, and 1 percent from gas flaring.

There have been many effects on Britain of global warming and climate change. Because statistics have been collected there since the 18th century, it has been easier to study the changes. The number of cold days has steadily decreased, with an average of 4 days per year above 68 degrees F (20 degrees C) for most of the period since 1772, but 26 days above 68 degrees F (20 degrees C) in 1995. Indeed, October 2001 was the warmest October in central England, with four of the five warmest years in the previous three and a half centuries being in the 1990s and early 2000s. One study has shown that oak trees have experienced earlier leafing as the climate gets warmer.

As well as rises in temperature, there have also been widespread floods, with that in October and November 2000 resulting in the flooding of some 10,000 houses at a cost of about $1.5 billion. This was the worst flooding in Britain since those in March to June of 1947, with the melting of a six-week snowpack, although some war damage to locks on canals leading into the River Thames made the floods worse than normal. Since then, there had been floods in 1968, 1993, and 1998, with those in 2000 following the wettest autumn since records were first collected in the late 1660s. Although floods have not been unknown in Britain—and the River Thames flooded again in 2003 and 2006—in June and July 2007 there were much more serious floods. These caused damage estimated at $3 billion, with Northern Ireland experiencing floods on June 12 and East Yorkshire and the Midlands being hit three days later.

Over the next five weeks, large parts of Berkshire, Gloucestershire, Oxfordshire, and South Wales were also inundated, with rainfall in June 2007 being twice the June average. Indeed, some areas of the country received the average monthly precipitation in one day. The worry has been that the floods, which took place twice a century on average, are now taking place every three to five years. The flooding would be much worse without the construction of the Thames Barrier in the 1970s, which has prevented any serious floods from happening in London since those in January 1928, March 1947, and 1968.

During the 1990s, a detailed survey of plant species in the country showed that the date of the first flowering of 385 British plant species had advanced by an average of 4.5 days when compared with the previous four decades. Flowering is particularly sensitive to the temperature in the month before the plant flowers, indicating that plants have become sensitive to the changes in temperature, with those that flower in spring being the most responsive. In terms of fauna, British birds have steadily expanded their ranges northward, with more birds that had previously only been found in the south of the country being spotted in northern England and Scotland. Over the last 25 years, some birds have expanded the northern margins of their ranges by about 12 mi. (19 km.). Another study of birds has shown that between the years 1971 and 1995, some 32 percent of the 65 species in the study have started laying eggs earlier—on average 8.8 days earlier—each year. In addition, frogs, toads, and newts have started spawning

Floods causing an estimated $3 billion in damages hit the United Kingdom in 2007, including this flooding in Sheffield.

between 9 and 10 days earlier than had been the case 20 years earlier.

The British government of John Major took part in the United Nations Framework Convention on Climate Change, signed in Rio de Janeiro in May 1992. The next government, that of Tony Blair, signed the Kyoto Protocol to the UN Framework Convention on Climate Change on April 29, 1998, ratifying it on May 31, 2002, with it entering into force on February 16, 2005.

The United Kingdom Climate Strategy introduced in 1994 had the objective of keeping greenhouse gas emissions of carbon dioxide, methane, and nitrous oxide at 1990 rates—carbon dioxide by incentives to business and home users to conserve energy; methane by reducing landfill through a landfill levy and a greater regulatory environment, as well as limiting methane emissions from coal production; and nitrous oxide through technological innovations in the manufacture of nylon. The introduction of three-way catalytic converters was planned to reduce carbon monoxide, especially from car exhausts, by up to 50 percent, and the reorganization of large power stations was expected to reduce nitrogen oxides by 35 percent.

In November 2000, the United Kingdom's Climate Change Policy, which was formulated following the United Nations Conference on Environment and Development, was formally launched. The United Kingdom was, in 2004, the eighth largest producer of carbon emissions, with the country being responsible for about 2.3 percent of the world's total coming from fossil fuels. The plan drawn up by the Blair government was not just to cut the emissions back to 12.5 percent less than the 1990 rate during the period from 2008 until 2012, as agreed by the Kyoto Protocol, but also to reduce them to 20 percent lower than the 1990 rate by 2010. The methods used by the British government to reduce carbon emissions largely hinged on encouraging business to improve its use of energy, to cut back on emissions from cars by providing better public transport, to promote energy efficiency in homes, and to get agriculture to reduce emissions.

As worry about the effects of global warming and climate change received much publicity in the British press, the Campaign against Climate Change was founded in 2001 to oppose the rejection of the Kyoto Protocol by U.S. president George W. Bush. Although it had small beginnings, on December 3, 2005, it did organize a large rally in London, and another took place on November 4, 2005. By this time, there were also a number of other pressure groups, including Stop Climate Chaos. Formed as a coalition of a number of other groups, including the Campaign against Climate Change, in September 2005, Stop Climate Chaos was also organizing protests. This was to lead to the I Count Campaign to try to get governments around the world to introduce measures to prevent world temperatures from rising more than 3.6 degrees F (2 degrees C).

On June 21, 2006, royal assent was given to a parliamentary bill that became the Climate Change and Sustainable Energy Act 2006, which, introduced to the British Parliament by Mark Lazarowicz, a Scottish Labour member of parliament, encourages microgeneration installations to reduce the use of large power stations and, as a result, reduce carbon dioxide emissions and fuel poverty, whereby some poor people had been unable to afford to heat their residences. The impetus from this led to the drafting of the Climate Change Bill, which was published on March 13, 2007, based heavily on the measures suggested in the I Count campaign. It aimed to reduce the United Kingdom's carbon emissions for 2050 to 60 percent of the level for 1990, with an intermediate target range of 26 to 32 percent by 2020. The bill was initially criticized for failing to include international aviation and shipping but quickly gained cross-party support, although it has not been passed into law. The British government has also been very keen on establishing a system of having a greenhouse gas allowance trading regime, although plans for this are still being drawn up.

SEE ALSO: Carbon Dioxide; Climate Change, Effects; Global Warming; Kyoto Protocol.

BIBLIOGRAPHY. *Climate Change: The United Kingdom Programme: 1st Report* (H.M. Stationary Office, 1994); Charles Furniss, "Dossier: Flooding in the UK," *Geographical* (v.78/6, June 2006); Robert Henson, *The Rough Guide to Climate Change* (Rough Guide, 2006); Mike Hulme and John Turnpenny, "Understanding and Managing Climate Change: The UK Perspective," *Geographical Journal* (v.170/2, June 2004); John McCormick, *Environmental Policy in the European Union* (Palgrave, 2001); Tim O'Riordan and Jill Jäger, eds., *Politics of Climate Change: A European Perspective* (Routledge, 1996); M. L. Parry, et al., *Review of the Potential Effects of Climate Change in the United Kingdom: Second*

Report (H.M. Stationary Office, 1996); *The Potential Effects of Climate Change in the United Kingdom* (Department of the Environment/H.M. Stationary Office, 1991); "United Kingdom—Climate and Atmosphere," www.earthtrends.wri.org (cited October 2007); Richard van der Wurff, *International Climate Change Politics: Interests and Perceptions—A Comparative Study of Climate Change Politics in Germany, the United Kingdom and the United States* (Universiteit van Amsterdam, 1997); Marcel Wissenburg, *European Discourses on Environmental Policy* (Ashgate, 1999); A. Wordsworth and M. Grubb, "Quantifying the UK's Incentives for Low Carbon Investment," *Climate Policy* (v.3, 2001); Farhana Yamin, ed., *Climate Change and Carbon Markets: A Handbook of Emissions Reduction Mechanisms* (London: Earthscan, 2005).

JUSTIN CORFIELD
GEELONG GRAMMAR SCHOOL, AUSTRALIA

United Nations

GLOBAL CLIMATE CHANGE is one of the most pressing issues of the 21st century. Because of the use of coal, oil, and gas for energy and the loss and degradation of forests, our planet is warming faster than at any time in the last several thousand years. We have already experienced warming temperatures, changing rainfall patterns, and sea-level rise. These disruptive forces have severe effects on economies, environment, and society of humankind. Nonetheless, the climate challenge may at the same time be viewed as an enormous opportunity for a significant economic change. It is quite evident that the United Nations (UN), by implementing a number of notable conventions and treaties, assumes a key role to play in a wide range of activities concerned with understanding, mitigating, and adapting to climate change. In a nutshell, the past few decades have seen a growing recognition of the importance of involvement of the UN with the complex scientific and technical issues related to global warming, climate change, and sustainable development.

Although it is rightly perceived that it is just a kind of global challenge the UN is uniquely positioned to address, it is also recognized that this is not a challenge for this world body alone. To handle the dilemma, it requires a truly concerted global effort—an initiative that draws together national governments, private sector, and civil society in one sustained push for change. This policy-relevant piece concludes by asserting the following forward-looking reflections that the UN, with its sensitivity and imagination, will be able to more successfully convey the urgency of the situation and send the following message to all of us: we should unite at any cost to save our beloved planet.

THE UN CONVENTIONS ON CLIMATE CHANGE

Regardless of the fact that the international scientists have drawn attention to the threats posed by global warming in the 1960s and 1970s, it took some years before the global community responded. In 1988, an Intergovernmental Panel on Climate Change (IPCC), an authoritative UN network of 2000 scientists, was created by the UN Environment Programme (UNEP) and the World Meteorological Organization. In 1990, this group presented a first assessment report that reflected the views of 400 scientists. The report indicated that global warming was real and urged that something be done about it. The findings of the panel prompted governments to create an international treaty, the UN Framework Convention on Climate Change (UNFCCC). By the standards of international agreements, the negotiation of the convention was rapid. It was ready for signature at the UN Conference on Environment and Development, more popularly known as the Earth Summit, held in Rio de Janeiro in 1992.

More comprehensively, the Convention on Climate Change, which entered into effect on March 21, 1994, sets an overall framework for intergovernmental efforts to tackle the challenge posed by climate change. It recognizes that the climate system is a shared resource, the stability of which can be affected by industrial and other emissions of carbon dioxide and other greenhouse gases (GHGs). Under this convention, which enjoys near universal membership, with 191 countries having ratified it, governments gather and share information on GHG emissions, national policies, and best practices; launch national strategies for addressing GHG emissions and adapting to expected effects, including the provision of financial and technological support to developing countries; and cooperate in preparing for adaptation to the effects of climate change. Notwithstanding, the recent UNFCCC report underscores the principal changes to patterns of investment and financial flows required to tackle climate change in the next quarter century. A major accomplishment of the convention, which is general and flexible in character, is that it admits that there is a problem.

A number of nations have recently approved an addition to the UNFCCC, the Kyoto Protocol, a second, more far-reaching international treaty on climate change that entered into force on February 16, 2005. The Kyoto Protocol has more powerful and legally binding measures that call for industrialized countries to collectively reduce emissions to 5 percent below 1990 levels between 2008 and 2012. In fact, these trends have been projected to accelerate over the recent years, including the Conference of the Parties of the UNFCCC, held in Bali (Indonesia) in December 2007, although the Vienna Climate Change Talks 2007 were attended by nearly 1,000 representatives from over 150 governments, business and industry, environmental organizations, journalists, and research institutions.

THE UN RESPONSE TO CLIMATE CHANGE ACTION

In 2007, climate change has indeed become one of the highly prioritized concerns for the UN, because there is now an overall understanding that the phenomenon will seriously affect the way the world operates in toady's challenging era of globalization, from healthcare and water issues to economic activity, humanitarian assistance, and the peace-building and security aspects. The UN has already demonstrated a long-standing commitment and responsibility toward resolving the cardinal environmental hazards encompassing reducing GHG emissions to limit future climate change and improving the capacity of the world's biodiversity and poorest communities to adapt to its inevitable effects. This universal organization has played a pivotal role in generating the scientific consensus, elevating the issue to the cover pages of the global media, and placing it on the in-tray of heads of state and government, as well as the chief executive officers of businesses and industries.

The UN, through the UNFCCC, helps to accelerate the take-up of clean and renewable energies to counter the climate change challenge. To be more specific, the Clean Development Mechanism (CDM) of the Kyoto Protocol permits developed countries to offset some of their emissions through clean and renewable energy projects and certain forestry schemes in developing countries. The CDM funds flowing from north to south will reach up to US$100 billion over the coming years. The high-technology industries and job opportunities are emerging in both developed and developing nations. China and India are currently homes to two of the biggest wind turbine and power companies. The foreign direct investment in renewable energy, driven in part by the UN-brokered climate treaties, is anticipated to top US$80 billion in 2007. The UN system is further endeavoring to nourish this process. For instance, the UNEP, in partnership with the UN Foundation and Asian Development Bank, piloted a project that has brought solar power to 100,000 people in India.

This sort of progress actually echoes the Millennium Development Goals adopted by the UN, as they relate to such areas as poverty alleviation, public health, basic education, and so on. Furthermore, the UN assists in harnessing the power of the carbon markets and evaluates the potential for forests to cope with challenge emanated from the climate change. The UN's Food and Agriculture Organization, for example, estimates that 13 million hectares of the world's forests are lost annually and that deforestation accounts for approximately 20 percent of the global GHG emissions.

The issue of climate change, along with such steps as the UN Global Compact, assists the restoration of nexus between the UN and other segments of society making up business and industry. Although an intriguing feature of recent months is a call by the private sector for international regulation across the globe, businesses appear eager to do their part if the ground rules are clear and comprehensive. In addition, many other welcoming initiatives are already being undertaken. The European Union, for instance, has agreed to a 20 percent emission reduction target, which will rise to 30 percent if other countries follow suit. The Group of Eight Summit in Germany has also affirmed the ongoing UN climate change process. It is demanding that governments and the UN deliver an international agreement to address this issue. Developing countries are also acting at the same time. In Brazil, efforts to counter deforestation in the Amazon have shown positive results, whereas China has recommitted itself to reduce its energy intensity by 20 percent.

In addition, the prime minister of India has recently ordered a review of his nation's GHG emissions.

The UN is looking at its own backyard as well. The Capital Master Plan for the refurbishment of the UN headquarters in New York is assessing how to factor green measures into the project, looking to make the structure a striking example of an ecofriendly building. This is part of a wider evaluation of how UN operations, from building to procurement of goods and services, can respond to the sustainability challenge. It is redundant to say that the UN World Tourism Organization (UNWTO) has developed an information-gathering Web resource consisting of data, studies, policy papers, videos, and other materials, as part of its effort to combat climate change. Although the tourism sector is expected to mitigate and adapt in the face of global warming and to explore and put in place more climate-friendly and climate-proof alternatives, the UNWTO service is a contribution to fostering the knowledge base and search for solutions to meet the climate challenge.

Finally, the UN is the only forum in which an agreement striving to reduce GHG emissions beyond 2012 could practically be brokered among the 190-plus countries with diverse outlooks and economies, but of a common atmosphere. Despite the fact that the warming of the Earth's atmosphere has by this time adversely affected fragile ecosystems and the livelihoods of poor people, this protest at the same time offers manifold alluring prospects for all of its member states.

FUTURE DIRECTIONS

The most recent findings of the IPCC have emphasized that the science on climate change is very clear. The panel report, issued in May 2007, has also unequivocally confirmed the warming of the climate system and linked it directly to human activities. The IPCC has outlined the likely effects of climate change in the near future if the international community fails to act. Although this grave threat is now beginning to receive the high attention it merits, it brings up an underpinning question about how the UN itself will meet this ultimatum, carrying serious implications for our planet and our future generations.

UN Secretary-General Ban Ki-moon, who has identified climate change as one of his top priorities since his

A moment of silence was observed for the victims of the Algiers bombing during the United Nations Climate Change Conference in Bali, Indonesia, December 3–14, 2007. The gathering brought together representatives of over 180 countries.

very first days in office, intends to take a leadership role in supporting efforts by the international community to address the problem by bringing world leaders together and ensuring that all parts of the UN system will contribute to the solution. The UN, as a global forum with broader participation, is best suited to forging a common approach to responding to climate change. Observing the imminence and severity of the problems posed by the accelerating changes in the global climate, the secretary-general has warned that we cannot go on this way for long, and that we cannot continue with business as usual. He also urged us to take joint action on a global scale to address climate change. It is true that there are numerous policy and technological options available to face this impending crisis, but we should have the political will to seize them.

Both developed and developing countries have to be able to reassess the big picture of what is required by identifying the key building blocks for an effective response to climate change. There is a consensus that the response needs to be global, with the involvement of all countries, and that it needs to give equal weight to adaptation and mitigation. The industrialized nations can do much more to reduce GHG emissions and encourage energy efficiency, and they can also support clean development in fast-growing economies and adaptation measures in countries facing the greatest hardships from climate change. In contrast, the developing countries need to be more engaged in addressing climate change while safeguarding their socioeconomic growth and poverty eradication. As for the UN as a whole, it should help the developing economies to finance and deploy energy efficient technologies and incorporate adaptation into the UNFCCC and related environmental organizations.

As an expert scientific panel has recently reported to the UN, to head off the worst of climate change, the governments must pour tens of billions of dollars more than they are into clean-energy research and enforce sharp rollbacks in fossil-fuel emissions. The UN itself must better prepare to help tens of millions of environmental refugees, and authorities everywhere should discourage new building on land less than 1 m. (39 in.) above sea level. The construction of climate-resilient cities may also be championed. The climate-resilient cities are identified as cities that produce low per capita emissions and that can manage weather-related events. Moreover, a climate-resilient city has a reliable supply of potable water, given that water will likely become scarcer as weather patterns change.

The UN is an intergovernmental organization, but it collects its strength and inspiration from the support of civil society worldwide. In fact, there are a number of wonderful illustrations of cohesive ties binding the UN to global civil society. Although these relationships date back to the earliest days of this global forum, they have truly come of age just recently. In spite of the truth that this has been possible because of a deliberate and sustained outreach effort on both parts, it also reflects the substantially expanded role of civil society organizations on the world stage. That is why today's UN relies on its partnership with the nongovernmental organization (NGO) community in virtually everything it does. Some research findings show that the nonstate actor can contribute effectively and efficiently, if it fosters awareness in developing inventive initiatives at the grassroots level, which inspire people to work toward a solution. Thus, one area in which the cooperative partnership between the UN and civil society is increasingly essential relates to the global challenge posed by climate change. Conceding that the NGOs have historically been at the forefront of the struggle to draw attention to the environment, and to push for action to protect it, they may be stimulated to shoulder their redoubled responsibility toward building grassroots support for a breakthrough, as well as common ground for fighting climate change.

Climate change is not just an environmental issue but one that has serious socioeconomic implications as well. Because this advanced issue requires the attention of many sectors such as finance, energy, transport, agriculture, and health, climate change should firmly be positioned in the broader sustainable development agenda. Climate change impinges on all countries, because its aftermaths know no boundaries. Hence, this issue must be addressed in the context of the international development agenda. To be more concrete, the actions on climate change must be integrated into development efforts and scientific research led by various parts of the UN family. This may include work to address investment flows and finance schemes relevant to the development of an effective and appropriate international response to climate change and increased support for adaptation and for involving industry leaders to encourage support from the private sector. In other words, multilateral and bilateral donors, regional development banks, and international investment flows into the developing countries ought to reflect adaptation in their investment decisions.

It is expected that the Bali summit, by building a climate of trust among all governments in that spirit, will determine future action on mitigation, adaptation, the global carbon market, and financing responses to climate change for the period after the expiry of the 166-member Kyoto Protocol, the current global framework for reducing GHG emissions, in 2012.

CONCLUSION

Combating climate change presents a remarkable opportunity to break with the past and to look anew at the way we operate, the way we do business, and the path we use to relate to each other, now and in the future. Nevertheless, as climate change is a global problem, it needs a long-term global solution. The UN's ultimate goal is to make a comprehensive agreement under the UNFCCC process. Such an agreement must tackle climate change on all fronts, including climate adaptation, disaster mitigation, clean technologies, global deforestation, and resource mobilization. The UN should be at the center of brokering a fair, equitable, and decisive climate change regime for the new millennium. Toward confronting the climate change challenge, the UN needs an international collaborative stride made by governments, the private sector, and civil society. If this comprehensive organization succeeds, it will not only change the world but will save humanity. The UN must lead the world to a universal consensus discussion and enroot a strategic policy guideline to be realized over the coming years; the faster we get at it, the easier it is going to be to adapt.

SEE ALSO: Adaptation; Clean Development Mechanism; Deforestation; Developing Countries; Energy Efficiency; European Union; Greenhouse Gases; Kyoto Protocol; Nongovernmental Organizations (NGOs); Refugees, Environmental; Tourism; United Nations Environment Programme (UNEP).

BIBLIOGRAPHY. Gary Braasch, *Earth Under Fire: How Global Warming is Changing the World* (University of California Press, 2007); Andrew E. Dessler and Edward A. Parson, *The Science and Politics of Global Climate Change: A Guide to the Debate* (Cambridge University Press, 2006); Charles J. Hanley, "Scientists Offer Climate Plan to UN," *The Washington Post* (February 27, 2007); Intergovernmental Panel on Climate Change, *Climate Change 2007: The Physical Science Basis* (Intergovernmental Panel on Climate Change, 2007); Brett M. Orlando, et al., *Carbon, Forests and People: Towards the Integrated Management of Carbon Sequestration, the Environment and Sustainable Livelihoods* (World Conservation Union, 2002); Nicholas Stern, *The Economics of Climate Change: The Stern Review* (Cambridge University Press, 2007); UN, *The Future in Our Hands: Addressing the Leadership Challenge on Climate Change* (United Nations Headquarters, 2007); UN, *United Nations Framework Convention on Climate Change* (United Nations Headquarters, 1992); United Nations Environment Programme, *Caring for Climate 2005: A Guide to the Climate Change Convention and the Kyoto Protocol 2005* (United Nations Environment Programme, 2005); United Nations Environment Programme, *Global Trends in Sustainable Energy Investment 2007* (United Nations Environment Programme, 2007); UN Foundation, *Confronting Climate Change: Avoiding the Unmanageable and Managing the Unavoidable* (UN Foundation with Sigma Xi, 2007); UN Framework Convention on Climate Change, *Technologies for Adaptation to Climate Change* (Secretariat of the United Nations Framework Convention on Climate Change, 2006).

MONIR HOSSAIN MONI
UNIVERSITY OF DHAKA

United Nations Development Programme (UNDP)

THE UNITED NATIONS Development Programme (UNDP) is the United Nations' (UN's) global development network that assists developing countries in managing and funding sustainable environmentally sensitive global and national development and, as such, is the world's largest multilateral development source. The UNDP was created (1965–71) by merging the Expanded Programme of Technical Assistance with the UN Special Fund. The UNDP currently works in 166 countries, with the UNDP Resident Representative in each generally serving as the resident coordinator of the UN development activities. The UNDP is an executive board within the UN General Assembly. The UNDP administrator is the third-ranking position in the UN, following the secretary-general and deputy secretary-general. Headquartered in New York City, the UNDP is funded entirely by voluntary contributions from member nations. The UNDP annually publishes local, regional, national, and global Human Development Reports. The UNDP concentrates on five development challenges: democratic governance, poverty reduction, crisis prevention and recovery, energy and environment, and HIV/AIDS.

UNDP GOALS AND INITIATIVES

The umbrella goal of the UNDP is to cut world poverty in half by 2015; that goal is supported by seeking to achieve in the same year eight subsidiary Millennium Development Goals (MDGs), so named because they were derived from the September 2000 UN Millennium Summit. The MDGs are to eradicate extreme poverty and hunger; achieve universal primary education; promote gender equality and empower women; reduce child mortality; improve maternal health; combat HIV/AIDS, malaria, and other diseases; ensure environmental sustainability; and develop the Global Partnership for Development endorsed at the March 2002 International Conference on Financing for Development in Monterrey, Mexico, and reaffirmed in August 2002 at the Johannesburg World Summit on Sustainable Development.

The goal of achieving environmental sustainability is rooted in the idea that developing and impoverished countries are those countries that are most damaged by environmental degradation and by the use of expensive polluting energy sources. The UNDP's goal of ensuring environmental sustainability concentrates on effective water governance, access to sustainable energy services, sustainable land management targeting desertification and land degradation, the conservation and sustainable use of biodiversity, and national/sectoral policies controlling emissions of ozone-depleting substances (ODSs) and persistent organic pollutant (POP) emissions. The UNDP seeks to accomplish these objectives by linking its developing country client states with environmentally sensitive development projects that produce long-term jobs. The UNDP

also helps these same countries develop and adopt policies that encourage and sustain these projects and assists in developing other governmental policies and action plans that might positively affect environmental sustainability. The UNDP also helps these countries develop the capacity to manage their environment, energy resource use, and sustainability while reducing poverty, sustaining their advancing development, and integrating the local communities and women into this management.

The UNDP promotes the effective use of the client state's water resources by developing policies and programs integrating the sustainable use of marine, coastal, and freshwater resources; adequate and accessible clean water sources; and the sanitation services necessary to sustain and improve these water resources. The UNDP supports these integrated water resources management programs by initiating and then requiring transboundary waters management within a water governance framework uniting local, national, and regional governmental entities.

The UNDP promotes access to sustainable clean energy services by supporting sustainable and integrated transboundary energy resource development and energy use targeted at reducing the poverty in the client state. The UNDP promotes the use of energy technologies that are climate change neutral or abating. The UNDP facilitates these programs by providing access to funding sources that include, but are not limited to, the Kyoto Protocol's Clean Development Mechanism and the Global Environment Facility (GEF) program. All of the funding sources accessible through the UNDP support projects that mitigate climate change and support indigenous sustainable livelihoods.

The UNDP promotes sustainable land management, contending that land degradation and desertification are two of the major causes of rural poverty in developing countries. The UNDP concentrates on educating and helping its client states and their composite communities maintain the integrity of the indigenous land-based ecosystems by developing or revising land governance policies that protect the land, sustain livelihoods, and mitigate or adapt to climate change. The UNDP also supports the creation of infrastructure projects that prevent or retard land degradation and desertification. These projects are funded for the most part by integrated multistakeholder relationships symbiotically uniting local, national, regional, and global entities.

The UNDP promotes conservation and sustainable biodiversity by helping client states and their composite communities maintain and develop the capacity to manage their indigenous biodiversity and ecosystems. This process not only seeks to sustain the indigenous biodiversity and ecosystems but also seeks to manage them so as to provide more food, fuel, sustainable livelihoods, and medicines in addition to better security and shelter. This process sometimes entails the development of clean water systems, improved disease control, and better preparedness for and reduced vulnerability to natural disasters. The UNDP helps these countries develop and then manage their agriculture, fisheries, forests, and other resources in a pro-poor approach oriented to developing marketable self-sustaining biotechnology.

The UNDP's Montreal Protocol and GEF programs are designed to support the reduction and elimination of ODSs and POP emissions on the national and sectoral levels while maintaining the economic competitiveness of the client states through alternative technologies and increasing indigenous capacity.

SEE ALSO: Clean Development Mechanism; Intergovernmental Panel on Climate Change (IPCC); Kyoto Protocol; Sustainability; United Nations.

BIBLIOGRAPHY. Stephan Klingebeil, *Effectiveness and Reform of the United Nations Development Programme (UNDP)* (Taylor & Francis, Inc., 1999); Craig N. Murphy, *The United Nations Development Programme: A Better Way?* (Cambridge University Press, 2006); United Nations, *Basic Facts about the United Nations* (United Nations, 2004); "United Nations Development Programme," www.undp.org (cited November 2007); United Nations Development Programme, *Generation, Portrait of the United Nations Development Programme* (United Nations Publications, 1995); United Nations Development Programme, *UNDP for Beginners: A Beginner's Guide to the United Nations Development Programme* (UNDP, 2006); "United States Committee for the United Nations Development Program," www.undp-usa.org (cited November 2007).

RICHARD MILTON EDWARDS
UNIVERSITY OF WISCONSIN COLLEGES

United Nations Environment Programme (UNEP)

THE UN ENVIRONMENT Programme (UNEP) coordinates all United Nations (UN) global and regional environmental activities, assists developing countries in implementing environmentally sound policies, encourages sustainable development through sound environmental practices, reviews the status of the global environment, seeks consensus in environmental policy, and alerts the global community and governments of new and emerging threats to the biosphere. The UNEP officially divides its responsibilities into seven divisions: Early Warning and Assessment; Environmental Policy Implementation; Technology, Industry, and Economics; Regional Cooperation; Environmental Law and Conventions; Global Environment Facility Coordination; and Communications and Public Information.

The UNEP grew out of the June 1972 UN Conference on the Human Environment, but its mandates and objectives are derived from UN General Assembly resolution 2997 (27) (December 15, 1972), as amended at the UN's 1992 Conference on Environment and Development, the 1997 Nairobi Declaration on the Role and Mandate of UNEP, and the Malmö Ministerial Declaration of May 31, 2000. The UNEP is headquartered in Gigiri, Nairobi, Kenya, and has six regional offices and a number of country offices all under the governance of UNEP's Governing Council. Fifty-eight member states allocated according to geographical regions and serving three-year terms make up the council. The council has the primary responsibility for the developing UN policies and programs on environmental issues. The council attempts to mediate differences and promote cooperation between UN member states. The UNEP secretariat employs 890 global staff members and is funded by UN member states. The UNEP's executive director, as of June 2006, was Achim Steiner. Steiner was preceded in the position by Dr. Klaus Töpfer, who followed Dr. Mostafa Kamal Tolba, who held the position from 1975 to 1992.

GOALS AND ACTIVITIES OF UNEP
UNEP activities span the spectrum of global environmental issues concerning the atmospheric, marine,

and terrestrial ecosystems. The UNEP develops international and regional meetings on environmental issues, promotes a synergy of science and policy on environmental issues, funds and implements development projects related to the environment, works with environmental nongovernmental organizations (NGOs), and is responsible for coordinating and implementing responses to climate change, especially when those changes relate to undeveloped countries with little funding. The UNEP cosponsors and coorganizes regional workshops on the common problems of climate change and possible response strategies in Africa, Latin America, the Caribbean, the Asian-Pacific Basin, and the former Soviet Union.

UNEP also sponsors the development of member state solar loan programs and plays a pivotal role in restoring the Shatt al Arab marshlands that were virtually destroyed by Iraq's Saddam Hussein when the Marsh Arabs sided with Iran in the Iran-Iraq War. The UNEP estimated in 2001 that the marshes were reduced to no more than 386 sq. mi. (1,000 sq. km.). Restoration of the marshlands began in 2003, following the end of organized Iraqi military resistance to the 2003 Anglo-American invasion of Iraq to oust Hussein, and by 2007, the marsh was restored to approximately 50 percent of the area it made up before the Iran-Iraq War. In a similar vein, the UNEP helped/helps create guidelines and treaties relating to transboundary air and water pollution and international trade in harmful chemicals.

The UNEP plays an integral role in creating criteria and indicators for assessing ecological and economic vulnerabilities to climate change and developing regional responses and adaptation. The UNEP is creating a handbook on economical agricultural adaptive strategies to climate change and has published a handbook on climate change communications. The UNEP also plans and implements the Climate Change Outreach Programme, promoting climate change awareness at the national level and assisting governments in developing adaptive response strategies. Its Assessment of Impacts of and Adaptation to Climate Change in Multiple Regions and Sectors project, funded by the Global Environment Facility, seeks to create the scientific and technical capacity in over 45 mostly African countries to respond to climate change.

The discovery of a hole in the ozone layer of the atmosphere in 1985, coupled with the earlier detection of the warming of the Earth (1980), gave impetus to the idea that one of the causes is human-induced climate change (anthropogenic climate change) from man-made, ozone-depleting gases. The United Nations and the World Meteorological Organization (WMO) responded by creating in 1988 the Intergovernmental Panel on Climate Change (IPCC) and tasking it with studying the hypothesized phenomenon on a comprehensive, objective, open, and transparent basis. The IPCC determined that the Earth had warmed over the last 150-year period and that that warming was in part caused by human activity. The executive summary of the report concluded that most of the observed global warming experienced in the last 50 years is a result of the increase in greenhouse gas concentrations. The IPCC and former U.S. vice president Al Gore were jointly awarded the Nobel Peace Prize on October 12, 2007, for their work on global warming.

The UNEP and the WMO provide the joint secretariat support to IPCC. The IPCC does not engage in research. Its assessments are based on peer-reviewed and published scientific/technical literature under guidelines set forth in the Principles Governing IPCC Work. The IPCC has three Working Groups and a Task Force. Working Group I assesses the scientific aspects of the climate system and climate change. Working Group II assesses the vulnerability of socioeconomic and natural systems to climate change, negative and positive consequences of climate change, and adaptive strategies. Working Group III assesses options for limiting greenhouse gas emissions and ways of mitigating climate change. The Task Force on National Greenhouse Gas Inventories is responsible for the IPCC National Greenhouse Gas Inventories Programme and developed the methodology for calculating national greenhouse gas inventories.

After the IPCCs Fourth Assessment Report was issued in February 2007, 46 nations, led by France, signed the Paris Call for Action seeking the replacement of the UNEP with a new and more powerful United Nations Environment Organization. The top four greenhouse gas emitters, the United States, China, Russia, and India, did not sign.

SEE ALSO: Climate; Global Environment Facility (GEF); Greenhouse Gases; Intergovernmental Panel on Climate Change (IPCC); United Nations; World Meteorological Organization.

BIBLIOGRAPHY. United Nations, *Basic Facts about the United Nations (United Nations*, 2004); United Nations Environment Programme, www.unep.org (cited November 2007); United Nations Environment Programme, *Natural Allies: UNEP and Civil Society* (United Nations Foundation, 2004); United Nations Environment Programme, "Organizational Profile," www.unep.org/PDF/UNEPOrganizationProfile.pdf (cited November 2007).

RICHARD MILTON EDWARDS
UNIVERSITY OF WISCONSIN COLLEGES

United States of America

THE UNITED STATES of America is the world's largest industrialized country and emitter of carbon dioxide. It is therefore widely regarded as the most significant contributor to global warming and climate change. U.S. climate change policies have never remained consistent, as they have tended to shift in accordance with the presidential administration in office. The current administration, led by George W. Bush, has come under particular scrutiny from the media, the scientific community, the general public, and other countries for its climate change policies.

The focus on climate change in the United States became particularly acute in the late 1980s and evolved out of concern about the growing damage to the ozone layer. For two decades, that problem had occupied the attention of scientists and policymakers. In the early 1970s, scientific researchers at the University of California in Irvine established clear evidence that chlorofluorocarbons (CFCs)—chemical compounds made up of fluorine, chlorine, and carbon—were damaging the ozone layer, the thin protective layer above the Earth's atmosphere. The increased solar radiation entering the atmosphere as a result of this damage had the potential to cause health problems such as skin cancers and cataracts. At the time, CFCs were widely used as refrigerants, cleaning solvents, and the basis for aerosol products.

In October 1976, the U.S. Environmental Protection Agency (EPA) prohibited companies from manufacturing CFC-based products unless they were essential and threatened to cancel the product registration of companies that failed to comply. The following year,

the agency ordered a gradual phase-out of CFCs from household products such as deodorants, hairsprays, and cleaners. In 1985, the United Nations Environment Programme (UNEP) created a framework convention for the protection of the ozone layer. It encouraged governments to adopt relevant measures to that end, devised a Conference of the Parties composed of governments that had ratified the convention, and appointed a UN secretariat to monitor and frame the actions of the Conference of the Parties. The creation of the framework convention paved the way for international negotiations over the regulation of CFCs. These negotiations, although initially hindered by energy interests in both the European Union (EU) and the United States, ultimately led to the Montreal Protocol of 1988, a worldwide treaty that mandated a staged reduction in the production and consumption of fully halogenated CFCs. The treaty was more lenient toward developing countries, as it gave them a 10-year grace period for the phase-out of CFCs and promised them technological assistance from industrialized nations in return for switching to CFC alternatives. The United States ratified the Montreal Protocol on April 21, 1988, and brought it into effect at the beginning of the following year.

During the next 18 months, the signatories of the Montreal Protocol realized that the original draft of the protocol did not go far enough in controlling CFC emissions. At the second Conference of the Parties in June 1990, delegates placed further restrictions on the production and use of CFCs and further reduced the acceptable level for CFCs in the atmosphere. The fourth Conference of the Parties, which took place two years later, led to further changes. For example, the United States had to limit its production of ozone-depleting substances in accordance with its own Clean Air Act. Over time, the Montreal Protocol virtually eliminated CFC emissions and significantly reduced the threat to the ozone layer. Ultimately, it was the model on which future environmental regulations were based.

INTERNATIONAL FOCUS ON CLIMATE CHANGE

At the same time as ozone depletion was occupying the attention of international policymakers, the issue of planetary climate change was also gaining ascendancy on the world stage. As a phenomenon, climate change was by no means new, having been identified by scientists in the late 1950s. However, in the 1980s, it became the focus of widespread public and political attention. In the summer of 1988, the United States experienced its hottest and driest summer since records began. Fire destroyed millions of acres of forest in western regions of the country, and in the south, barges and small boats became stranded as the Mississippi River began to recede. The Midwest experienced prolonged periods of drought, and across the country crop yields declined severely. On June 23, the temperature in Washington, D.C., reached a record 101 degrees F (38 degrees C).

In response to the extreme weather conditions, Senator Timothy Wirth (D-CO) convened a U.S. Senate hearing on the issue of climate change. Paul Revere, a scientist for NASA, confirmed at the hearing that extra carbon was building in the atmosphere and heating the Earth. His colleague James E. Hansen similarly asserted that the planet was experiencing its highest temperatures of the 20th century. Both men used the phrase "the greenhouse effect" to describe global warming. They also claimed that global warming was occurring as a result of human activity, especially the combustion of fossil fuels such as coal and oil.

Just days after the Senate hearing ended, politicians, officials, and scientists gathered in Toronto, Canada, for an environmental conference. The delegates proposed a 20 percent reduction in global carbon dioxide emissions by 2005, although they based this proposal less on firm economic and scientific analysis than on a desire to show their commitment to solving the climate change problem. Later that year, in December, the UNEP and the World Meteorological Organization jointly created the Intergovernmental Panel on Climate Change (IPCC), an organization designed to bring together climate information from around the world.

THE DIVISIVE ISSUE OF GLOBAL WARMING

At a 1989 summit meeting of the world's seven largest industrial democracies, it was apparent that the EU and United States were becoming increasingly divided over global warming. West European countries thought that the evidence about climate change was irrefutable, whereas George H.W. Bush, the American president, argued that the matter required further research. A number of factors accounted for

the disparity between the two sides, particularly the fact that the United States was experiencing higher and faster population growth than Europe, and thus would be more adversely affected by limitations on carbon emissions and fuel consumption. In 1990, the IPCC released its first report, which reflected the general consensus of scientists that global warming was a serious issue.

The 1992 UN Conference on Environment and Development, at Rio de Janeiro in Brazil, represented the first step toward international regulation of climate change. Attended by governments from around the world, the conference resulted in a treaty called the Framework Convention on Climate Change (FCCC), which sought to limit greenhouse gases at a level that would not interfere with the planet's natural climate system. It also established a new Conference of the Parties, a forum in which nations would meet and establish environmental standards. The United States, still under the leadership of George H.W. Bush, promptly ratified and implemented the treaty, as did the majority of governments at the conference. Industrialized countries also used the conference as an opportunity to adopt a program called Agenda 21, which promised development assistance to developing countries on the condition they took steps to adopt more environmentally sound policies.

In January 1993, recently elected U.S. president William J. Clinton assumed control of the White House. The vice president in the new administration, Albert Gore, had published a book the previous year called *Earth in the Balance*, which he hoped would alert the general public to the fragility of the natural environment. On Earth Day in April of that year, Clinton, in adherence with the stipulations laid down in the FCCC, publicly announced that he aimed to reduce carbon dioxide emissions to their 1990 level by the year 2000. He also called for a tax on energy consumption, although strong opposition from the Democratic-controlled Congress thwarted this proposal.

The IPCC released its second report in 1996. The report stated that human activities were having a clear effect on climate change. Although some scientists were still skeptical about the findings, the majority agreed that global warming had become a major problem. In light of the evidence presented in the report, Conference of the Parties delegates sought further preventative measures. At the Second Conference of the Parties in Geneva that year, Timothy Wirth, now undersecretary for global affairs in the Clinton Administration, announced that the United States was willing to limit its greenhouse gas emissions if other countries were prepared to do the same. Delegates at the conference began to draw up a protocol for the framework convention that they could sign at the next Conference of the Parties, scheduled for Kyoto, Japan, in 1997. The most important aspect of the draft protocol was its proposal to reduce carbon emissions to a set level between 2008 and 2012.

As part of his ongoing efforts to underline his commitment to solving the global climate crisis, Clinton gave a speech at a special meeting of the UN in New York in June 1997. He proposed the development of new technology, the introduction of strategies such as emissions trading, and the adoption of measures that would protect the environment without impeding economic growth. He followed up this speech by holding a press conference with Gore at the White House the following month. At the conference, he reiterated his government's commitment to tacking climate change. The U.S. Senate, which had come under Republican control in the midterm elections, was becoming increasingly uneasy at Clinton's rhetoric, and in late July it adopted a resolution that warned the president to exercise caution at the forthcoming Kyoto conference. Known as the Byrd-Hagel Resolution, after its two main sponsors, the resolution advised Clinton not to accept terms that would jeopardize the American economy or place less responsibility for climate change on developing nations. Although opposition in Congress was mounting, Clinton stated at a meeting of the National Geographic Society in October that he supported reducing U.S. emissions to their 1990 level between 2008 and 2012. He neglected to explain, however, that the plan might have negative economic consequences for the United States.

INTERNATIONAL RESPONSES

The third Conference of the Parties took place in Kyoto, Japan, in November 1997 and was attended by 10,000 government officials from across the globe, as well as scores of lobbyists, observers, and media representatives. During the proceedings, the United States reiterated Clinton's plan to reduce carbon dioxide emissions to their 1990 level during the first commitment period, between 2008 and 2012. Most of the European

countries felt that this measure was inadequate, and they demanded more substantial action. The dissension between Europe and America thwarted an agreement over emissions policy, and only a last-minute trip to Japan by Gore resulted in a compromise between the two sides. Gore said that the United States would reduce its carbon dioxide emissions to 7 percent below the 1990 level during the first commitment period, scheduled for 2008 to 2012, and the Europeans agreed to cut their emissions to 8 percent below the 1990 level during the same time frame. Overall, the Kyoto text only committed the 38 industrialized countries in attendance at the conference to reducing their emissions. It placed no obligation on developing nations, who argued that a binding emissions treaty would hinder their path to full industrialization.

The treaty, although open for signatures at the end of the conference, was by no means complete. A number of issues remained unresolved, including emissions trading and carbon dioxide sinks. Emissions trading, a policy strongly favored by the United States, would allow countries that had improved on their emissions targets to sell the difference to countries at risk of not meeting their targets. Many European countries were unfamiliar with this policy and were reluctant to adhere to it without full knowledge of its consequences. Australia, Japan, and the United States also advocated that carbon dioxide sinks—reservoirs such as forests and oceans that absorb carbon and prevent it from escaping into the atmosphere—should be included as part of countries' emissions targets. Clinton signed the treaty in 1998 but did not present it to the Senate for ratification.

The sixth Conference of the Parties opened at The Hague in the Netherlands in late November 2000. It was somewhat overshadowed by events in the United States, where the result of the November 7 presidential election was still in dispute because of ongoing recounts in the state of Florida. The key purpose of the conference was to establish the regulations for the implementation of the Kyoto Protocol. For the previous three years, dissension and conflicting interests had prevented the protocol from coming into force. The Umbrella Group—a federation composed of the United States, Australia, Canada, and Japan and supported by energy companies and oil-rich countries such as Kuwait—argued that carbon sinks should be allowed to count as part of countries' reduction targets. The Umbrella Group also hoped to include a provision for emissions trading in the Kyoto regulations.

In its *Third Assessment Report*, published at the beginning of 2001, the IPCC further emphasized the role of humans in the climate change process. Discrediting the idea that global warming stemmed from natural phenomena such as volcanic activity and changes in the sun's radiation, the IPCC argued that the combustion of fossil fuels such as coal, oil, and natural gas was the only reasonable cause of climate change. Moreover, it attributed global warming to an average temperature increase of just 1 degree over the previous century and warned that if fossil fuel consumption continued without restraint, the planet would experience an additional 3 degrees to 10 degrees of warming by the year 2100. The panel was confident, though, that there was still enough time to halt the process of climate change.

U.S. WITHDRAWL FROM THE KYOTO PROTOCOL

The same month the IPCC report was published, George W. Bush entered the White House, and he moved quickly to limit his country's responsibility for reducing carbon dioxide emissions. On March 28, he announced that he was going to withdraw the United States from the Kyoto Protocol. He justified his decision by arguing that the protocol would have a harmful effect on the American economy and that it unfairly placed most of the burden for reducing carbon dioxide emissions on industrialized nations. Dick Cheney, the vice president in the Bush Administration, openly repudiated the protocol. As an alternative, Bush proposed an extensive program of scientific research and technological innovation. To that end, he instructed the National Academy of Sciences (NAS) to carry out an extensive study on the impact of the greenhouse effect. Bush's actions prompted outrage among the signatories of the Kyoto Protocol. Many of the EU member countries said that the United States had no right to reject the validity of the protocol.

The NAS released its findings in May 2001. The report reiterated that climate change was a very real phenomenon, caused predominantly by human actions.

The Intergovernmental Panel on Climate Change warned that if fossil fuel consumption continued without restraint, the planet would experience an additional 3 to 10 degrees of warming by the year 2100.

It concurred that there had been an average temperature increase of 1 degree F during the 20th century, and estimated that temperatures would increase between 2.5–10 degrees F (1.4–5.5 degrees C) during the 21st century. It further predicted that sea levels could rise as much as 3 ft. (1 m.) by 2100, resulting in adverse social consequences such as homelessness, starvation, and growing numbers of environmental refugees. Even in light of clear evidence about the harmful consequences of carbon emissions, the Bush Administration continued to circumvent the issue of climate change.

A NEW U.S. PLAN

The same month as the NAS published its report, Vice President Dick Cheney's energy task force announced a national plan to develop new sources of energy. Cheney publicly announced that the plan would secure America's energy needs for the long-term future, but the plan did not place any emphasis on conserving energy or making energy consumption more efficient. An environmental group later discovered that representatives from the energy industry had exerted a large degree of influence over the task force. A few weeks later, Bush revealed his plans for a U.S. Climate Change Initiative, which would support climate change research and manage the distribution of funding for such research.

An American delegation was present at the seventh Conference of the Parties in Bonn in July 2001, despite the United States' repudiation of the Kyoto treaty. Paula Dobriansky, the head of the delegation, agreed that it would not participate in any discussions pertaining to the treaty. However, the United States still made its presence felt at the conference. Dobriansky announced to a bemused audience that Bush was committed to halting the process of climate change and then went on to reiterate the administration's opposition to the Kyoto Protocol. She said that the United States would only stop other countries from adhering to the protocol if such actions harmed the country's interests. The conference was also notable for the fact that the EU finally consented to the idea of carbon sinks, agreeing to let countries include sinks as part of their carbon reductions. The EU also agreed that there should be no limit on emissions trading. Both provisions were included in the Bonn Agreement, which also stipulated that countries would be responsible for additional emission reductions in the second commitment period if they missed their initial targets.

In 2002, Bush announced that he aimed for an 18 percent reduction in the intensity of greenhouse gas emissions by 2012. He proposed to achieve this through measures such as consumer information campaigns, new mandatory regulations, and partnerships with energy firms. He argued that his plan demonstrated his government's commitment to meet the terms of the United Nations Framework Convention on Climate Change. However, many scientists and politicians pointed out that Bush's plan only sought to reduce the intensity, and not the quantity, of greenhouse gas emissions. The scientific community predicted that if the United States continued to follow the same plan in the coming decades, it could potentially cause greenhouse gas emissions to increase 30 percent above their 1990 level by the year 2030. In addition to that plan, the Bush Administration and the energy industry established a new energy project called FutureGen in 2003. The purpose of the project was to build an advanced facility for generating power through the gasification of coal and the sequestering of carbon emissions. The facility, which is currently under construction, is scheduled to open in 2012.

Although Bush began to acknowledge in 2003 that climate change stemmed from human actions, his initiatives did not extend much beyond those that were already in force. He was reelected in 2004, and by the following year the United States was still heavily reliant on coal-burning power plants for its energy. In 2005, coal-generated energy represented 32 percent of all the energy generated across the country. Around 23 percent of the nation's energy came from natural gas, 18 percent from natural gas and petroleum, and 10 percent from nuclear generation. According to figures compiled by the UN Statistics Division, the United States had the highest level of carbon dioxide emissions in the world in 2004. It emitted the equivalent of 7 billion tons of carbon dioxide, by far the highest rate of any country in the world, and the same amount of greenhouse gases as 2.6 billion people living in 151 developing nations. China, the world's fastest-growing economy, was in second place, having emitted the equivalent of just over 4 billion tons of carbon dioxide. However, the difference in carbon emissions between the two countries was far greater when broken down to an average figure for each member of the population. American emissions amounted to an average of 23.92 tons of emissions per person, whereas China's emissions only averaged out to 3.36 tons of emissions for every member of the population.

Many prominent American environmental groups continued to campaign vigorously for the United States to join the Kyoto Protocol but were hindered by a powerful industrial lobby. However, more than half of the states in the union had started their own initiatives to reduce greenhouse gas emissions. In California in 2003, then-governor Gray Davis stipulated that car manufacturers had to reduce the emissions level of all vehicles sold in the state. The California Air Resources Board established new emissions standards for cars that came into effect at the end of 2005. Davis's successor, Arnold Schwarzenegger, also took a firm stance on the environment.

Despite the Bush Administration's claims that the United States cannot adhere to the Kyoto Protocol, scientific evidence has proven that more restrictive policies on carbon emissions will not have a detrimental effect on American economic growth. During the energy crisis of 1973 to 1986, the country actually improved its energy efficiency: its economy grew by 35 percent, and its energy output remained at a consistent level. Robert Ayers, an ecological pioneer, estimates that 19 of every 20 units of energy in the United States are wasted and advocates capturing those wasted units to cut America's energy consumption.

In recent years, there have been clear indicators about how global warming has been affecting the United States. Eighteen of the warmest years on record occurred between 1980 and 2006; 2005 was officially the hottest year ever recorded. Ten of the 12 strongest hurricanes on record occurred in 2005. Many of those hurricanes had a direct effect on the United States, particularly Hurricane Katrina, which devastated the city of New Orleans and parts of Louisiana and killed hundreds of people. In Montana, the number of glaciers at Glacier National Park is rapidly falling. In 1910, the park had 150 glaciers, and that number is now less than 30.

Evidence about how climate change will affect the United States in the future has been growing. Peer-reviewed data show that the country could lose up to 14,000 sq. mi. (36,260 sq. km.) of territory as a result of global warming. The NAS predicts a 3-ft. (1 m.) rise in sea levels by 2100, and the Environmental Protection Agency estimates that the acreage available for cultivation in Maryland and Pennsylvania could drop as much as 43 percent in the coming decades. Scientists estimate that Glacier National Park's glaciers will have disappeared by 2030. They are also predicting a major reduction in winter snowpack in the Pacific Northwest and Rocky Mountain regions, which would not only harm the economy of those regions by eradicating the conditions necessary for winter sports such as skiing and snowboarding but would also threaten the drinking water supplies, drawn from melting snow, of millions of people. Other possible consequences of global warming for America include higher levels of aridity in the Southwest and Great Plains, increased flooding along major river basins, more wildfires, and the spread of disease and illness. In the past few years, the effect of climate change has become a prominent issue in popular American culture. Both the fictional disaster movie *The Day After Tomorrow* and Gore's documentary *An Inconvenient Truth* graphically depict what will happen to the United States, and other parts of the world, if temperatures continue to rise unabated.

SEE ALSO: Bush (George H.W.) Administration; Bush (George W.) Administration; California; Carbon Emissions; Carbon Footprint; Carbon Sinks; Carter Administration; Climate Action Network (CAN); Clinton Administration; Department of Defense, U.S.; Department of Energy, U.S.; Department of State, U.S.; Emissions Trading; Framework Convention on Climate Change; Global Warming; Gore, Albert Jr.; Greenhouse Effect; Greenhouse Gases; Intergovernmental Panel on Climate Change (IPCC); Kyoto Mechanisms; Kyoto Protocol; Louisiana; Montreal Protocol; Refugees, Environmental; Toronto Conference; United Nations Environment Programme (UNEP); World Meteorological Organization.

BIBLIOGRAPHY. J.W. Anderson, *How Climate Change Policy Developed: A Short History* (Resources for the Future, 2006); Marc Allen Eisner, *Governing the Environment: The Transformation of Environmental Regulation* (Lynne Rienner Publishers, 2007); Tim Flannery, T*he Weather Makers: How Man Is Changing The Climate and What It Means for Life on Earth* (Atlantic Monthly Press, 2005); *Fourth Climate Action Report to the UN Framework Convention on Climate Change* (U.S. Department of State, 2006); Al Gore, *An Inconvenient Truth: The Planetary Emergency of Global Warming and What We Can Do About It* (Rodale Books,

2006); Raymond J. Kopp, *Recent Trends in U.S. Greenhouse Gas Emissions: An Introductory Guide to Data and Sources* (Resources for the Future, 2006); Mark Lynas, *High Tide: The Truth About Our Climate Crisis* (Picador, 2004); A. Barrie Pittock, *Climate Change: Turning Up the Heat* (Csiro Publishing, 2005); James Gustave Speth, *Red Sky At Morning: America and the Crisis of the Global Environment* (Yale University Press, 2004); Mike Tidwell, *The Ravaging Tide: Strange Weather, Future Katrinas, and the Coming Death of America's Coastal Cities* (Free Press, 2006); United Nations Statistics Division, "Climate Change: Greenhouse Gas Emissions," www.unstats.un.org/unsd/environment/air_greenhouse_emissions.htm; Norman J. Vig and Michael G. Faure, *Green Giants: Environmental Policies of the United States and the European Union* (MIT Press, 2004).

RICHARD FRY
WAYNE STATE UNIVERSITY

University Corporation for Atmospheric Research

THE UNIVERSITY CORPORATION for Atmospheric Research (UCAR) is a nonprofit institution that has a mission to "support, enhance, and extend the capabilities of the university community, nationally and internationally; understand the behavior of the atmosphere and related systems and the global environment; and foster the transfer of knowledge and technology for the betterment of life on Earth." It was founded in the year 1960 and is based in Boulder, Colorado. The UCAR research lab maintains an affiliated nature preserve.

UCAR collaborates with universities to manage the National Center for Atmospheric Research (NCAR) and the UCAR Office of Programs. The goal of these three organizations is "Understanding our changing Earth system, Educating about the atmosphere and related sciences, Supporting a global community of researchers, and Benefiting society through science and technology."

The establishment of UCAR began in the 1950s, when faculty representatives from 14 universities met to discuss the need for supporting the atmospheric sciences, as well as enhancing the study of these sciences. These faculty members realized the research potential in an institution that could foster collaborations and maintain personnel that one university on its own would not have the resources to do. Thus, the NCAR was founded with support from the U.S. National Science Foundation. UCAR was formally established in 1960 to manage NCAR and foster Earth systems science. Earth systems science investigates not only the atmosphere as an entity but also its relations with the Earth's oceans and lands, as well as with the sun.

The universities and institutions that work with UCAR are member universities, international affiliates, or academic affiliates. Member universities must be North American universities offering doctoral degrees in the atmospheric and related sciences. International affiliates are international universities that grant equivalent degrees, and academic affiliates are North American universities that award predoctoral degrees in similar fields. Some of the member universities include the University of Alaska, Columbia University, Drexel University, the University of Illinois at Urbana-Champaign, the University of Missouri, Old Dominion University, Saint Louis University, and the University of Wyoming. In addition, there are private-sector members who participate in UCAR by funding projects, assisting with technology, collaborating with UCAR research, or acting on UCAR governance boards.

NCAR researchers and technical staff frequently collaborate with the UN Intergovernmental Panel on Climate Change (IPCC) to write or review reports of the latter institution. On October 11, 2007, the Nobel Peace Prize was awarded to former U.S. vice president Al Gore and the IPCC for their work evidencing the human effect on global warming.

Earlier that year, the IPCC had published its fourth periodic assessment of climate change; these assessments began in the year 1991. Assessments take into account numerous pieces of climatological data, analyze past climate patterns, and predict future patterns for local and global environments.

UCAR also manages an Office of Education and Outreach (EO), which communicates with the public, and especially youths, about the necessity and intrigues of working in Earth systems science.

SEE ALSO: Clinton Administration; Colorado; Columbia University; National Center for Atmospheric Research;

United Nations; University of Alaska; University of Illinois.

BIBLIOGRAPHY. Al Gore, *An Inconvenient Truth: The Planetary Emergency of Global Warming and What We Can Do About It* (Rodale Books, 2006); Al Gore, *An Inconvenient Truth: The Crisis of Global Warming* (Viking Juvenile, 2007); Intergovernmental Panel on Climate Change, *Climate Change 2007—The Physical Science Basis: Working Group I Contribution to the Fourth Assessment Report of the IPCC* (Climate Change 2007) (Cambridge University Press, 2007).

CLAUDIA WINOGRAD
UNIVERSITY OF ILLINOIS AT URBANA-CHAMPAIGN

University Corporation for Atmospheric Research Joint Office for Science Support

THE JOINT OFFICE for Science Support (JOSS) of the University Corporation for Atmospheric Research (UCAR) is a group of professional and skilled technical and administrative specialists whose mission is to serve and support the scientific community. JOSS headquarters is located in Boulder, Colorado. The office receives funding from the National Science Foundation and the National Oceanic and Atmospheric Administration (NOAA), as well as from other U.S. agencies, private sources, and international organizations. Before 2005, JOSS was divided into two groups: the Field Operations and Data Management (FODM) group and the Program Support Operations (PSO) group. In 2005, the FODM staff moved to NCAR's Earth Observing Laboratory, and PSO/JOSS remained in UCAR UOP (Office of Programs) and retained the name JOSS. Many of the events sponsored by JOSS have to do with climate change and its effect on human existence. JOSS also collaborated closely with the Global Atmospheric Research Program (GARP) for its alpine (1979) and monsoon (1982) experiments in Switzerland and India, respectively.

JOSS collaborates with scientists and research managers to plan and conduct scientific programs in the most productive and cost-effective ways. JOSS supports planning efforts, research programs, field experiments, and data management activities worldwide. The office offers a wide range of services. It can be a consultant both for individual investigators and for research managers and funding agency officials planning extensive geophysical field experiments and monitoring projects.

JOSS also helps researchers to establish their own network of professional relationships with key people in other institutions and agencies, the U.S. Department of State, and other governments. JOSS aids researchers in planning and conducting meetings, workshops, and conferences. The office has an extensive breadth of knowledge and many years of experience in most regions of the world and advises others about research activities in the United States and elsewhere.

The office team organizes field trips through detailed budgeting, site surveys, and logistical and operational support. It takes care of establishing staff and operations centers for the field project and directs daily operations. JOSS supplies ground support for aircraft, ships, radar, and other observing platforms and systems and hires and trains local workers. On average, JOSS provides various support services for over 475 scientific events and more than 2,500 travelers every year. Supported meeting size has ranged from 12 to 1,200 participants. These international and domestic meetings are relevant to scientists and governmental agencies because they are the initial planning stages of future research or the sharing of information and opinions within the scientific community and government. JOSS-supported meetings have also provided a wide range of topics related to climate change.

For example, JOSS supported a large meeting of Native Americans from across the United States. Representatives from various tribes came together to share their opinions about how climate affects their own lives and how they are adapting their old ways into the modern world with regard to climate issues. JOSS has also supported several large, politically significant meetings in Washington, D.C., designed to help atmospheric scientists share their knowledge, research findings, and predictions with agencies that have the power to influence government decisions.

From 1995 to 2001, JOSS took an active part in the Indian Ocean Experiment (INDOEX). The experiment started from the assumption that regional consequences of global warming depend critically on the potentially large cooling effect of another pollutant, known as aerosols. These aerosols scatter sunlight back to space and cause a regional cooling effect, thus causing uncertainty in predicting future climate. The Indian Ocean Experiment attempted to address the complex influence of aerosol cooling on global warming by collecting in situ data on the regional cooling effect of sulfate and other aerosols.

The project's goal was to study natural and anthropogenic climate forcing by aerosols and feedbacks on regional and global climate. International Global Change Research Program considered this issue to be of critical importance. INDOEX measured long-range transport of air pollution from south and southeast Asia toward the Indian Ocean during the dry monsoon season from January to March 1999. Surprisingly high pollution levels were observed over the entire northern Indian Ocean toward the Intertropical Convergence Zone at about 6 degrees S.

Agricultural burning and especially biofuel use were shown to enhance carbon monoxide concentrations. Fossil fuel combustion and biomass burning caused a high aerosol loading. The growing pollution in the region gave rise to extensive air quality degradation, which, the experiment pointed out, had local, regional, and global implications, including a reduction of the oxidizing power of the atmosphere. JOSS was also involved in GARP experiments, as well as in the establishment of the international research institute on El Niño and its consequences.

JOSS provides administrative, travel, and logistical support for several institutions that operate in the field of climate change and global warming, such as the Intergovernmental Panel on Climate Change, the International Research Institute for Climate and Society, the Office of the U.S. Global Change Research Program, the NOAA Climate Program Office, and the Inter-American Institute for Global Climate Change Studies. It also produces Reports to the Nation on Our Changing Planet, a series of general interest monographs that provide science-based information on issues regarding climate and global change for the general public.

SEE ALSO: Aerosols; Global Warming; Greenhouse Effect; Greenhouse Gases; Kyoto Protocol.

BIBLIOGRAPHY. INDOEX—The Indian Ocean Experiment, http://www-indoex.ucsd.edu; Joint Office for Science Support, http://www.joss.ucar.edu/index.html.

Claudia Winograd
University of Illinois at Urbana-Champaign

University of Alaska

THE UNIVERSITY OF Alaska system has three main campuses (University of Alaska Fairbanks (UAF), University of Alaska Anchorage, and University of Southeast) and 15 satellite campuses. However, beyond individual researchers at various campuses, all four global warming or climate change research centers, institutes, and groups are found at the UAF campus. The four centers are the Alaska Climate Research Center (ACRC), the International Arctic Research Center (IARC), Center for Global Change and Arctic System Research, and the Alaska Center for Climate Assessment and Policy (ACCAP).

Under the direction of the Geophysical Institute, the ACRC conducts secondary data gathering, storage, and report analysis. The ACRC is funded by the State of Alaska under Title 14, Chapter 40, Section 085. The IARC is a joint venture between Japan and the United States. It is an international focused research center that includes 20 research groups and over 60 international scientists. The IARC mission is to determine whether climate change is manmade or natural, what the data points are needed to make this determination, and what the possible effects of climate change are.

In 2007, a national and international controversy sprang out of the IARC when Director Syun-Ichi Akasofu claimed the 2005 peer-review Internatioal Panel on Climate Change (IPCC) report was methodologically flawed. Dr. Akasofu claimed the IPCC did not have an anthropogenic (natural) greenhouse gases control to account for their assertion of a 0.6 degrees man-made climate change over the past 100 years. Professor Akasofu also claimed that CO_2 and other greenhouse gases are not responsible for climate change the researchers would have included anthropogenic greenhouse gases. In a public response to criticism from the academic community, Akasofu noted that, "Since I am not a climatologist, all the data presented in my note are found in papers and books published in the past; that is why I do not want to publish my note as a paper in a professional journal." Although not a climatologist, Akasofu learned speculation on climatology has widely circulated among antiglobal warming groups.

The Center for Global Change and Arctic System Research was founded in 1990 with the goal of fostering interdisciplinary Arctic and sub-Arctic research in Arctic biology, atmospheric chemistry, climatology, engineering, geophysics, hydrology, natural resources management, social sciences, and marine sciences to better understand global change in the Arctic. There are two subgroups and a 1,042-page Arctic research study by the Center for Global Change and Arctic System Research affiliates. One subgroup is the Globe Learning and Observations to Benefit the Environment (GLOBE), founded in 1997. GLOBE is an international group that develops hands-on environmental science curriculum for K–12 students, teachers, and scientists. Another subgroup, founded in 1994, is the University of Alaska-National Oceanic and Atmospheric Administration (UA-NOAA) Cooperative Institute for Arctic Research (CIFAR). CIFAR is one of 13 national university-based NOAA institutes.

It focuses on atmospheric and climate research/ modeling. The UA-NOAA studies focus on arctic haze, marine science, fisheries, and sea ice research. The last subgroup is a research study called the Arctic Climate Impact Assessment (ACIA). The ACIA report was prepared for the Fourth Arctic Council Meeting in Reykjavik, Iceland, in November 2004. Some of the findings from the report note that over the past 50 years, the mean surface air temperature has increased 3.6–5.4 degrees F (2–3 degrees C) and late summer ice decreased by 15–20 percent over the past 30 years; in addition, between 1961 and 1998, North American glaciers lost 108 cu. mi. of ice.

The ACCAP was founded in 2006. The ACCAP mission is to determine the biophysical and socioeconomic effect of climate change within Alaska and to improve regional, local, and Alaskan ability to create policies that address the changing climate. The ACCAP works in affiliation with NOAA-Regional Integrated Sciences and Assessments, Institute of Northern Engineering, International Arctic Research Center, the Institute for Socioeconomic Research at the University of Alaska Anchorage, and the Alaska Climate Research Center.

SEE ALSO: Alaska; Climate Change, Effects; Education.

BIBLIOGRAPHY. S. Akasofu, www.iarc.uaf.edu/highlights /2007/akasofu_3_07/index.php (cited April 24, 2007); Alaska Center for Climate Assessment and Policy, www. uaf.edu/accap/; Alaska Climate Research Center, www. climate.gi.alaska.edu/index.html; Arctic Climate Impact Assessment, "Impacts of a Warming Arctic, Arctic Climate Impact Assessment," 2004; Arctic Monitoring and Assessment Programme, 1998; "Arctic Pollution Issues: A State of the Arctic Environmental Report," www.amap.no/ (under scientific papers); Center for Global Change and Arctic System Research, www.cgc.uaf.edu/; Intergovernmental Panel on Climate Change, www.ipcc.ch; International Arctic Research Center, www.iarc.uaf.edu/; World View of Global Warming, www.worldviewofglobalwarming.org/ index.html.

Andrew Hund
University of Alaska, Anchorage

University of Arizona

THE UNIVERSITY OF Arizona (UA or U of A) is a land-grant and space-grant public institution of higher education and research located in Tucson. The University of Arizona was the first university in the state of Arizona, founded in 1885, when Arizona was still a territory. UA includes Arizona's only allopathic medical school. In 2006, total enrollment was 36,805 students. UA embraces its threefold mission of excellence in teaching, research, and public service. Now in its second century of service to the state, UA has become one of the nation's top 20 public research institutions.

The UA graduate school offers several programs of study in atmospheric and environmental science. The Department of Atmospheric Sciences offers programs leading to the master of science and doctoral degrees. Research is conducted through the Institute of Atmospheric Physics in areas such as climate and global change, land-atmosphere interaction, convective processes, atmospheric dynamics, radiative transfer, remote sensing, atmospheric aerosols, atmospheric chemistry, cloud and precipitation physics, lightning, atmospheric electricity, weather forecasting, and numerical weather prediction. The Department of Soil, Water, and Environmental Science (SWES) offers graduate work leading to M.S. and Ph.D. degrees in soil, water, and environmental science. Two tracks are offered, environmental science and soil and water science. In addition to the major, each Ph.D. student must complete a minor, which can be intra- or interdepartmental. Many, if not most, SWES graduate students enroll in several non-SWES courses as part of their program. This reflects the multidisciplinary characteristics of the SWES program in general, and also that a minor is required for all Ph.D. students.

The most frequently studied outside courses are in chemistry, chemical and environmental engineering, hydrology and water resources, and microbiology. With an M.S. or Ph.D. in environmental science, students will be prepared for careers in business and industry, governmental agencies, educational institutions, and private consulting firms. Many Ph.D. students obtain faculty positions at colleges and universities. The Environmental Studies Laboratory was created in 1999 to provide facilities and equipment for the research of past, present, and future environmental variability and change. Current research encompasses paleoclimate studies across the globe, wildfire studies in southwestern North America, and relationships between climate and society across multiple spatial and temporal scales. The laboratory is just one of several complementary investigative groups of environmental and earth system sciences in the Department of Geosciences and at UA. The laboratory is involved in a number of efforts aimed at improving the quality of and participation in K–12 education.

Research carried out by staff and students at the Environmental Studies Laboratory and their forerunners at the University of Colorado and Columbia University has centered around two broad themes. The oldest concentrates on paleoenvironmental science and its application toward understanding the full range of environmental variability, with a focus on climate and ecology. More recently, the laboratory has worked on improving connections between environmental sciences and society, with the specific goal of increasing the scientific basis of environmental decision making. Both themes involve a conscious effort to work across broad temporal scales and over spatial scales that extend from local to global. By working on projects in many key systems around the world, the hope is to craft a better understanding of the global system.

SEE ALSO: Arizona; Climate Change, Effects; Weather; Soils.

BIBLIOGRAPHY. Department of Geosciences Studies Laboratory, www.geo.arizona.edu; Intergovernmental Panel on Climate Change, www.ipcc.ch; University of Arizona, www.arizona.edu.

FERNANDO HERRERA
UNIVERSITY OF CALIFORNIA, SAN DIEGO

University of Birmingham

THE UNIVERSITY OF Birmingham is an English university in the city of Birmingham. Founded in 1900 as a successor to Mason Science College, with origins dating back to the 1825 Birmingham Medical School, the University of Birmingham was arguably the first so-called red brick university. It currently has over 18,000 undergraduate and over 11,000 postgraduate students. The University of Birmingham has an international reputation for excellence in research and teaching in environmental science, engineering, and policy. There are currently around 150 academic staff actively investigating scientific, technical, and socioeconomic aspects across a broad range of environmental disciplines, including the management of freshwater resources, environmental restoration, sustainable use of natural materials, pollution control, waste management, management of natural hazards, and human health.

The Center for Environmental Research and Training plays a key role in providing a focus for the university's environmental expertise. It acts as a gateway for external organizations, enabling this expertise to be made more widely accessible, and provides a mechanism for the promotion and management of interdisciplinary research within the university. The Institute for Energy Research and Policy, founded in 2005, focuses on research into the economics of energy policy, energy efficiency, wind power, and hydrogen-based energy systems. The Centre for Environmental Research and Training and the School of Geography Earth and Environmental Sciences are internationally recognized in many areas of climate change research. The center seeks to enhance the university's international reputation for environmental research and teaching by advancing partnerships with institutions overseas, such as in Poland, Malaysia, Thailand, and the West Indies.

In October 2006, as part of an ongoing series of initiatives to reduce greenhouse gas emissions from university activities, a new climate change website was launched. The website's purpose was to provide advice and information on the solutions that the university as a whole can provide to the challenge of climate change. In addition, extra effort has been made to implement more energy-efficient practices.

The University of Birmingham is also one of a select group of universities that are committed to carbon management with the Carbon Trust to actively cut their carbon emissions and so minimize the long-term effects of climate change. The Carbon Trust has designed a university-focused program, which will provide technical and change management support to help the sector realize carbon emissions savings. The primary focus of the work is to reduce emissions under the control of the university, such as academic, accommodation, and leisure buildings and vehicle fleets. Practical support will be given in areas such as identifying carbon-saving opportunities, developing an emissions reduction implementation plan, providing analysis software, and offering workshop support for staff and senior management training.

SEE ALSO: Carbon Trust; Energy Research; Environmental Policy.

BIBLIOGRAPHY. Climate Change Action at Birmingham, www.climatechange.bham.ac.uk; Institute for Energy Research and policy, www.ierp.bham.ac.uk; University of Birmingham, www.bham.ac.uk.

FERNANDO HERRERA
UNIVERSITY OF CALIFORNIA, SAN DIEGO

University of California, Berkeley

THE UNIVERSITY OF California, Berkeley, is the premier public research university in the United States, with 97 percent of its academic programs being among the top 10 in the country. Commonly referred to as UC Berkeley, Berkeley, and Cal, the university's academic excellence is sustained by a $2.46 billion endowment. Berkeley was founded in 1868 and is the oldest of the 10 University of California campuses. During the 1930s, the leadership of university president Robert Sproul helped Berkeley to establish itself as a leading research university, and by 1942 the American Council on Education ranked Berkeley second only to Harvard in the number of distinguished academic departments. A reorganization of the University of California system in 1952 resulted in the naming of Clark Kerr as the first chancellor for the Berkeley campus. Since then, there have been nine other chancellors; the current chancellor is Robert Birgeneau, who has filled this role since 2004.

Berkeley is a comprehensive university offering over 7,000 courses in 130-plus academic departments organized into 14 colleges and schools, offering nearly 300 degree programs. The university awards over 5,500 bachelor's degrees, 2,000 master's degrees, 900 doctorates, and 200 law degrees each year. With 33,558 students and 1,950 faculty, the student-faculty ratio is 17 to 1—among the lowest of any major university. Berkeley is the most selective school in the UC system and is one of the most selective universities in the country. For the 2006–07 academic year, 4,157 freshmen matriculated at Berkeley, from an applicant pool of just under 41,750 applicants. Graduate admissions vary by department, although in 2006 the university's doctoral programs admitted 1,058 students from a pool of 14,263 applicants.

Collectively, Berkeley's 32 libraries tie with University of Illinois for the fourth largest academic library system in the United States, surpassed only by the U.S. Library of Congress, Harvard, and Yale. In 2003, the Association of Research Libraries ranked Berkeley as the top public university library in North America and third among all universities. As of 2006, Berkeley's library system contains over 10 million volumes and maintains over 70,000 serial titles.

The scholarly achievements and excellence of the faculty and alumni have helped to build and maintain

Berkeley's excellent reputation. Berkeley scientists invented the cyclotron, discovered the antiproton, isolated the polio virus, created the Unix computer operating system, and discovered numerous transuranic elements including seaborgium, plutonium, berkelium, lawrencium, and californium. During World War II, Ernest Orlando Lawrence's Radiation Laboratory at Berkeley contracted with the U.S. Army to develop the atomic bomb, and Berkeley physics professor J. Robert Oppenheimer was named scientific head of the Manhattan Project in 1942. Berkeley faculty have a no less distinguished record in fields outside the physical sciences: they include 221 American Academy of Arts and Sciences Fellows, 83 Fulbright Scholars, 28 MacArthur Fellowships, 384 Guggenheim Fellows, 87 members of the National Academy of Engineering, 132 members of the National Academy of Sciences, 3 Pulitzer Prize winners, and 92 Sloan Fellows. Berkeley counts 61 Nobel laureates among its faculty, researchers, and alumni—the sixth most of any university in the world; 20 have served on its faculty.

Berkeley's reputation for student activism was forged in the 1960s. With the end of World War II and the subsequent rise of student activism, the California Board of Regents succumbed to pressure from the student government and ended compulsory military training at Berkeley in 1962. Then, in 1964, an impromptu response by students to the university's ban on campus political activity led to the beginning of the Free Speech Movement and, ultimately, the freedom of expression by students. This movement grew during the protests against U.S. involvement in the Vietnam War during the 1960s and early 1970s. Today Berkeley has over 700 established student groups, nearly 100 of which are political. There is also a strong sense of public service among Berkeley graduates. For example, Berkeley sends the most students to the U.S. Peace Corps of any university in the nation.

As part of its academic excellence, Berkeley is also a leader in environmental research. There are over 100 individual undergraduate and graduate programs at Berkeley that focus on the environment, in addition to dozens of top research centers. The university is also active in research concerning global warming and climate change, with several centers involved in research and advocacy on these and related issues. For example, the University of California Climate Change Center was established in 2003 by the California Energy Commission to undertake a broad program of scientific and economic research on climate change in California. The center has sites at both the Berkeley and San Diego campuses of UC. The Berkeley center, based at the Goldman School of Public Policy, focuses on economic and policy analysis, whereas the site in San Diego (at the Scripps Institute of Oceanography) focuses on physical climate modeling. Several other departments on the Berkeley campus are involved in the work of the center including the Department of Agricultural and Resource Economics, the Department of City and Regional Planning, the Department of Civil and Environmental Engineering, the Graduate Group in Energy and Resources (ERG), and the Environmental Energy Technologies Division of the Lawrence Berkeley National Laboratory.

The ERG is an interdisciplinary academic unit of the Berkeley campus that was created in 1973 to develop, transmit, and apply critical knowledge to enable a future in which human material needs and a healthy environment are mutually and sustainably satisfied. ERG conducts programs that include graduate teaching and research on issues of energy, resources, development, human and biological diversity, environmental justice, governance, global climate change, and new approaches to thinking about economics and consumption. The University of California Energy Institute (UCEI) is a multicampus research unit of the University of California system begun in 1980, whose mission has been to foster research and educate students and policymakers on energy issues. The Center for Global Metropolitan Studies (GMS) is a campus initiative to foster interdisciplinary collaboration to investigate and address problems and opportunities posed by global metropolitan growth and change through research. The Berkeley Institute of the Environment (BIE) was established in 2005 and brings together and helps enhance diverse campus programs and research units by making research tools and understanding accessible across disciplinary lines to address complex environmental problems, while fostering collaboration and thinking about critical environmental problems. The Institute of Transportation Studies (ITS) is one of the world's leading centers for transportation research, education, and scholarship. Research areas include transportation sustainability, future urban transit systems, and environmental effects. The UC Berkeley Transportation Sustainability Research Center (TSRC) was formed in 2006 to combine the research forces of the five aforementioned centers, institutes, and groups (ERG, UCEI, GMS, BIE, and ITS). The TSRC is a multicampus unit that supports research, education, and outreach.

Other campus activities related to global warming include the research and advocacy work of the College of Engineering. For example, on August 2, 2007, the college released a blueprint for fighting global warming by reducing the amount of carbon emitted when transportation fuels are used in California. This low-carbon fuel standard is designed to stimulate improvements in transportation fuel technologies and is expected to become the foundation for similar initiatives in other states, as well as nationally and internationally.

There is also a Chancellor's Advisory Committee on Sustainability that promotes environmental management and sustainable development on the Berkeley campus. The mission of the committee is to engage the campus in an ongoing dialogue about reaching environmental sustainability and to integrate environmental sustainability with existing campus programs in education, research, operations, and public service. The committee is charged with advising the chancellor on matters pertaining to the environment and sustainability as it directly relates to the university.

Berkeley's tradition of student political action has also merged with global climatic change issues. For example, in March 2007, students organized the California Campus Climate Challenge Summit to learn about global warming, climate change, and methods for influencing policy change via student activism. Such activities also highlight the social justice aspects of issues concerning global warming and the desire by some to create an environmental and social movement to help raise awareness about this issue. Also, Berkeley alumna Sissel Waage has just coedited a book to address issues of climate change and global warming. The book features a wide array of authors ranging from activists to scholars to students, who each discuss what the average person can do to turn their private concerns into public action.

SEE ALSO: Carbon Footprint; Climate Change; Education.

BIBLIOGRAPHY. *Brief History of the University of California, Berkeley,* www.berkeley.edu/about/history/; Edward B. Fisk, *Fiske Guide to Colleges 2005,* 21st ed. (Sourcebooks, 2005); Jonathan Isham and Sissel Waage, eds., *Ignition: What You Can Do to Fight Global Warming and Spark a Movement* (Island Press, 2007); Eric Owens, *America's Best Value Colleges* (Princeton Review, 2004).

Michael Joseph Simsik
U.S. Peace Corps

University of Cambridge

THE UNIVERSITY OF Cambridge (Cambridge University), located in Cambridge, England, is the second-oldest university in the English-speaking world and has a reputation as one of the world's most prestigious universities. The university grew out of an association of scholars in the city of Cambridge that was formed in 1209 by scholars leaving Oxford after a dispute with the townsfolk there. The universities of Oxford and Cambridge are often jointly referred to as Oxbridge. In addition to their cultural and practical associations as a historic part of English society, the two universities also have a long history of rivalry with each other. The University of Cambridge currently has 31 colleges, of which three admit only women (New Hall, Newnham, and Lucy Cavendish). The remaining 28 are mixed, with Magdalene being the last all-male college to begin admitting women, in 1988. Two colleges admit only postgraduates (Clare Hall and Darwin), and four more admit mainly mature students or graduate students (Hughes Hall, Lucy Cavendish, St. Edmund's, and Wolfson).

The other 25 colleges admit mainly undergraduate students but also allow postgraduates following courses of study or research. Although various colleges are traditionally strong in a particular subject, for example, Churchill has a formalized bias toward the sciences and engineering, the colleges all admit students from just about the whole range of subjects, although some colleges do not take students for a handful of subjects such as architecture or history of art. It is noteworthy that costs to students (accommodation and food prices) vary considerably from college to college. This may be of increasing significance to potential applicants as government grants decline in the next few years.

The Department of Land Economy offers a three-year honors undergraduate degree program, as well as postgraduate studies in environmental policy. The main focus of the program is land and environmental protection. Up to 40 students are admitted each year. The University of Cambridge offers a variety of taught courses relating to the environment, whether as part of an undergraduate degree course or a postgraduate course. Climate change research at the University of Cambridge ranges from long-term climate modeling using geological data to in situ measurements of the present composition of the Earth's atmosphere, at all scales from global climate models to study of the dynamic chemical processes controlling the present state of the atmosphere. Significant expertise in atmospheric sciences and earth sciences provides a strong foundation for research into limiting the effects of human activity on the Earth's climate, mitigating the effects of climate change, assessing the effects of climate change on human health, and designing economic policies and technologies for climate change mitigation.

The Cambridge Environmental Initiative was launched in December 2004. The primary mission of this initiative is to facilitate and support interdisciplinary environmental research within the University of Cambridge, promote the university's external profile in environmental research, and encourage new environmental research initiatives within the university.

The Cambridge Center for Climate Change Mitigation Research is an interdisciplinary research center, focusing on climate change mitigation at the local and global levels. The main objective of this collaboration is to foresee strategies, policies, and processes to mitigate human-induced climate change, which are effective, efficient, and equitable, including understanding and modeling transitions to low-carbon energy-environment-economy systems. The center houses global, European, and U.K. research teams adopting a common set of conventions and protocols that are networked with multidisciplinary teams working in different countries.

In March 2007, former U.S. vice president Al Gore delivered a climate change training program to 200 of the United Kingdom's top leaders from business, government, media, education, and civil society. The program brought together leaders committed to communicating and taking action on climate change

across the United Kingdom and internationally. The program, hosted by the University of Cambridge, was designed to help trainees understand the issues and think critically about solutions and actions required to address climate change issues.

SEE ALSO: Cambridge Environmental Initiative; Oxford University.

BIBLIOGRAPHY. Cambridge Environmental Initiative, www.cei.group.cam.ac.uk; The Climate Project, www.cpi.cam.ac.uk; University of Cambridge, www.cam.ac.uk.

FERNANDO HERRERA
UNIVERSITY OF CALIFORNIA, SAN DIEGO

University of Colorado

THE UNIVERSITY OF Colorado at Boulder (CU) is the flagship university of the University of Colorado system. CU has produced a number of astronauts, Nobel laureates, Pulitzer Prize winners, and other notable individuals in their fields.

The Department of Atmospheric and Oceanic Sciences (ATOC) is an interdisciplinary program that provides an educational and research environment to examine the dynamic, physical, and chemical processes that occur in the atmosphere and the ocean. A major theme is the establishment of a physical basis for understanding, observing, and modeling climate and global change. At the undergraduate level, approximately 2,000 students are pursuing baccalaureate degrees in environmental studies, environmental biology, environmental engineering, environmental law, geological sciences, geography, environmental policy, and other subjects. Several hundred graduate students are also pursuing advanced degrees involving research on environmental topics at CU.

ATOC is coordinated with the environmental program at the University of Colorado. Interdisciplinary education and research opportunities exist with the hydrology program and the environmental policy program. Interdisciplinary research opportunities also exist with the Cooperative Institute for Environmental Studies, the Institute for Arctic and

Alpine Research, the Center for Complexity, and the Laboratory for Atmospheric and Space Physics. Graduate students, research staff, and faculty work together on a wide range of research projects. ATOC has extensive computer facilities and laboratories in remote sensing, chemistry, and hydrodynamics. The presence of leading laboratories in the environmental sciences in Boulder, including the National Center for Atmospheric Research and the National Oceanic and Atmospheric Administration Environmental Research Laboratories, provides additional opportunities for a rich educational experience.

Climate research at the University of Colorado is driven by the goals and broad objectives that have been articulated by the World Climate Research Program and the U.S. Global Change Research Program, which are to develop the fundamental scientific understanding of the climate system and climate processes that is needed to determine to which extent climate can be predicted, as well as the extent of man's influence on climate. The program encompasses studies of the global atmosphere, oceans, sea and land ice, and the land surface, as well as their coupling. To achieve these goals, climate research at the University of Colorado plans to design and implement observational and diagnostic research activities that will lead to a quantitative understanding of significant climate processes, including the transport and storage of heat by the ocean; the exchange of heat, moisture, and momentum between atmosphere, ocean, and sea ice; and the interaction among cloudiness, radiation, the land surface, and the global hydrological cycle. Research in global and regional models capable of simulating the present climate and, to the extent possible, of predicting climate variations on a wide range of space and timescales are also being actively pursued.

CU is known internationally for its interdisciplinary research on a variety of global change issues. Working closely with Boulder-based federal research laboratories, CU research centers such as the Cooperative Institute for Research in the Environmental Sciences, the Institute for Arctic and Alpine Research, and the Laboratory for Atmospheric and Space Physics have successfully brought together a variety of academic

The University of Colorado at Boulder was founded in 1876 at the base of the Rocky Mountains. The natural beauty of the area draws many students interested in pursuing degrees in environmental studies and policies.

disciplines to work on global change issues, which cross traditional disciplinary and departmental lines. On the policy side, the Natural Resources Law Center and the Environmental Policy Program have brought together lawyers, economists, historians, and political scientists to study global change and other environmental issues. Unique interdisciplinary education programs in climate change are being developed at the University of Colorado through a partnership of the Department of Atmospheric and Oceanic Sciences and the Environmental Studies Program, overcoming the disciplinary and departmental barriers that exist in many university programs.

The program at the University of Colorado proves a unique combination of disciplinary depth and inter- and multidisciplinary breadth necessary for students who plan to work in this area. Current research topics include El Niño and tropical climate variability, polar climate; polar regional climate modeling, World Data Center for Glaciology, Institute for Arctic and Alpine Research, Program in Arctic Regional Climate Assessment, sea ice remote sensing, sea ice modeling, land/atmosphere interactions, land surface model, boreal forest dynamics model, Land-Atmosphere CO_2 exchange, paleoclimate, Past Global Change Group/Institute for Arctic and Alpine Research, impact of clouds and aerosols on climate, aerosol modeling research group home pages, and global climate modeling.

SEE ALSO: Colorado; El Niño and La Niña; National Center for Atmospheric Research (NCAR).

BILBLIOGRAPHY. Climate and Global Dynamics, www.cgd.ucar.edu; National Center for Atmospheric Research, www.ncar.ucar.edu; University of Colorado, www.colorado.edu.

FERNANDO HERRERA
UNIVERSITY OF CALIFORNIA, SAN DIEGO

University of Delaware

THE UNIVERSITY OF Delaware (UD) is the largest university in Delaware. The main campus is located in Newark, with satellite campuses in Dover, Wilmington, Lewes, and Georgetown. Approximately 16,000 undergraduate and 3,000 graduate students attend this university annually. Although UD receives public funding for being a land-grant, sea-grant, space-grant, and urban-grant state-supported research institution, it is also privately chartered. The university's endowment is currently valued at about $1.2 billion. In 2007, UD was ranked 15th nationally in *Kiplinger's Personal Finance* magazine list of the 100 best public institutions of higher education. The University of Delaware was also ranked 15th best value for in-state students and 10th best value for out-of-state students. Seven academic colleges confer degrees at UD.

The College of Marine and Earth Studies is one of the seven colleges at UD. The undergraduate component of the college is currently housed within the Department of Geological Sciences, where students can major in earth science education or geology. Geology majors can undertake concentrations in paleobiology or coastal and marine geoscience. Students may also major in environmental science, collaboration between the Departments of Geological Sciences, Geography, and Biological Sciences. Environmental science majors can select several concentrations including the geological environment or the marine environment. There are also marine studies courses that are open to undergraduate students. These range from introductory classes for nonscience majors to advanced programs for science and engineering majors. The college offers graduate programs (master's and doctoral degrees) in geology, oceanography, and marine studies. The College of Marine and Earth Studies brings the latest advances in technology to bear on ocean, Earth, and atmospheric research, as well as on teaching.

Graduate study in climatology involves exposure to a wide range of research methods that can be used to help solve climate-related environmental problems. Faculty and graduate student research interests span the range of climatology and include climatic modeling, synoptic climatology, atmospheric dynamics, climate dynamics, physical climatology, water-budget climatology, paleoclimatology, climatic geomorphology, glaciology, global climate change, human effects on climate, and climatic influences on society, particularly on human health and socioeconomic activity.

Climatology courses offered at UD include Geography 612: Physical Climatology, Geography 620:

Atmospheric Physics, Geography 623: Atmospheric Dynamics, Geography 651: Microclimatology, Geography 652: Seminar in Climatology, Geography 653: Synoptic Climatology, Geography 655: Water Budget in Environmental Analysis, Geography 657: Climate Dynamics, Geography 681: Remote Sensing of Environment, Marine Studies 809: The Ocean and Climate Variation, and Geography 855: Climatological Research.

SEE ALSO: Climate Change, Effects; Glaciology; Paleoclimates.

BIBLIOGRPHY. Climate Change-Global Warming, co2.cms.udel.edu; University of Delaware, www.udel.edu.

FERNANDO HERRERA
UNIVERSITY OF CALIFORNIA, SAN DIEGO

University of East Anglia

THE UNIVERSITY OF East Anglia (UEA) is a leading campus university located in Norwich, England. It was founded as part of the British government's New Universities Program in the 1960s. The university is a member of the 1994 group of leading research-intensive universities. Academically, it is one of the most successful universities founded in the 1960s, consistently ranking among Britain's top higher-education institutions. It was 19th in the Sunday Times University League Table 2006, and joint first for student satisfaction among mainstream English universities in the 2006 National Student Survey. Furthermore, the university was ranked 57th in Europe and one of the top 200 universities in the world in the 2006 World University Rankings undertaken by the Shanghai Jiao Tong University.

The School of Environmental Sciences is one of the longest-established, largest, and most fully developed Schools of Environmental Sciences in Europe. The School of Environmental Sciences offers undergraduate and graduate degree programs with an emphasis in a number of studies. The school is also engaged in research in a number of areas from atmospheric sciences to marine sciences. The school has been ranked high in research and teaching.

The Climatic Research Unit (CRU) is widely recognized as one of the world's leading institutions concerned with the study of natural and anthropogenic climate change. The unit consists of a staff of around 30 research scientists and students and has developed a number of the data sets widely used in climate research, including the global temperature record used to monitor the state of the climate system, as well as statistical software packages and climate models. The aim of the Climatic Research Unit is to improve scientific understanding in three areas: past climate history and its effect on humanity, course and causes of climate change during the present century, and prospects for the future.

The Climatic Research Unit is part of the School of Environmental Sciences, with close links to other research groups within the department, such as the Centre for Social and Economic Research on the Global Environment. The unit undertakes collaborative research with institutes throughout the world on a diverse range of topics and is coordinating or contributing to a number of networking activities. The CRU participates in both pure and applied research, sponsored almost entirely by external contracts and grants from academic funding councils, government departments, intergovernmental agencies, charitable foundations, nongovernmental organizations, commerce, and industry. Alongside its research activities, the unit has an educational role through its contribution to formal teaching with the School of Environmental Sciences (most notably, the M.Sc. in climate change) and various forms of in-service training, including postgraduate education. It is regarded as an authoritative source of information on both the science and policy aspects of climate change by the media and maintains a high public profile. The unit staff have published many peer-reviewed articles as well as editing various scientific journals, newsletters, and bulletins.

SEE ALSO: Climate Change, Effects; United Kingdom.

BIBLIOGRAPHY. Climate Research Unit, www.cru.uea.ac.uk/cru; Intergovernmental Panel on Climate Change, www.ipcc.ch; University of East Anglia, www1.uea.ac.uk/cm/home.

FERNANDO HERRERA
UNIVERSITY OF CALIFORNIA, SAN DIEGO

University of Florida

ORIGINALLY FOUNDED IN 1853 as the East Florida Seminary, the institution later renamed the University of Florida (UF) moved to its current location in Gainesville in 1906. The largest and oldest university in the state, UF resides on 2,000 acres and provides educational opportunities for over 46,000 students. As the flagship institution of higher learning in the state, UF has acted as a steward in addressing global warming and climate change issues through its sustainability initiatives.

In 1994, UF president Charles Young signed the Talloires Declaration in an effort to convey support for actions to reduce environmental degradation and the depletion of natural resources. Three years later, members of the UF community initiated a grassroots campaign called Greening UF to engender a sense of environmental stewardship on campus. During 2000, the need for an administrative presence to develop sustainability initiatives was realized in the creation of an Office of Sustainability, housed within the College of Design, Construction, and Planning. In 2001, UF took another step along the path to greater campus sustainability by mandating that Leadership in Energy and Environmental Design (LEED) criteria be followed on all new building and renovation endeavors. UF's own Rinker Hall became the first building in the state of Florida to earn LEED Gold certification in 2004 and the newly renovated Library West achieved Gold status in 2007.

The call for a campus-wide Office of Sustainability was made by the Student Senate in 2004, and in fall 2005, a national search was initiated for a director of a centralized Office of Sustainability. The search for a director of the Office of Sustainability concluded in 2006 with the hiring of Dedee DeLongpré, a former development director for schools in California and program administrator for Fauna and Flora International. UF galvanized support for statewide sustainable action by hosting the inaugural Florida Campus and Community Sustainability Conference in October 2006.

Though UF began supporting major environmental initiatives over a decade ago, much of the university's recent push to enhance its stewardship has taken place during the tenure of President Bernie Machen (2004–present). In 2005, UF became the first university to be recognized as a Certified Audubon Cooperative Society. Speaking of this accomplishment and of sustainability during National Campus Sustainability Day on October 25, 2005, President Machen described the challenges and opportunities that the university has faced. President Machen espoused a full-spectrum approach to increasing sustainability by addressing the university's effect on the environment through green building, decreasing energy dependence, and reducing waste disposal to zero by 2015.

A year later, at the Florida Campus and Community Sustainability Conference, President Machen took on a much more direct tone regarding the issue of climate change, which he declared to be an urgent problem. In his speech, he described how UF would tailor its sustainability goals for 2007 specifically to decrease the university's emissions to combat climate change. President Machen outlined the progress made toward this end and also described new programs that UF would implement. As an academic institution, UF has already made strides to integrate sustainability into the classroom, as evidenced by the over 100 courses, 10 programs, and over 20 research entities relating to sustainability that are offered by the university.

In terms of engaging climate change from a supply chain perspective, President Machen talked of UF's effort to reduce emissions by buying locally produced food for use in dorm cafeterias and by implementing a sustainable purchasing policy. President Machen also detailed the university's energy-reduction strategy, which involves using power produced by renewable sources, turning down heat and lights in many campus buildings during vacation breaks, and purchasing more fuel-efficient fleet vehicles. Following the announcement of these environmental actions, President Machen reinforced UF's dedication to addressing climate change by being the first to sign the American College and University Presidents Climate Commitment at the end of 2006.

Through its multifaceted sustainability initiatives and commitment to emissions and energy reductions at all levels, UF has emerged as an environmental steward in the state of Florida and among universities across the United States. With the aid of strong administrative backing and the support of an environmentally conscious student body and faculty, UF provides a prime example of how a large educational institution can foster widespread social responsibility while minimizing its own negative effect on the environment.

SEE ALSO: Florida; Green Buildings; Sustainability.

BIBLIOGRAPHY. *ULSF—University Leaders For A Sustainable Future—The Talloires Declaration*, www.ulsf.org/programs_talloires.html; University of Florida, www.ufl.edu.

JOSHUA CHAD GELLERS
COLUMBIA UNIVERSITY

University of Hawaii

THE UNIVERSITY OF Hawaii, also known as UH, is a public, coeducational college and university system that confers associate, bachelor's, master's, doctoral, and postdoctoral degrees through three university campuses, seven community college campuses, an employment training center, three university centers, four education centers, and various other research facilities distributed across six islands throughout the state of Hawaii. All schools of the UH system are accredited by the Western Association of Schools and Colleges. UH at Manoa is the flagship institution of the system. It is well respected for its programs in Hawaiian/Pacific studies, astronomy, east asian languages and literature, asian studies, second language studies, linguistics, ethnomusicology, medicine, and law. UH education centers are located in more remote areas of the state, supporting rural communities via distance education.

A total of 616 programs are offered throughout the UH system, with 123 devoted to bachelor's degrees, 92 to master's degrees, 53 to doctoral degrees, three to first professional degrees, four to postbaccalaureate degrees, 115 to associate's degrees, and various other certifications. The Department of Oceanography offers a bachelor's degree program in global environmental science. This program emphasizes the study of Earth and Earth's physical, chemical, biological, and human systems. Students learn to investigate natural as well as economic, policy, and social systems and their response to and interaction with the Earth system.

The School of Ocean and Earth Science and Technology (SOEST) was established by the Board of Regents of UH in 1988 in recognition of the need to realign and further strengthen the excellent education and research resources available within the university. SOEST brings together four academic departments, three research institutes, several federal cooperative programs, and support facilities of the highest quality in the nation to meet challenges in the ocean, Earth, and planetary sciences and technologies.

The Hawaii Natural Energy Institute (HNEI), located on the campus of UH, is a one of the many research institutes housed by the SOEST. HNEI has become an acknowledged international leader in the energy field and has broadened its expertise to encompass the development of technologies that will enable us to tap our oceans for energy, food, minerals, and other resources. The institute's responsibilities include conducting and supporting basic research, managing research facilities and laboratories, demonstrating the applications of its work, and investigating the social, environmental, and financial effect of energy- and marine-related activities. Researchers at HNEI have a long history of investigating technology solutions, such as renewable energy systems and carbon sequestration, to reduce greenhouse gas emissions. Since 1989, HNEI has been a major participant in an international effort to study the feasibility of sequestering carbon dioxide in the deep ocean. Laboratory studies conducted in a novel large deep ocean

Divers clean the outside of SeaStation 3000, an open-ocean aquaculture project operated by the University of Hawaii.

simulator designed and built by HNEI have provided extensive data on the breakup of liquid carbon dioxide jets and fundamental information on carbon dioxide hydrate formation. HNEI cooperates in these endeavors with faculty from UH; federal, state, and local governments; private industry; public utilities; foreign governments; community groups; and universities and research institutes throughout the world.

The International Pacific Research Center is a climate research center housed by SOEST. This center was founded to gain greater understanding of nature and the causes of climate variation in the Asia-Pacific region, to determine whether such variations are predictable and to discover how global climate change affects the region.

SEE ALSO: Hawaii; National Science Foundation; Pacific Ocean.

BIBLIOGRAPHY. Hawaii Natural Energy Institute, http://www.hnei.hawaii.edu; International Pacific Research Center, http://www.iprc.soest.hawaii.edu; School of Ocean and Earth Science and Technology, http://www.soest.hawaii.edu.

FERNANDO HERRERA
UNIVERSITY OF CALIFORNIA, SAN DIEGO

University of Illinois

THE UNIVERSITY OF Illinois is a system of public universities. It consists of three campuses: Urbana-Champaign, Chicago, and Springfield. The governing body of the three campuses is the board of trustees. The campus at Urbana-Champaign is known as the U of I and UIUC, whereas the Chicago campus is known as UIC, and the Springfield campus UIS. The largest university in the Chicago area, UIC has 25,000 students, 15 colleges, including the nation's largest medical school, and annual research expenditures exceeding $290 million. Playing a critical role in Illinois healthcare, UIC operates the state's major public medical center and serves as the principal educator of Illinois' physicians, dentists, pharmacists, nurses, and other healthcare professionals.

The modern UIC was formed in 1982 by the consolidation of two University of Illinois campuses: the Medical Center campus, which dates back to the 19th century, and the comprehensive Chicago Circle campus, which in 1965 replaced the two-year undergraduate Navy Pier campus designated to educate returning veterans. UIC's student body is recognized as one of the nation's most diverse, and the students reflect the global character of Chicago.

The School of Earth, Society, and Environment brings the resources of the Departments of Atmospheric Science, Geology, and Geography at the University of Illinois together to study Earth systems. The school awards bachelor of science degrees in atmospheric science, geography, and geology, as well as in the interdisciplinary major Earth Systems, Environment, and Society (ESES). The school also actively pursues advanced research and programs of graduate studies. At the Department of Atmospheric Sciences, students can pursue master of science (M.S.), and doctor of philosophy (Ph.D.) degrees. The University of Illinois is deeply involved in climate research at many levels, both within the department and in interdisciplinary research across the campus.

Faculty and students in the department carry out a number of research projects directed toward understanding climate variability and climate change. Students work with professors to conduct research to help explain the interactions between climate and the biosphere, ocean, and human activities, as humans alter the cycles of greenhouse gases such as carbon and methane. Students use global climate models to make projections of future changes under various plausible economic scenarios and simulate important geophysical processes, such as the past, present, and possible future behavior of the Atlantic thermohaline circulation.

Students also conduct research on how energy, water, and carbon are transported between the land surface and the atmosphere in systems ranging from agroecosystems to rainforests. Quantifying and understanding the causes of climate change is one of the greatest challenges of our time. The ESES major is an academic, liberal arts, and sciences degree. Students interested in a getting a more applied degree might be interested in the College of Agricultural, Consumer, and Environmental Sciences Natural Resources and Environmental Sciences (NRES) program, which offers a wide range of NRES and horticulture courses, from plant propagation to wildlife ecology.

Courses offered by UIC in the area of climate change include: ATMS 447: Climate Change Assessment, ATMS 448: Climate and Climate Change, ATMS 300: Weather Processes, ATMS 401: Atmospheric Physics, ATMS 402: Principles of Atmospheric Dynamics, ATMS 403: Weather Forecasting, ATMS 449: Biogeochemical Cycles and Global Change, ATMS 491: Topics in Atmospheric Sciences, GEOG 415: Physical Climatology, and IB 440: Plants and Global Change.

SEE ALSO: Climate Change, Effects; Illinois.

BIBLIOGRAPHY. Atmospheric Sciences, www.atmos.uiuc.edu/research; Climate Research Group, www.crga.atmos.uiuc.edu; University of Illinois, www.uillinois.edu.

FERNANDO HERRERA
UNIVERSITY OF CALIFORNIA, SAN DIEGO

University of Kentucky

THE UNIVERSITY OF Kentucky, also referred to as UK, is a public, coeducational university located in Lexington. Founded in 1865, the university is the largest in the commonwealth by enrollment, with 27,209 students. The university as a whole has been ranked the 19th-best public research university based on the scholarly activity of faculty. The university features 16 colleges, a graduate school, 93 undergraduate programs, 99 master's degrees, 66 programs in Ph.D.s and other doctoral degrees, and four professional programs.

The Department of Earth and Environmental Sciences at UK offers B.S., B.A., M.S., and Ph.D. degrees in geology, a minor in geology, and administers a topical major (B.S.) in environmental sciences. The department currently includes 10 regular faculty members, one lecturer, and 14 other faculty members who hold adjunct appointments. Approximately 40 graduate students and 35 undergraduates are enrolled annually. The study of environmental issues such as groundwater quality, waste disposal associated with the extraction of Earth resources, and climate change require understanding and application of several disciplines, including the fields of geology, chemistry, biology, agronomy, hydrology, and engineering. An environmental scientist must have a diverse background in all the natural sciences to develop creative solutions to environmental problems. The topical major in environmental sciences is intended to provide the breadth of scientific training needed to develop such solutions. At the same time, it provides flexibility for the student and adviser to build a curriculum tailored to the student's specific interests.

Courses with environmental applications offered at UK include: GEO 251: Weather and Climate, GLY 585: Hydrogeology, NRC 301: Resource Management and Conservation, NRC 359: Global Positioning Systems, NRC450G: Biogeochemistry, NRC 555 GIS: and Landscape Analysis, and PLS 366: Fundamentals of Soil Science.

Research within the Department of Earth and Environmental Sciences is funded by grants and contracts from the National Science Foundation, the Department of Energy, the PRF, and other federal, state, and industrial sources. Areas of graduate research include tectonics, hydrogeology, sedimentary geology, geochemistry, petrology, geophysics, and coal geology.

SEE ALSO: Climate Change, Effects; Global Warming.

BIBLIOGRAPHY. College of Agriculture, www.ca.uky.edu; Kentucky Geological Survey, www.eurekalert.org; University of Kentucky, www.uky.edu.

FERNANDO HERRERA
UNIVERSITY OF CALIFORNIA, SAN DIEGO

University of Leeds

THE UNIVERSITY OF Leeds is a major teaching and research university in Leeds, West Yorkshire. Leeds is a leading research institution and one of the largest in the United Kingdom with over 32,000 full-time students. Established in 1904, it is one of the six original civic universities, and in 2006 it was ranked second for the number of applications received. Leeds is now among the top 10 universities for research in the United Kingdom and is internationally acknowledged as a center of excellence in a wide range of academic and professional disciplines. Many of its research initiatives cross traditional subject boundaries, and Leeds currently promotes projects through 58 interdisciplinary centers and seven research schools. The university recognizes the importance of the environment and has established policies and programs to tackle environmental issues at a local and global level. Business in the Environment has ranked the University of Leeds as the top university in the region for environmental management and performance. The university also won the 2006 Green Gown Award for Waste Management and the 2007 Award for Continuous Improvement.

The School of Earth and Environment was formed in 2004 from the merger of the previous schools of Earth Sciences and of Environment. The school currently has over 60 academic staff, 35 support staff, and over 50 postdoctoral research fellows and associates. It is one of the largest schools in the United Kingdom that focuses on a multidisciplinary approach to understanding environment: the school studies the Earth from its core to its atmosphere and examines the social and economic dimensions of sustainability. The school offers both three-year BSc and four-year undergraduate degrees in geological, geophysical, and environmental sciences. The school also offers graduate degrees in Earth, atmospheric sciences, and sustainability research. The University of Leeds has recognized Earth and environmental systems science as one of its 12 gold research peaks of excellence. The school also has extensive ongoing research projects and collaborates with members of other departments and institutes such as mathematics and applied physical sciences and engineering, the Earth and Biosphere Institute, and the Earth, Energy, and the Environment Institute. This research grouping receives the second-highest research income nationally from the Natural Environment Research Council through competitive grant proposals.

The Earth and Biosphere Institute is a group of internationally recognized scientists with interests in the effects of biotic and environmental changes on a spectrum of time and space scales, from short term to geological, and from nanoscale to global. Research initiatives are ongoing within the University of Leeds and with groups from outside the university. The Earth, Energy, and Environment Institute is a multidisciplinary institute housed at the University of Leeds. The purpose of the institute is to provide integrated research solutions, knowledge transfer, and training in this energy-related area on all levels, from the regional to the global.

SEE ALSO: Education; European Union; United Kingdom.

BIBLIOGRAPHY. Climate Change Research Group, www.geography.leeds.ac.uk; School of Earth and Environment, www.see.leeds.ac.uk; University of Leeds, www.leeds.ac.uk.

FERNANDO HERRERA
UNIVERSITY OF CALIFORNIA, SAN DIEGO

University of Maine

THE UNIVERSITY OF Maine, established in 1865, is the flagship university of the University of Maine system. It is located in Orono, just outside Bangor, one of Maine's largest cities. Also known as UMaine, the university has an enrollment of over 11,000 students, making it the largest university in the state. The college was the fourth to be established in Maine, after Bowdoin, Bates, and Colby. Originally intended as an agricultural college, it also places a large emphasis on engineering and the sciences. The University of Maine is a major research institute composed of six separate and distinct colleges offering 184 areas of study. The Department of Earth Sciences offers graduate degree programs focusing on climate change in conjunction with the Climate Change Institute.

The Climate Change Institute (formerly the Institute for Quaternary and Climate Studies) is an interdisciplinary research unit organized to conduct research and graduate education focused on the variability of the Earth's climate, ecosystems, and other environmental systems and on the interaction between humans and the natural world. Institute investigations cover the Quaternary period, a time of numerous glacial/interglacial cycles and abrupt changes in climate, ranging in time from the present to nearly 2 million years ago. Research activities include field, laboratory, and modeling studies that focus on the timing, causes, and mechanisms of natural and anthropogenically forced climate change and on the effects of past climate changes on the physical, biological, chemical, social, and economic conditions of the Earth. Institute research is supported by grants from a variety of sources including the National Science Foundation, the National Oceanic and Atmospheric Administration, NASA, and endowments from the Bingham Trust and the Dan and Betty Churchill Exploration Fund.

The Climate Change Institute offers an M.S. degree program with potential for a continuing Ph.D. program through Earth Sciences or an independent degree through the graduate school. Students in the program investigate human and global change, past, present, and future. This program involves in-depth documenting of past and present and predicting future environmental changes and human cultures through the primary disciplines of climatology, archaeology, glaciology, glacial geology, ecology, history, marine geology, and modeling. Over 40 faculty and staff of the institute provide multidisciplinary strength in the fields of climatology, archaeology, glaciology, glacial geology, geochemistry, ecology, history, remote sensing, and marine geology.

The institute's internationally recognized researchers have contributed to the scientific literature in multiple areas of study including understanding controls on the climate system, identifying and exploring abrupt behavior in the climate system, detecting and documenting human-source pollution in the atmosphere, exploring changes over time in the behavior of major atmospheric circulation systems such as El Niño/Southern Oscillation and the North Atlantic Oscillation, studying climate-induced changes in the distribution and abundance of plants and animals, detecting and documenting changes in civilization resulting from climate change, and identifying changes in the extent of ice sheets and mountain glaciers and their effect on the environment.

SEE ALSO: Greenhouse Effect; Maine.

BIBLIOGRAPHY. Climate Change Institute, www.climate change.umaine.edu; Department of Environment, www.maine.gov; University of Maine, www2.umaine.edu.

Fernando Herrera
University of California, San Diego

University of Maryland

THE UNIVERSITY OF Maryland in College Park was founded in 1856 as the Maryland Agricultural College and renamed in 1920 as the University of Maryland. The College Park main campus operates within a group of state-supported institutions of higher education in Maryland. The system's research and service components include the University of Maryland Biotechnology Institute in College Park and the Center for Environmental and Estuarine Studies, with laboratories at Horn Point, Solomons, and Frostburg.

The meteorology/oceanography program is a graduate department program with an expanded scope of activity to include the dynamic system of atmosphere in relation to oceans, land, and life. Though the department does not offer an undergraduate major, they do offer an undergraduate minor in meteorology and oceanography. Advantages to the university programs include proximity to nearby federal agencies such as NASA and the National Centers for Environmental Prediction, as well as the department's partnerships with NASA and the National Oceanic and Atmospheric Administration (NOAA), allowing for close collaboration. Graduate students in the department may have the opportunity to conduct research through these partnerships. An example of these partnerships is the Joint Global Change Research Institute—a university-based collaboration between the University of Maryland and the U.S. Department of Energy's Pacific Northwest National Laboratory or the Earth System Science Interdisciplinary Center, a joint center between the university's departments of meteorology, geology, and geography and the Earth Sciences Directorate at the NASA Goddard Space Flight Center.

The meteorology department conducts research in a broad range of areas, including atmospheric chemistry, climate studies, glaciology, numerical weather prediction, and remote sensing, with a range of activities including fieldwork, remote sensing, and numerical modeling from pole to pole and from the troposphere to the mesosphere. Current funded climate research in particular focuses on global change (including Earth system modeling and analysis and modeling of climate hesitative to greenhouse gas concentrations), atmospheric and oceanic reanalyses, hydroclimate studies, ocean-atmosphere interaction, monsoons, extratropical interannual variability, clouds and radiation, and NWP methods in climate modeling.

Founded as a joint collaboration between the Departments of Atmospheric and Oceanic Science and Geology and Geography at the University of Maryland and the Earth Sciences Directorate at the NASA/Goddard Space Flight Center, ESSIC (the Earth System Science Interdisciplinary Center) is based at the university. ESSIC's focus is to study and understand the dynamics of human activities on atmosphere, ocean, land, and biosphere components of the Earth. ESSIC also administers the Cooperative Institute for Climate Studies, sponsored by the NOAA National Satellite, Data, and Information Services and the NOAA National Centers for Environmental Prediction.

ESSIC's research centers on climate variability and change, atmospheric composition and processes, the global carbon cycle (including terrestrial and marine ecosystems/land use/cover change), and the global water cycle. ESSIC combines primary research of within-system observations with remote sensing and predictive models to forecast global changes and potential regional impact.

The University of Maryland Enterprise Campus, or M Square, is a 124-acre research park adjacent to the University of Maryland/College Park Metro Station. NOAA's National Center for Weather and Climate Prediction is one of the anchor tenants of M Square to take advantage of proximity to the University of Maryland—a leading center for climate research and numeric weather forecasting, which is developing major new partnerships with federal agencies in the areas of earth science, remote imaging, climate change, and energy research.

In addition to academics, the Center for Integrative Environmental Research (CIER) is based at the university. CIER is dedicated to creating a comprehensive understanding of the complex environmental challenges facing society and to developing valuable tools to inform policy and investment decision making. To support resilient flourishing natural and human systems, a truly integrative approach is needed. These multidisciplinary processes combine insights from the physical, engineering, natural, social, and health sciences and stimulate active dialogue across the science-society divide. CIER researchers and graduate students collaborate at global, national, regional, and local scales to explore issues with and across two major sustainability challenges: society's use of material and energy and urban environmental change.

The University of Maryland is setting an example for public and private sectors, as the president of the university joined other college presidents and chancellors around the country in taking a community leadership role and modeling ways to minimize global warming emissions and integrate sustainability into the curriculum and university environment, via membership in the Leadership Circle of the American College and University Presidents Climate Commitment.

SEE ALSO: Education; Maryland; National Oceanic and Atmospheric Administration (NOAA).

BIBLIOGRAPHY. Department of Atmospheric and Oceanic Science, The University of Maryland, http://www.atmos.umd.edu; Earth System Science Interdisciplinary Center, "About ESSIC," http://www.essic.umd.edu.

LYN MICHAUD
INDEPENDENT SCHOLAR

University of Miami

THE UNIVERSITY OF Miami, founded in 1925, is a private, coeducational teaching and research institution noted for its extensive study-abroad programs, marine science institute, and medical center. The university confers bachelor's, master's, doctoral, and professional degrees in a broad range of academic disciplines including the arts and sciences and marine science. The main campus is in Coral Gables, Florida, and the Rosenstiel School of Marine and Atmospheric Science (RSMAS) is located on Virginia Key on Biscayne Bay.

The university's undergraduate programs related to climate change include ecosystem science and policy and geological sciences. Ecosystem science and policy covers a range of environmental issues from various perspectives. The geological sciences program is taught in the College of Arts and Sciences; the major includes themes of Earth origins, environmental preservation, global dynamics, and internal and surface processes. A special aspect of the geological sciences program is an option for a five-year B.S./M.S.

program coordinated with RSMAS to complete the graduate portion of the education.

The Rosenstiel School was established in 1940 on the Coral Gables campus and moved to Virginia Key on Biscayne Bay in 1957 after outgrowing its initial location. RSMAS is now a 16-building complex and home to a museum with a collection of approximately 400,000 invertebrate specimens, the National Institute of Environmental Health Sciences Marine and Freshwater Biomedical Sciences Center, NSF/NIEHS Oceans and Human Health Center, and the Division of Marine and Atmospheric Chemistry. Research facilities include advanced technology computing and laboratories with precision instruments (mass spectrometer, X-ray spectrograph, gas chromatograph, and a scanning electron microscope).

RSMAS's Division of Marine and Atmospheric Chemistry is a graduate program offering master's and doctoral degrees in chemical processes of the atmosphere and chemical processes of the ocean and hydrology. Fieldwork plays a major role, and the department's small size ensures student access to faculty, staff, and research opportunities.

The teaching/research facilities include chemical laboratory instrumentation, a laboratory onboard a cruise ship for collecting marine and atmospheric data, a 90-ft. research catamaran, and a simulation wind and wave tank for observation of specific air/sea interactions. RSMAS's library holds an extensive marine science collection.

Research opportunities are diverse as a result of having a faculty with a wide range of expertise and research interests. Faculty members from the six divisions at RSMAS (Applied Marine Physics, Marine and Atmospheric Chemistry, Marine Affairs and Policy, Marine Biology and Fisheries, Marine Geology and Geophysics, and Meteorology and Physical Oceanography) are working on research related to changing climate impact. A sampling of climate-related studies include the Southeast Climate Consortium (using advanced forecasting tools for agriculture, forestry, and water resources management in the southeastern United States), water resource management in the drought-prone state of Ceara in northeast Brazil (human response to climate change), and climate and fisheries (how changes in sea-level pressure, sea surface temperature patterns, and global mean temperature affect fish stocks).

The Cooperative Institute for Marine and Atmospheric Studies (CIMAS) is a research institute of the University of Miami located in the RSMAS. CIMAS serves as a mechanism to bring together the research resources of the university with those of the National Oceanic and Atmospheric Administration (NOAA) to perform research for understanding the Earth's oceans and atmosphere within the context of NOAA's mission.

The scientific activities in CIMAS are organized under broad research themes. The themes are topics of climate variability, fisheries dynamics, regional coastal ecosystem processes, human interactions with the coastal environment, air-sea interactions and exchanges, and integrated ocean observations; their scientific objectives are guided by NOAA's Strategic Plan and its specific goals in the context of the research activities and expertise resident in the university and the local Miami laboratories of NOAA.

The University of Miami is setting an example for public and private sectors, as when the president of the university joined other college presidents and chancellors around the country in taking a community leadership role and modeling ways to minimize global warming emissions and integrate sustainability into the curriculum and university environment, via membership in the American College and University Presidents Climate Commitment.

SEE ALSO: Atlantic Ocean; National Oceanic and Atmospheric Administration (NOAA); Oceanography.

BIBLIOGRAPHY. Edward B. Fiske, *Fiske Guide to Colleges 2005*, 21st ed. (Sourcebooks, 2004); University of Miami, "Research at Rosenstiel School," www.rsmas.miami.edu.

Lyn Michaud
Independent Scholar

University of Michigan

THE UNIVERSITY OF Michigan (also known as Michigan, or UM) was established in 1817 and today is one of the premier research universities in the United States. The university has a strong reputation in all academic fields, with more than 70 percent of

Michigan's 200 major programs, departments, and schools ranking in the top 10 nationally. The university has one of the largest annual research expenditures of any university in the United States (nearly $800 million in 2006), and the university's endowment was valued at $5.65 billion also in 2006, making it the ninth-largest endowment in the United States and the third largest among public universities. The university has over 6,200 faculty members, 73 of whom are members of the National Academy and 400 of whom hold an endowed chair in their discipline. The university consistently leads the nation in the number of Fulbright Scholars and has matriculated 25 Rhodes Scholars, while having produced seven Nobel Prize winners.

The Michigan library system comprises 19 libraries with 24 collections, totaling 8.13 million volumes, with 177,000 volumes being added each year. The campus libraries include the Gerald R. Ford Presidential Library. Over 40,000 students attend UM, including 25,555 undergraduate and 14,470 graduate students in 600 academic programs. Students come from all 50 states and more than 100 countries. One quarter of all undergraduates are members of ethnic minority groups. The university also has a strong social conscience. It was on the steps of the Michigan Student Union on October 14, 1960, that President John F. Kennedy proposed the concept of the Peace Corps. Lyndon B. Johnson's speech outlining his Great Society program also occurred at Michigan.

Over 300 Michigan faculty members are engaged in research on environmental issues. Likewise, there are a number of institutes and centers at the university that are actively engaged in research concerning global warming and climatic change, as well as related issues such as sustainability and alternative energy. For example, the UM Climate Change Consortium is an initiative of the School of Engineering that runs an electronic forum about climate change. The School of Engineering is also renowned for its Solar Car Team, which placed first in the American Solar Challenge four times and third in the World Solar Challenge three times. The University of Michigan Transportation Research Institute provides leadership in interdisciplinary transportation-related research and also has experts conducting research on energy efficiency and emissions. The Frederick A. and Barbara M. Erb Institute for Global Sustainable

Enterprises is affiliated with the School of Business and fosters global sustainable enterprise through interdisciplinary research and education initiatives. Even the operations of the university itself strive to be environmentally sensitive, with UM Waste Management Services encouraging recycling on campus. The university operates a 13,000-acre (53 sq. km.) biological station in the northern Lower Peninsula of Michigan—one of only 47 biosphere reserves in the United States.

The university also houses the Michigan Memorial Phoenix Energy Institute, which focuses on identifying and using secure, affordable, and sustainable energy sources. The institute uses public policy, economics, business, and social sciences to lay the foundation for successful implementation of scientific and technological achievements. Created in 1947 as a memorial to UM grads who gave their lives in World War II, the institute originally sought to explore peaceful ways to use atomic energy for the benefit of humankind. Then in 2004 the regents of the university broadened the charter of the Phoenix Project beyond atomic energy to encompass interdisciplinary research on and education about the development of energy policies that promote world peace, responsible use of the environment, and economic prosperity.

The UM Center for Sustainable Systems is located within the School of Natural Resources and Environment (SNRE) and focuses on the identification of systems-based approaches to sustainability through creative and effective teaching and research. SNRE also hosted in 2007 a national summit on climate change as part of the Clinton Global Initiative. The purpose of the summit was to bring together global leaders and key stakeholders to examine what is known about the composite regional effect of climate change and what management and policy options can help regions deal with changes in average conditions, as well as with extreme events.

SEE ALSO: Alternative Energy, Solar; Climate Change, Effects; Education.

BIBLIOGRAPHY. Facts and Figures, University of Michigan Office of Budget and Planning, http://sitemaker.umich.edu/obpinfo/facts_figures; Edward B. Fiske, *Fiske Guide to Colleges 2005*, 21st ed. (Sourcebooks, 2004); Howard H.

Peckham, *The Making of the University of Michigan 1817–1992* (University of Michigan Press, 1994).

MICHAEL J. SIMSIK
U.S. PEACE CORPS

University of Nebraska

THE UNIVERSITY OF Nebraska in Lincoln, Nebraska (UNL), founded in 1869, is part of the state-supported University of Nebraska system. The discipline of ecology began at the University of Nebraska, and the campuses are botanical gardens and arboreta. The university confers bachelor's, master's, doctoral, and professional degrees in a variety of academic and professional disciplines, including agricultural sciences, animal sciences, health sciences, atmospheric sciences, earth and space sciences (which includes the former departments of geology and geophysics), and environmental sciences. Research facilities include the Cedar Point Biological Station, the Center for Great Plains Studies, and the Engineering Research Centers.

Within the university, the College of Agricultural Sciences and Natural Resources encompasses 27 programs of study and two preprofessional programs ranging from animal science, fisheries and wildlife, and agribusiness to professional golf course management. The areas related to climate change and the environment include Environmental Restoration Science, environmental studies, natural resource and environmental economics, and applied climate sciences.

Environmental restoration science is the study of soil (for producing crops and supporting groundwater purification and the conversion of waste material). The program combines classroom work with out-of-class experiences (internships and fieldwork), with an emphasis on soil in an environmental context. In addition to frequent opportunities for faculty and staff contact, students may be involved in the Soil and Water Resources Club, Agronomy Club, Soil Judging Team, or the Lincoln Chapter of the Soil and Water Conservation Society.

Environmental studies provides interdisciplinary study encompassing a variety of academic departments in the College of Agricultural Sciences and Natural Resources and the College of Arts and Sciences to study environmental and social sciences, global issues, and environmental protection. In addition to academic work across academic disciplines, the Environmental Resource Center provides access to environmental information, internship opportunities, and a study area. Students may be involved in a variety of environmentally related student groups such as Ecology Now, the Soil and Water Resources Club, and the Wildlife Club.

Natural Resource and Environmental Economics brings together natural sciences with economics, law, and other social sciences with a focus on assessing public policies regulating the use of natural resources and environmental amenities.

The Applied Climate Sciences (ACS) program (formerly called the Department of Agricultural Meteorology) is a graduate program bringing together a variety of academic disciplines for academic study and applied research on climate effects and variability; drought monitoring, mitigation, and planning; environmental biophysics; global climate change; High Plains climate; micrometeorology; and severe weather, among others.

Within the program are the bio-atmospheric interactions specialization (to understand interactions between the atmosphere and the biosphere) and the agricultural meteorology specialization (to understand bioatmospheric interaction and agriculture).

The facilities include laboratory, office, and classroom facilities to support research, teaching, and outreach regarding natural resources; the Agro-Meteorology Laboratory at the University of Nebraska Agricultural Research and Development Center near Mead; measuring stations for the National Atmospheric Deposition Program and the U.S. Department of Agriculture Ultraviolet B Monitoring Network; the High Plains Climate Center and its Automated Weather Data Network; and the National Drought Mitigation Center.

A sampling of research opportunities related to climate in the ACS programs includes climate variations, climate modeling, micrometeorology, remote sensing, and climate effects.

UNL's School of Natural Resources hosts the Great Plains Regional Center of the National Institute for Global Environmental Change. The focus of the center's research is environmental change in the grasslands of Colorado, Iowa, Kansas, Minnesota, Missouri, Montana, Nebraska, North and South Dakota, and Wyoming. Because of the agricultural value of crops and livestock produced in this region, even minor climate changes could have a major effect on the nation and the world. The data collected are also valuable for reference with regard to global regions with semiarid grassland ecosystems similar to the Great Plains.

SEE ALSO: Climate Change, Effects; Education; Nebraska.

BIBLIOGRAPHY. University of Nebraska, "College of Agricultural Sciences and Natural Resources," www.casnr.unl.edu; University of Nebraska–Lincoln, "Applied Climate Sciences," www.snr.unl.edu/acs.

LYN MICHAUD
INDEPENDENT SCHOLAR

University of New Hampshire

THE UNIVERSITY OF New Hampshire in Durham, New Hampshire, was founded in 1866 as the College of Agriculture and Mechanic Arts. The school became the University of New Hampshire (UNH) in 1923 and is the main campus of the state's university system. Academic programs cover a variety of disciplines such as the arts and sciences, humanities, business, engineering, education, and the health professions. Research facilities included Anadromous Fish and Aquatic Invertebrate Research Laboratory, Jackson Estuarine Laboratory, Coastal Marine Laboratory, Jere A. Chase Ocean Engineering Laboratory, and Shoals Marine Laboratory.

Courses, programs, and departments interconnect on the subject of climate. The Climate Change Research Center (CCRC) is dedicated to the study of the Earth's atmosphere. Set up for graduate-level instruction, the CCRC also provides information materials (interactive displays) to various organizations and provides support to K–12 teachers through lesson guides and teacher training. Several center faculty also have appointments to the Departments of Earth Science, Natural Resources, Chemistry, and Engineering and work with undergraduates in the Research & Discover internship and Undergraduate Research Opportunity Program.

CCRC research focuses on understanding the fundamental properties of the atmosphere and how they have been affected by human activities and will continue to be so in the future, with areas of interest including air quality and climate, airborne sciences, biosphere-atmosphere exchange, halogen chemistry, ice course and air-snow exchange, new england climate assessment, and organic aerosols.

Courses taught by CCRC faculty are specific to the research being conducted in the department: Introduction to Atmospheric Science (fundamental principles and dynamics of the earth's atmosphere), Global Atmospheric Chemistry (relationship between atmospheric chemistry, climate, and global change), Atmospheric Aerosol and Precipitation Chemistry (processes determining the chemical and physical characteristics of atmospheric aerosol particles and precipitation), Measurement Techniques in Atmospheric Chemistry (instrumental methods used in atmospheric chemistry and biogeochemical research), Earth System Science (components, interactions, and concepts for characterizing the Earth's integrated system), and Regional Air Quality (measurement programs to examine air quality in New England and other global regions).

Earth Sciences is a multidisciplinary department including geology, oceanography, hydrology, ocean mapping, and geochemical systems specialization. Students have the option of choosing among the different disciplines in creating a program specific to interest and, in addition, have many different research opportunities from local research to field work at sites including Antarctica, Greenland, the Pacific and Indian oceans, Mexico, China, the Himalayas, Indonesia, Pakistan, and the western United States. Research in earth sciences covers all the Earth system components and relies on connections with researchers and faculty in natural resources.

Natural Resources and Earth Systems Science is an interdepartmental graduate program offering only a doctoral degree in natural resources and environmental studies or Earth and environmental science. The program relies on interdisciplinary work to understand and manage the environment with many possible options for study including atmospheric science and ethical and policy issues.

The UNH/National Oceanic and Atmospheric Administration Cooperative Institute for Coastal and Estuarine Environmental Technology (CICEET) researches coastal resource management. Working with a network of colleagues throughout the United States, CICEET's focus is ecosystem health and coastal resiliency to find technology to address coastal challenge as well as providing usable information for people to have continued access to clean water and healthy coasts.

UNH hosted the Global Analysis, Integration, and Modeling International Project Office of the International Geosphere-Biosphere Program until termination of the program in 2004. The decade-long project researched existing models. The more-than-decade-long project provided the foundation for the Global Carbon Project and created the platform for new integrative modeling activities.

The UNH is setting an example for public and private sectors; the president of the university joined other college presidents and chancellors around the country in taking a community leadership role and modeling ways to minimize global warming emissions and integrate sustainability into the curriculum and university environment with membership in the American College and University Presidents Climate Commitment. Since 1997, the University Office of Sustainability—the oldest endowed sustainability program in U.S. higher education—has brought together all members of the university community in integrating sustainability throughout UNH (in curriculum, operations, and research and engagement).

SEE ALSO: International Geosphere-Biosphere Program (IGBP); National Oceanic and Atmospheric Administration (NOAA); Sustainability.

BIBLIOGRAPHY. Climate Change Research Center, "Research," www.ccrc.unh.edu/research; International Geosphere-Biosphere Program, "Global Analysis, Integration and Modelling—GAIM," www.igbp.kva.se; University of New Hampshire Graduate Program, "Natural Resources," "Earth Sciences," www.unh.edu.

Lyn Michaud
Independent Scholar

University of Oklahoma

THE UNIVERSITY OF Oklahoma (OU), located in Norman, was founded in 1890. The university confers bachelor's, master's, doctoral, and professional degrees in a variety of academic and professional disciplines. The university is home to the College of Atmospheric and Geographic Sciences, containing the School of Meteorology and the Department of Geography.

The School of Meteorology will celebrate its fiftieth anniversary in 2010. The school is housed in the National Weather Center building, which also is home to 12 academic, research, and operational meteorology organizations. With over 320 undergraduate and 80 graduate students, OU has the largest meteorology program in the United States, and the department offers a Bachelor of Science in Meteorology degree or a Professional Meteorology master's degree. Undergraduate students also have the opportunity to participate in an international exchange program featuring the universities of Reading in England, Monash in Australia, and Hamburg in Germany.

The School of Meteorology performs shared research with the National Oceanic and Atmospheric Administration's National Severe Storms Laboratory and participates in a variety of national projects. In February 2008, the National Weather Center in Norman, Oklahoma, hosts the first U.S.–China Symposium on Meteorology: Mesoscale Meteorology and Data Assimilation.

The Department of Geography provides for the study of human interactions with climate, earth structures, and natural resources and the effect these interactions have on human culture. The department offers bachelor's, master's, and doctoral degrees. Faculty expertise includes cultural, historical, political, and economic geography; applied physical geography; and geographic information science. Research is conducted in climatology, natural resources, human effects on the planet, and cartography.

In addition to academics, the College of Atmospheric and Geographic Sciences hosts several research centers. The Center for the Analysis and Prediction of Storms includes the Center for Collaborative, Adaptive Sensing of the Atmosphere and the Linked Environments for Atmospheric Discovery. The mission of the center is the development of techniques to predict severe weather conditions using on-site and remote sensing systems.

The Cooperative Institute for Mesoscale Meteorological Studies was established in 1978 between OU and the National Oceanic and Atmospheric Administration. In addition to linking scientific expertise from academia, the center has collaborative relationships with all the National Oceanic and Atmospheric Administration units in Norman. Research programs include basic and convective mesoscale research, climate change monitoring and detection, and climatic effects. Additional research involves improving technology for the prediction of weather events.

Under legislative mandate, the Oklahoma Climatological Survey (OCS) was established in 1980 to collect, analyze, interpret, and provide information on climate and weather data to residents of Oklahoma. The OCS conducts research on climate change effects and provides data and support to the state climatologist.

Under joint administration by OU and Oklahoma State University, the Oklahoma Mesonet monitors weather and soil conditions at over 100 automated observing stations.

The Center for Spatial Analysis brings together various academic disciplines to study and apply geospatial science and technology.

Focused on strengthening the research of environmental scientists and facilitating relationships between the scientific community and the public, the Environmental Verification and Analysis Center participates in earth science work.

A network opportunity for teachers of geography, the Oklahoma Alliance for Geographical Education relies on affiliation with the National Geographic Society along with its associated state geographic alliances, as well as geography organizations from around the country.

The opportunity for Oklahomans to learn about and support NASA's mission research programs, including science and geography, along with other academic disciplines falls under the auspices of the Oklahoma NASA Space Grant Consortium. The consortium includes partnerships with Oklahoma universities, a science museum, a cooperative extension service, state and local governments, and other interested institutions.

Beyond academics and research, the president of the University of Oklahoma joined other college presidents and chancellors around the United States to take a community leadership role and to model ways to minimize global warming emissions and integrate sustainability within the curriculum and university operations and environment, with membership in the American College and University Presidents Climate Commitment.

SEE ALSO: National Aeronautics and Space Administration (NASA); National Oceanic and Atmospheric Administration (NOAA); University of Reading.

BIBLIOGRAPHY. School of Meteorology, "Cooperative Institute for Mesoscale Meteorological Studies," *Meteorogram* (Spring 2007); University of Oklahoma, Department of Geography, http://www.geography.ou.edu; University of Oklahoma, School of Meteorology, http://www.weather.ou.edu.

Lyn Michaud
Independent Scholar

University of Reading

THE UNIVERSITY OF Reading is located in Reading, England. With a foundation made by joining the School of Art begun in 1860 and the School of Science begun in 1870, Christ Church of Oxford established an extension college in 1892. Funding and expansion ensured success and led to the University of Reading being given a royal charter in 1926. The University of Reading has several educational centers conducting research in a variety of disciplines in the arts, humanities, sciences, and social sciences, including environmental sciences and meteorology.

The environmental science degree investigates the components of the Earth system, along with providing a scientific understanding of current and future environmental issues and their management. Within the major, students may choose a pathway in consultation with a course adviser. The pathway options include earth and atmosphere (interactions between the solid Earth, the Earth's surface, the oceans, and the atmosphere through geology, soil science, and meteorology), earth systems and environmental change (changes in rates, causes, and extents of change; the contribution of human activity; and how that activity can be managed appropriately), and habitat management.

The Department of Meteorology was founded in 1965, and offers a range of courses for both undergraduates and postgraduates. The teaching facilities include a lecture hall, laboratories for practical experimentation, and a meteorological enclosure for observations and experimentation. For teaching purposes and general access, current weather data and satellite pictures stream into the department.

Active research in regional weather systems, data assimilation, global atmospheric modeling, and global circulation and climate are ongoing. Established research and teaching links with the European Centre for Medium Range Weather Forecasts and the U.K. Meteorological Office provide additional learning opportunities and access to state-of-the-art equipment.

In addition to academics, the University of Reading is the home of the National Centre for Atmospheric Science (NCAS) Climate @ Reading Programme, the Environmental Systems Science Centre (ESSC), and the Data Assimilation Research Centre (DARC).

Based within the Department of Meteorology is the NCAS Climate @ Reading Programme. NCAS was called the Centre for Global Atmospheric Modeling from May 1990 until April 2006. NCAS Climate @ Reading provides scientific expertise, technical support, and training to address the following key challenges: increasing knowledge of climate variability and change on various timescales, using advanced computer technology for simulations; establishing the usefulness and accuracy of climate prediction models; and integrating various specialties for creating adaptability or mitigation of climate change.

NCAS Climate @ Reading is part of the Climate Programme charged with carrying out climate change research into how the climate will change by the end of the century, while taking into account natural variation and finding ways to predict climate change for the benefit of society. NCAS carries out this mission by integrating dynamics of the global climate system with earth system modeling, improved prediction, and ever-improving computer technology.

ESSC is a research center that is not involved in undergraduate teaching. It is funded as such by the Natural Environment Research Council (NERC). ESSC is one of the focal centers in the country for the study of remotely sensed data; the work involves coupling data with computer models of environmental processes. The wide range of interests of researchers parallels the research interests of NERC: meteorology, oceanography, geology, and hydrology. ESSC forms part of the School of Mathematics, Meteorology, and Physics.

Research interests are largely driven by the strategy of NERC, and so address problems relevant to issues within topics of global environmental change, hazard assessment and mitigation, natural resources, waste, and pollution. Research includes land surface processes, marine science, meteorology, and others.

DARC is funded under the NERC Earth Observation Centre of Excellence Initiative. The meteorology department holds the directorship and partners with Rutherford Appleton Laboratory and the universities of Cambridge, Edinburgh, and Oxford. The focus of DARC is theory development research and studying applications of data assimilation methods. This work involves the highly integrated combination of various Earth observation data sources with information available from sophisticated Earth system models and the European Space Agency Enlist satellite.

In addition to many collaborative partnerships worldwide, the University of Reading is one site of an international exchange program with the University of Oklahoma.

SEE ALSO: Climate Change, Effects; Education; University of Oklahoma.

BIBLIOGRAPHY. Department of Meteorology, "About Us," www.met.reading.ac.uk; University of Reading, "Department of Meteorology Excellence in Weather and Climate Research Since 1965," Research Summary (April 2006); University of Reading, www.rdg.ac.uk.

LYN MICHAUD
INDEPENDENT SCHOLAR

University of St. Gallen (Switzerland)

THE UNIVERSITY OF St. Gallen, Graduate School of Business Administration, Economics, Law, and Social Sciences (HSG) is located in St. Gallen in northeastern Switzerland. The town of St. Gallen is a commercial center with roots in the early 7th century and a hermitage established by the Irish missionary Saint Gall. The hermitage became a Benedictine abbey and noted educational center in the 8th century.

On May 28, 1898, the Cantonal Parliament of the Canton of St. Gallen issued a resolution to establish a school for transport, commerce, and administration; this academy opened in 1899. The school continued under various names until being established as the University of St. Gallen. For the 2000/2001 educational year, the curriculum was revised to offer bachelor's and master's degrees.

HSG remains true to its original mission to provide an education easily translated into real life. Topics of study include business administration, economics, law, and social sciences leading to bachelor's, master's, and doctoral academic degrees. More programs are being offered in English, drawing many international students. In addition, executive education is offered by HSG to provide educational opportunities to people in business.

The Institute for Economy and the Environment (IWOe-HSG) was founded on October 1, 1992, at the University of St. Gallen to explore the global intersections of the natural world with the forces of economy and society. IWOe-HSG identifies potential problems and researches options to determine practical solutions. These issues cover a wide range of topics regarding renewable energy, financing renewable and sustainable options as well as addressing issues of concern like landscapes for the use and production of electricity from wind farms. To perform this work, the institute maintains a network with international businesses, other educational institutions, and research organizations.

In general, the institute analyzes conditions necessary for an environmentally sound market economy and corporate sustainability management. Furthermore, it imparts the findings to practical business situations to ensure that the work it does is relevant and has an effect on real-life situations. The foundation of the IWOe-HSG was supported by Oikos (the environmental student organization at the HSG) and the Oikos Foundation for Economy and Environment. IWOe-HSG is a nonprofit organization that receives support from the university and from private and nonprofit companies. At pres-

ent, the institute's primary research projects include culture and sustainability, management systems for corporate sustainability, sustainability marketing and sustainable consumption, life cycle assessment, environmental finance and entrepreneurship, and renewable energy technologies.

In 1968, five students from the University of St. Gallen chose to make a difference by fostering dialogue between groups instead of joining the protests of many of their peers. They set up the International Students' Committee and formed a symposium designed to foster cross-generational dialogue in a neutral environment. Each year in May since 1970, the St. Gallen Symposium has been held at the university as a forum for international dialogue on the connections among business, politics, and society. Attendees include business professionals (corporate and entrepreneurial), policymakers, academic professionals, and students from the University of St. Gallen in Switzerland for integrative discussions on sustaining business and society in general. The topic for the symposium held in 2007 was "The Power of Natural Resources."

The International Students' Committee organizes the St. Gallen Symposium under the auspices of the St. Gallen Foundation for International Studies (which serves in an advisory and supervisory capacity). In addition to the students from St. Gallen an office (International Students' Committee Harvard) at Harvard University in the United States fosters international cooperation, with Harvard students supporting the St. Gallen member students. The team of students designs and plans the symposium from concept and content through to marketing and implementation. The team also determines key themes and interacts with a network of selected speakers from business, politics, and society, along with maintaining contacts with other leading international universities.

SEE ALSO: Climate Change, Effects; Harvard University.

BIBLIOGRAPHY. Institute for Economy and Environment, "News," www.iwoe.unish.ch; University of St. Gallen, "St. Gallen Symposium," www.st.gallen-symposium.org.

Lyn Michaud
Independent Scholar

University of Utah

THE UNIVERSITY OF Utah, founded in 1850, is a public, coeducational institution located in Salt Lake City. The university confers bachelor's, master's, doctoral, and professional degrees in a variety of academic and professional disciplines. The university offers a focus on environmental science, and the university's College of Mines and Earth Sciences is home to the Departments of Meteorology and Geology and Geophysics.

The Department of Meteorology focuses on offering a background for meteorology and related environmental science careers with an understanding of system processes involving weather and climate. The program's emphasis on mountain weather and climate (as well as the complexities involved in forecasting weather on complex terrain) stems from Salt Lake's location on the western slope of the Wasatch Mountains (a range in the Rocky Mountains). Integrating math, physics, and chemistry and a broad range of academic opportunities with dynamic, physical, and synoptic meteorology prepares students for employment as meteorologists with the National Weather Service or for a broad range of careers with government and private employers in meteorology or environmental sciences.

The department's small size ensures student access to faculty, staff, and research opportunities. The faculty has diverse expertise and research interests in a broad range of observational, modeling, and theoretical studies. Current interests include tropical convection and hurricanes; boundary layer modeling; fire weather prediction and fire modeling; mountain meteorology; weather analysis and prediction; parameterization, remote sensing, and modeling of clouds; aerosol physics and

Water levels have been steadily decreasing in Lake Powell, which borders Utah and Arizona. The university's environmental program focuses on environmental policy and how human actions affect the natural dynamics.

air pollution; numerical modeling, data assimilation, and predictability; and climate change.

The department's teaching facilities on the main campus of the University of Utah include an audiovisual teaching laboratory also serving as a classroom, research equipment, and computing resource. A mountain meteorology laboratory is located about a mile away from the primary meteorology department.

Active research currently being done within the department by faculty members and funded by governmental and private organizations includes active research in clouds, aerosols, and climate; numerical weather prediction; mountain weather and climate; tropical convection and storms; and climate variability and change.

The Department of Geology and Geophysics offers both undergraduate and graduate degree programs. Undergraduates may declare majors in geoscience (with emphases in geology, geophysics, or environmental geoscience), geological engineering, or Earth science teaching. The four graduate programs are in geology, geophysics, geological engineering, and environmental engineering.

The Earth sciences overlap and integrate with a broadening range of disciplines. With a specific understanding of specialized processes and of fitting them into the broader system framework, students and researchers both gain a better understanding of earth science systems and the internal functional processes of geophysics, geobiology, and geochemistry. A sampling of projects and themes of special interest to those studying climate change includes environmental geology, paleoclimatology, geothermics, marine geology, and groundwater and surfacewater hydrology. Recent investigations of special note undertaken by University of Utah faculty, students, and collaborators were borehole thermal gradients as long-term records of earth surface paleotemperatures and global warming and Antarctic drill holes as recorders of Tertiary paleoclimate change.

The Environmental Studies Program at the University of Utah focuses on environmental policy and exists not as a single field of study but as a curriculum to provide an understanding of environmental systems and how human actions affect the natural dynamics. Students have the option of choosing from three different perspectives: biology and natural sciences, humanities and aesthetics, or

human behavior/policy/decision making. In addition to academics, the University of Utah is home to the National Oceanic and Atmospheric Administration Cooperative Institute for Regional Prediction (CIRP). Established in 1996, and building on an applied research program started in 1991 for faculty and students to work in cooperation with local and regional forecasters, CIRP conducts research to improve regional forecasting and knowledge of western U.S. meteorology.

Benefits of this research to weather professionals, researchers, and the general public relate to improved access and quality of data related to weather prediction. Various programs within CIRP provide access to surface observations incorporated from numerous sources (surface stations around the United States), analyses of compiled data, specific information for the fire weather community, and extended forecasts. CIRP also provides workshops.

SEE ALSO: National Oceanic and Atmospheric Administration (NOAA); Paleoclimates; Utah.

BIBLIOGRAPHY. Cooperative Institute for Regional Prediction, www.met.utah.edu/jhorel/cirp; Department of Geology and Geophysics, www.earth.utah.edu; University of Utah, "About Environmental Studies at the University of Utah," www.envst.utah.edu/about.htm; University of Utah Meteorology Department, www.met.utah.edu.

LYN MICHAUD
INDEPENDENT SCHOLAR

University of Washington

THE UNIVERSITY OF Washington, located in Seattle, was founded in 1851. The university confers bachelor's, master's, doctoral, and professional degrees in a variety of academic and professional disciplines including atmospheric sciences and Earth and space sciences (includes the former Departments of Geology and Geophysics) and environmental sciences.

Founded in 1947, the Department of Atmospheric Sciences focuses on teaching excellence incorporating a wide range of topics including weather forecasting, global warming, air quality, mountain weather, marine weather, El Niño, the ozone hole, ice ages, and the weather of Mars. Students who choose atmospheric science majors are prepared for a range of career options (weather forecasting, environmental meteorology, broadcast weather, and a broad range of careers with government and private employers in meteorology or environmental sciences).

The teaching facilities include a map room (for viewing weather data including radar and satellite imagery and forecast models) and an instruments laboratory (for learning about various types of observational instruments and computer interfacing by taking local observations).

The department's internship program encourages students to explore a variety of career paths while gaining valuable experience in addition to academic credit and possible stipends. The internships range from National Weather Service Forecast Offices to television stations, U.S. Forest Service, the Northwest Avalanche Center, the Pacific Marine Environmental Lab, and other laboratories and businesses. The faculty has diverse expertise and research interests, with research opportunities available to students.

The Department of Earth and Space Sciences integrates geology, math, physics, biology, and chemistry into a program to further the understanding of Earth structure, processes, and history and solar system structure, processes, and histories. The faculty areas of research are aimed toward predicting the future conditions of Earth, studying the geologic record, observing conditions, and modeling the present state. A sampling of projects of special interest to those studying climate changes includes glaciology, Quaternary research, climate, and paleoclimate.

The Program on the Environment offers interdisciplinary environmental education and is a focal point for information exchange on environmental education and research opportunities. The Earth Initiative encourages innovative partnerships to address environmental and natural resource challenges. By focusing on problem-specific environmental issues in the Pacific Northwest and beyond, the initiative brings together faculty, students, and community partners to create collaborative research, teaching, and scholarship.

The Program on the Environment, established in 1997, exists not as a separate department but as a collaborative network linking all university departments in providing environmental education. The programs provide education in natural sciences; social sciences; law, policy, and management; and ethics, values, and culture.

In addition to academics, the University of Washington is home to the Climate Impacts Group (CIG), which integrates climate science with public policy. With a focus on the U.S. Pacific Northwest, the group performs research on the consequences of climate change and provides information to policy makers on preserving regional resources vulnerable to climate changes.

By bringing together various disciplines to examine past climate and stressors, to analyze patterns of and predictable climate variations, as well as to determine past climate change effects, CIG extrapolates possible future responses to climate change in the Pacific Northwest's vulnerable areas.

CIG makes recommendations for approaches of human-induced changes on natural resources and overall climate effect. By developing close connections and maintaining networks with government, private, and North American tribal groups, as well as the agencies in charge of water, forests, fisheries, and coastal resources, the group ensures that their information is useful, informative, and transformational to the daily activities of appropriate resource management.

The University of Washington is setting an example for public and private sectors. The president of the university joined other college presidents and chancellors around the country in taking a community leadership role and modeling ways to minimize global warming emissions and integrate sustainability into the curriculum and university environment, with membership in the Leadership Circle of the American College and University Presidents Climate Commitment. By doing so, the university is building on past successes in developing plans for reducing energy consumption with the Environmental Stewardship Advisory Committee and Policy on Environmental Stewardship adopted in 2004, joining the UPASS program (bus pass) to reduce single-occupancy vehicle trips to campus, and acting as a founding partner of the Seattle Climate Partnership. In addition, the university has pledged to reduce greenhouse gas emissions by 2012 to 7 percent below the levels in 1990.

SEE ALSO: National Oceanic and Atmospheric Administration (NOAA); Washington.

BIBLIOGRAPHY. University of Washington, "About the Climate Impacts Group," www.cses.washington.edu/cig; University of Washington, "Schools and Centers," www.washington.edu.

LYN MICHAUD
INDEPENDENT SCHOLAR

Upwelling, Coastal

COASTAL UPWELLING OCCURS when water along a coastline flows offshore and deeper water—usually relatively cool, rich in nutrients, and high in partial pressure of carbon dioxide—flows upward to fill its place. Upwelling areas are notable for their effect on carbon cycling, as upwelling not only brings dissolved inorganic carbon to the surface, where it is released into the atmosphere, but also stimulates phytoplankton blooms that further remove some of that carbon through photosynthesis; a small percentage of this bloom also sinks in the form of organic matter (organic carbon) to deep water and becomes buried in sediment, creating a long-term carbon sink. There is considerable interest among carbon-cycle scientists regarding the reciprocal interactions between upwelling systems and climate change. Although such upwelling can in principle occur along any coastline, marine or freshwater, some marine coastlines (e.g., Peru, the western United States, northwest Africa, and southwest Africa) are renowned for their annual upwelling events that are the source of major blooms of diatoms and dinoflagellates, which become the base for extensive marine food webs and coastal fishing industries.

In the past several decades, major research programs have developed around the influence of coastal upwelling ecosystems on ocean carbon cycling and atmospheric carbon dioxide, how natural climate change (such as glacial-interglacial cycles) has affected coastal upwelling and associated biological productivity over a range of timescales, and how human-induced climate change is affecting coastal upwelling rates and timing and the associated fisheries.

Carbon dioxide exchange between coastal surface water and the atmosphere varies considerably in time and space. Because the pattern is complicated and dynamic relative to the number of direct measurements, considerable uncertainty lingers regarding the net carbon flux through the system over the course of a year. In general, outgassing occurs near the coastline, where upwelled water outcrops at the surface. This water is often rich in carbon dioxide arising from the respiration of organisms ingesting organic matter that sank from the surface to deeper water (which may be the sea bottom along the continental shelf). As upwelled water moves from shore, phytoplankton bloom in response to dissolved nitrogen, phosphorus, and other nutrients and begin to use up some of the dissolved inorganic carbon, reducing the partial pressure of carbon dioxide.

Because this process occurs over a period of several weeks, the rate of uptake of dissolved inorganic carbon also changes through time, so that net outgassing will occur early in an upwelling event, gradually changing to net ingassing. Much of the phytoplankton is recycled in the surface layer, prolonging the bloom, but some of the nutrients and carbon escape the system through the fecal material of heterotrophs feeding on the phytoplankton. The nutrients of remineralized organic matter that sink may come to the surface in future years through upwelling, or the organic matter may sink below the depth of upwelled water into the deep sea or get buried in sediment. The latter two processes can take carbon out of the atmosphere for thousands or millions of years, respectively. Although these processes occur in other aquatic areas, enough of the global ocean carbon flux in a given year occurs through coastal upwelling zones to affect atmospheric carbon dioxide.

The strength and direction of surface winds that drive coastal upwelling vary over a broad spectrum of timescales. Changes in global heat retention through time affect the potential for temperature gradients that influence wind speed, and the distribution of land masses and topographic features such as mountains affect coastal shape, coastal currents, sea level and coastal profile, and atmospheric circulation patterns. Temperature and precipitation patterns and sealevel, among other variables, affect nutrient distribution in the oceans. All of these affect upwelling strength, biological productivity, carbon burial, and net effect on the global carbon cycle. Much research has been dedicated to understanding upwelling changes during glacial-interglacial cycles, tracking responses to changed wind speeds and to lowered sea level, and therefore steeper coastal profiles. Other research has examined how to predict occurrences of upwelling in, for example, the Mesozoic, under the assumption that upwelling is responsible for the accumulation of some petroleum deposits.

Both models and empirical observations of several coastal upwelling areas, such as off the coast of California and northwest Africa, suggest that atmospheric warming is leading to greater rates of upwelling. This increase is driven by a greater land-ocean temperature gradient and therefore greater wind speeds. This can lead both to greater outgassing of carbon dioxide (if not balanced by increased productivity) and loss of certain fish that cannot maintain their population position as a result of higher offshore current velocities.

SEE ALSO: Benguala Current; Oceanic Changes; Peruvian Current; Upwelling, Equatorial.

BIBLIOGRAPHY. John A. Barth, et al., "Delayed Upwelling Alters Nearshore Coastal Ocean Ecosystems in the Northern California Current," *Proceedings of the National Academy of Sciences* (v.104/10, 2006); F.P. Chavez and T. Takahashi, "Coastal Oceans," *The First State of the Carbon Cycle Report [SOCCR]* (November 2007); H.V. McGregor, et al., "Rapid 20th-Century Increase in Coastal Upwelling off Northwest Africa," *Science* (v.315/5812, 2007).

ROBERT M. ROSS
PALEONTOLOGICAL RESEARCH INSTITUTE

Upwelling Equatorial

UPWELLING EQUATORIAL (UE) is upward water's motion in the upper layer of the Equatorial Ocean and occurs when a persistent easterly wind is blowing over the equatorial zone. Maximum upward velocity in the UE occurs just at the equator.

UE is a result of the permanent divergence of a westward surface south equatorial current in the narrow equator vicinity, forced by the southeast trade wind. Divergence of westward current at the equator is caused by the change of sign of the Coriolis force between the Northern and Southern hemispheres. As a consequence of divergence, the upper thermocline becomes shallower at the equator. Strong permanent equatorial divergence also causes an intense entrainment of more cold water from the thermocline into the upper mixed layer. This leads to cooling of the upper mixed layer. As a result, the sea surface temperature is about 1.8 degrees F (1 degree C) lower in the equator vicinity than in the interior Equatorial Ocean outside of it.

Pure UE occurs in the narrow vicinity of the equator, just within the divergent zone. Because of the slope of equatorial thermocline in a zonal direction (the thermocline is deeper in the western equatorial Atlantic and Pacific oceans than in the eastern) and the generation of coastal upwelling in the eastern Equatorial Oceans, UE manifestation, as relatively cold surface water, is more pronounced in the upper layer of the eastern Equatorial Oceans. Therefore, the cooler sea surface water looks like a long and thin equatorial tongue spreading from the eastern Equatorial Oceans. There is also quite high biological activity in this relatively cold tongue.

The thickness of the UE is restricted by the upper boundary of the equatorial undercurrent because the eastward current is accompanied by equatorial convergence and, hence, downward water's motion. That is why this thickness varies from about 328 to 656 ft. or 100 or 200 m. (in the western Equatorial Atlantic or Pacific Ocean, respectively) to 33 to 66 ft. or 10 to 20 m. (in the eastern Equatorial Atlantic and the Pacific Ocean).

UE is a quite persistent phenomenon in the Atlantic and Pacific oceans because the westward surface south equatorial current occurs there in the equator's vicinity almost throughout the entire year. However, UE intensity varies from season to season and from year to year. Seasonally, it is at a maximum in the Equatorial Atlantic and Pacific, when the south equatorial current intensifies, following the seasonal cycle of the southeast trade wind (with some delay that does not typically exceed a month); that is, in boreal late summer to early fall. Interannular variations of UE are mostly to the result of the El Niño/La Niña phenomena, especially in the Pacific Ocean. Just before El Niño developing (i.e., an anomalous warming of the upper layer in the Equatorial Pacific), the southeast trade wind dramatically weakens, and UE is over. In contrast, during La Niña (a cold episode in

Cold-water upwelling in the Gulf of Tehuantepec: This image of the Isthmus of Tehuantepec in Mexico shows sea surface temperatures observed by the Moderate Resolution Imaging Spectroradiometer (MODIS) on NASA's Aqua satellite.

the Equatorial Pacific Ocean), UE is strongly developed as a result of an anomalous intensification of the southeast trade wind and, hence, the south equatorial current. Interannular variability of the UE in the Equatorial Atlantic follows Pacific variability with some delay, which is typically not more than a few months. However, the magnitude of interannular UE variations in the Atlantic Ocean is not as large as in the Pacific Ocean. A seasonal cycle prevails in the Equatorial Atlantic, where the magnitude of the seasonal UE variations is two to three times bigger than the interannular ones.

In the Indian Ocean, UE (as a persistent phenomenon) occurs only in boreal winter, when the northeast monsoon has been developing. The UE is most pronounced in the western part of this basin. Seasonal UE variability is at maximum in the Indian Ocean. Interannular UE variability in the Indian Ocean is controlled by the Indo-Ocean Dipole, which is an inherent Indo-Ocean mode interrelated with the Pacific interannular variability (i.e., El Niño/La Niña phenomena), as can be seen in the recent results of Alexander Polonsky and coauthors and of Swadhin Behera and Toshio Yamagata.

Low-frequency (decade-to-decade) variability of the southeast trade wind or the northeast monsoon generates quasi-equilibrium UE variations. A more (or less) intense southeast trade wind and northeast monsoon leads to more (or less) intense UE.

SEE ALSO: Atlantic Ocean; Equatorial Undercurrent; Indian Ocean; Mixed Layer; Pacific Ocean; Thermocline; Trade Winds; Upwelling, Coastal.

BIBLIOGRAPHY. Swadhin Behera and Toshio Yamagata, "Influence of the Indian Ocean Dipole on the Southern Oscillation," *Journal of the Meterological Society of Japan* (v.81/1, 2003); Eric Kraus, ed., *Modelling and Prediction of the Upper Layers of the Ocean, Proceedings of a NATO Advanced Study Institute* (Pergamon Press, 1977); Alexander Polonsky, Gary Meyers, and Anton Torbinsky, "Interannual Variability of Heat Content of Upper Equatorial Layer in the Indian Ocean and Indo-Ocean Dipole," *Physical Oceanography* (v.21/1, 2007).

ALEXANDER BORIS POLONSKY
MARINE HYDROPHYSICAL INSTITUTE, SEVASTOPOL

Uruguay

URUGUAY IS LOCATED in southern South America, making a small wedge between Argentina and Brazil. Most of the country is covered by rolling grasslands, though it is crossed by several river systems and has a coastline on the Atlantic Ocean. Uruguay has no mountain ranges to buffer it from weather systems, making it susceptible to rapid weather changes. Droughts and periodic flooding are common. Climate change is expected to have some initial benefits for the livestock industry, but its long-term effects are unclear. Current climate models predict that Uruguay will see temperature increases of 2 degrees F (1.1 degrees C) by 2050 and 3.4 degrees F (1.9 degrees C) by 2100. Some models also indicate that there will be increased precipitation in both the summer and winter seasons, although there is some disagreement in these models.

Most of Uruguay is covered with evergreen grasslands perfect for the raising of cattle and sheep, which is a major part of the national economy. The grasslands may at first benefit from increased temperatures and higher atmospheric concentrations of CO_2, but it is not clear at which point the positive becomes a negative.

Sea-level rise is a concern along the Atlantic coastline. The capital city of Montevideo, home to about 45 percent of the population, sits on the coast, and much of the country's industrial infrastructure is concentrated around Montevideo Bay. A 1997 study predicted a rise in sea level of between 1.6–3.2 ft. (0.5–1.0 m.) by 2100. This would destroy much of the high-value real estate along the coast and severely damage the sewage and water purification systems of the city.

Uruguay has a population of just 3.5 million people. It has low population density, relatively high per capita income, and a well-developed economy. It is not a significant contributor to global carbon emissions; in fact, it is one of only a handful of countries that is carbon neutral, removing as much carbon from the atmosphere as it releases. In 1998, CO_2 emissions were about 1,800 metric tons per capita. Ninety-two percent of emissions come from liquid fuel use, and 8 percent from cement manufacture. Uruguay was one of the first countries to submit a greenhouse gas inventory to the United Nations Framework Convention on Climate Change, of which it is a signatory, and it has made regular reports on its progress.

Uruguay's government has moved aggressively to institute sustainable practices since the 1980s. In 1982, it instituted a law on soil management that has led to the sequestration of 1.8 million metric tons of carbon annually for the past 20 years; the Forest Protection Act of 1987 has increased the size of forest plantations from 200 to 6,500 sq. km. (77 to 2,510 sq. mi.), and the country has had a cumulative net carbon sequestration of 27.4 million metric tons.

SEE ALSO: Atlantic Ocean; Carbon Dioxide; Climate Change, Effects; Drought; Floods; Global Warming.

BIBLIOGRAPHY. Mario Bidegain and Daniel Panario, "Climate Change Effects on Grasslands in Uruguay," *Climate Research* (v.9, 1997); "Development and Climate Change in Uruguay: Focus on Coastal Zones, Agriculture and Forestry," www.oecd.org/dataoecd/42/7/32427988.pdf (cited November 1, 2007).

HEATHER K. MICHON
INDEPENDENT SCHOLAR

U.S. Global Change Research Program

THE U.S. Global Change Research Program (USGCRP) supports research on the interactions of natural and human-induced changes in the global environment and their effects on society. The USGCRP began as a presidential initiative in 1989 and was codified by Congress in the Global Change Research Act of 1990, which mandates development of a coordinated interagency research program. Participants in the USGCRP include the Agency for International Development, Department of Agriculture, Department of Commerce, National Oceanic and Atmospheric Administration, Department of Defense, Department of Energy, Department of Health and Human Services, National Institutes of Health, Department of State, Department of Transportation, Department of the Interior, U.S. Geological Survey, Environmental Protection Agency, National Aeronautics and Space Administration, National Science Foundation, and the Smithsonian Institution.

The Office of Science and Technology Policy, the Office of Management and Budget, and the Council on Environmental Quality supervise the activities on behalf of the Executive Office of the President. Since its establishment, the USGCRP has funded research and activities, together with several other national and international science programs, to document important aspects of the sources and lifetimes of greenhouse gases. The program has also established extensive space-based systems for global monitoring of climate and ecosystem. Its researchers have started to analyze the complex issues of various aerosol species that have a significant effect on the climate. They have also furthered understanding of the global water and carbon cycles and made major progress in computer modeling of the global climate.

The USGCRP carries out research in several focus areas. The research in atmospheric composition focuses on how human activities and natural phenomena affect the atmosphere and on how those changes relate to important issues such as climate change and ozone layer depletion. The research in this area aims to develop a framework for observation that will provide decisionmakers with sound scientific information both in the United States and abroad. The research

carried out in this area has contributed to the approval of laws and international treaties that protect the national and global environment. In addition, it has shown connections between global change and ozone depletion, as well as air-quality degradation at local, regional, and global levels. Scientists have been able to improve predictive models on climate change by taking into account the interconnectedness of these factors. This area of study has also tackled the important problem of atmospheric aerosols (particulate matter). Similar to greenhouse gases, aerosols have increased greatly in their atmospheric concentration since the Industrial Revolution and cause changes to the energy balance of the planet. Unlike greenhouse gases, however, aerosols can have either a warming or a cooling influence on the climate.

Since 2002, the Global Change Research Program and the administration's U.S. Climate Change Research Initiative (CCRI) established the Climate Change Science Program (CCSP) to allow the United States and the global community with scientific knowledge to manage risks and opportunities of change in the climate and the related environmental systems. Every year the CCSP produces the report "Our Changing Planet," which supplements the president's budget for that fiscal year. The preview of the report for year 2008 unmistakably states that "climate research conducted over the past several years indicates that most of the global warming experienced in the past few decades is very likely due to the observed increase in greenhouse gas concentrations from human activities. Research also indicates that the human influence on the climate system is expected to increase." Because of these findings, the report recommends that society should be "equipped with the best possible knowledge of climate variability and change so that we may exercise responsible stewardship for the environment, lessen the potential for negative climate impacts, and take advantage of positive opportunities where they exist."

The scientists working in climate variability and change have made significant progress in identifying the causes of climate change. USGCRP projects in this area have also attempted to incorporate this new knowledge into models for the prediction of future climate variability and for the exploration of the effects of human activities on climate. A new generation of climate models improved representations of physical processes, as well as increased resolution. Despite these improvements, there are still significant uncertainties associated with certain aspects of climate models.

Research on the global carbon cycle has attempted to quantify the extent of the dynamic reservoirs and fluxes of carbon within the Earth system. Researchers have also tried to predict how carbon cycling might change and be managed in future years. This research should supply guidelines on how to achieve an appropriate balance of risk, cost, and benefit.

The Global Water Cycle was selected as a focus area in USGCRP research, as the cycle plays a critical role in the functioning of the Earth system, and an incorrect understanding of such a cycle is one of the main sources of uncertainty in climate prediction. The water cycle includes all the complex physical, chemical, and biological processes that are necessary for ecosystems and that influence climate. The research in this area aims at developing responses to the consequences of water cycle variability. The changing ecosystems area is motivated by the awareness that global change can affect the structure and functioning of ecosystems in complex ways. The role of ecosystems research is to assess the potential effects of global change on ecosystems to help society respond effectively to such change. Research focuses on modifications in ecosystem structure and functioning and on potential alterations in the frequency and intensity of climate-related disturbances that may have significant effects on society.

The area of land use and land cover change acknowledges that land use and land cover are connected to climate. They can be the determining factors in the exchange of greenhouse gases between the land surface and the atmosphere, the radiation balance of the land surface, and the exchange of sensible heat between the land surface and the atmosphere. Researchers in this area analyze how changes in land use and land cover contribute to climate change and variability.

The research in human contributions and responses is based on the assumption that decision makers and other interested citizens need reliable scientific data to make decisions and actions that address the risks of and opportunities for changes in climate and related systems. The results of these USGCRP researchers are intended to shape public debates about climate-related issues. They should point at ways to reduce climate change and to adapt to climate variability.

In addition to these focus areas, the USGCRP also has a series of cross-cutting activities. The Observing and Monitoring the Climate System interagency working group develops research in the planning and operation of observing systems, including several new Earth-observing satellites, suborbital systems, surface networks, reference sites, and process studies. These can supply reliable data on the Earth's climate system and can be helpful in the prevention of natural disasters. USGCRP sections and researchers are all involved in communications initiatives to improve public understanding of climate change research and to make scientific findings more accessible to different audiences. Finally, the interagency working group of International Research Cooperation works with major international scientific organizations on behalf of the U.S. government and the scientific community.

SEE ALSO: Carbon Dioxide; Climate Change, Effects; Department of Defense, U.S.; Department of Energy, U.S.; National Aeronautics And Space Administration (NASA); National Oceanic and Atmospheric Administration (NOAA).

BIBLIOGRAPHY. Preview of Our Changing Planet Fiscal Year 2008, www.usgcrp.gov/usgcrp/Library/ocp2008preview/OCP08preview-intro.htm; U.S. Global Change Research Program, www.usgcrp.gov.

Luca Prono
University of Nottingham

Utah

THE EARTH'S CLIMATE has changed since the Industrial Revolution, and there is substantial evidence that man is influencing the change through greenhouse gas production. Utah's contribution to global climate change is significant and comes mostly from carbon dioxide (CO_2) emissions from electricity production, mainly from coal consumption in power plants. Other leading contributors to Utah's greenhouse gas emissions are transportation, oil and gas production, and emissions from livestock and microbial processes. By 1993, Utah's per capita greenhouse gas emissions were twice the global average, and Utah's total emissions were 1.2 percent of the global total. Baseline CO_2 emissions for Utah are projected to be 54 percent above 1990 levels by 2010 and 95 percent above 1990 levels by 2020, mainly as a result of its growing population.

The effect of climate change on Utah is likely to be significant as well. The western United States is warming faster than the global average, and Utah's average temperature for the last decade (1996–2006) was 2 degrees F (1 degrees C) warmer than the state's average over the past 100 years. This trend is causing earlier springs, less snowfall and more rain, earlier plant blooms, and a shorter frost season, although Utah's mountain snowpack has shown no long-term changing trends. Furthermore, climate models project that, even if greenhouse gas emissions stabilize, global average temperatures will continue to increase for centuries, and current and future emissions will affect Earth's temperature more over the next 100 years than in the past 100 years. The effect is projected to be even greater on Utah. For example, a climate model based on a 2.5-times atmospheric CO_2 increase by 2100 projects that Utah's mean annual temperature will increase by approximately 8 degrees F (4.4 degrees C), whereas the global average will increase by 5 degrees F (2.7 degrees C).

Higher temperatures will likely mean earlier springs and less water storage in Utah. Earlier Utah springs will cause rainfall instead of snowfall. This will replenish the reservoirs, but the mountain snowpacks will receive less snowfall and melt earlier in the season, and the increased temperatures will increase evaporation in the reservoirs. The combination of these factors could jeopardize the state's already-taxed water supply. Climate model projections show that there may be heavier episodes of precipitation, but there also will be longer time spans between them.

This weather pattern may induce a higher chance of flash flooding and a larger amount of runoff to the rivers and reservoirs, instead of steady groundwater replenishment through more frequent, but less intense, rainfall. In addition, there will likely be ecological impacts from warming rivers and lakes—animal and insect populations may migrate, and insect migration times may no longer be synchronized with the plants they pollinate. Utah's agricultural economy may also be affected. With a 2 to 5 degree

F increase, Utah's crops may thrive, but with the decreased water supply and infrequent rainfall, more frequent and severe droughts may counter the positive effects of a slight warming. Beyond an increase of a few degrees, crops may no longer endure the heat, and Utah's agriculture would need to change to accommodate the increased temperature.

GRASSROOT EFFORTS

Statewide, grassroots campaigns designed to educate the general public on the state of the science on climate change and its potential effect are neither common nor well received in Utah. However, over the past several years, state officials of this politically conservative state have become increasingly aware and proactive in climate discussions and policymaking. Recent policies have focused on energy efficiency in the form of corporate and personal tax credits for renewable energy, sales tax exemptions for renewable energy, and grants and loan programs for clean fuels and vehicle technology.

In May 2007, Utah was the sixth state to sign the Western Region Climate Action Pact, an agreement between states designed to set standards for reducing greenhouse gas emissions under a market-based program. In line with this new pact and Utah's focus on energy efficiency, Governor Jon Huntsman established a goal for increased energy efficiency in all state facilities of 20 percent by 2015. To aid in reaching this goal and in developing future policies, the governor formed the Blue Ribbon Advisory Council (BRAC). BRAC is a council of atmospheric, climate, and environmental scientists within Utah who advise the governor on the state of science on climate change. These scientists work alongside politicians and economists to evaluate the potential effect of policies designed to control and mitigate future climate change in Utah.

The scientific and political communities recognize that there is sound scientific evidence of human-induced climate change in Utah. This, in combination with the projections of future climate change and the concern over future water availability in the region, has spurred local and state leaders to take a close look at current climate initiatives and future policies. Although some of the general populace and politicians of Utah remain skeptical, many state officials are embracing advice from atmospheric scientists and climatologists and are making efforts to decrease Utah's footprint on global and regional climate change.

SEE ALSO: Carbon Dioxide; Climate Models; Greenhouse Gases.

BIBLIOGRAPHY. *Climate Change and Utah: The Scientific Consensus* (Blue Ribbon Advisory Council, 2007); *Current Utah Clean Energy Policies* (Utah Stakeholder Working Group on Climate Change, 2007); *Final Utah Greenhouse Gas Inventory and Reference Case Projections: 1990–2020* (Utah Center for Climate Strategies, 2007).

SUMMER RUPPER
RACHELLE HART
BRIGHAM YOUNG UNIVERSITY

Utah Climate Center

THE UTAH CLIMATE Center aims to disseminate climate data and information and to use expertise in atmospheric science to interpret climate information in an accurate and original way for the public. The mission includes the design of new products to meet the present and future needs of agriculture, natural resources, government, industry, tourism, and educational organizations in Utah and the intermountain region. It is part of Utah State University and has been recognized as a state center by the American Association of State Climatologists.

Much of climate information requested from the center comes from published records and computerized databases. Published records in print form extend through 2006 and are available at the National Climatic Data Center (NCDC). Many original historical data records for Utah for the 19th and early 20th centuries have been transferred to and archived in the Utah State University Merrill-Cazier Library. The Utah Climate Center serves as an official repository for both published climate data records and official publications from the NCDC, encompassing several decades, as part of an official agreement with the NCDC.

The weather and climate data provided by the Utah Climate Center are elaborated with the cooperation of the National Oceanic and Atmospheric Administration, the Federal Aviation Administration, and other federal, state, and local authorities. The Utah Climate Center strives to provide quality climate data.

In the past, the Utah Climate Center has provided information through paper via the postal service, via fax, or via electronic mail, but it now uses a GIS search facility. The Utah Climate Center gathers and archives climatic data from 22 networks throughout the state. It also monitors and compiles information from networks used by the Forest Service, the University of Utah, the Department of Agriculture, the Bureau of Land Management, the Natural Resource Conservation Service, and others. Data such as maximum and minimum temperature, precipitation, evaporation and evapotranspiration (a measure of water lost from the soil as a result of transfer to plants, rather than straight evaporation), and solar radiation are collected from various sites. The center collects 57,000 pieces of information from hundreds of locations each day. The earliest data at the UCC date back to the 1870s and 1880s, although most of the stations were put in place in the 1900s.

The center was initially timid in assessing global warming. In 1998, then-director Donald T. Jensen stated that any signals of global warming in Utah "have been lost in the noise of other temperature fluctuations. ... It's hard to find any real evidence here because temperatures here have always bounced up and down." However, with the passing years, the center has expressed views that are more attuned to the general opinion on global warming. Robert Gillies, the present director of the center, has pointed out that "the massive and growing scientific evidence has convinced the atmospheric science community that climate change is occurring, and is the result of human activities, specifically the release of greenhouse gases."

SEE ALSO: Carbon Dioxide; Climate Change, Effects; Greenhouse Gases; Utah.

BIBLIOGRAPHY. Robert Gillies, "Letter to the Editor: Statement on Climate Change From the Utah Climate Center," *Hard News Café,* www.hardnewscafe.usu.edu/archive/feb2007/020707_climateletter.html; Utah Climate Center, www.climate.usurf.usu.edu/.

LUCA PRONO
UNIVERSITY OF NOTTINGHAM

Uzbekistan

UZBEKISTAN IS LOCATED in central Asia, south of Kazakhstan. It is one of the world's two doubly-landlocked countries, being totally surrounded by other landlocked countries. The terrain is mostly rolling desert, with about 10 percent of the land lying within fertile river valleys. Poor conservation practices during the Soviet era had already caused enormous environmental damage, and many believe that this damage will exacerbate the effects of climate change in coming years.

The Aral Sea, which Uzbekistan shares with neighboring Kazakhstan, is one of modern history's great environmental disasters. The inland sea is over 5 million years old and used to be the world's fourth largest lake; today it is the world's eighth largest and is shrinking fast. During the Soviet era, massive amounts of water were diverted to irrigate crops, both from the Aral itself and from its feeder rivers. At the same time, it was used as a dumping ground for pesticides, raw sewage, and even nuclear waste.

Over the last two decades, 90 percent of the lake's source flow has dried up. Its surface area has decreased by 50–60 percent, and it has lost 80 percent of its volume. Salinity has increased from 10 grams per liter in 1960 to over 45 grams per liter in 2000. Many fish populations within the lake have long since died out. The lake has shrunk so much that it has actually split into separate parts, which had to be reconnected with a man-made channel.

Climate change will have an equally negative effect on the Aral. The glaciers that form an important part of the Aral Sea Basin shrank by 34 percent between 1960 and 2000. Runoff from mountain snow packs and annual rainfall has also decreased. The Amudarya River, a critical irrigation source, has decreased in flow to the point that it no longer reaches its outlet on the Aral.

Climate models expect the country's two main climate zones, the dry/tropical and moderate zones, to shift 93 to 124 mi. (150 to 200 km.) to the north by 2100. Air temperature is expected to increase by between 2.7–3.6 degrees F (1.5–2 degrees C). In the initial phases, the shift in climate zones and increased temperature, along with higher atmospheric concentrations of CO_2, might help the country's agricultural output, but with water sources already strained, that increase will be shortlived. Warmer temperatures and less water will damage the fertility of the grasslands, increase desertification, and reduce the country's livestock population.

Uzbekistan has a population of about 28 million people and is not a significant contributor to global carbon emissions. In 1998, per capita emissions were about 4,600 metric tons. About 80 percent of these emissions came from the release of gaseous fuels, 16 percent from liquid fuel sources, and 5 percent from cement manufacturing and solid fuel use.

The Uzbeki government has worked to study the potential effects of climate change and to raise public awareness of the problem. It is currently formulating plans to change damaging agricultural and industrial practices and move toward more sustainable growth.

SEE ALSO: Carbon Dioxide; Climate Models; Greenhouse Gases.

BIBLIOGRAPHY. Climate School, United Nations Environment Programme, "Uzbekistan and Climate Change Problem," www.climate.uz/en/section.scm?sectionId=6889&contentId=6944 (cited November 1, 2007); Embassy of the Republic of Uzbekistan in Israel, "Uzbek Experts Discuss Scenarios of Climate Change," www.embuzisr.mfa.uz/modules.php?op=modload&name=News&file=article&sid=577 (cited November 1, 2007); UNO Tashkent, "Climate Change Issues Gain Uzbek Media Attention," www.uzbekistan.unic.org/index.php?option=com_content&task=view&id=67&Itemid=73 (cited November 1, 2007).

HEATHER K. MICHON
INDEPENDENT SCHOLAR

Validation of Climate Models

THE CLIMATIC SYSTEM is constituted by four intimately interconnected subsystems—atmosphere, hydrosphere, cryosphere, and biosphere—which evolve under the action of macroscopic driving and modulating agents, such as solar heating, Earth's rotation, and gravitation. The climate system features many degrees of freedom, which make it complicated, and nonlinear interactions taking place on a vast range of time-space scales accompanying sensitive dependence on initial conditions, which makes it *complex*. The climate is defined as the set of statistical properties of the observable physical quantities of the climatic system.

The evaluation of the accuracy of numerical climate models and the definition of strategies for their improvement are crucial issues in the Earth system scientific community. On one hand, climate models of various degrees of complexity constitute tools of fundamental importance to reconstruct and project in the future the state of the planet and to test theories related to basic geophysical fluid dynamical properties of the atmosphere and of the ocean, as well as of the physical and chemical feedbacks within the various subdomains and between them. On the other hand, the outputs of climate models, and especially future

climate projections, are gaining an ever-increasing relevance in several fields, such as ecology, economics, engineering, energy, and architecture, as well as for the process of policymaking at a national and international level. Regarding influences at the societal level of climate-related finding, the effects of the fourth assessment report of the Intergovernmental Panel on Climate Change (IPCC4AR) are unprecedented.

The validation or auditing—overall evaluation of accuracy—of a set of climate models is a delicate operation that can be decomposed in two related, albeit distinct, procedures. The first procedure is the intercomparison, which aims at assessing the consistency of the models in the simulation of certain physical phenomena over a certain time frame. The second procedure is the verification, the goal of which is to compare the models' outputs with corresponding observed or reconstructed quantities. Difficulties emerge because we always have to deal with three different kinds of attractor: the attractor of the real climate system, its reconstruction from observations, and the attractors of the climate models. Depending on the timescale of interest and on the problem under investigation, the relevant active degrees of freedom (mathematically corresponding to the separation between the slow and fast manifolds) needing the most careful representation change dramatically. For relatively short timescales (below 10 years), the atmospheric degrees of freedom are active, whereas the other subsystems can be considered to be essentially frozen. For longer timescales (100–1,000 years), the ocean dominates the dynamics of climate, whereas for even longer timescales (over 5,000 years), the continental ice sheet changes are the most relevant factors of variability. Therefore, the scientific community has produced different families of climate models, spanning a hierarchical ladder of complexity, each formulated and structured for specifically tackling a class of problems.

COUPLED GLOBAL CLIMATE AND REGIONAL CLIMATE MODELS

Here, whereas most considerations are quite general, we mainly refer to the coupled global climate models (GCMs) and regional climate models (RCMs) currently used for the simulation of the present climate and for the analysis of the climate variability up to centennial scales. In these models, whereas the dynamical processes of the atmosphere and of the hydrosphere are represented within a wide framework of numerical discretization techniques applied to simplified versions of thermodynamics and Navier-Stokes equations in a rotating environment, the continental ice sheets are typically taken as fixed parameters of the system. In contrast, the so-called subscale processes, which cannot be explicitly represented within the resolution of the model, are taken care of through simplified parameterizations.

Several crucial processes, such as radiative transfer, atmospheric convection, microphysics of clouds, land-atmosphere fluxes, ice dynamics, and eddies and mixing in the ocean, as well as most of those controlling the biosphere evolution, undergo severe simplifications. With time, the formulation of the GCMs has developed through refinements to the spatial resolution, ameliorations of numerical schemes, and improved parameterizations, as well as through the inclusion of a larger and larger set of processes, such as aerosol chemistry and interactive vegetation, which are relevant for the representation of the system feedbacks and forcings.

In addition, limited-area climate modeling faces the mathematical complication of being a time-varying boundary conditions problem, as RCMs are nested into driving GCMs. Therefore, RCMs tend to be enslaved on timescales, depending on the size (and position) of

their domain, and in principle, the balances evaluated over the limited domain are constrained at all times. Therefore, climate reconstructions and projections performed with an RCM can critically depend on the driving GCM. Other more technical issues arise from the delicate process of matching the boundary conditions at the models' interface, where rather different spatial and time grids have to be joined.

Model results and approximate theories can be tested only against past observational data of nonuniform quality and quantity, essentially because of the space and the timescales involved. The available historical observations sometimes feature a relatively low degree of reciprocal synchronic coherence and individually present problems of diachronic coherence, as a result of changes in the strategies of data gathering with time, whereas proxy data, by definition, provide only semiquantitative information on the past state of the climate system. Extensive scientific effort is aimed at improving the quality and quantity of the climatic databases. In particular, the best guess of the atmospheric state of roughly the last 50 years has been reconstructed by two independent research initiatives, through the variational adaptation of model trajectories to all available meteorological observations, including the satellite-retrieved data, producing the so-called reanalyses.

Given all the above mentioned difficulties, as well as the impossibility, because of the entropic time arrow, of repeating world experiments, the validation of GCMs is not epistemologically trivial, as the Galilean paradigmatic approach cannot be followed. Validation has to be framed in probabilistic terms, and the choice of the observables of interest is crucial for determining robust metrics able to audit effectively the models. Recently, the detailed investigation of the behavior of GCMs has been greatly fostered and facilitated, as some research initiatives have been providing open access to standardized outputs of simulations performed within a well-defined set of scenarios. Relevant examples to be mentioned are the project PRUDENCE (RCMs) and the PCMDI/CMIP3 initiative (GCMs included in the IPCC4AR).

One aim—from the end-user's viewpoint—is checking how realistic the modeled fields of practical interest are, such as surface temperature, pressure, and precipitation. In these terms, current GCMs feature a good degree of consistency and realism when considering present climate simulation, and they basically agree on short-term climate projections down to seasonal averages on continental scales. When decreasing the spatial or the temporal scale of interest, the signal-to-noise ratio of climatic signals—both observative and model generated—typically decreases, so that the validation of GCMs in control runs and climate change simulations becomes more difficult, even if improvements are observed over time by state-of-the-art models. Statistical and dynamical—provided by nested RCMs—downscaling of climatological variables enlarges the scopes of model validation. In particular, RCMs provide a better outlook on small-scale and nonlinear processes, such as surface-atmosphere coupling, precipitation, and effects of climate change on the biosphere.

However, the above-mentioned quantities can hardly be considered climate state variables, whereas strategies for model improvement can benefit from understanding the differences in the representation of the climatic machine among GCMs. The comparison of the statistical properties of bulk quantities defin-

ing the climatic state, such as top-of-the-atmosphere energy fluxes, tropospheric average temperature, tropopause height, geopotential height at various pressure levels, tropospheric average specific humidity, and ocean water structure, allows the definition of global metrics that constitute robust diagnostic tools. Moreover, to capture the differences in the representation of specific physical processes, it is necessary to use specialized diagnostic tools—process-oriented metrics—as indexes for model reliability.

Examples of these metrics are major features of atmospheric variability, such as tropical and extratropical cyclones; detailed balances, such as water vapor convergence over continents or river basins; teleconnection patterns, such as El Niño–Southern Oscillation or Madden–Julian Oscillation; or oceanic features, such as the overturning circulation and the Antarctic current intensity. The latter approach may be especially helpful in clarifying the distinction between the performance of the models in reproducing diagnostic and prognostic variables. Even if improvement

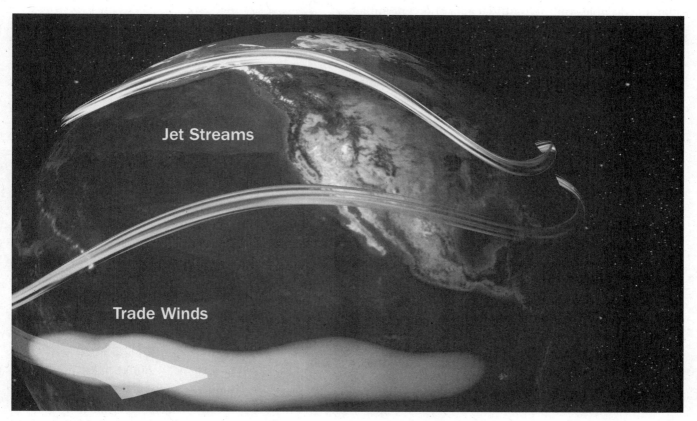

A NASA diagram of a strong El Niño striking surface waters in the Pacific Ocean. Warm water anomalies develop (indicated by shape at bottom) and westerly winds weaken, allowing the easterly winds to push the warm water against the South American coast.

is ongoing and promising, in these more fundamental metrics describing the climatic machine, current GCMs do not generally feature a comparable degree of consistency and realism at a quantitative level, and further investigations on basic physical and dynamical processes are needed.

Because the goal of a climate model is to reproduce the most relevant statistical properties of the climate system, the structural deficiencies, together with an unavoidably limited knowledge of the external forcings (uncertainties of the second kind) limit intrinsically the possibility of performing realistic simulations, especially affecting the possibility of representing abrupt climate change processes. The uncertainties of the initial conditions (uncertainties of the first kind), constituting, because of the chaotic nature of the system, probably the most critical issue in weather forecasting, are not in principle so troublesome—assuming that the system is ergodic—when considering the long-term behavior, where "long" is evaluated with respect to the longest timescale of the system. Nevertheless, to avoid transient behaviors, which may induce spurious trends in the large-scale climate variables on the multidecadal and centennial scales, it is crucial to initialize efficiently the slowest dynamical component of the GCMs, namely, the ocean. The validation of GCMs requires considering such uncertainties and devising strategies for limiting their influence when control run and, especially, climate change experiments are performed.

As for taking care of possible issues related to initial conditions, often an ensemble of simulations, where the same climate model is run under identical conditions from a slightly different initial state, allows a more detailed exploration of the phase space of the system, with a better sampling—on a finite time—of the attractor of the model. Some climate models have recently shown a rather encouraging ability to act as weather forecasting models, thus featuring encouraging local, in-phase space properties. Although such an ability gives evidence that short timescales' physical processes are well-represented, it says little on the overall performances when statistical properties are considered.

The structural deficiencies of a single GCM and the stability of its statistical properties can be addressed, at least empirically, by applying Monte Carlo techniques to generate an ensemble of simulations, each characterized by different values of some key uncertain parameters characterizing the global climatic properties, such as the climate sensitivity. Therefore, in this case, sampling is performed by considering attractors that are parametrically deformed.

To describe synthetically and comprehensively the outputs of a growing number of GCMs, recently it has become common to consider multimodel ensembles and to focus the attention of the ensemble mean and the ensemble spread of the models, taken respectively as the (possibly weighted) first two moments of the models outputs for the considered metric. Then information from rather different attractors is merged. Although this procedure surely has advantages, especially for GCMs intercomparison, such statistical estimators should not be interpreted in the standard way—with the mean approximating the truth and the standard deviation describing the uncertainty—because such a straightforward perspective relies on the (false) assumptions that the set is a probabilistic ensemble, formed by equivalent realizations of given process, and that the underlying probability distribution is unimodal.

SEE ALSO: Abrupt Climate Changes; Atmospheric Component of Models; Atmospheric General Circulation Models; Biogeochemical Feedbacks; Chaos Theory; Climate; Climate Models; Climate Sensitivity and Feedbacks; Climatic Data, Historical Records; Climatic Data, Proxy Records; Climatic Data, Reanalysis; Intergovernmental Panel on Climate Change (IPCC); Modeling of Paleoclimates; Ocean Component of Models.

BIBLIOGRAPHY. Isaac M. Held, "The Gap Between Simulation and Understanding in Climate Modeling," *Bulletin of the American Meteorological Society* (v.86/11, 2005); Intergovernmental Panel on Climate Change, *Climate Change 2007: The Physical Science Basis. Contribution of Working Group I to the Fourth Assessment Report of the Intergovernmental Panel on Climate Change* (Cambridge University Press, 2007); Valerio Lucarini, "Towards a Definition of Climate Science," *International Journal of Environment and Pollution* (v.18/5, 2002); Valerio Lucarini, Sandro Calmanti, Alessandro Dell'Aquila, Paolo M. Ruti, and Antonio Speranza, "Intercomparison of the Northern Hemisphere Winter Mid-Latitude Atmospheric Variability of the IPCC Models," *Climate Dynamics* (v.28/7-8, 2007); Josè P. Peixoto and Abraham H. Oort, *Physics of Climate* (American Institute of Physics, 1992); Barry Saltzmann, *Dynamic Paleoclimatology* (Academic Press, 2002).

Valerio Lucarini
University of Bologna

Vanuatu

VANUATU IS A country located in the South Pacific Ocean, composed of an archipelago of 83 volcanic islands with a population around 220,000. It has been a global leader in the development of renewable resources for sustained energy.

Vanuatu is governed by a parliamentary democracy, with the prime minister, elected by Parliament, being the leader of the government.

Its economy is primarily agricultural, with 80 percent of the population engaged in agricultural activities. Coconut is by far the most important cash crop, making up more than 35 percent of the country's exports, with petroleum fuels making up the majority of its imports. Interestingly, Vanuatu is a tax haven and international financial center. Despite this, Vanuatu is still a country in need of substantial foreign aid.

Vanuatu is also member of Alliance of Small Island Developing States, which is a progressive force in the United Nations for initiating climate change protocols.

In 1999, coral reef scientists reported massive coral reef death directly attributed to global warming. This, along with the high price of importing oil, led Vanuatu to commence switching from reliance on fossil fuels to becoming dependent on renewable energy economies based on hydrogen fuel technologies.

Thus, in September 2000 at the Eco-Asia conference, the chairman of the conference, Vanuatu Parliament member and former prime minister Maxime Carlot Korman challenged top scientists in the field to find realistic solutions for sustainable resources.

Korman proposed that his country would be the testing site for these solutions because of a variety of social and economic issues. First of all, the country had an electricity consumption of around 35 GWh (by comparison, in 1998, California used 570,000 GWh), all of which produced by fossil fuels, which is an amount that could feasibly be replaced by renewable resources. Second, this project would need to be funded by private investors, and Vanuatu's status as a tax haven provided incentive. Finally, the combination of reducing the country's dependence on foreign oil at the same time it was bringing in foreign investors would help the country reduce its need for outside aid.

The plan was to primarily use geothermal and solar energy to produce hydrogen to be used in hydrogen-cell-powered generators. In addition, this would be supplemented by directly converting wind and water currents into energy and storing them in batteries. The endpoint was to stop importing fossil fuels by 2010 and by 2020, to eliminate remaining internal combustion engines. Some privately funded projects currently being undertaken include Pacific Energy installing solar panels to supply energy to schools, Unelco using Vanuatu's natural resources and producing a coconut oil-diesel fuel blend, and the VANREPA research company installing wind turbines and solar panels to generate electricity for local schools and a solar still to desalinize ocean water.

Although it appears imminent that, as of 2007, Vanuatu appears to be heading toward failing to meet its objective of independence of foreign oil by 2010, the country has made tremendous strides in becoming a green nation. Despite this, Vanuatu was listed as the world's happiest nation in 2006 by the Happiest Planet Index, a New Economics Foundation index based on human well-being and environmental impact.

SEE ALSO: Climate Change, Effects; Global Warming.

BIBLIOGRAPHY. "An Index of Human Well-Being and Environmental Impact," www.happyplanetindex.org (cited August 2007); Seth Dunn, "Islands of Hope in Hydrogen: Clean Hydrogen Energy is Just Around the Corner... If Governments Have the Guts to Take on the Oil Barons," *New Internationalist* (December 1, 2002); Gavin Evans, "Alternative Fuel Grows on Trees: South Pacific Islands Rely on Coconuts to Cut Diesel Imports," *International Herald Tribune* (January 4, 2006); Jolanka K. Fisher and Timothy P. Duane, *Trends in Energy Consumption, Peak Demand and Generating Capacity in California and Western Grid 1977–2000* (University of California Energy Institute, 2001); "Presentation to EcoAsia 2000," www.vanuatu-government.gov.vu (cited September 2007).

Robert Klever
Independent Scholar

Venezuela

VENEZUELA IS A major oil-producing state and a founding member of the Organization of Petroleum Exporting Countries (OPEC). Its leaders resist global efforts to accelerate cuts in carbon dioxide emissions, and the country's low-cost oil supply to Latin America and the Caribbean countries may stall regional transition to alternative energy sources. Venezuela has been criticized for encouraging energy inefficiency with oil subsidies, but new state environmental programs promote conservation.

Venezuela has proven crude oil reserves of 80,012 million barrels, and the state-owned Petróleos de Venezuela (PDV) is one of the world's largest oil companies. Most of the nation's oil exports enter the United States. PDV's subsidiary Citgo refines the crude oil in Texas. In addition to Venezuela's petroleum trade

Angel Falls in Venezuela is the world's highest free-falling, freshwater waterfall, with a drop of 2,648 feet.

with the United States, its Chinese oil investments are growing. Venezuela recently signed energy agreements guaranteeing petroleum to many Latin America and Caribbean countries. These pacts include subsidized oil, an exchange of goods and services for oil, and interest-deferred financing for oil purchases.

There are many uncertainties about the potential effect of climate change on Venezuela. As a precautionary method, the development of agricultural varieties resistant to drought and adverse climate conditions has been recommended. Flooding is likely in other areas. In December 1999, Venezuela experienced its highest monthly rainfall in a century. Massive landslides and flooding led to the deaths of more than 30,000 people. The risk of increased mortality from diseases with mosquito vectors such as yellow fever increases after floods, and malaria has been documented to increase in the country's coastal regions after the onset of El Niño.

Although the country has been criticized for contributing to global warming, some conservation measures are in place. Venezuela ratified the UN Framework Convention on Climate Change in 1994 and the Kyoto Protocol in 2005. In 2006, President Hugo Chavez launched a reforestation program to plant 100 million trees. These trees will be intercropped with cacao and coffee to give farmers an incentive to abandon environmentally destructive farming methods. Chavez also launched an energy efficiency program that promotes improved light bulbs and natural gas, wind, and solar power. Venezuela's large reserves of natural gas remain largely untapped.

Venezuela is among the top 20 countries in terms of endemism, and more than 200 protected areas cover in excess of 70 percent of the nation. Its diverse climatic and biogeographical regions cover a range of elevations, and there are 1,740 mi. (2,800 km.) of coastline, including vast mangrove swamps and numerous islands. There has been a documented retreat of glaciers in the Sierra Nevada range, however, and the glacier on Pico Bolivar may completely disappear during the next decade.

A climate change mystery occurs in Venezuela: methane builds up over the country at night. Scientists that researched this peculiar phenomenon have recently suggested, amid some skepticism, that the methane is being released from certain plants growing in the savanna.

SEE ALSO: El Niño and La Niña; Floods; Oil, Consumption of; Oil, Production of.

BIBLIOGRPHY. Luis Jose Mata and Carlos Nobre, *Impacts, Vulnerability and Adaptation to Climate Change in Latin America* (UNFCCC, 2006); Petroleum of Venezuela, www.pdvsa.com/ (cited June 2007); Quirin Schiermeier, "The Methane Mystery," *Nature* (v. 442, August 2006).

MARY FINLEY-BROOK
UNIVERSITY OF RICHMOND

Vermont

KNOWN AS THE Green Mountain State, Vermont underwent widespread reforestation following farm abandonment in the mid-1800s. In addition to the existence of several land trusts, the Green Mountain Club has protected more than 55 mi. (88 km.) along a hikers' Long Trail. There is a strong state environmental movement, with local groups and chapters of national organizations. Vermont hosts one of the nation's leading environmental law and policy programs at the Vermont Law School. The state government has invested in energy efficiency and joined regional efforts to reduce greenhouse gas (GHG) emissions. Climate change and global warming concern Vermonters because of the economic revenue the state gains from nature tourism, particularly during the autumn foliage and winter skiing seasons.

In spite of some local variation, the northeastern United States has experienced a temperature increase of 1.8 degrees between 1899 and 2000. The average length of the annual growing season has been extended by eight days, and bloom dates have changed. Scientists have documented quicker thawing of lakes, known as ice-in and ice-out dates, as well as earlier runoff from mountains.

Data have suggested that maple production from the sugar maple (*Acer saccarum*), the Vermont state tree, may be vulnerable to climate shifts. Under several climate change models, this species may entirely shift out of the United States. Studies suggest that seven of 80 eastern tree species may decline by as much as 90 percent in the next century if temperature changes continue at current rates.

In terms of GHG emissions, Vermont's energy portfolio is one of the greenest in the nation. Vermont Yankee Nuclear Power Plant, online since 1973, provides approximately 35 percent of the state's energy requirements. Its license is scheduled to expire in 2012, but petitions are in place for renewal and expansion of production. There is some local opposition to both proposals.

Energy from Hydro-Quebec, an extensive series of dams across Quebec Province in Canada, provides a third of Vermont's energy needs. Although hydropower is a renewable energy source, the creation of the project's dams has been controversial, in that it caused flooding of indigenous lands. Research on dam building also suggests that methane may be released from the decomposition of flooded forests. Releases are expected to be considerably higher in tropical ecosystems, but more research is needed to determine the exact levels. Fossil fuels are also used in the creation and maintenance of dam structures.

Nearly one-third of Vermont's energy is created from a mixture of in-state renewable energy resources. The Cow Power program uses manure to create energy and provide income to dairy farmers. Efficiency Vermont is the nation's first statewide provider of energy efficiency services. The program is operated by an independent, nonprofit organization under contract to the Vermont Public Service Board. It provides technical advice, financial assistance, and design guidance to make homes, schools, and businesses more energy efficient. The program is funded by a charge on users' electric bills.

In 2006, Governor Jim Douglas announced a public-private partnership that will invest $20 million in energy efficiency. He also extended state energy efficiency programs. In addition to $8 million in weatherization grants available to low-income Vermonters each year, Douglas' program also created private capital to provide $20 million no- or low-interest loans to assist homeowners and small businesses to make buildings more fuel efficient.

Vermont has joined regional climate change mitigation initiatives. The state joined seven others in 2004 in a legal suit against several energy companies to require them to reduce carbon dioxide emissions and control GHG outputs. Vermont is part of the Regional Greenhouse Gas Initiative's cap-and-trade program for power plants, aimed at reducing carbon emissions. Six New England states and five eastern Canadian provinces adopted a joint Climate Change Action Plan to reduce GHG emissions.

SEE ALSO: Energy Efficiency; Energy, Renewable; Greenhouse Gases; Vermont Law School.

BIBLIOGRAPHY. Christa Farrand Case, "Climate Change Could Sour US Maple Sugaring," *Christian Science Monitor*, April 6, 2005; Vermont Office of the Governor, www. vermont.gov/governor/ (cited June 2007).

MARY FINLEY-BROOK
UNIVERSITY OF RICHMOND

Vermont Law School

VERMONT LAW SCHOOL'S (VLS's) environmental law program is ranked number one in the nation by *U.S. News & World Report*. Students at VLS participate in a variety of legal and environmental study programs, many of which focus on climate change and related policies. The Environmental Law Center at VLS includes several specialized institutes and clinics that are involved in various aspects of climate change research and litigation. At present, VLS is home to 612 students, participating in four different degree programs: Juris Doctor (JD), Master of Studies in Environmental Law (MSEL), Master of Laws in Environmental Law (LLM), and Master of Laws in American Legal Studies (LLM) for international students.

VLS is known not only for its academic commitment to reducing the effects of global climate change but also for putting this commitment into action in its day-to-day operations. The U.S. Green Building Council awarded Debevoise Hall, VLS's main administration and academic building, with a Leader in Energy Efficiency and Design Silver Certification. The renovated 1893 school building features an energy-efficient, compact fluorescent lighting system that is linked to sensors that turn off the lights when no one is present. A cross-flow heat exchanger and desiccant wheel provide temperature- and humidity-adjusted airflow year-round. Debevoise Hall and the adjacent classroom building Oakes Hall (constructed in 1998) both feature composting toilets and other water conservation measures. In Oakes Hall, these measures have decreased the projected water consumption rate from 15 gallons used per person, per day, to 15.5 gallons used per day, total, for the entire facility.

VLS contains several academic centers that support climate change research, litigation, and problem solving. The Institute for Energy and the Environment is VLS's premier research institution on energy policy and law. Professor Michael H. Dworkin, former chair of the Vermont Public Service Board (1999–2005), directs the institute. The institute provides conferences and produces scholarly publications on topics such as biofuels and related policy recommendations; net-metering legislation; energy efficiency and the need for demand-side regulation; and health and environmental costs of burning fossil fuels.

The Land Use Institute, directed by Professor L. Kinvin Wroth, seeks to address the legal and planning issues surrounding current forms of land use. This includes sustainable design and ecology planning, siting of new energy facilities, revisions to permit composition and process, and the scope of eminent domain.

The Climate Legacy Initiative, under the directorship of Professor Burns Weston, has been established to clarify and advocate the legal rights, including human rights, of present and future generations relative to the harms likely to result from global warming both in the United States and internationally.

VLS also offers an Environmental and Natural Resources Law Clinic, the *Vermont Journal of Environmental Law*, the Environmental Tax Policy Institute, and an environmental summer session. Because global climate change is one of the world's most expansive environmental challenges, each of these programs are working to address global warming–related issues and inspire future problem solvers. In all, VLS promotes a commitment to bettering the environment through education.

SEE ALSO: Education; Energy; Green Design; Policy, US; Regulation; Vermont.

BIBLIOGRAPHY. "America's Best Graduate Schools in 2008" *U.S. News and World Report*, April 2, 2007; Department of Energy Efficiency and Renewable Energy, Building Technologies Program, Building Database, www.eere.energy.

gov/buildings/database/energy.cfm?ProjectID=93; Vermont Law School, www.vermontlaw.edu.

ELLEN J. CRIVELLA
VERMONT LAW SCHOOL

Vienna Convention

THE VIENNA CONVENTION for the Protection of the Ozone Layer came into being in March 1985. The convention was developed initially out of recognition of the international concern over the loss of the ozone in the stratosphere. Subsequently, the Montreal Protocol in 1987 strengthened the convention, which was further adjusted and amended on June 29, 1990.

The concern over ozone was linked to attention to the observed phenomena of climate change. Eventually, the issue of the ozone appears to have been stabilized, and the issue of climate change evolved into a larger international concern.

The attention to the ozone issue relied on the scientific understanding of the effects of chlorofluorocarbons in the context of the atmosphere. As a result of the interest in the subject, scientific research invested in developing increasingly complex models of the atmosphere to more accurately address climate issues.

Over time, concern over the issue of the loss of ozone in the stratosphere declined, as international efforts appeared to have improved the condition of the stratospheric ozone layer. Ozone levels in the troposphere continue to be a contributor to local pollution.

As the concern over stratospheric ozone declined, the broader concern over climate change that was noted in the convention continued and assumed a proportionately larger role. The improved science used to address the ozone issues then also became valuable in addressing the more complex climate change matter.

The development of the convention and climate change regulations through the United Nations Environment Programme has been closely associated with the science and models of the atmosphere. The science was addressed in related conferences and workshops including the Villach Conference in 1985. Eventually, the convention led to the formulation of the UN Framework Convention on Climate Change.

The preamble to the convention indicates the generally accepted views of the international community. This preamble mentions the relationship between the environment, its modification, and the potentially harmful effects on human health.

The awareness noted in the preamble is an acknowledgement that human activity can and does modify the environment. It also permits the recognition of the association between modifying the environment and its effect on human health itself. This awareness then establishes the possibility of regulating human activity.

Human activity is also associated with relevant scientific and technical considerations. There is then a recognized need for research and systematic observations into the phenomena. The use of science and technology is an important part of the convention and its implementation.

The intended result of the convention is twofold. First, the twin concerns are to protect human health and the environment. The particular focus then is the concern about adverse effects to both humans and the environment, which result specifically from modifications of the stratospheric ozone layer.

The convention is not concerned with all modifications of the ozone layer but only those that have adverse effects on human health and the environment. In fact, any modification of the ozone layer that has a positive effect on human health and the environment would be favorable. Under the definitions in Article 1, "adverse effects" refer to changes in the physical environment or biota, including changes in climate. The relevant changes are those that have significant deleterious effects on human health, on ecosystems whether natural or managed, and on "materials useful to mankind."

To achieve the intended result, the preamble raises some significant considerations. The reference to principle 21 of the Declaration of the UN Conference on the Human Environment highlights the context for the convention, which includes the sovereignty of each country, also known as a state, over its own domain and resources. It also underlines the responsibility of each such state not to cause damage beyond its national jurisdiction or to other states.

In such circumstances, a state's performance can then be evaluated in the international context against its own sovereign jurisdiction and its effect beyond its boundaries.

The sovereignty of each state also necessarily involves the recognition of cooperation between states. It is through cooperation and action that states can then be brought within the scope of international law and the Charter of the United Nations.

The preamble discloses its relatively recent formulation when it includes nonsovereign entities in its discussions. It generally acknowledges international and national organizations, whose work and studies are pertinent to the scope of the convention. An example of such an organization noted in the convention is the World Plan of Action on the Ozone Layer of the UN Environment Programme.

Another significant feature of the convention is the attention to developing countries in the global effort. The preamble generally recognizes the circumstances and particular requirements of developing countries. This recognition eventually leads to the more complex form of differentiated contributions to the total human activities and their consequences.

The agency responsible for the Convention is the Ozone Secretariat of the UN Environment Programme. In 1988, the assessment panel process was initiated under the Montreal Protocol, Article 6, and four panels were established. These are the panels for scientific, environmental, technology, and economic assessments. In 1990, the Panels for Technical Assessment and for Economic Assessment were merged into the Technology and Economic Assessment Panel.

Under Article 6 of the convention, a Conference of the Parties was established. The Conference of the Parties to the Convention, in its decision VCV/2 of its fifth meeting, acknowledges and encourages the collaboration of the three assessment panels with other entities. These entities link the convention with other international entities involved in addressing climate change issues. These other entities named in this decision are the Intergovernmental Panel on Climate Change, the Subsidiary Body on Scientific, Technical and Technological Advice under the UN Framework Convention on Climate Change, the International Civil Aviation Organization, and the World Meteorological Organization.

These entities were in addition to those recognized in Article 6, which included the World Health Organization and the International Atomic Energy Agency. Within limits, it also allowed for any state not a party to this convention and for any body or agency, whether national or international, governmental or nongovernmental, to be represented at the Conference of the Parties by observers.

The convention is specifically referred to in the preamble of the UN Framework Convention on Climate Change, signed in 1992, which came into force in 1994.

SEE ALSO: Atmospheric Composition; Climate Change, Effects.

BIBLIOGRAPHY. Dale S. Bryk, "The Montreal Protocol and Recent Developments to Protect the Ozone Layer," *Harvard Environmental Law Review* (vol.15/1, 1991); W. Lang, H. Neuhold, and K. Zemanek, eds., *Environmental Protection and International Law* (, 1991); M. Paterson, *Global Warming and Global Politics* (Routledge, 1996); Social Learning Group, Massachusetts Institute of Technology, *Learning to Manage Global Environmental Risks, Volume 1: A Comparative History of Social Responses to Climate Change, Ozone Depletion, and Acid Rain* (MIT Press, 2001); Edith Brown Weiss, "International Law," *Encyclopedia of World Environmental History* (2003).

LESTER DE SOUZA
INDEPENDENT SCHOLAR

Vietnam

THIS SOUTHEAST ASIAN country, with a long coastline on the South China Sea, has a land area of 128,065 sq. mi. (331,689 sq. km.), with a population of 87,375,000 (2006 est.) and a population density of 655 people per sq. mi. (253 people per sq. km.). The largest city and the former capital of South Vietnam, Ho Chi Minh City (formerly Saigon) has a population of 3,525,300, making it the 49th largest in the world. Some 17 percent of the land in Vietnam is arable, with 4 percent being under permanent cultivation and 1 percent used as meadow or pasture. In addition, 30 percent of the country is forested, with an increasing logging industry operating in the country.

As there are three major rivers in Vietnam—the Red River, the Pearl River, and the Mekong—there has been a heavy investment in hydropower. In 1964, at a speech at Johns Hopkins University, U.S. President Lyndon B. Johnson offered a hydroelectric scheme for the Mekong River that was quickly dubbed the TVA for the Mekong. It was rejected by Ho Chi Minh, and the Vietnam War, which was escalating at that time, led to massive bombing of parts of the country, destruction of much of the infrastructure of both North Vietnam and South Vietnam, and the defoliation of some parts of the jungle. Following the end of the war in 1975 and the nation's reunification in the following year, the country was desperately poor. In spite of this, heavy government investment in hydropower then results in it now accounting for 59.3 percent of the country's electricity production, with the rest coming from fossil fuels. Electricity production accounts for 25 percent of the carbon dioxide emissions in the country, with 38 percent from transportation and 30 percent from manufacturing and construction.

During the 1990s, Vietnam became a major tourist destination, and this has led to a significant rise in the rate of per capita carbon dioxide emissions, from 0.3 metric tons per person in 1990 to 1.18 metric tons per person by 2004. In 1998, 48 percent of carbon dioxide emissions in the country was from liquid fuels, 37 percent from solid fuels, 8 percent from gas flaring, and 7 percent from cement manufacturing.

The Vietnamese government of Vo Chi Cong took part in the United Nations Framework Convention on Climate Change, signed in Rio de Janeiro in May 1992. The Vietnamese government signed the Kyoto Protocol to the UN Framework Convention on Climate Change on December 3, 1998, which was ratified on September 25, 2002, with it entering into force on February 16, 2005.

SEE ALSO: Climate Change, Effects; Deforestation; Tourism.

BIBLIOGRAPHY. Jos Frijns, Phung Thuy Phuong, and Arthur P.J. Mol, "Ecological Modernisation Theory and Industrialising Economies: The Case of Viet Nam," (*Environmental Politics* (v.9/1, Spring 2000); Michael J. McRae, "Tam Dao—Sanctuary Under Siege," *National Geographic* (v.195/6, June 1999); Thomas O'Neill, "Mekong River," *National Geographic* (v.183/2, February 1993); World Resources Institute, "Vietnam—Climate and Atmosphere," www.earthtrends.wri.org (cited October 2007).

ROBIN S. CORFIELD
INDEPENDENT SCHOLAR

Villach Conference

THE CONFERENCE HELD in Villach, Austria, from October 9 to 15, 1985, was the result of the continuing work of several international entities. In some views, the background to the conference and the starting point for internationally cohesive attempts to understand the issues related to the stratospheric ozone layer depletion and climate change have been traced to the UN Conference on Human Development in Stockholm in 1972. The technical, scientific understanding of the possibility of human-induced effects on the ozone layer and of climate change developed in the diplomatic context as complementary efforts linked through a reliance on common research initiatives.

A World Climate Conference held in Geneva in 1979 continued the efforts of the 1972 UN Conference on Human Development and led to the World Climate Programme. The World Meteorological Organization, the UN Environment Programme, and the International Council of Scientific Unions collaborated to hold a series of workshops that have come to be known as the Villach Conference.

The UN Environment Programme Ad Hoc Working Group of Legal and Technical Experts for the Elaboration of a Global Framework Convention for the Protection of the Ozone Layer was established by decision of the UN Environment Programme Governing Council decision 12/14, part I. The first part of the fourth session of the working group was held at the Palais des Nations, Geneva, from October 22 to 26, 1984.

The Working Group was also informed of the collaborative effort of the UN Environment Programme, the World Meteorological Organization, and the International Council of Scientific Unions to hold a major scientific conference to assess the carbon dioxide/ozone and climate question in October 1985 at Villach, Austria, with the support of the government of Austria.

Located at the intersection of the ozone and climate change issues, the Villach Conference became immediately significant to the Working Groups of both UN endeavors and to the overall development of international initiatives to address atmospheric environmental issues. A workshop on chlorofluorocarbons at Villach initiated the processes that led to a protocol to the Vienna Convention, which had been signed earlier that year in March 1985. The same conference augmented the UN Environment Programme's role in determining the assessment of the greenhouse gas/climate issue. In this regard, the October 1985 Villach Conference was significant.

The participants at the Villach Conference were a small group of environmental scientists and research managers in nongovernmental organizations. The dominant contribution of these participants was their expertise in climate modeling. With accreditation from recognized scientific institutions such as the International Institute for Applied Systems Analysis and from Harvard University, the results of this series of workshops carried considerable weight internationally. Even before the conference, a majority of the scientists who attended had publicly advocated what they considered an imperative to respond to a perceived threat to planetary climate stability within a strategy that was consistent with sustainable development, which was eventually incorporated in the Brundtland Report. Through affiliations with nongovernmental institutions, the scientists had improved modeling techniques that they relied on and that led them to generally agree with the conclusions and recommendations of the conference.

It was at the Villach Conference that the scientific community present arrived at an initial consensus as to the technical features of greenhouse gases, the depletion of the stratospheric ozone layer, and the chemical reactions that were relevant.

The general conclusion of the scientists and participants at the conference was that they could anticipate an unprecedented rise of global mean temperature in the first half of the 21st century. The scale and actual increase in global mean temperature was expected to be higher than any rise in the record of the planet's history. To mitigate the perceived events, the participants recommended a strategy that relied on technical and science-based research to establish target emission or concentration limits. In doing so, they sought to regulate the rate of change of global mean temperature within specific parameters.

Another result of the work by the participants at Villach was the establishment of the Advisory Group on Greenhouse Gases in 1986. This group was established to ensure continued academic and public interest in the effect of rising levels of greenhouse gases on the ozone layer and on climate change. The Advisory Group on Greenhouse Gases was jointly sponsored by the World Meteorological Organization, the UN Environment Programme, and the International Council of Science Unions.

The Brundtland Report, published in 1987, popularized sustainable development and advocated the development of a low-energy economy. This publication included a section on energy authored by Professor Gordon Goodman, who was by then a prominent member of the Advisory Group on Greenhouse Gases.

Work by the Advisory Group on Greenhouse Gases subsequently led to the 1988 Conference on the Changing Atmosphere: Implications for Global Security in Toronto, Canada, which called for 20 percent reductions in CO_2 emissions. The Advisory Group followed that up by preparing the Meeting of Legal and Policy Experts in February 1989 in Ottawa, which recommended an umbrella consortium to protect the atmosphere. The 1988 Toronto Conference then led to the establishment of the Intergovernmental Panel on Climate Change, with the mandate for continuing international research on climate change phenomena.

SEE ALSO: Atmospheric Composition; Climate Change, Effects.

BIBLIOGRAPHY. Dean Edwin Abrahamson, ed., *The Challenge of Global Warming* (Island Press, 1989); Matthew Paterson, *Global Warming and Global Politics* (Routledge, 1996); T. Skodvin, *Structure and Agent in the Scientific Diplomacy of Climate Change: An Empirical Case Study of Science-Policy Interaction in the Intergovernmental Panel on Climate Change* (Kluwer, 2000).

LESTER DE SOUZA
INDEPENDENT SCHOLAR

Virginia

THE COMMONWEALTH OF Virginia is a southern Atlantic state on the eastern seaboard of the United States of America. Virginia is the 35th largest state in the United States, with an area of over 42,000 sq.

mi. (108,780 sq. km.) and an estimated population of about 7.6 million. Virginia's state capital is Richmond, located in central Virginia, and the largest city is Virginia Beach, on the coast. Cities in the commonwealth are considered independent and function in the same manner as counties. Virginia's geography is divided into six different regions, from the tidewater along Chesapeake Bay on the eastern coast, to the Appalachian Plateau in the West. Virginia's climate is considered mild for the United States but can be quite variable because of significant topographic differences across the state and the influence of the Atlantic Ocean. The regions east of the Blue Ridge, as well as the southern part of the Shenandoah Valley, are considered a humid subtropical climate (Koppen climate classification *Cfa*). In the mountainous areas between the Allegheny and Blue Ridge mountains in the west, the climate becomes humid continental (Koppen climate classification *Dfa*).

Virginia was founded on May 13, 1607, at Jamestown, the first permanent English settlement in North America, and is one of the original 13 colonies of the American Revolution. Virginia joined the Union on June 25, 1788, and was the 10th state to ratify the U.S. Constitution. More U.S. presidents (eight) have come from Virginia than from any other state. The current and 70th governor of Virginia is Democrat Tim Kaine, who assumed office in January 2006. Virginia's geography varies considerably across the state, including the Shenandoah Valley, bordered by the Allegheny and Blue Ridge mountains in the west; the rolling hills of the central piedmont; and the flat eastern coastal plain extending from the piedmont to the Atlantic Ocean. Elevations in Virginia range from sea level to nearly 6,000 ft. at the summit of Mt. Rogers.

The state is divided into six distinct geographical regions: the ridge and valley between the Blue Ridge Mountains to the east and the Appalachian and Allegheny plateaus to the west; the Shenandoah Valley, located within the ridge and valley; the Blue Ridge Mountains between the ridge and valley to the west and the piedmont region to the east; the foothills between the piedmont and the Blue Ridge Mountains; the piedmont between the Blue Ridge Mountains to the west and the tidewater region to the east; and, the tidewater between the fall line to the west and the Atlantic east, including the eastern shore.

The southern area of Virginia's Shenandoah Valley is considered a humid subtropical climate.

Virginia has five distinct climate regions: tidewater, piedmont, northern Virginia, western mountains, and southwestern mountains. Climate varies significantly across the state because of differences in topography, the influence of the Atlantic Ocean and the Gulf Stream in the east, and the complex pattern of rivers and streams. Much of the precipitation that Virginia receives results from storms associated with low-pressure (cyclone) frontal systems—warm and cold fronts that typically move from west to east across the state, curving northwestward as they reach the Atlantic Ocean. Virginia also experiences occasional tropical storm activity, most often in early August and September, and typically through the venue of the mouth of the Chesapeake Bay. The tropical storms can provide significant amounts of precipitation. In recent years, development trends in northern Virginia extending out from Washington, D.C., have created an urban heat-island effect characteristic of other large urban areas.

In Virginia, emissions of carbon dioxide, one of the primary greenhouse gases that alter the Earth's climate, have risen by over 30 percent since 1990, driven by economic growth, development patterns, and increased transportation trends. Human-induced climate change is expected to have many effects on Virginia's weather, wildlife, food production, and water supplies. Sea level in the mid-Atlantic region is estimated to rise several inches in the coming decades, threatening low-lying areas and coastal developments. Changes in precipitation and temperature regimes have the potential to disrupt agriculture and forestry. Although at present it is rare for a major hurricane to threaten the Virginia coast, hurricanes make the coastal area of Virginia vulnerable. Increased tropical storm activity that might result from climate change would carry the threat of damage to Virginia's communities.

Virginia has begun to address the issue of climate change in several ways. Virginia is one of several states to have completed, in 2007, a comprehensive Climate Action Plan, manifested as the Virginia Energy Plan. The process of developing a climate action plan includes identifying cost-effective opportunities by which a state may reduce emissions of the greenhouse gases that alter climate. However, unless clear policies are stipulated, climate action plans cannot ensure real reductions in emissions. Virginia adopted a voluntary renewable portfolio standard in 2007, and in May 2007, Governor Tim Kaine announced that Virginia had joined the Climate Registry, a state-sponsored initiative to standardize methods to record and measure greenhouse gas emissions.

SEE ALSO: Atlantic Ocean; Climate; Gulf Stream; United States.

BIBLIOGRAPHY. Commonwealth of Virginia, www..va.us; Commonwealth of Virginia Department of Mines, Minerals, and Energy, *The Virginia Energy Plan* (Commonwealth of Virginia, 2007); Bruce P. Hayden and Patrick J. Michaels, *Virginia's Climate*, (University of Virginia Climatology Office, 2007).

Karin Warren
Randolph College

Volcanism

MEMBERS OF THE scientific community by and large concur that the Earth is undergoing a change in climate and that global warming is occurring at an increasing rate. In fact, scientific modeling suggests that Earth will experience an increase in temperature during the next 100 years at a pace up to four times greater than that in the previous 100 years. To a large extent this acceleration in the late 20th century is attributed to carbon dioxide (CO_2) emissions generated by human activity. Carbon dioxide acts like a glass barrier over the Earth, preventing heat from leaking into the environment, and thus creating a greenhouse effect. Human activity has counteracted global warming to a small degree by emitting aerosols in the environment, causing global cooling.

Volcanism is another contributing factor to climate change. Each volcano affects the climate based on its location and the nature and extent of an eruption.

Similar to human activity, volcanism leads to both global warming and cooling. The effect of volcanism on climate depends on the interaction between the sun's heat and the volcanic debris. In essence, scientists believe that ongoing volcanic eruptions have maintained the Earth's temperate climate for millions of years. Circumstantial evidence suggests that volcanism can influence short-term weather patterns in addition to having an effect on long-term climate change.

Volcanic dust blown into the atmosphere can remain for months and produce temporary cooling, the degree to which is dependent on the volume of dust, and the duration of which is dependent on the size of the dust particles. The strength of gases can vary greatly among volcanoes. Water vapor is typically the most abundant volcanic gas, followed by carbon dioxide and sulfur dioxide. Other principal volcanic gases include hydrogen sulfide, hydrogen chloride, and hydrogen fluoride.

Volcanoes that discharge great quantities of sulfur compounds affect the climate more significantly than those that release only dust. In fact, the greatest volcanic effect on the Earth's short-term weather patterns is caused by sulfur dioxide gas. Sulfur dioxide can form sulfuric acid aerosols that reflect the sun's heat and trigger cooling of the Earth's surface for as long as two years. Scientific literature frequently refers to the drop in global temperature after the eruption of Mount Pinatubo, Philippines, in 1991 and the very cold temperatures leading to crop failures and famine in North America and Europe for two years following the eruption of Mt. Tambora, Indonesia, in 1815 as examples of this effect. Furthermore, sulfuric dioxide can form acid rain, a combination of sulfuric acid and nitric acid. Acid rain, which is also caused by the burning of fossil fuels, is a critical environmental problem that can

Yasur volcano in Vanuatu is one of the world's most active volcanoes, erupting dozens of times a day for about 800 years.

injure lakes, streams, and forests and the inhabitants of these ecosystems.

In addition, volcanoes discharge water and carbon dioxide in large quantities in the form of atmospheric gases, and can absorb and retain in the atmosphere heat radiation emanating from the ground. Estimates suggest that water makes up to 99 percent of gas in volcanic expulsions. This short-term warming of the air leads the water to become rain within a matter of hours or days and the carbon dioxide to dissolve in the ocean or to be absorbed by plants. The majority of the heat energy connected to global warming exists in the ocean. If the oceanic depth at which heat is stored is decreased, then global temperature increases are expected to be greater than predicted.

Volcanic eruptions combined with man-made chlorofluorocarbons (CFCs) also can contribute to ozone depletion. CFCs were developed in the early 1930s and were used in industrial, commercial, and household applications, such as refrigeration units and aerosol propellants, because they were nontoxic and nonflammable and met a number of safety criteria. In February 1992, however, following evidence that CFCs contributed to depletion of the ozone layer, the United States announced the phase-out of the production of CFCs by December 1995. Members of the Montreal Protocol in 1992 followed suit and agreed to an accelerated phase-out by the end of 1995 as well.

The ozone layer, which rests in the stratosphere and begins at 7.5 mi. (12 km.) above the Earth, is a shield that protects living beings from ultraviolet-B (UV-B), the sun's most harmful UV radiation. In high doses, UV-B can lead to cellular damage in plants and animals. Scientists believe that global warming will lead to a weakened ozone layer, because as the Earth's surface temperature rises, the stratosphere will get colder, slowing the natural repairing of the ozone layer. Decreased ozone in the stratosphere results in lower temperatures.

Unlike ozone depletion created by man-made CFCs, which will take decades to repair, theories indicate that repair of damage to the ozone layer caused by volcanoes occurred as volcanic activity diminished. Recent mathematical models intimate that a volcanism in Siberia of great proportion significantly depleted the ozone layer.

Finally, hydrogen fluoride gas can concentrate in rain or on ash particles, contaminating grass, streams, and lakes with excess fluorine. Excess fluorine in grass and water supplies can poison the animals that eat and drink at contaminated sites, eventually causing fluorisis, which destroys bones.

SEE ALSO: Forced Climate Variability; Oceanic Changes; Stratosphere; Sulfur Dioxide.

BIBLIOGRAPHY. Australian Greenhouse Office, Department of the Environment and Heritage, "Is Global Warming Primarily Due to Solar Variability?" www.greenhouse.gov.au/science/hottopics/pubs/topic6.pdf; Robert C. Balling, *Environmental Geosciences* (v.7/4, 2000); Ryan C. Bay, Nathan Bramall, and P. Buford Price, "Bipolar Correlation of Volcanism With Millennial Climate Change," *Proceedings of the National Academy of Sciences of the United*

States of America (2007); Center for Educational Technologies, Wheeling Jesuit University, "Exploring the Environment: Volcanoes," www.cotf.edu/ete/modules/volcanoes/vclimate.html; Richard V. Fisher, "Volcano Information Center," www.volcanology.geol.ucsb.edu/gas.htm; Terrence M. Joyce, "Observations on Global Warming," www.whoi.edu/page.do?pid=12457&tid=282&cid=14087; Kenneth A. McGee, Michael P. Doukas, Richard Kessler, and Terrence M. Gerlach, "Impacts of Volcanic Gases on Climate, the Environment, and People," U.S. Geological Survey Open-File Report 97-262, www.pubs.usgs.gov/open-file/of97-262/of97-262.html; Ian Plimer, "The Past is the Key to the Present: Greenhouse and Icehouse Over Time," *Institute of Public Affairs Review*, (v.55/1, 2003); University of Sheffield, www.shef.ac.uk/mediacentre/2007/; U.S. Geological Survey, "Volcanic Eruptions Temporarily Reduce the Effects of Global Warming," www.hvo.wr.usgs.gov/volcanowatch/2005/05_11_03.html.

Robin K. Dillow
Rotary International Archives

Von Neumann, John (1903–57)

JOHN VON NEUMANN is a Hungarian-born American mathematician who contributed important theories to all the different branches of the discipline. His discoveries influenced quantum theory, automata theory, economics, and defense planning. Von Neumann was one of the founders of game theory and, along with Alan Turing and Claude Shannon, was one of the conceptual inventors of the stored-program digital computer. He is mostly remembered in the field of global warming for his elaboration of general circulation models and for his 1955 article in *Fortune*, in which he stated that "microscopic layers of colored matter spread on an icy surface, or in the atmosphere above one, could inhibit the reflection-radiation process, melt the ice, and change the local climate." This is one of the earliest conceptualizations of the problem of global warming.

John Von Neumann was born János Neumann in Budapest, Hungary, on December 28, 1903, into a wealthy and completely assimilated Jewish family. His father was a banker, and his mother originally came from a family who made a fortune selling farm equipment. John was a child prodigy and was initially educated by private tutors in mathematics and foreign languages. In 1911, he entered Budapest's most prestigious school, the Lutheran Gymnasium. When Bela Kun established his revolutionary government, Von Neumann's family fled Hungary and briefly emigrated to Austria. Kun's government failed after only five months. Because it was mainly composed by Jews, Von Neumann's family, no matter how hostile to the regime, was blamed, together with many other Jews, for the brutality of the revolutionary government. In 1921, Von Neumann completed his education at the gymnasium and his father strongly advised him to take up a career in business and not in mathematics, where Von Neumann's talent was already obvious. His father was afraid that mathematics would not allow Von Neumann to lead a wealthy and comfortable existence. As a compromise, it was decided that Von Neumann would study both chemistry and mathematics. In spite of the strict quotas for Jewish entry to university, Von Neumann succeeded in attending the University of Budapest for mathematics and the University of Berlin for chemistry. He later switched to the Swiss Federal Institute in Zürich, from where he graduated with a degree in chemical engineering in 1925. The following year, he obtained a doctorate in mathematics from the University of Budapest.

Von Neumann quickly gained a reputation in set theory, algebra, and quantum mechanics. In particular, his paper "An Axiomatization of Set Theory" (1925) was read and appreciated by the famous mathematician David Hilbert. Because of Hilbert's interest, Von Neumann worked at the University of Göttingen in 1926 and 1927. Von Neumann was then employed as a *Privatdozent* ("private lecturer") at the Universities of Berlin (1927–29) and Hamburg (1929–30). At a time when Europe was characterized by political unrest and totalitarianism, he was invited to visit Princeton University in 1929 to lecture on quantum theory. He was then hired at Princeton as visiting professor the following year, although teaching was not one of his strongest assets, and when the Institute for Advanced Studies was founded there in 1933, he was appointed to be one of the original six professors of mathematics, a position he retained for the rest of his life. In 1930, Von Neumann married Mariette Koevesi. They had one child, Marina, who later became an economist. The couple separated amicably in 1937, and the following year Von Neumann married his childhood sweetheart Klara Dan, with whom he remained until his death. After his appointment at the Institute for Advanced Studies, Von Neumann resigned from all his German academic positions. Being a Jew, he could not work for a Nazi state. He prophetically stated that if Hitler remained in power, it would ruin German science for a long time.

During the World War II, Von Neumann worked for the Manhattan Project at the invitation of its director Robert Oppenheimer. His expertise in the nonlinear physics of hydrodynamics and shock waves was instrumental in the design of the Fat Man atomic bomb dropped on the Japanese port of Nagasaki. Von Neumann argued against dropping the bomb in Tokyo on the Imperial Palace. In the postwar years, Von Neumann continued to work as a consultant to government and industry. Starting in 1944, he contributed important insights for the development of the U.S. Army's ENIAC computer, initially designed by J. Presper Eckert, Jr., and John W. Mauchly. In a crucial contribution, Von Neumann modified the ENIAC to run as a stored-program machine. Von Neumann did not invent the computer itself, but he invented the software that made it run. The ENIAC machine was a combination of electronic hardware and punch-card software that allowed it to be employed for a variety of uses including weather forecast. Von Neumann then campaigned to build an improved and faster computer at the Institute for Advanced Study. The institute's machine, which began operating in 1951, used binary arithmetic where the ENIAC had used decimal numbers. Von Neumann's publications on computer design (1945–51) caused a clash with Eckert and Mauchly, who wanted to patent their contributions, and paved the way for the independent construction of similar machines around the world. Von Neumann's intuition for a single-processor, stored-program computer became the accepted standard.

As a consultant for RAND Corporation, Neumann was given the task of planning a nuclear strategy for the U.S. Air Force. In this capacity, he was a strong advocate of nuclear weapons, and he became one of President Dwight Eisenhower's top advisers on his nuclear deterrence policy. Von Neumann was diagnosed with bone cancer in 1955. In spite of his dete-

riorating health, he continued to work, and in 1956 he was honored with the Enrico Fermi Award. He converted to Roman Catholicism just before his death on February 8, 1957, in Washington, D.C.

SEE ALSO: Climate Models; Education.

BIBLIOGRAPHY. William Aspray, "The Mathematical Reception of the Modern Computer: John von Neumann and the Institute for Advanced Study Computer," in Esther R. Phillips, ed., *Studies in the History of Mathematics*, vol. 26 (Mathematical Association of America, 1987); Salomon Bochner, "John von Neumann," *Biographical Memoirs*, vol. 32 (National Academy of Sciences, 1958); J. Dieudonné, "Von Neumann, Johann (or John)," in Charles C. Gillespie, *Dictionary of Scientific Biography* (Charles Scribner's Sons, 1981); H. S. Tropp, "John von Neumann," in Anthony Ralston and Edwin D. Reilly, Jr., *Encyclopedia of Computer Science and Engineering* (Van Nostrand Reinhold, 1983); Spencer Weart, *The Discovery of Global Warming* (Harvard University Press, 2004).

LUCA PRONO
UNIVERSITY OF NOTTINGHAM

Vostok Core

RUSSIA'S VOSTOK STATION is located in east Antarctica. Vostok Station holds the record for lowest temperature ever recorded at minus 129 degrees F (minus 89 degrees C). Soviet researchers began deep drilling at the Vostok Station in 1980. The ice cores brought to the surface in segments provide information (chemistry, structure, inclusions) about climate conditions similar to tree ring samples. The information from air bubbles allows for measurement of the atmospheric concentration of greenhouse gases (carbon dioxide, methane, nitrogen, helium, sodium, and organic carbon). Besides presenting an extraordinary human effort spanning two decades in one of the most inhospitable places on earth, the drilling at Vostok has produced one of the richest scientific treasure troves of all time. Previously, analysis revealed tracking between carbon dioxide and temperature and that the magnitude of the carbon dioxide swings could account for the magnitude of the temperature swings.

The first hole drilling stopped in 1985 because of problems. A second hole drilled with French-Russian cooperation produced an ice core 2,083 m. long, or 1.33 mi. With a climate record of 160,000 years, drilling on this hole ended in 1990. A third hole was drilled with collaboration among Russia, France, and the United States. The drilling reached a depth of 2.25 mi. (3.6 km.) and in January 1998 produced the deepest ice core recovered at the time (now exceeded by the European Project for Ice Coring in Antarctica)—11,886 ft. (3,623 m.) deep, containing a climate record of 420,000 years, for a total of four climate cycles. Drilling stopped at this depth because the researchers were recovering accretion ice refrozen to the bottom of the glacier, indicating the presence of an underlying lake, and did not want to put themselves in danger from the release of pressurized lake water or risking contamination of the lake. Researchers have found microbes in the glacial ice from the Vostok core and four times more in the glacial-accretion ice transition, suggesting that the underground lake contains microbes and organic carbon.

Polar snowfall can be preserved in annual layers within an ice sheet to provide a climate record. These layers can be studied to develop an accurate picture of the climate history extending over the time periods (the deepest Vostok core extends over a 400,000 year time frame). Impurities (volcanic debris, sea salt, organic material, and interstellar particles) are also deposited with snow, making those layers distinctive.

Air bubbles trap gases in the ice and allow for testing to determine the air's composition at distinctive periods in the climate record. Water pockets may also become trapped the deeper the ice core is, and closer to the underlying rock or water. Researchers can determine the composition of water in comparison with heavy water isotopes to indicate environmental temperature; cold periods are those with moisture being removed from the atmosphere.

Studies on the second Vostok core showed a correlation between carbon dioxide and temperature over the past 160,000 years and provided evidence linking climate change with the greenhouse effect. The trapped air bubbles provided gas isotopes and, when compared with the temperature variations, matched up to show that greenhouse gases were the primary driver of climate change over time. The climate variation of ice ages also matched the solar records.

Initial Vostok studies, when combined with later research, provide an inclusive representation of the multiple factors involved in climate change by using a multidisciplinary approach to climate change research by using astronomical tables, chemistry, and physics. The Vostok ice cores indicate periods of ice ages, contain gases for comparison with temperature changes, and highlight the last ice age of 8 degrees F (4.4 degrees C) cooler than the present, taking place about 18,000 years ago.

Vostok's cores have provided significant evidence of greenhouse gas variations driving climate change and have provided information for the modeling of future climate changes in relation to greenhouse gas concentrations. The third Vostok core, recovered in 1998, provides additional confirmation and extends the historical record through the four most recent glacial cycles, showing that increased concentrations of greenhouse gases have forced the temperature higher and can be compared with the geological record of the same time frames.

The Vostok ice core has become the standard for creating timescales from cores recovered from other parts of the world. Researchers are able to plot the isotopes of their samples with similar isotope ratios of the known sample to provide an accurate time period reference.

SEE ALSO: Greenhouse Gases; Milankovitch, Milutin; Paleoclimates.

BIBLIOGRAPHY. Mark Bowen, *Thin Ice Unlocking the Secrets of Climate in the World's Highest Mountains* (Henry Holt, 2005); Mason Inman, "The Dark and Mushy Side of a Frozen Continent," *Science* (July 6, 2007); World Data Center for Paleoclimatology, "Vostok Ice Core Data," www.ncdc.noaa.gov.

LYN MICHAUD
INDEPENDENT SCHOLAR

Walker, Gilbert Thomas (1868–1958)

GILBERT THOMAS WALKER is the British physicist and statistician who first described the phenomenon of Southern Oscillation, a coherent interannular fluctuation of atmospheric pressure over the tropical Indo-Pacific region that produces wind anomalies. This was part of Walker's project to determine the connections between the Asian monsoon and other climatic fluctuations in the global climate in an effort to predict unusual monsoon years that cause drought and famine to the Asian sector. Although he was not a meteorologist by education, Walker greatly advanced the study of global climate with his discovery.

Gilbert Thomas Walker was born on June 14, 1868, in Rochdale, Lancashire. He was the eldest son and fourth child in a large family of eight. Soon after his birth, his family relocated to Croydon, where his father became the borough's chief engineer. From 1876 on, Gilbert attended Whitgift School, and in 1881 he won a scholarship to St. Paul's School. He already excelled in mathematics from these early years and passed the London matriculation in 1884. However, he did not stay in London for his degree, opting instead for Trinity College, Cambridge, where he enrolled in 1886,

thanks to a scholarship. In 1891, he was elected a Trinity Fellow. He received an M.A. in 1893, and two years later, he was appointed a lecturer in mathematics. The heavy work necessary to attain these successes eventually took their toll on Walker's health, which broke down in 1890, forcing the mathematician to spend the next three winters in Switzerland. As a result of his precarious health, Walker did not publish many significant papers in the following years, but his 1899 work "Aberration and Some Other Problems Connected With the Electromagnetic Field" earned him the prestigious Adams Prize from Cambridge University. Walker was elected a fellow of the Royal Society in 1904, the same year that he received a Sc.D. from Cambridge University.

In the summer of 1903, Walker resigned his academic positions to become assistant to Sir John Eliot, who was the meteorological reporter to the government of India and director-general of Indian Observatories. The choice of Walker as special assistant was surprising, as Walker was not a meteorologist but a mathematician. At the end of 1903, Eliot retired, and Walker became the sole person responsible for the Indian Meteorological Department. He continued Eliot's quest for professional individuals to become members of his staff. He made prestigious appointments including J.H. Field, J. Patterson, and G.C.

Simpson, who later became directors of meteorological services in India, Canada, and the United Kingdom, respectively. From the beginning of his appointment, Walker devoted his research to the problems of monsoon and, in 1909, published his first meteorological papers. In 1908, Walker also gave lectures at the University of Calcutta, which were then published in 1910 by Cambridge University Press. Walker married May Constance Carter in 1908, and the couple had a son and a daughter.

Walker's interest in the monsoon resulted from the famine that the absence of rains had caused in India in 1899. Walker soon understood that he could not tackle meteorological problems through mathematical analysis and tried to develop more empirical techniques. He called his methodology seasonal foreshadowing, rather than weather forecasting, as the phrase indicated a vaguer prediction. Walker calculated statistical delay correlations between antecedent meteorological events both within and outside India and the subsequent behavior of the Indian monsoon. He was one of the first scientists to establish relationships between apparently separate events. The sets of relationships that he established, subsequently called Walker circulation in his honor, create a system resembling a global heat engine influencing the world's climate. The Walker circulation works like a swing in which warm, moist air rises in the western Pacific, becomes drier at high elevation, and displaces eastward, where heavy air sinks and returns westward. The phenomenon thus creates high air currents moving from the west to the east and, at the same time, east-to-west trade winds near the ocean surface. Global warming theorists have predicted that the rising temperatures will eventually slow down this mechanism.

Walker retired as director of the Indian Meteorological Department in 1924 and became a professor of meteorology at the Imperial College of Science and Technology in London. There, Walker continued his research into global weather and simultaneously carried out laboratory experiments in physics to study the convection of unstable fluids, particularly in its applications to cloud formations. He retired from Imperial College in 1934 and moved to Cambridge, where he lived until 1950. Although retired, Walker remained an active researcher and served as the editor of the *Quarterly Journal of the Royal Meteorological Society* from 1934 to 1941. He died in Coulsdon,

Surrey, on November 4, 1958. Although he had only mixed success in his original goal, the prediction of monsoonal failures, Walker conceived theories that allowed his successors to move beyond local observation and forecasting toward comprehensive models of climate worldwide

Throughout his distinguished career, Walker was a member of prestigious societies and received numerous honors. He was elected a fellow of the Royal Meteorological Society in 1905 and served as president of the society in 1926 and 1927. He was also vice president of the society three times and was an ordinary member of council in 1925 and from 1935 to 1939. He was awarded the society's Symons Gold Medal in 1934. While working in India, Walker was the president of the Royal Society of Bengal and president of the Indian Science Congress. He became a Companion of the Order of the Star of India in 1911 and was knighted on the king's birthday in 1924. Walker was also an honorary fellow of Imperial College and a fellow of the Royal Astronomical Society.

SEE ALSO: Climate Models; Southern Oscillation.

BIBLIOGRAPHY. Gilbert Walker, "On the Meteorological Evidence for Supposed Changes of Climate in India," *Indian Meteorological Memoirs* (v.21/1, 1910); J. M. Walker, "Pen Portrait of Sir Gilbert Walker, CSI, MA, ScD, FRS," *Weather* (v.52/7, 1997).

LUCA PRONO
UNIVERSITY OF NOTTINGHAM

Walker Circulation

THE WALKER CIRCULATION is an atmospheric system of air flow in the equatorial Pacific Ocean. The trade winds across the tropical Pacific flow from east to west: air rises above the warm waters of the western Pacific, flows eastward at high altitudes, and descends over the eastern Pacific. A weaker Walker circulation (in the reverse direction) occurs over the Indian Ocean.

Sir Gilbert Walker assumed the post of director-general of the observatory in India following catastrophic famines in the late 1800s resulting from a general failure of the South Asian monsoon. In an effort to predict the monsoons, Walker undertook an investigation into the regional climate system. Over time, he recognized that the monsoonal system extended to a panoceanic scale. Walker observed that an inverse relation of atmospheric pressures at sea level generally existed between the two sides of the Pacific Ocean. A high-pressure phase in South America was usually accompanied by low pressure in the western Pacific and vice versa—the Southern Oscillation (SO). The generally accepted measure of the SO is the inverse relationship between surface air pressure at Darwin, Australia, and Tahiti (stations used by Walker); indeed, the SO is normally identified by the SO Index (SOI), that is, the difference in atmospheric pressure at sea level between these stations (Tahiti minus Darwin). The greater the SOI, the greater the intensity of the trade winds. Historically, the SO has exhibited a more or less cyclical pattern, in that it weakens or reverses every few years.

On the basis of data collection initiated during the International Geophysical Year in 1957 to 1958, in the 1960s Jacob Bjerknes of the University of California described the general nature of the mechanism linking the system. He extended the horizontal picture of the SO vertically by theorizing that to complete the system of the trade winds and atmospheric air pressure, there needed to be a countercirculation of air from west to east at high altitudes, descending over the eastern Pacific.

Atmospheric circulation is intimately coupled with the movement of water in the tropical Pacific. At the surface, the trade winds initiate a westward flow of surface water across the Pacific Ocean, producing an increase of sea level in the western Pacific of approximately 40 cm. The equatorial heating of this water produces high seasurface temperatures (SSTs) in oceanic waters near Indonesia. The resulting low atmospheric pressure and evaporation fuels the pan-Pacific upper-atmospheric circulation characterized by convection (low atmospheric pressure) in the west and subsidence (high atmospheric pressure) in the east. This is termed the Walker circulation in honor of Sir Gilbert.

The Walker circulation is closely connected to oceanic upwelling off the coast of South America. Fluctuations in the circulation are closely linked to El Niño and La Niña events—together termed the El Niño/SO system (ENSO). A weakening or reversal of the Walker circulation is closely linked with the El Niño phenomenon, with warmer-than-average SSTs in the eastern Pacific as upwelling diminishes. In contrast, the opposite phase, a particularly strong Walker circulation, produces a La Niña event, with cooler SSTs caused by increased upwelling. Interannular switches in the dipole are linked to global-scale changes in patterns of weather. Several explanations for the variation in the Walker circulation have been hypothesized, but the nature of the mechanisms initiating the change in phase has not been fully identified.

Evidence suggests that the Walker circulation may have been weakening since the mid-19th century. However, there is a high degree of uncertainty concerning the potential effects of climate change on the Walker circulation. Transient warming may dominate before the ocean has had a chance to reach equilibrium. In the short term, warming may occur more quickly in the western Pacific, thereby enhancing the circulation. In contrast, atmospheric models have indicated that climate change will, as part of a weakening of the entire tropical circulation, lead to a general decrease in the strength of the Walker circulation. The specific mechanisms involved are not fully known, and projections remain rather speculative and are a focus of intense research.

SEE ALSO: El Niño and La Niña; Pacific Ocean; Southern Oscillation; Trade Winds; Walker, Gilbert Thomas.

BIBLIOGRAPHY. Ana Cristina Ravelo, "Walker Circulation and Global Warming: Lessons From the Geologic Past," *Oceanography* (v. 19/4, December 2006); Gabriel A. Vecchi, Brian J. Soden, Andrew T. Wittenberg, Isaac M. Held, Ants Leetmaa, and Matthew J. Harrison "Weakening of Tropical Pacific Atmospheric Circulation Due to Anthropogenic Forcing," *Nature* (v.441, May 4, 2006).

MICHAL BARDECKI
RYERSON UNIVERSITY

Washington

WASHINGTON STATE IS 71,300 sq. mi. (184,666 sq. km.), with inland water making up 1,553 sq. mi. (4,022 sq. km.), coastal water making up 2,535 sq. mi. (6,565 sq. km.), and access to territorial water making up 666 sq. mi. (1,725 sq. km.). Washington's average elevation is 1,700 ft. (518 m.) above sea level, with a range in elevation from sea level on the Pacific Ocean to 14,410 ft. (4,392 m.) at the peak of Mt. Rainier. Western Washington lies on the Juan de Fuca Plate, with overriding by the North American Plate.

Washington is fairly warm and is kept that way by ocean currents including frequent rain and the rain shadow effect. Moist air streams up slopes of mountains, and rainfall or snowfall is increased. However, when the air, depleted of much of its moisture, begins to descend down the slopes, the temperatures warm, clouds dissipate, and very little precipitation falls. Western mountains take Pacific moisture that flows onshore all year long. The western slopes of the Olympic Mountains have a true rainforest, but the rain shadow on the eastern slope means it receives less than 15 in. of rain per year. The highest temperature recorded in the state

is 118 degrees F (48 degrees C) in Ice Harbor Dam on August 5, 1961, and the lowest temperature recorded in the state is minus 48 degrees F (minus 44 degrees C) in Mazama and Winthrop on December 30, 1968. Mount St. Helens, an active volcano, spouted volcanic ash into the atmosphere in 1980. In 1993, the year of the flood in Iowa, Washington recorded the coolest summer on record in Spokane.

The state supports a population of over 6 million people. Major industries include agriculture, with the major products being apples, beef, milk, timber, and wheat; manufacturing computers, food, machinery, and paper products; and mining coal, gold, sand, and gravel.

The Columbia River is the second largest river in the country based on volume. The Grand Coulee Dam retains some water from spring runoff for summer use. It is the largest concrete structure at 55 ft. (16.7 m.) tall. It holds 24 generators supplying 65 mil-

Mount St. Helens in Washington is an active volcano, and had an enormous eruption at 8:32 in the morning on May 18, 1980. The debris blasted down nearly 230 sq. mi. (596 sq. km.) of forest and buried much of it beneath volcanic mud deposits.

lion kilowatts of electricity. This electricity is carried throughout the west and east to Chicago.

Commercial logging started in the 1800s and has claimed 90 percent of the forests that once grew in the Pacific Northwest. Logging in the national forests since World War II has divided mountainsides and river valleys into checkerboards of clear-cut and uncut forest, creating vulnerable areas to succumb to environmental pressures. More than half of the remaining untouched forest areas in Olympic National Forest are slated for cutting during the next 50 years, as is 69 percent of the old growth in Oregon's Siuslaw National Forest. In 1990, Congress voided all court injunctions brought to stop cutting of ancient trees on lands administered in Washington and Oregon by the Bureau of Land Management and the Forest Service and removed the right of citizen and conservation groups to seek further injunctions in any future cuts planned.

The Climate Impacts Group at the University of Washington is working to further understanding of the patterns and predictability of regional climate variability, the influence of climate variation on the Pacific Northwest, and providing strategies to prepare for climate change.

Washington is a member of the Western Regional Climate Action Initiative, along with Arizona, California, New Mexico, Oregon, Utah, and the Canadian provinces of British Columbia and Manitoba. This cooperative group works together to identify, evaluate, and implement ways to reduce greenhouse gas emissions on the regional level.

The effects of climate change that are already being experienced by Washington include weather changes, reduced summer water supply dependent on winter snowpack, and rising water levels along the coastline. Because of these effects, Washington is taking steps to prepare for climate change and reduce human-induced contributions to global warming. In February 2007, the governor signed an executive order establishing goals for reduction in climate pollution, increases in jobs, and reductions in expenditures on imported fuel. Because the United States has not ratified the Kyoto Protocol calling for greenhouse gas emission reduction, some states and regions have taken voluntary initiatives to reduce these emissions on their own. Washington's goal is to cut emissions by 20 percent by 2050, with reduced auto emissions,

use of renewable fuels, green building standards, and energy efficiency; passing a clean renewable energy initiative; and adopting a CO_2 emissions performance standard for electric generating units.

Not only is Washington proactive on reduced emissions but they provide education and call for public support by using more energy-efficient transportation (public transportation, ride sharing, walking, or bicycling), improving home energy efficiency (insulation, energy-efficient appliances, and fluorescent lighting), and encouraging the planting of trees and plants.

Though climate models show no precipitation change, a change in temperature of 2 degrees to 3 degrees may mean less snowfall. An expected decrease this century of 60 percent in Cascade snowpack—even in the most reassuring of global warming scenarios—stands to have devastating consequences in the Pacific Northwest. The Columbia River's Grand Coulee Dam has made it possible for Washington, along with other Pacific and mountain states, to increase in population. A survey of nearly 600 snowfields in the Sierra Nevada, the Rocky Mountains, and the cascades of Washington and Oregon shows that 85 percent of them have lost volume since the 1950s. A higher incidence of wildfires resulting from increasing drought levels, as well as rising sea levels, could displace people from their homes.

The associated costs include those affiliated with fighting wildfires, flood damage, lost revenue from tourism, increase in water pricing because of shortages and drought conditions, and increased costs for healthcare related to poor air quality and increased infection.

With rising temperatures, variable precipitation, and rising sea levels, the quality of the water supply in Washington could be compromised, with an increasing risk of water-borne infections. The potential for agricultural disruption also could lead to risks of malnutrition, and increasing heat waves could cause more heat-related illness.

SEE ALSO: Deforestation; Drought; Floods; University of Washington; World Health Organization.

BIBLIOGRAPHY. Mark Bowen, *Thin Ice: Unlocking the Secrets of Climate in the World's Highest Mountains* (Henry Holt, 2005); Climate Impacts Group, "Climate Change," http://www.cses.washington.edu; Roger L. DiSilvestro, *Fight for Survival Audubon Perspectives* (Wiley, 1990); Washington State Department of Ecology

"Impacts of Climate Change on Washington's Economy," http://www.ecy.wa.gov; World Health Organization, *Climate and Health Fact Sheet* (WHO, July 2005).

LYN MICHAUD
INDEPENDENT SCHOLAR

Washington, Warren (1936–)

WARREN WASHINGTON IS an African-American meteorologist and atmospheric scientist whose research focuses on the development of computer models that describe and predict the Earth's climate. Washington was one of the first developers of atmospheric computer models in the early 1960s at the National Center for Atmospheric Research in Boulder, Colorado. These computer models use the basic laws of physics to predict the future states of the atmosphere. Because these equations are extremely complex, it is almost impossible to solve them without a powerful computer system. Washington's book *An Introduction to Three-Dimensional Climate Modeling* (1986), coauthored with Claire Parkinson, is a standard reference for climate modeling. In his subsequent research, Washington worked with others to incorporate ocean and sea ice physics as part of a climate model. Such models now involve atmospheric, ocean, sea ice, surface hydrology, and vegetation components. Washington is the director of the Climate and Global Dynamics Division of the National Center for Atmospheric Research (NCAR), in Boulder, Colorado. He has advised the U.S. Congress and several U.S. presidents on climate-system modeling, serving on the President's National Advisory Committee on Oceans and Atmosphere from 1978 to 1984.

Washington was born in Portland, Oregon, in 1936. His father, Edwin Washington, Jr., wanted to be a schoolteacher, but in the 1920s, it was impossible for African Americans to be hired as teachers in Portland public schools. Thus, Edwin was forced to work as a waiter in Pullman cars to support his family. His wife Dorothy Grace Morton Washington became a practical nurse after Warren and his four brothers grew up. Washington's interest in scientific research was apparent from an early age, and he was encouraged by his high school teachers. When Washington had to choose what to do after high school, his counselor advised him to attend a business school rather than college. However, his ambition was to be a scientist, so he enrolled at Oregon State University, where he earned his bachelor's degree in physics in 1958. During his undergraduate years, Washington became interested in meteorology while working on a project at a weather station near the campus. The project involved using radar equipment to follow storms as they came in off the coast. Because of his growing interest in meteorology, Washington began a master's degree in meteorology at Oregon State, graduating in 1960. He then began a Ph.D. at Pennsylvania State University, graduating in 1964, thus becoming one of only four African Americans to receive a doctorate in meteorology.

Washington began working for NCAR in Boulder, Colorado, in 1963 and has remained associated with that institution throughout his career. His research at the center has attempted to describe patterns of oceanic and atmospheric circulation. Washington has contributed to the creation of complex mathematical models that include the effects of surface and air temperature, soil and atmospheric moisture, sea ice volume, various geographical traits, and other factors on past and current climates.

Washington's research has helped further our present understanding of the greenhouse effect. He has contributed to determining the process in which excess carbon dioxide in the Earth's atmosphere causes the retention of heat, thus leading to what is known

Warren M. Washington is the head of the Climate Change Research Section, National Center for Atmospheric Research.

as global warming. Washington's research also shed light on other mechanisms of global climate change. In interviews and statements, Washington has made it clear that he firmly believes that global warming is to the result of human actions: "For researchers in climate science, the question of whether or not climate change is attributable to human activity was put to rest several years ago with our DOE-supported simulations showing that the only way to duplicate the sharp increase of the global average temperature observed in the late 20th century was to include human generated greenhouse gases in the simulations. When the same simulation was run without the human-generated greenhouse gas increases, the model simulations show that the Earth would be in a slight cooling trend with the natural forcings of volcanic and solar activities. For us, that was the smoking gun for human-induced climate change." He has thus pleaded for climate science to rise in priority as a science problem for American administrations. He has also claimed that the Department of Energy has a particular responsibility to help find solutions for the global warming problem. According to Washington, "as the impacts of climate change become more apparent with increased severity of heat waves, droughts, water shortages, and more severe hurricanes, there will be more emphasis on understanding how we can better mitigate and adapt to the changes." The meteorologist has suggested that the Department of Energy study the carbon footprint and effect of various technology paths for the production of energy. He has also argued for an increased focus on what strategies to use to mitigate climate change and to find a long-term stabilization for carbon dioxide and other greenhouse gases in the atmosphere.

Throughout his career, Washington has published over 100 professional articles about atmospheric science, scientific textbooks, and an autobiographical volume. He has also served as a member and a director of prestigious institutes and commissions. Washington was appointed the director of the Climate and Global Dynamics Division at NCAR in 1987. In 1994, he was elected president of the American Meteorological Society. He is a fellow of the American Association for the Advancement of Science and a member of its board of directors, a fellow of the African Scientific Institute, a distinguished alumnus of Pennsylvania State University, a fellow of Oregon State University,

and founder and president of the Black Environmental Science Trust, a nonprofit foundation that encourages African-American participation in environmental research and policymaking. From 1974 to 1984, Washington served on the President's National Advisory Committee on Oceans and Atmosphere. In 1995, he was appointed by President Bill Clinton to a six-year term on the National Science Board. In 1997, he was awarded the Department of Energy Biological and Environmental Research Program Exceptional Service Award for Atmospheric Sciences in the development and application of advanced coupled atmospheric-ocean general circulation models to study the effects of anthropogenic activities on future climates.

SEE ALSO: Climate Models; Education.

BIBLIOGRAPHY. "Advanced Computing for Understanding and Adapting to Climate Change. Interview with Warren Washington," http://www.scidacreview.org/0702/html/interview.html; Warren Washington, *Odyssey in Climate Modeling, Global Warming and Advising Five Presidents* (Lulu.com, 2007).

LUCA PRONO
UNIVERSITY OF NOTTINGHAM

Waves, Gravity

ATMOSPHERIC GRAVITY WAVES are generated by atmospheric disturbances such as storm fronts, strong wind shears, and flow over mountains and play a key role in coupling the lower and upper atmosphere, causing a redistribution of momentum and energy from the troposphere and lower stratosphere into the upper atmospheric regions of the middle stratosphere, mesosphere, and lower thermosphere. Gravity waves trigger convection and induce mixing and transport of atmospheric chemicals such as ozone. Gravity waves typically form within or near the back edge of a precipitation shield. The strongest upward motions of gravity waves occur just following the surface pressure trough and lead to maximum precipitation rates just ahead of the ridge.

Atmospheric gravity waves can occur at all altitudes in the atmosphere and are important for the transport of energy and momentum from one region of the atmosphere to another, to initiate and modulate convection and subsequent hydrological processes, and to inject energy and momentum into the flow. When the gravity wave breaks, the resulting turbulence mixes atmospheric chemicals. These wave-breaking processes occur globally and affect climate of the mesosphere and stratosphere.

These mesoscale–regional scale processes have global significance because of their accumulative effects from the global distribution of various wave sources. The primary challenges to observational, numerical, and analytical studies are how to better quantify gravity wave excitation as it is related to various tropospheric processes, the global distribution of the wave sources, their propagation and breaking, and the multiscale interactions involving gravity waves.

The difficulty in producing the observed Arctic climate change in models may be a result of not including gravity waves in the models. The most likely energy source mechanisms are latent heat release in deep convection and shear instability, in which waves can extract energy from the jet stream when vertical wind shear is sufficiently strong to reduce the Richardson number below 0.25. Alternatively, wave energy loss can be prevented by an efficient wave duct, which appears to be the most prevalent of the three mechanisms described.

Gravity waves are maintained by wave-ducting processes requiring a layer of static stability (the duct depth near the surface), no critical levels (wind moving in the direction of the wave at the same speed) in the lower stable layer that would absorb the wave's energy, and a reflecting layer above the stable layer to keep the wave from losing its energy.

Gravity waves can affect an existing cloud pattern in several ways as they propagate: through modulating the cloud pattern, with the development of wave cloud formations, the wave and cloud can propagate in tandem with little effect on the overall cloud pattern. Convection can generate a broad spectrum of waves, ranging from short-period waves excited by the development of convective cells along a thunderstorm gust front to large wavelength disturbances resulting from the release of latent heat in a thunderstorm complex.

The challenge of including gravity waves in global climate models stems from the resolution ability of computers; with increasing computer power, more complex equations over smaller distances can be resolved to examine gravity waves. Current models often use one of the available gravity wave drag parameters and assume a fixed gravity wave source for proper representation of turbulence on the small scale.

The vast spatial and temporal extent of gravity waves has important implications for the atmosphere from the mesoscale to the global scale and poses a stiff challenge to improving weather and climate predictions at all ranges.

An important part of the National Center for Atmospheric Research mission is to understand the coupling of the lower and upper atmosphere through dynamical, chemical, and radiative processes. Sudden stratospheric warming involves dynamical changes on vastly different scales from the troposphere to the lower thermosphere, and thus provides us an opportunity to understand the coupling process. Further wave source sensitivity studies and observations will help to define gravity wave sources and behavior.

SEE ALSO: Climate Models; Jet Streams; National Center for Atmospheric Research.

BIBLIOGRAPHY. Marvin A. Geller, Hanli Liu, Jadwiga H. Richter, Dong Wu, and Fuqing Zhang, "Gravity Waves in Weather, Climate, and Atmospheric Chemistry: Issues and Challenges for the Community," Gravity Wave Retreat, Boulder, Colorado, June 2006; Steven Koch, Hugh D. Cobb III, and Neil A. Stuart, "Notes on Gravity Waves—Operational Forecasting and Detection of Gravity Waves," *Weather and Forecasting* (June 1997); The Earth and Sun System Laboratory, "Gravity Waves," http://www.essl.ucar.edu.

Lyn Michaud
Independent Scholar

Waves, Internal

AN INTERNAL WAVE is a wave that develops below the surface of a fluid along changes in density. With increased depth, this change may be gradual, or it may occur abruptly at the interface. Similar to the transmission of energy by wind along the surface of the ocean, the density interface beneath the ocean surface transmits energy to produce internal waves. The greater the difference in density between the two fluids, the faster the wave will move.

Internal waves may have a variety of causes, including the tidal pull of water, wind stress, and energy put into the water by moving vessels. Year-round internal waves caused by tidal forces carry between 30 and 50 percent of their energy away from their source. Seasonal (stormy winter months) internal waves caused by wind and storms carry at least 15 to 20 percent of the energy input from their source.

Internal waves reach greater highs (above 100 m., or 328 ft.) from a smaller energy input than do the waves resulting from large energy input at the ocean's surface. This is because they move along interfaces with less density difference than between the ocean surface and the atmosphere.

Internal waves eventually run out of energy and break, similar to surface waves. When they break in the deep ocean, they create turbulence. Where they create this turbulence, heat can be transferred from the upper ocean and stored in the deep ocean, with the exception of the Arctic Ocean, which is warmer in deeper water than on the surface. In the Arctic, turbulence transfers heat from the deep ocean to the surface.

Beneath the surface of the ocean exists a unique weather and climate resulting from the fluctuating currents driving wind in the atmosphere and from deep waves with similar patterns to those on the surface of the ocean—but unable to be seen from the surface—created by the ocean movement caused by the tides.

Internal tide is an internal wave created from the back-and-forth flows of water over geographic features at lower depths. These internal waves radiate away in the form of tide current carrying energy away from the source.

Internal waves are of special interest to climate modeling researchers because heat transfer is one of the roles of the ocean in regulating and changing climate. For example, the amount of heat transferred from the deep ocean in the Arctic will affect the amount of floating sea ice above. In the same way heat from the sun at the equator is transferred to the atmosphere, the upwelling of water enhances the global ocean circulation; otherwise, the depths of the ocean would be much colder and retain nutrients deeper down and away from supporting life or

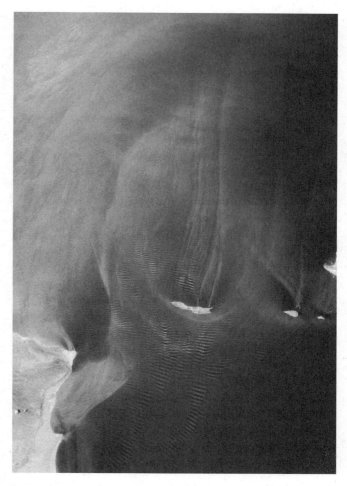

Internal waves occur beneath the water's surface. The reflection of light makes it possible to see them from high altitudes.

the ability for continuance of the carbon dioxide cycle absorption and release.

Internal waves mix and redistribute heat, salt, and nutrients in the oceans; mixing is accomplished more easily in water having uniform density. Most of this mixing occurs where internal waves break, overturning the density stratification of the ocean and creating patches of turbulence. Scientists have observed that the internal wave rates of dissipation and the redistribution of heat and nutrients is 10 percent less near the equator than at midlatitudes. This ocean dynamic would have to be accounted for in climate models.

Researchers are studying internal waves, as determining where internal wave energy might be high and where it might be low will help researchers distinguish between fluctuations in the data record originating from ocean currents and fluctuations resulting from internal waves/tides. Thus, oceanographic instruments can be deployed in oceanographically interesting locations where scientists can quantify the vertical redistribution of heat or assess the potential contribution to climate change and variability.

SEE ALSO: Ocean Component of Models; Oceanic Changes; Oceanography.

BIBLIOGRAPHY. Michael C. Gregg, Thomas B. Sanford, and David P. Winkel, "Reduced Mixing From the Breaking of Internal Waves in Equatorial Waters," http://apl.washington.edu; Robert Hallberg, "Reply," *Journal of Physical Oceanography* (v.31, July 2001); Sandra Hines, "Internal Waves Appear to Have the Muscle to Pump up Mid-Lats," *Science* (June 2003).

LYN MICHAUD
INDEPENDENT SCHOLAR

Waves, Kelvin

KELVIN WAVES AFFECTING weather and climate occur in both the oceans and the atmosphere. These low-frequency, gravity-driven waves propagate vertically and parallel to boundaries (e.g., equator, coastline, air masses, and topography). Kelvin waves are nondispersive and carry energy from one point to another. The height or amplitude of a Kelvin wave is highest near the boundary where it propagates; the wave height decreases as the wave moves farther away from the boundary. In the Northern Hemisphere, the waves propagate anticlockwise, and in the Southern Hemisphere, the waves propagate clockwise. The flow of the Kelvin wave balances pressure perpendicular to the boundary by the forces of gravity and the Coriolis effect.

Kelvin waves in the ocean form as coastal waves or equatorial waves, both of which are caused by external forces—often a shift in the trade winds or resulting from temperature variations—and the water inside the Kelvin waves is usually a few degrees warmer than the surrounding water. Kelvin waves may be called external (or barotropic) Kelvin waves if the ocean is homogenous, and internal (or baroclinic) Kelvin waves if the ocean is stratified.

In the Northern Hemisphere, the equatorial waves propagate parallel to the equator and to the east, and the coastal Kelvin waves propagate in a counterclockwise direction, using the coastline for direction. These waves can be between 5 and 10 cm. high and hundreds of km. wide. Kelvin waves tend to move quickly, with a typical speed of approximately 250 km. (155 mi.)/day and can cross the Pacific in approximately 2 months.

The tidal cycles can cause Kelvin waves by the mechanism of a progressive tide wave moving from open ocean into and out of a narrowed body of water. Because of the Earth's rotation, resulting in an anticlockwise direction of current flow inside the channel, flood tides will be greater on the right side of the channel.

The effect on climate results from the Kelvin waves causing a variation in the depth of the oceanic thermocline (the boundary between warm waters in the upper ocean and cold waters in the deep ocean). Because of this variation, Kelvin waves can be used to predict and monitor El Niño activity. In comparison with Rossby waves, which carry water back toward the western Pacific and take as long as a decade to move from the eastern Pacific to the western Pacific, the faster-moving Kelvin waves carry warm water eastward in approximately 2 months.

In the atmosphere near the equator, Kelvin waves travel eastward and may propagate upward to higher altitudes. The formation of Kelvin waves is triggered by mountains, thunderstorm updrafts, and anything that interrupts the normal flow of stable air. The trigger forces the air upward, and the stable air sinks by gravity instead of just returning to normal; when air sinks farther, it causes the wave motion. Kelvin waves cannot happen in unstable air because the motion would just allow the movement of air to continue upward to higher altitudes.

Kelvin waves may propagate in the lower and upper stratosphere and mesosphere. In the lower stratosphere, the eastward-moving Kelvin wave is associated with periods of 10 to 30 days, and in the upper stratosphere, the Kelvin wave is associated with periods of 5 to 7 days. In the mesosphere, the Kelvin wave is associated with periods of 3 to 4 days. These Kelvin waves transport energy and eastward momentum upward and contribute to the maintenance of the eastward flow.

For predicting weather and future climate change, climate models including the COMMA-LIM model can reproduce the Kelvin waves to correlate with how the Kelvin wave acts in the atmosphere. The wave action can also interact with other wave types and the flow of air masses.

SEE ALSO: Climate Models; El Niño and La Niña; Waves, Gravity; Waves, Rossby.

BIBLIOGRAPHY. Y. Hayashi, D.G. Golder, and J.D. Mahlman, "Stratospheric and Mesospheric Kelvin Waves Simulated by the GFDL 'SKYHI' General Circulation Model," *Journal of Atmospheric Sciences* (v.41, 1984); "Kelvin Waves," www. oc.nps.navy.mil; Tony Phillips, "A Curios Pacific Wave," www.science.nasa.gov.

LYN MICHAUD
INDEPENDENT SCHOLAR

Waves, Planetary

PLANETARY-SCALE WAVES HAVE their origin relating to the Earth's shape and rotation; the waves are so large that some of them wrap around the whole Earth and can be observed in the atmosphere through the meandering of the jet stream. A long wave or planetary wave is a weather system that circles the world, with one to three waves forming a looping path around the Earth at any given time and displacing air north and south. Planetary waves have ridges (the high points) and troughs (the low points). Warmer upper air is associated with an increasing number of waves or stronger waves.

Planetary waves form in the lowest part of the atmosphere, called the troposphere, and propagate upward, transferring energy into the stratosphere and heating polar air between 9–18 degrees F (5–10 degrees C). Because of a larger landmass, with the majority of the highest mountains and land-sea boundaries in the Northern Hemisphere, planetary waves form more strongly in the Northern Hemisphere. Once the wave dissipates, the polar air begins to cool. In the Southern Hemisphere, landforms also produce planetary waves, although they are weaker there because there are fewer tall mountain ranges and vast open ocean surrounding Antarctica. The warming of the Arctic stratosphere suppresses ozone destruction. Ozone exists in the lower level of the stratosphere and is caused by sunlight splitting the oxygen molecules at cooler temperatures, with less ozone destruction at warmer temperatures.

The Himalayas and other land features create the planetary atmospheric waves that serve to decrease the formation of an ozone hole at the northern pole and therefore limit solar ultraviolet radiation exposure in the Arctic. Climate change could open ozone holes in the Arctic; in the spring of 1997, weak planetary waves created conditions that formed a small ozone hole over the Arctic. The chemistry of ozone destruction requires very cold air temperatures in the stratosphere, and because of planetary wave action, the Arctic stratosphere stays warmer than the Antarctic stratosphere.

In contrast, researchers announced in 1992 that El Niño weather changes and a large number of planetary waves in the atmosphere had caused shrinking of the Antarctic ozone hole, with the ozone hole in September 2002 being half the size it was in 2000. Large-scale weather patterns (similar to the semipermanent area of high pressure) generate more frequent and stronger planetary waves. If the waves are more frequent and stronger as they move from the surface to the upper atmosphere, they warm the upper air. Because ozone breaks down more easily with colder temperatures, the warmer the upper air around the "polar vortex," or rotating column of winds that reach into the upper atmosphere where the protective ozone layer is, the less ozone is depleted.

Researchers working in Esrange, Sweden, studied the main features of planetary waves and variability of the semidiurnal tide, with planetary wave periods observed by meteor radar. They focused observation on 5-, 8- to 10-, 16-, and 23-day planetary waves by meteor radar measurements in the mesosphere and lower thermosphere. In the winter, when the planetary waves are significantly amplified, a very strong periodic variability of the semidiurnal tide is observed as well. This result indicates that the most probable mechanism responsible for the periodic tidal variability during winter is in situ nonlinear coupling between tides and planetary waves. They established a correlation between the planetary wave and semidiurnal tide and secondary waves with frequency, phase, and vertical wavenumber (wavelength) correlation.

The influence of planetary waves on global system dynamics with airflow and temperature distribution include the indirect effect of upper air patterns on lower air patterns through feedback linking all layers of the atmosphere.

Planetary waves (also called Rossby waves) form in the ocean and affect ocean circulation over longer periods of time—from one to 10 years.

SEE ALSO: Waves, Gravity; Waves, Rossby.

BIBLIOGRAPHY. Patrick L. Barry and Tony Phillips, "Planetary Waves Break Ozone Holes: Huge Planet-Girdling Atmospheric Waves Suppress Ozone Holes Over Earth's Northern Hemisphere," www.science.nasa.gov; Dora V. Pancheva and Nicholas J. Mitchell, "Planetary Waves and Variability of the Semidiurnal Tide in the Mesosphere and Lower Thermosphere Over Esrange (68°N, 21°E) During Winter," www.agu.org.

LYN MICHAUD
INDEPENDENT SCHOLAR

Waves, Rossby

NASA RESEARCHERS AT the Goddard Institute for Space Studies describe Rossby Waves as "slow-moving waves in the ocean or atmosphere, driven from west to east by the force of Earth spinning." These are naturally occurring phenomena first recognized in 1939 by a Swedish-American meteorologist named Carl-Gustav Rossby. These waves, which are found in both the atmosphere and the oceans, are important mechanisms for the redistribution of energy around the globe. In three sections, this essay describes the formation of atmospheric Rossby Waves, oceanic Rossby Waves, and the connection between Rossby Waves and global climate change.

ATMOSPHERIC ROSSBY WAVES

This eponymously named phenomenon was first identified as atmospheric oscillations that occurred in the mid-latitudes in the northern and southern hemispheres. In Europe and North America, people typically experience a Rossby wave as a large cold front plunging southward. The jet stream, guised as a tongue of cold air, dips southward as a large tropical air mass moves northward. The interaction between these air masses, affected by the coriolis affect that intensifies at lower latitudes, generates our changing weather on a day-to-day and week-to-week basis. During televised weather reports, we see North American Rossby waves as the large-scale oscillations of clouds moving from west to east across a continent.

OCEANIC ROSSBY WAVES

Scientists later identified a similar phenomenon at work in the water of all ocean basins. Researchers discovered that oceanic Rossby waves represent a mechanism by which the ocean responds to significant atmospheric "forcing" or wind-related disruption. Rossby waves disperse the atmospheric energy across ocean basins and can be measured through satellite imagery. Because of the impact of the earth's axial rotation and the Coriolis Effect, the oceanic Rossby waves tend to spiral away from the equator in both the northern and southern hemispheres.

ROSSBY WAVES AND CLIMATE CHANGE

Scientists are increasingly interested in Rossby Waves because of the possible connection between these atmospheric and oceanic waves and global warming. An understanding of atmospheric Rossby Waves enabled researchers to effectively study long term temperature fluctuations and provide concrete evidence that global warming was occurring.

Rossby waves can affect entire ocean basins. They also tend to move from the eastern part of the Pacific and Atlantic Oceans towards the west on either side of the equator. A complementary Kelvin wave moves in the opposite direction from west to east along the equator. The multiple axes along which the ocean moves has the ability to disrupt oceanic circulation. Researchers propose that global warming will generate stronger weather events with greater frequency. Because oceanic Rossby waves transmit atmospheric disruptions, the theory is that as storms occur more frequently, the wavelength and frequency of Rossby waves will also change—with a potentially disruptive impact on ocean currents such as the Gulf Stream. A disrupted Gulf Stream could cause cooling at higher latitudes in the North Atlantic.

Researchers also see a connection between changes in Rossby Waves and the intensity of El Niño and La Niña ocean surface water temperature fluctuations in the Pacific Ocean. Also known as the El Niño-Southern Oscillation (ENSO), sea surface temperature increases during El Niño events and decreases in La Niña events can alter or intensify the monsoonal rainfall and hurricane patterns in North and South America. The challenge for climatologists and oceanographers is to understand how disrupted Rossby waves are a cause and consequence of global warming. The broader point is that global climate change is a complex process that involves the interplay between large scale atmospheric and oceanic processes operating at multiple scales.

SEE ALSO: Coriolis Force; El Niño and La Niña; Jet Streams; Rossby, Carl-Gustav; Waves, Gravity; Waves, Kelvin.

BIBLIOGRAPHY. James Hansen, et al., "GISS Surface Temperature Analysis. Global Temperature Trends: 2005 Summation," Goddard Institute for Space Studies, http://data.giss.nasa.gov/gistemp/2005/ (cited February 2008); James Hansen and Sergej Lebedeff, "Global Trends of Measured Surface Air Temperature," *Journal of Geophysical Research* (v.92, 1987); David Herring, "Earth's Temperature Tracker," Goddard Institute for Space Studies, http://www.giss.nasa.gov/research/features/temptracker/ (cited December 2007); Graham Quartly, et al., "Rossby Waves: Synergy in action," *Philosophical Transactions: Mathematical, Physical and Engineering Sciences* (v.361/1802, 2003); Saenko, Oleg, "Influence of Global Warming on Baroclinic Rossby Radius in the Ocean: A model Intercomparison," *Journal of Climate* (v.17, 2005); James Speth, *Red Sky at Morning: America and the Crisis of the Global Environment* (Yale University Press, 2005); Gabriel Vecchi, et al., "Weakening of Tropical Pacific Atmospheric Circulation Due to Anthropogenic Forcing," *Nature* (v.441, 2006).

CHRISTOPHER D. MERRETT
ILLINOIS INSTITUTE FOR RURAL AFFAIRS
WESTERN ILLINOIS UNIVERSITY

Weather

WEATHER IS THE physical condition or state of the atmosphere at any given time. It is what is happening in the atmosphere at any time or over any short period of time. If there were no atmosphere, there would be no weather. The principal elements of weather are temperature, pressure, winds, moisture, and precipitation. Thus, weather of any place is the sum total of its temperature, pressure, winds, moisture, and precipitation conditions for a short period of a day or a week. Temperature expresses intensity of heat. Unequal distribution of temperature over the Earth's surface causes differences in atmospheric pressure, which causes winds. Moisture is present in the atmosphere as water vapor, often condensed into clouds. It may be precipitated in the form of rain, hail, sleet, or snow. The capacity of air to gather and retain water vapor is largely dependent on its temperature. The higher the temperature, the greater the capacity of air to hold moisture. On cooling, the air is not able to retain all the moisture it gathers while warm. This leads to condensation and precipitation.

Weather is not synonymous with climate. Weather changes from day to day, whereas climate is something more stable, commonly defined as the average weather. Climate is the composite weather conditions over a considerable period of time. Weather conditions can change suddenly. Today may be warm and sunny, tomorrow may be cool and cloudy. Weather conditions include clouds, rain, snow, sleet, hail, fog, mist, sunshine, wind, temperature, and thunderstorms. Weather is driven by the heat stored in the Earth's atmosphere, which comes from solar energy. When heat is moved around the Earth's surface and in the atmosphere because of differences in temperature between places, this makes winds. Winds form part of larger weather systems, the most powerful of which is the hurricane. Other weather features like the thunderstorm also develop because of the movement of heat in the atmosphere. Some thunderstorms result in tornadoes. The Earth's water cycle plays an important role in the development of many weather features such as dew, fog, clouds, and rain.

Weather can be described using terms such as wet or fine, warm or cold, windy or calm. The science of studying weather is called meteorology. Meteorologists measure temperature, rainfall, air pressure, humidity, sunshine, and cloudiness, and they make predictions and forecasts about what the weather will do in the future. This is important for giving people advance notice of severe weather such as floods and hurricanes. Temperature is measured with a mercury thermometer in degrees Celsius. Temperature is the hotness or coldness of an object. Rainfall

Thunderclouds contain a great amount of energy, and the currents of air are strong enough to split apart the raindrops that are forming. This builds up an electric charge, released as lightning. The sound of thunder is the effect of the lightning strike on the air.

is usually measured by collecting what falls in rain gauges and is expressed as a depth of water that has fallen, in millimeters. Wind can be observed with a weather vane, but to measure its speed, more technical equipment is needed. Alternatively, the Beaufort scale can be used to make a judgment of the strength of the wind by observing how it affects objects outdoors, such as trees. The relative humidity and dew point temperature of the air can be determined by making measurements with a hygrometer and reading a table of numbers.

The purpose of a weather map is to give a graphical or pictorial image of weather to a meteorologist. As a forecasting tool, weather maps allow a meteorologist to see what is happening in the atmosphere at virtually any location on earth. Complex three-dimensional models of weather systems can be made by collecting weather data at multiple levels in the atmosphere. Computers then compile that information to produce the pictures that weather scientists analyze.

In the early days of meteorology, these pictures were all drawn by hand.

WEATHER PREDICTION AND FORECASTING

Weather affects virtually everyone daily. Human beings live largely at the mercy of the weather. It influences our daily lives and choices and has an enormous effect on corporate revenues and earnings. Weather can be predicted to some degree by observing the state of the sky and the wind. Weather is measured, and forecasts are usually released and used to make important decisions about travels, timing, and so on. Weather forecasts can save lives, reduce damages to property, reduce damages to crops, and tell the public and the global community about expected weather conditions. A forecast is basically predicting how the present state of the atmosphere will change with time. This involves plotting weather information on special charts. Weather radar and satellites are now also used to help predict the weather. Weather forecasters measure the weather so they can forecast it. Temperature, rainfall, wind, clouds, sunshine, and air pressure are measured all over the country, and the information is plotted on special charts. These weather charts have been used by forecasters for many years to predict what the weather will do over the next few days. Weather presenters often show simplified charts on television. Sophisticated equipment is used to help forecast the weather: weather radar can help to show where it is raining over a country, whereas satellites are used to reveal cloud cover and the development of large weather systems. Procedures for collecting and taking the observations are determined by the World Meteorological Organization.

There are a variety of forecasting techniques. The easiest of the techniques is called persistence. In this technique, tomorrow's weather is said to be same as today's weather. Local factors that should be considered when forecasting include clouds and snow. Clouds during the day will decrease the maximum temperature expected. Without the clouds, higher maximum temperatures would be obtained. With snow, the surface stays colder during the day, as less short-wave radiation is absorbed. During the night, radiational cooling effectively cools the surface. If there is some wind, the wind could create mechanical turbulence, which will mix down warmer air to the surface. Hence, winds keep the minimum nighttime temperature warmer than in calm conditions. By looking at cloud movement at different levels, one can infer the type of temperature advection that may be occurring, and therefore atmospheric stability.

Synoptic weather analysis and forecasting began after the telegraph made instantaneous long-distance communications possible after 1850. The invention of radio allowed the extension of the weather observation net over the oceans, and this was one of its first important uses. Every ship became a weather station, radioing reports at regular intervals to a central office. This great advance occurred by 1915, and weather prediction was of great service in Would War I. Significant theoretical advances were made during the middle of the 20th century, especially in the properties of the upper atmosphere, elucidated by such indirect methods as sound propagation and meteor trails.

THE SUN, THE AIR, AND THE WEATHER

The state of the air is important in weather studies. The elements of weather are air temperature, precipitation, cloud cover and sunshine, wind speed and direction, and air pressure. The air in the atmosphere is a mixture of gases consisting of nitrogen, oxygen, water vapor, carbon (IV) oxide, and some rare gases. The amount of water vapor is very important, as it gives us clouds and rain. Clouds are seen in the sky every day. They come in all shapes and sizes and bring with them all sorts of weather. A cloud is simply a visible mass of tiny water droplets that have formed because the air has become too cold at that height to store all its water as invisible vapor. This usually happens when warmer air near the ground is cooled down by rising higher in the atmosphere.

Different types of clouds can be described as they are viewed from the ground, using different terms. A cloud's name generally reflects the height at which it forms, as well as its general shape. The three main types of clouds are cirrus, cumulus, and stratus. Cirrus clouds are wispy in appearance and resemble horsetails. They are formed almost entirely of tiny ice crystals. Cumulus clouds look like fluffy balls of cotton wool. These can sometimes grow much larger, becoming cumulonimbus clouds, which bring heavy bursts of rain during thunderstorms. Cumulus clouds are clouds of vertical development and may grow upward dramatically under certain circumstances. Another type of cumulus cloud is the altocumulus cloud, which sometimes resembles fish scales. They sometimes have dark, shadowed undersides. Stratus clouds are layer clouds that form near the ground and make the weather very grey and dreary, and sometimes rainy. Stratus clouds form flat layers or uniform sheets. Only a fine drizzle can form from stratus clouds because there is no vertical development.

Air usually contains some water in the form of moisture called dew. The water is hardly seen, as it is like a gas. Humid air contains more moisture than dry air, but when the temperature of air falls, its ability to hold moisture decreases. If the temperature drops low enough, air can become saturated, even if it was originally much drier at the warmer temperature. At this point, excess moisture begins to condense, forming small water droplets on the ground called dew. As an air mass cools, it can hold less and less water vapor. If it cools down enough, it reaches a point at which the water vapor present in the air mass represents the amount needed to saturate an air mass at the lower temperature. The temperature at which saturation occurs is the dew point. The dew point depends on the amount of water vapor in the air. In winter, the temperature may fall below freezing. If dew has formed on the ground, it will freeze, forming white crystals called frost.

When the temperature of air close to the ground falls low enough, dew will form. If a larger layer of air is cooled, the condensation of excess moisture in the air forms a mist of tiny water droplets known as fog. Fog is common in the autumn and winter and usually forms when there is little or no wind to disturb it. It is also more common in hollows and valleys, where the air tends to be a little colder because it is heavier and sinks down into these places. Fog is least likely on hilltops, unless low clouds have descended to cover them.

The major driving force behind the weather is the sun. Energy from the sun is stored in the atmosphere as heat. When this heat is moved around, it makes the weather. The equator is much hotter than the poles because it receives much stronger sunlight. There is more heat stored in the atmosphere nearer the equator. Heat, however, likes to flow from warm to cold temperatures. More heat is also stored nearer the ground. This makes the air lighter than that above it, and it rises. When air rises, its temperature falls, which makes weather features such as clouds and rain.

SEVERE WEATHER

The greater part of Europe often experiences bad weather, particularly in winter. Bad weather usually comes in from the Atlantic Ocean with weather systems called depressions—regions of low pressure, strong winds, and rain. In the tropics, bad weather is much less common, but when it strikes, it can be devastating. Large tropical storms, which usually develop toward the end of summer, are called hurricanes. The most common place for hurricanes to form is in the Caribbean. Here, seawater temperature is high because the sunlight is strong, and a lot of heat is stored there. Under the right conditions, a storm will develop, which, with sufficient energy, will become a hurricane. Viewed from a satellite, a hurricane appears almost circular, with clouds spiralling out from a small center. On the ground, the weather in the center may be fairly calm, with clear skies, but

as the hurricane moves over, the weather can become very nasty, with winds of over 100 mi. (161 km.) per hour that are strong enough to tear roofs off houses.

Heavy rain, dark black clouds, and lightning are evidence of a thunderstorm. Thunderstorms are not nearly as strong as hurricanes, but they can be damaging, particularly if large hailstones fall out from their clouds. Thunderclouds are known scientifically as cumulonimbus clouds. Thunderstorms are more common in summer because they need a lot of energy to form. The energy comes from the heating of the ground and the surface air by the sun. If this heating is strong enough, air heated near the ground will rise up a long way into the atmosphere because it is lighter than the air around it. Warmer air is lighter than colder air. As the air rises up, it becomes colder. Moisture in the air begins to condense out as clouds, in the same way as fog forms on a calm, cool night. In thunderclouds, however, the energy is much greater, and the currents of air are strong enough to split apart the raindrops that are forming. This builds up an electric charge, which, when released, is seen as lightning. The sound of thunder is the effect of the lightning strike on the surrounding air. When rain or hail begins to fall from a thundercloud, it is usually very heavy, but it generally lasts no more than 30 minutes. Sometimes, however, the death of one thunderstorm may lead to the development of another, and the bad weather may continue for several hours.

A rapidly spiralling column of air is called a tornado. Large thunderstorms develop because there is an awful lot of energy stored in the atmosphere. Some large thunderstorms give birth to tornadoes. The tornado is usually very small in comparison to the thunderstorm, but it can wreak terrible havoc across the small area over which it moves. At a distance, a tornado is seen as a rapidly rotating funnel or spout of air, usually colored grey because of clouds and earth debris caught up inside it. The strongest tornadoes can have wind speeds of over 250 mi. (402 km.) per hour—and sometimes over 300 mi. (483 km.) per hour.

Most water on earth is in the oceans. A little, however, is contained by air in the atmosphere. The water in the atmosphere is usually not seen except when it rains, as it is in the form of moisture or vapor. Water enters the atmosphere by evaporating from the surface of the oceans, lakes, and other liquid water bodies. At higher levels in the atmosphere, the air is colder, and moisture begins to condense out as fine droplets, which we see as clouds. When conditions are right on the ground, the same process forms fog. Eventually, water in clouds forms rain, hail, sleet, or snow, which falls back to the ground. This movement of water from the earth to the atmosphere and back again is called the water cycle. The water cycle is responsible for much of the world's weather.

SEE ALSO: Atmospheric Composition; Clouds, Cirrus; Clouds, Cumulus; Clouds, Stratus Rain; Thunderstorms.

BIBLIOGRAPHY. P. Gore, *Basic Introduction to Weather* (Georgia Perimeter College, 2005); Introduction to Weather, www.ace.mmu.ac.uk/eae/Weather/Older/Weather_Introduction; R. Oblack, "Your Guide to Weather" (2007); E. S. Rubin and C. I. Davidson, *Introduction to Engineering & the Environment* (The McGraw-Hill Companies, 2001).

AKAN BASSEY WILLIAMS
COVENANT UNIVERSITY

Weather World 2010 Project

DEVELOPED BY THE Department of Atmospheric Sciences (DAS) at the University of Illinois Urbana-Champaign (UIUC), the Weather World 2010 project (WW2010) is a WWW framework for integrating current and archived weather data with multimedia instructional resources, using new and innovative technologies. The project focuses on many different manifestations of global warming such as hurricanes, clouds and precipitation, and El Niño. The accuracy of the instructional resources on the website has been reviewed by professors and scientists from the DAS at the UIUC and at the Illinois State Water Survey.

Weather World 2010 is the result of a long process dating back to January 1993, when Dr. Mohan Ramamurthy and programmer John Kemp created the Weather Machine. This resource allowed users to view weather images through a gopher server. At the same time, Steve Hall started to devise instructional modules in HyperCard for use by the educationally motivated Collaborative Visualization (CoVis) Project, a project that strives to promote project-based science learning. The advent of the first Web browser Mosaic marked the birth of a new medium for the exchange of information on the internet. As a result, efforts were made to place both weather and educational resources into HTML format. This conversion transformed the original Weather Machine products into the Daily Planet, which was created in 1994. This first attempt to realize a World Wide Web product was a success, and it soon became a popular site for many to browse.

Meanwhile, Mythili Sridhar and Steve Hall converted the HyperCard-based educational modules to HTML so that they could be used on the Web. The people working on the project soon became aware of the necessity of integrating weather data with explanatory and educational material. This led to the creation of the CoVis-sponsored Electronic Textbook, later named the Online Guide to Meteorology, in 1995. This server became an extremely useful location for users to learn about weather with archived data for examples. The CoVis also provided the Geosciences Web Server, a useful resource allowing users large collections of links to all sorts of weather information, as well as educational material.

In the summer of 1995, the team of researchers released the Weather Visualizer, which became one of the most successful attempts to integrate real-time weather and instructional material. It allowed users to customize their own weather maps and to get explanations about difficult or technical terms. Seeing the rising popularity of HTML and newer technologies such as Java, fall 1995 witnessed the creation of the Image Animator and the Interactive Weather Report—a couple of the first Java weather tools on the Web. Thanks to real-time instant access to weather data using Java applets, these tools allowed an increased level of educational interactivity.

The first discussions of a dynamic framework for hypermedia and CD-Web interactivity started to take place in the early months of 1996. Such discussion was prompted by the difficulties experienced by CoVis teachers in accessing the large volume of information within the educational modules. Often, their connections were too slow to effectively use the resources in their classrooms. The desire to improve and redesign the existing resources, including developing better graphical interfaces and navigation systems, prompted Steve Hall and Dave Wojtowicz to begin the construction of the early form of Weather World 2010 in May 1996. This project soon grew into a much larger task than was initially conceived: rather than a simple improvement, this new project was to provide a new framework to support the pre-existing modules and allow users access to weather data and information. In October 1996, the project was presented to the students, faculty, and staff of the Department of Atmospheric Sciences at the University of Illinois. The project drew on the different expertise of the many volunteers that it managed to involve, and in return, it provided them with a chance to learn HTML. Since October 1996, nearly 20 people have volunteered for a variety of tasks: writing scripts to generate image products, developing helper sections to describe images and fundamental concepts, creating instructional modules, and participating in design input and performance testing.

Initially, the team of WW2010 spent considerable time browsing the internet, finding the most appealing features of available websites, and incorporating some of these features into the server. Although Java and other new technologies are opening up exciting possibilities, the researchers of WW2010 remain concerned with allowing easy access to large sets of data and instructional materials and efficient navigation. WW2010 is an example of how new and innovative technologies can be used to visualize some of the effects of climate change and global warming on our lives.

SEE ALSO: Climate Change, Effects; Global Warming; University of Illinois; Weather.

BIBLIOGRAPHY. Mike Harris, *Understanding Weatherfax: A Guide to Forecasting the Weather from Radio and Internet Fax Charts* (Sheridan House, 2005); Weather World 2010 Project, http://ww2010.atmos.uiuc.edu/(Gh)/home.rxml.

LUCA PRONO
UNIVERSITY OF NOTTINGHAM

Western Boundary Currents

WESTERN BOUNDARY CURRENTS are intense jet currents at the western periphery of large-scale oceanic gyres in the World Ocean. As was shown in the pioneer paper of Henry Stommel in 1948, they are the result of two causes: the so-called β-effect (this term

has arisen from traditional representation of Coriolis force, f, in the following form: $f = f_0 + \beta y$, where f_0 is a Coriolis parameter at a definite latitude; in other words, the β-effect is to the result of the spherical form of the Earth turning around its axis), and low conservation of absolute vortex for the oceanic motions.

Oceanic gyres are forced by horizontally inhomogeneous large-scale wind fields (or wind vorticity). For instance, in the North Atlantic Ocean, anticyclonic subtropical gyre is situated under northeastern trade wind and midlatitude westerly wind as a result of clockwise wind vorticity, whereas north tropical cyclonic gyre is a result of anticlockwise wind vorticity between the Intertropical Convergence Zone and the northeastern trade wind. Currents in the western part of each gyre are more intense than in the eastern part because this is dictated by low conservation of absolute (relative plus planetary) vortex.

Each particle moving northward (southward) gets an additional (loses) planetary vorticity as a result of the spherical form of the Earth. In the clockwise gyre, this should be compensated for by the increasing relative negative vorticity, that is, by the intensification of clockwise rotating. In the anticlockwise gyre, this should be compensated by the increasing of relative positive vorticity, that is, by the intensification of anticlockwise rotating. In both cases, this leads to intensification of currents in the western periphery of the basin. In the eastern part of gyres, all particles move in the opposite direction, in comparison with the western part. It leads to the weakening of circulation in the eastern gyres' end.

The β-effect may be also understood in terms of Rossby waves. Actually, long, nondispersive Rossby waves carry (kinetic) energy from the east to the west within each gyre. After their reflection from the western boundary of the basin, the short dispersive Rossby waves are generated and move to the east. However, the short Rossby waves are dissipated in the relatively narrow vicinity of the near-coastal zone as a result of their shortness and dispersive properties, which lead to more affective realization of dissipative processes. Thus, the kinetic energy of the planetary Rossby waves is accumulated in the vicinity of the western periphery of the gyres.

In fact, western boundary currents (especially in the Atlantic Ocean) are also controlled by thermohaline factors. The β-effect affects thermohaline circulation and causes the intensification of the thermohaline currents in the western part of the basin. Deep thermohaline currents in the North Atlantic Ocean (generating in the region of sinking of Deep Atlantic Ocean Water and spreading at the depths between 1.5 and 2.5 mi., or 2.5 and 4 km.) are southward, whereas compensative thermohaline currents in the upper baroclinic layer (between the surface and 0.6 and 1.2 mi., or 1 to 2 km.) are northward. As a result, the wind-driven northward Western Boundary Currents in the clockwise gyres intensify, and southward ones in the anticlockwise gyres weaken as a result of meridional thermohaline circulation.

The most intense Western Boundary Currents in the Northern Hemisphere are the Gulf Stream, Labrador Current, North Brazil Current (Atlantic Ocean), Kuroshio (Pacific Ocean), and Somali Current (Indian Ocean). The velocity in these currents' axes reaches or even exceeds 6.5 ft. (2 m.) per second. Detailed analysis of the structure and origins of Western Boundary Currents (e.g., the Gulf Stream) was done by Henry Stommel in 1958 and 1966.

Western boundary currents in the North Atlantic Ocean carry up to about 100 Sv (1 Sverdrup = 10^6 cu. m. per second) of water in the upper baroclinic layer. The wind vorticity accounts for about 30 to 60 Sv (30–60 multiplied by 10^6 cu. m. per second). The average power of source of Deep Atlantic Ocean Water is about 20 Sv (20 multiplied by 10^6 cu. m. per second). So, the joint effect of wind vorticity and meridional thermohaline circulation can explain up to 80 percent of observed transport of the western boundary currents in the North Atlantic Ocean.

Western boundary currents are meandered jets that generate the intense mesoscale eddies, the so-called rings. The typical horizontal size of rings is about 62 mi. (100 km.), and orbital velocity is 3.3 to 6.5 ft. (1 to 2 m.) per second. Rings trap the water in their central part and carry it with a typical speed of about a few centimeters per second. The lifetime of the rings may reach 4 years. Thus, mesoscale eddies account for a significant portion of volume transport in the vicinity of Western Boundary Currents. Recirculation of the Gulf Stream is one of the integral manifestations of mesoscale effects.

SEE ALSO: Coriolis Force; Gulf Stream; Intertropical Convergence Zone; Kuroshio Current; Stommel, Henry; Thermohaline Circulation; Waves, Planetary; Waves, Rossby; Wind-Driven Circulation.

BIBLIOGRAPHY. Henry Stommel, "The Westward Intensification of the Wind-Driven Ocean Currents," *Transactions of the American Geophysical Union*, (v.29/2, 1948); Henry Stommel, *The Gulfstream. A Physical and Dynamical Description* (Cambridge University Press, 1958).

ALEXANDER BORIS POLONSKY
MARINE HYDROPHYSICAL INSTITUTE, SEBASTOPOL

Western Regional Climate Center

THE WESTERN REGIONAL Climate Center (WRCC), based in Reno, Nevada, and inaugurated in 1986, is one of six regional climate centers in the United States. The regional climate center program is administered by the National Oceanic and Atmospheric Administration. Specific supervision is provided by the National Climatic Data Center of the National Environmental Satellite, Data, and Information Service.

The mission of the Western Regional Climate Center is to disseminate climate data and information of the highest standards pertaining to the western United States; promote better use of this information in policymaking; carry out applied research related to climate issues; and improve the coordination of climate-related activities at the state, regional, and national levels. The center receives queries from lawyers, media, insurance companies, different businesses, teachers and students, contractors, the Forest Service, state and local government, and individuals interested in weather observation.

The data collected by the center include daily climate observations for a digital period of record (6,781 stations, about 2,608 now active), summarized monthly climate data (5,240 stations), hourly precipitation data (1,937 stations), upper air soundings recorded twice a day (about 50 stations), surface airway hourly observations (over 1,800 stations nationwide). In addition, the center provides access to these databases: remote automatic weather station, historic lightning data though 1996, access to Natural

Resources Conservation Service SNOTEL (SNOwpack TELemetry), and other western databases.

The WRCC coordinates the work of federal resource management agencies and western committees and commissions. It liaises with other centers and programs such as the National Climate Data Center in Asheville, North Carolina; regional climate centers; state climatologists and state climate programs; the Climate Analysis Center, Washington D.C.; and the National Weather Service. Its main areas of research include the effects of climate variability in the western United States, the quality control of western databases, the relation of El Niño/Southern Oscillation to western climate, the climatic trends and fluctuations in the West, geographic information systems, and remote sensing. It is also the home of the regional climatologist.

Although the WRCC is concerned with some of the most evident phenomena of global warming, such as the effect of El Niño on western climate, the climatologists at the center do not agree on a single definition of global warming. For example, in July 2007, Jim Ashby, one of the WRCC climatologists, challenged the credibility of weather data collected by weather stations, as they are often moved. Even moving the weather station just few hundred yards could make a difference in temperature and moisture, Ashby maintained, creating a situation in which data belonging to different areas are compared. This comparison of inhomogeneous data would be, according to Ashby, the same as comparing apples and oranges. Ashby has also stressed that even stations that remain in the same place can have changing circumstances that alter weather readings.

Atmospheric readings can be altered by the surroundings (the presence of trees favors cooler temperatures, whereas new buildings and more roads and rooftops retain more heat). Ashby takes Reno, where the Western Regional Climate Center is situated, as a classic example of a place where weather is being changed by urbanization. Average low temperatures in town have risen about 10 degrees (5 degrees C) in the last 20 years. "Most of the warming is due to the fact Reno is growing like a weed," Ashby argued. "The weather station used to be out in a field somewhere. Now it's surrounded by asphalt." This line of reasoning comes dangerously close to that of global warming contrarians, who say that all the moves and changes at weather stations have affected the validity of the climate record.

Another climatologist at the center, Kelly Redmond, has instead pointed out the visible effects of global warming on the vegetation of the region. Western states are heavily dependent on snowpack for their flora. Melting snow provides three-fourths of the water in streams. Over the past 35 years, temperatures across the region have risen from 1–3 degrees F (0.5–1.6 degrees C), causing the snow to melt as much as three weeks earlier and leading to widespread droughts.

SEE ALSO: Climate Change, Effects; Nevada; Weather.

BIBLIOGRAPHY. Associated Press, "West Getting Burned by Global Warming," www.heatisonline.org/contentserver/objecthandlers/index.cfm?id=4659&method=full; Robert Krier, "Weather Station Moves Add Degree of Difficulty," www.signonsandiego.com/news/nation/20070709-9999-1n9station.html; Western Regional Climate Center, www.wrcc.dri.edu/wrccmssn.html.

LUCA PRONO
UNIVERSITY OF NOTTINGHAM

West Virginia

WEST VIRGINIA HAS the 37th largest population of America's 48 contiguous states and is the 20th largest emitter of carbon dioxide. The past century has seen an average temperature increase of 1.1 degrees F (degrees C) in Charleston, the state's capital, and an average precipitation increase of 10 percent across the state. Scientists predict that if the climate continues to rise, there could be an increase in the number of thunderstorms and extremely hot days. By 2100, average temperatures in winter, spring, and summer could be 3 degrees F (1.6 degrees C) warmer than at present, and autumn temperatures could be 4 degrees F (2.2 degrees C) higher. Precipitation could also be significantly higher.

Rising temperatures have the potential to alter the state's natural environment, particularly the topography of its forests. The exact effect of a warmer climate on forests, which constitute 97 percent of the state's land mass, is unknown. If average rainfall decreases in the coming years, forests could shrink by as much as 10 percent, and grassland will appear where trees once stood. In contrast, a combination of warmer temperatures and increased rainfall could support the growth of tree species that thrive in such conditions, such as oaks and pines, and create much denser forests. Changes to the state's forests would have significant implications for the plant and animal species that rely on forests for their survival. Hares, red squirrels, and the northern flying squirrel are particularly at risk, as their habitats are concentrated in mountain forests, thus making it impossible for them to move to higher altitudes. Significant climate change over the next century could potentially eradicate some or all of West Virginia's high-altitude forests. The state's most expansive forest, the Monongahela Forest, is one such example.

The state's water supply, much of which originates from groundwater such as lakes and rivers, could also be affected by hotter weather. Much of the state lies at a high altitude, and an earlier spring would cause water levels to rise as a result of melting snow. Higher water levels may cause problems such as flooding, property damage, and erosion. Summer droughts, which already have an effect on water supplies, could cause rivers and lakes to dry up entirely, cutting off water to a substantial portion of the state's population. At present, nearly all the rural population and around half the urban population are supplied by groundwater. Receding water reserves could exacerbate problems such as acid and sludge drainage from abandoned coal mines and increase concentrations of bacteria and feces in the water supplied to homes and businesses. Low water supplies would also affect the tourist industry by limiting the opportunities for sports such as white-water rafting.

A warmer climate in West Virginia could also change patterns of farming and agriculture. Less rainfall would reduce soil moisture and create the need for irrigation—a practice that would divert water from homes and communities. Alternatively, it could push agriculture to the northern reaches of the state. More sun and less rain could increase yields of hay, the state's main crop, by as much as 30 percent, but at the same time it would destroy arable pasture used for livestock grazing. In turn, livestock farming would be less profitable, because animals would not yield as much meat.

If climate change continues, it will create health risks for the inhabitants of West Virginia. Heat-related deaths and illnesses would likely increase, particularly among elderly people living alone. The increase in ground-level ozone that would result from warmer temperatures could potentially lead to more incidences of respiratory illnesses such as asthma and impaired lung function. Ticks and certain species of insect would be more able to thrive and spread diseases such as dengue fever in a warmer climate. Mosquitoes would be especially adept at surviving and would likely spread malaria, a disease that does not naturally exist in the state, and California encephalitis, a lethal disease that causes neurological damage. Higher rainfall, another possible consequence of climate change, could facilitate the transmission of water-borne illnesses such as giardia, crypto-sporidia, and bacterial gastroenteritides.

The state faces specific policymaking challenges to address the possible consequences of climate change.

SEE ALSO: Agriculture; Forests; Rain; Rainfall Patterns.

BIBLIOGRAPHY. The Earth Institute, Columbia University, "Next Generation Earth: West Virginia," www.nextgenerationearth.org (cited December 2007); Environmental Protection Agency, *Climate Change and West Virginia* (U.S. Environmental Protection Agency, Office of Policy, 1998).

RICHARD FRY
WAYNE STATE UNIVERSITY

Wind-Driven Circulation

WIND-DRIVEN CIRCULATION (WDC) refers to ocean currents initiated and propelled by winds blowing across the surface of the ocean. Wind-driven circulation is part of the complex system of energy and heat redistribution that helps to draw tropical heat away from equatorial regions to moderate the earth's climate.

WIND GENERATION

Wind is caused by the uneven solar heating of the Earth's surface, which creates areas of high and low atmospheric pressure. Pressure differentials cause air to circulate from areas of high to low pressure. Solar heating is greatest near the equator, causing a region of low pressure. Rising air moves northward as cooler air from high pressure, mid-latitude regions moves towards the equator. The Coriolis Effect, the rotational action of the earth spinning on its access, deflects the winds towards the east north of the equator and to the west south of the equator. This explains why prevailing winds in the mid-latitude northern hemisphere come from the west.

OCEANIC CIRCULATION

An understanding of wind generation helps to explain the operation of ocean currents. There are two general types of ocean circulation and both result from solar heating: (1) WDC, and (2) thermohaline circulation (THC). Differential solar heating generates wind. Wind blowing across the surface of the ocean causes the water to move. The direction of the ocean current is determined in part by the direction of the prevailing winds at a given latitude. WDC is responsible for currents near the ocean's surface.

THC drives deep-water ocean currents and vertical movement of water. THCs occur as water density increases, either through cooling as water moves towards the poles, or because of increased salinity. In both cases, the THC is responsible for carrying warmer surface water to the ocean depths.

There is an interplay, then, between the WDCs and THCs that is exemplified in the Gulf Stream that moves warm water from the Caribbean in a northeasterly direction, along the eastern seaboard of North America, towards northern Europe. As WDCs carry water away from the tropics, the water cools and becomes more saline. It consequently becomes more dense, which causes it to sink along a THC. This mechanism whereby warm surface water is transported towards the poles, descends as it becomes cooler, to be re-circulated southward as cool deepwater currents is sometimes referred to as the North Atlantic conveyor.

GLOBAL WARMING AND THE INTERPLAY OF WDCS AND THCS

The Gulf Stream starts as a WDC because it gets its initial energy from the trade winds. Global warming could affect the Gulf Stream in two ways. The first, and more widely reported mechanism occurs because global warming could melt arctic ice, decreasing sea salinity. This will in turn hinder the ability of the water to descend as part of the THC. The end result is the cooling of northern latitudes.

Global warming could also affect the initial forcing of the Gulf Stream. Some researchers suggest that global warming will disrupt and weaken the trade winds. Diminished trade winds means less energy driving the Gulf Stream northward, also contributing

to a cooler climate at high latitudes. This could cause dramatic cooling in northern Europe which needs the North Atlantic Conveyor to moderate its climate. The paradox is that global warming could ultimately cause regional cooling in some parts of the world by affecting both the THCs and WDCs.

SEE ALSO: Coriolis Force; El Niño and La Niña; Jet Streams; Trade Winds; Winds, Easterlies; Winds, Westerlies.

BIBLIOGRAPHY. Oleg Saenko, et al., "On the Response of the Oceanic Wind-Driven Circulation to Atmospheric CO_2 Increase," *Climate Dynamics* (v.25, 2005.); J.R. Toggweiler, and Joellen Russell, "Ocean Circulation in a Warming Climate," *Nature* (v.451, 2008).

Lyn Michaud
Independent Scholar

Winds, Easterlies

WINDS ARE DEFINED by their origins. The "easterly" descriptor refers to winds with an easterly zonal component (coming from the east). These include northeasterly and southeasterly winds. Easterly winds occur at all atmospheric scales, including local and synoptic. However, the term "easterly winds" generally refers to large-scale belts of winds operating within the global circulation of the atmosphere.

In the atmosphere's general circulation, two distinct bands of easterly winds exist: the trade winds and the polar easterlies. They are found at low and high latitudes, respectively, and arise from the dynamics of air flow among pressure systems.

Among the most consistent winds on earth, the trade winds (or trades) are part of the Hadley cell circulation found from approximately 0 degrees to 30 degrees north and south of the equator. Air rises near the equator as a result of a combination of convection spurred by intense solar radiation and the low-level convergence of wind in a circumpolar zone called the Intertropical Convergence Zone. As the rising air approaches the tropopause, it turns. At approximately 30 degrees, the air subsides, or sinks, resulting in the belt of persistent subtropical highs. Air diverges out of these anticyclones and flows toward the equatorial low. This outflow of air gives rise to the trades. As the air moves toward the equatorial low, it is deflected as a consequence of the Coriolis force (or effect). This deflection results in northeasterly trade winds in the Northern Hemisphere and southeasterly trades in the Southern Hemisphere. The trade winds are also referred to as tropical easterlies, particularly when the associated vertical wind shear is large.

The polar easterlies, a belt of winds found from approximately 60–90 degrees in each hemisphere, constitute another component of the general circulation. These winds originate from dynamics similar to those of the Hadley cell. Thermally driven high pressure exists at the poles; similarly, a circumpolar zone of low pressure is found near 60 degrees. As air flows from the polar high to the subpolar low (as a result of the presence of a pressure gradient), it is deflected. This deflection leads to a belt of easterly winds in both Northern and Southern hemispheres.

Rising air over the western tropical Pacific, with sinking air over the eastern tropical Pacific, comprises the Walker air current. The trades, as part of the Walker circulation, push warm surface waters toward the west.

Any weakening of the trades, then, would disrupt this oceanic transport of water. This dynamic is similar to an El Niño, in which the trade winds slow and even break down. Recent research suggests that climate changes, specifically global warming, would weaken or slow the trade winds. Indeed, the Walker circulation has already diminished by 3.5 percent since the 1800s. This slowing is projected to continue, and much of this change is attributed to anthropogenic activity.

The culprit in the slowing trade winds is the balance between evaporation and precipitation. To maintain a balance, the atmospheric absorption of moisture must be balanced by its release by precipitation. Water vapor, transported west by the trade winds, is precipitated out over the western Pacific. Warmer temperatures, then, spur the absorption of additional water vapor by enhancing evaporation rates. However, precipitation rates increase more slowly. To balance these processes, wind flow must decrease.

Weakened trade winds would result in numerous consequences, including the disruption of normal weather and climate patterns, as well as suppressed oceanic upwelling and potentially reduced biological productivity. The latter would have important economic ramifications, particularly for the Pacific fishing industries.

SEE ALSO: El Niño and La Niña; Walker Circulation; Winds, Westerlies.

BIBLIOGRAPHY: C. Donald Ahrens, *Meteorology Today*, 8th ed. (Thompson Brooks/Cole, 2007); Peter J. Robinson and Ann Henderson-Sellers, *Contemporary Climatology*, 2nd ed. (Prentice Hall, 1999); Gabriel A. Vecchi, Brian J. Soden, Andrew T. Wittenberg, Isaac M. Held, Ants Leetma, and Matthew J. Harrison, "Weakening of Tropical Pacific Circulation as a Result of Anthropogenic Forcing," *Nature* v.441/4, May 2006).

PETRA A. ZIMMERMANN
BALL STATE UNIVERSITY

Winds, Westerlies

THE WESTERLIES ARE the prevailing winds in the middle latitudes blowing from the subtropical high pressure toward the poles. The westerlies originate as a result of pressure differences between the subtropical high-pressure zone and the subpolar low-pressure zone. The westerlies curve to the east because of the Coriolis effect, caused by the Earth's rotation. In the Northern Hemisphere, the westerlies blow predominantly from the southwest, whereas in the Southern Hemisphere, they blow predominantly from the northwest. The equatorward boundary is fairly well defined by the subtropical high-pressure belts, whereas the poleward boundary is more variable. The westerlies can be quite strong, particularly in the Southern Hemisphere, where less land causes friction to slow them down. The strongest westerly winds typically occur between 40 degrees and 50 degrees latitude.

Winds transport heat from warmer areas to cooler areas and help the Earth to maintain equilibrium of its thermal environment. In the midlatitude, the westerlies play a big role in the weather and atmospheric circulation in the middle latitudes. They transport warm, moist air to polar fronts and are also responsible for the formation of extratropical cyclones. In winter, they collect warm, moist air from over the oceans, move it to the cooler continents, and bring heavy rainfall to the areas like the northwest coast of the United States. In summer, they collect hot, drier air from over the continents and move it to the oceans.

Does global warming influence westerlies? A recent study of westerlies in the Southern Hemisphere shows that the westerlies are shifting southward, toward Antarctica. No conclusion has been made yet, however. Some scientists believe that recent observations are related to global warming, but others think of them as a part of natural variations. North Atlantic Oscillation (NAO) is one indicator that shows the relationship between global warming and westerlies. NAO is calculated by the difference in pressure between the permanent low-pressure system located over Iceland (Icelandic Low) and the permanent high-pressure system located over the Azores (Azores High). Global warming can reduce the difference in pressure between the two places. At a high NAO index, a large pressure difference between the two places induces stronger westerlies flow. Storm tracks advance northward, and Europe experiences milder winters but more frequent rainfall in central Europe and nearby. At a low NAO index, with suppressed westerlies, storm tracks move more toward the Mediterranean, which results in colder winters in Europe and southern Europe and in North Africa receiving more storms and higher rainfall.

El Niño–Southern Oscillation (ENSO) is another indicator. In winters of El Niño years, the polar jet stream in the Northern Hemisphere moves farther poleward and brings warmer winter weather to the northeastern part of the United States. In the winter of 2006 to 2007, the warming induced was about 9 degrees F (5 degrees C), which was as much as five times the typical air temperature increase compared with a warming in a typical El Niño year. Changes in both surface and upper-level westerlies resulting from El Niño patterns can also influence the development, intensity, and track of hurricane over the tropical Atlantic Ocean. In fall 2006, El Niño strengthened the upper-level westerlies, increased wind shear, and discouraged tropical cyclogenesis over the tropical Atlantic. Whether or not global warming is behind these stronger El Niño patterns is still being researched. A recent climate model (Joellen L. Russell et al., 2007) indicates that westerlies influence the temperature of the Southern Ocean. According to the model, the southward movement of the Southern Hemisphere westerlies in recent years transfers more heat and carbon dioxide into the deeper waters of the Southern Ocean. This poleward shift of the westerlies has intensified the strength of the westerlies near Antarctica. The pattern could slow down global warming somewhat but also induce ocean levels to rise in Antarctica.

How global warming influences the westerlies still remains in question. The recent observation, however, suggests that global warming brings noticeable change in the westerlies.

SEE ALSO: Climate Change, Effects; Winds, Easterlies.

BIBLIOGRAPHY: C. Donald Ahrens, *Meteorology Today*, 8th ed. (Thompson Brooks/Cole, 2007); Peter J. Robinson and Ann Henderson-Sellers, *Contemporary Climatology*, 2nd ed. (Prentice Hall, 1999).

JONGNAM CHOI
WESTERN ILLINOIS UNIVERSITY

Wisconsin

WISCONSIN IS 65,498 sq. mi. (169,639 sq. km.) in size, with inland water making up 1,830 sq. mi., (4,740 sq. km.), and Great Lakes Coast water of 9,358 sq. mi., (24,237 sq. km.). Wisconsin's average elevation is 1,050 ft., (320 m.) above sea level, with a range in elevation from 579 ft., (176 m.) above sea level on Lake Michigan to 1,951 ft., (595 m.) at Timms Hill. Wisconsin is customarily divided into two major natural regions: the Central Lowland (low-lying area and swings in a broad belt across the southern two-thirds of Wisconsin) and the Superior Upland (higher elevation, forest cover, and numerous small glacial lakes). A low-lying and partially swampy plain, known as the Lake Superior Lowland, occupies the areas along the southern shore of Lake Superior. Wisconsin's rivers drain into either the Mississippi River system or the Great Lakes–St. Lawrence system. The Mississippi River forms part of the border with Minnesota and the entire border with Iowa. Wisconsin has many lakes and a few reservoirs.

Weather in Wisconsin varies greatly from winter to summer, with bitter cold and lots of snow in winter and warm and humid summers. This type of climate is called temperate continental. Three different types of air masses affect Wisconsin. The continental polar air masses from the northwest bring bitter cold, dry weather in winter. The maritime tropical gulf air mass from the Gulf of Mexico brings high humidity and heat in summer. The maritime polar Pacific air mass can bring more moderate weather straight from the west any time during the year. Wisconsin's weather is also affected by the Great Lakes on its northern and eastern sides by their moderating of the weather in the coastal areas, which are warmer in winter and cooler in summer than the rest of the state. Residents of shoreline cities get lake-effect snow (perhaps several inches, while 30 mi. (48 km.) inland, the sky is sunny and nobody needs to shovel). The annual precipitation is 31 in (79 cm.). The southern counties get about 30 in. (76 cm.) of snow each winter, and the annual snowfall in Iron County to the north can be as much as 100 in. (254 cm.). The highest temperature recorded in the state is 114 degrees F (45.5 degrees C) on July 13, 1936, and minus 55 degrees F (13 degrees C), on February 4, 1996, is the lowest temperature recorded in the state.

Wisconsin is a Corn Belt state, and corn is its major crop. Other leading crops are soybeans, hay, sweet corn, potatoes, cranberries, and oats. Major industries include dairy products, corn, hay, and beef. Wisconsin takes advantage of Lake Michigan for commercial fishing, though it is only a small industry in the state. Important species are whitefish, lake trout, perch, chub, alewife, and carp. Forestry is a minor industry, with most hardwood cut going to plywood and veneer manufac-

Canadian geese seeking refuge in the comparatively warm water temperature during -20°F weather in Mequon, Wisconsin.

ture; the pulp and paper industry consumes much of the softwood harvest. Wisconsin's mineral output is limited to stone, sand and gravel, copper, lime, lead, and talc.

Wisconsin experienced a sample of possible effects of climate change with the Great Flood of 1993, when rain and snow runoff raised the Mississippi River along with other rivers and small streams. Madison reported 21.49 in. (54.5 cm.) of rain. Signs of global warming are apparent throughout the Great Lakes. Over the past 150 years, the average extent of ice cover on many of Wisconsin's lakes has continuously declined.

Although climate models vary on the amount of temperature increase during the 21st century, ranging from 8–7 degrees F (4.4–3.8 degrees C), the average summer water level in the Great Lakes could drop 1.5–8 ft. (0.4–2.4 m.) by the end of the century, and extreme 100-year floods (named because they happen once every 100–200 years) could begin to occur on a much more frequent basis. With extreme floods and higher temperatures, Wisconsin could experience more drought conditions as a result of increased evaporation, and reduced soil moisture may force farmers to rely more on irrigation.

Possible effects from increased temperatures include decreased water supplies; changes in food production, with agriculture improving in cooler climates and decreasing in warmer climates (Wisconsin farmers could see an increase in demand for corn-based ethanol); changes in rain pattern to downpours, with the potential for flash flooding and health risks of certain infectious diseases from water contamination or disease-carrying vectors such as mosquitoes, ticks, and rodents; warmer temperatures can cause heat-related illnesses and lead to higher concentrations of ground-level ozone pollution causing respiratory illnesses, especially in cities with smog, like Milwaukee.

On the basis of energy consumption data from EIA's State Energy Consumption, Price, and Expenditure Estimates, released June 1, 2007, Wisconsin's total CO_2 emissions from fossil fuel combustion in 2004 were 107.05 million metric tons, made up of contributions from commercial (5.67 million metric tons), industrial (16.67 million metric tons), residential (10.02 million metric tons), transportation (30.57 million metric tons), and electric power (44.13 million metric tons).

Wisconsin established the Task Force on Global Warming in April 2007 to understand emissions impacts, determine modeling scenarios, and define the sensitivities that will need to be run to establish reasonable ranges of outcomes for confident decision making by Wisconsin legislators. Wisconsin joined the Climate Registry, a voluntary national initiative to track, verify, and report greenhouse gas emissions, with acceptance of data from state agencies, corporations, and educational institutions beginning in January 2008.

Wisconsin lawmakers have enacted a minimum renewable electricity standard that requires utilities to provide 22 percent of the state's power from renewable sources by 2011. To meet this goal, the Wisconsin Public Service Corporation installed 14 wind turbines in Kewanee County that are expected to provide enough electricity for 3,600 homes and generate tax revenue. The National Renewable Energy Laboratory also estimates Wisconsin as having the solar potential to provide electricity generation.

The Wisconsin Department of Natural Resources is the agency responsible for carrying out state conservation programs. Reforestation, soil erosion, and wildlife management are the principal concerns of conservationists in Wisconsin.

SEE ALSO: Climate Change, Effects.

BIBLIOGRAPHY. Forestry and Agriculture Work and Department of Natural Resources, www.wisconsin.gov; Alfred N. Garwood, *Weather America Latest Detailed Climatological Data for Over 4,000 Places and Rankings* (Toucan Valley Publications, 1996); Walter A. Lyons, *The Handy Weather Answer Book* (Visible Ink, 1997); National Wildlife Federation, "Global Warming and Wisconsin" (April 24, 2007).

LYN MICHAUD
INDEPENDENT SCHOLAR

Woods Hole Oceanographic Institution

THE WOODS HOLE Oceanographic Institution (WHOI), a world-renowned private, not-for-profit organization founded in 1930, is the largest independent oceanographic research institution in the United States. WHOI is committed to higher education and scientific research that furthers the understanding of the world's oceans and their role within the Earth's ecosystem as a whole. In addition, WHOI disseminates its research findings and information to the public and policymakers to foster understanding and decisionmaking for the greater good of society. At any given moment, more than 350 projects are in progress around the world, representing a wide range of scientific inquiries, in which collaboration and creativity are highly valued and encouraged.

WHOI recruits distinctly qualified scientists, engineers, staff, and students and provides an interdisciplinary and flexible setting in which students, as future scientists and engineers, can thrive. The research activities of the WHOI are centered in five departments: applied ocean physics and engineering, encompassing ocean acoustics, observation systems, and immersible vehicles; biology, with specialties in biooptical studies, pollution effects in the sea, and the behavior of marine animals; marine chemistry and geochemistry, focusing on chemical analyses and modeling of ocean processes; physical oceanography, including the examination of the geography and physics of the ocean currents; and geology and geophysics, including the study of the oceans' role in past climate change.

WHOI's research efforts are enhanced by four ocean institutes, established in 2000, which address the concerns of members of the general public and policymaking bodies and make research findings available as expeditiously as possible. The Coastal Ocean Institute encourages pioneering, interdisciplinary experiments, and field missions to increase knowledge and understanding about basic ocean processes. The Ocean Life Institute sponsors studies of the oceans' organisms and processes to understand the evolution of life and adaptability of species to their natural surroundings.

The Ocean and Climate Change Institute, among other undertakings, supports research about the effect of greenhouse gases on the ocean and the effect of ocean dynamics that may cause large, sudden climate shifts. WHOI's interest in research and technology that focus on the possible regional and global effects of changes in the Arctic on circulation and climate, including the effects on fisheries and ecosystems beyond the Arctic, was advanced in 2006 with the receipt of a grant to establish the Clark Arctic Research Initiative. The initiative is intended to make resources available to support individual and collaborative focused research based in the Arctic region through 2007.

The WHOI's fourth institute, the Deep Ocean Exploration Institute, encourages multidisciplinary research endeavors throughout WHOI and advocates the development of deep-sea technology, including vehicles such as *Alvin,* the Deep Submergence Vehicle owned by the U.S. Navy and operated by WHOI. In addition to *Alvin,* the National Deep Submergence Facility, funded by the federal government and located at WHOI, oversees the operation of the remotely operated vehicle *Jason/Medea* and the robotic underwater vehicle *ABE.* The year 2008 will be one of significant change as the *ABE* is replaced by the *Sentry,* which can dive to greater depths and has longer deployment capability than its predecessor.

Of even greater note is the anticipated retirement in 2008 of *Alvin,* which carries two scientists and a pilot as deep as 3 mi. (4.8 km.) for 6- to 10-hour dives. The submersible vehicle, which has made more than 4,000 dives since 1964, can reach almost two-thirds of the ocean's floor, moving at speeds up to 1.5 m.p.h. *Alvin* has remained a state-of-the-art vehicle because it has been disassembled, inspected, and reassembled every 3 to 5 years. Every part of *Alvin* has been replaced at least once in the vehicle's lifetime, but additional refurbishing will unlikely enable the sub to do more than it does now. *Alvin's* replacement vehicle will be funded by the National Science Foundation and will reach greater depths than *Alvin,* reaching almost the complete ocean floor. WHOI has indicated that operating both submergence vehicles is cost-prohibitive. *Alvin* leaves behind a rich history that has included locating a lost hydrogen bomb in the Mediterranean Sea, photographing the Titanic, and exploring deep-sea hydrothermal vents, where it gathered information about 300 life-forms that were previously unknown.

The WHOI's research fleet, which in addition to its submergence vehicles includes one of the United States' newest research vessels, the Navy-owned *Atlantis,* among its ships, provides students in the joint program offered by WHOI and the Massachusetts Institute of Technology (MIT) incomparable opportunities at research and learning. The MIT/WHOI Joint Program ranks among the leading graduate marine science programs in the world. In addition,

WHOI 's Geophysical Fluids Dynamic (GFD) Program offers eight to 10 new graduate fellows each year the unique experience of participating in an intensive 10-week interdisciplinary research program. The students present a lecture and prepare a written report for inclusion in the GFD Proceedings Volumes.

Students who have fulfilled the requirements of a Ph.D. program may be awarded one of several postdoctoral appointments: the scholar, fellow, or investigator. Postdoctoral scholarships are awarded for 18 months in the fields of oceanography, biology, chemistry, engineering, geology, geophysics, mathematics, meteorology, physics, and biology. In addition, professionals in law, social sciences, or natural sciences may apply for Marine Policy Fellowships, which focus on the examination of maritime conflicts. The WHOI also appoints postdoctoral investigators to positions that fall within the parameters of existing research contracts or grants.

In addition to its premier graduate and postdoctoral programs, WHOI provides opportunities for undergraduate students to gain experience through its Summer Student and Minority Fellowship programs, working in partnership with scientists and engineers on a wide range of scientific topics.

WHOI also grants a limited number of undergraduate students and certain advanced high school students the chance to participate in WHOI's education programs as Guest Students for up to 1 year. Middle school teachers benefit from participation in profession development workshops presented by WHOI scientists and engineers at the WHOI Exhibit Center, which highlights the institution's research programs and vessels.

The WHOI, in partnership with the Marine Biological Laboratory (MBL), an international research center for biology, biomedicine, and ecology, makes a critical contribution to the everyday operation of the MBLWHOI Library. The library, acclaimed for having one of the world's largest collections, both print and electronic, in biomedicine, oceanography, marine biology, and ecology, meets the daily needs of WHOI and MBL scientists and associated researchers. Researchers and students affiliated with the MIT/WHOI Joint Program, the Boston University Marine Program, the Sea Education Association, the National Marine Fisheries Service, the National Ocean and Atmospheric Administration, and the United States Geological Survey also make use of the library's services and resources.

The MBLWHOI Library houses the Data Library and Archives, which, through a wide-ranging collection of administrative records, oral papers and histories, personal papers and diaries, photographs, film, video, instruments, and other documents, makes the history of the WHOI accessible to WHOI's scientific community. The library encourages access to the collection by other institutions and researchers, but its services are restricted to those compatible with the needs of WHOI and MBL. The MBLWHOI Library offers an impressive collection of online resources including the *Alvin* dive log database, the *Alvin* ocean floor photos, digital photos from the WHOI archives, a searchable database of nearly 4,000 films and videotapes in the library's collection, WHOI oral history project, a database providing links to data files from research cruises, and images and descriptions about WHOI ships.

The WHOI receives funding from private contributions and endowment income, but the majority of its funding is generated by peer-reviewed grants and contracts from government agencies, including the National Science Foundation.

In August 2007, a joint venture led by WHOI and including the Scripps Institution of Oceanography at the University of California, San Diego, and Oregon State University's College of Oceanic and Atmospheric Sciences, was awarded a $97.7 million contract by the Joint Oceanographic Institutions in support of the National Science Foundation's Ocean Observatories Initiative. Charity Navigator, an independent evaluator of charities in the United States, assigned its highest ranking of four stars to WHOI—one of 1,300 charities to be recognized as outperforming most charities in its cause. Evaluations are based on annual financial information each charity provides to the U.S. Internal Revenue Service.

SEE ALSO: Massachusetts Institute of Technology; Navy, U.S.; Oceanic Changes; Oceanography.

BIBLIOGRAPHY. Jean Thilmany, "Searching Deeper," *Mechanical Engineering* (v.128/7, 2006); Woods Hole Oceanographic Institution, www.whoi.edu/page.do?pid=8118 (cited January 2008).

ROBIN K. DILLOW
ROTARY INTERNATIONAL ARCHIVES

World Bank

THE WORLD BANK (or the Bank) was established on December 27, 1945, following the ratification of the Bretton Woods agreement. The World Bank was conceived of in July 1944 at the United Nations Monetary and Financial Conference to provide development assistance to facilitate the reconstruction of Europe following World War II. Since then, the Bank has provided financial assistance to developing countries following natural disasters and humanitarian emergencies to facilitate postconflict rehabilitation and economic liberalization and development.

The organization of the World Bank consists of two agencies of the five that make up the World Bank Group (WBG): the International Bank for Reconstruction and Development (IBRD) and the International Development Association (IDA). The Bank consists of 185 member countries, all of whom are shareholders, represented by a board of governors—the ultimate policymakers of the Bank. The board of governors consists of member countries' ministers of finance or development, and it meets annually. The governors delegate specific duties to 24 on-site executive directors.

The five largest shareholders—France, Germany, Japan, the United Kingdom, and the United States—each appoint one executive director, and the remaining member countries are represented by 19 other executive directors, thus making up the 24. The president of the Bank, currently Robert B. Zoellick, is responsible for chairing the meetings of the board of directors and for overall management of the Bank. The president of the Bank serves a renewable five-year term. The president is, by tradition, a U.S. citizen nominated by the president of the United States—the bank's largest shareholder. The presidential nominee is confirmed by the board of governors.

The World Bank's activities are focused on the reduction of global poverty, especially on the achievement of the Millennium Development Goals (MDGs), goals calling for the elimination of poverty and the implementation of sustainable development. The constituent parts of the Bank, the IBRD and the IDA, achieve their aims through the provision of low- or no-interest loans and grants to countries with little or no access to international credit markets. The Bank is a market-based, nonprofit organization, using its high credit rating to make up for the low interest rate of loans.

MISSION AND FUNDING

The Bank's mission is to aid developing countries and their inhabitants achieve the MDGs through the alleviation of poverty by developing an environment for investment, jobs, and sustainable growth, thus promoting economic growth, and through investment in and empowerment of the poor to enable them to participate in development. The World Bank focuses on four key factors necessary for economic growth and the creation of a business environment: capacity building (strengthening governments and educating government officials), infrastructure creation (implementation of legal and judicial systems for the encouragement of business, the protection of individual and property rights and the honoring of contracts), development of financial systems (the establishment of strong systems capable of supporting endeavors from micro credit to the financing of larger corporate ventures), and combating corruption (eradicating corruption to ensure optimal effect of actions).

The Bank obtains funding for its operations primarily through the IBRD's sale of AAA-rated bonds in the world's financial markets. Although this does generate some profit, the majority of the IBRD's income is generated from lending its own capital. The IDA obtains the majority of its funds from 40 donor countries, who replenish the bank's funds every 3 years, and from loan repayments, which then become available for relending.

The Bank offers two basic types of loans: investment loans and development policy loans. The former are made for the support of economic and social development projects, whereas the latter provide quick disbursing finance to support countries' policy and institutional reforms. Although the IBRD provides loans with a low interest rate (between 0.5 and 1 percent for a standard Bank loan), the IDA's loans are interest free. The project proposals of borrowers are evaluated for their economical, financial, social, and environmental aspects to ensure that they are viable before any amount of money is distributed.

The Bank also distributes grants for the facilitation of development projects through the encouragement of innovation, cooperation between organizations, and participation of local stakeholders in projects.

IDA grants are predominantly used for debt burden relief in the most indebted and poverty-struck countries, amelioration of sanitation and water supply, support of vaccination and immunization programs for the reduction of communicable diseases such as HIV/AIDS and malaria, support to civil society organizations, and the creation of initiatives for the reduction of greenhouse gases.

THE WORLD BANK AND CLIMATE ISSUES

With respect to its work addressing issues of global warming, the World Bank has created and funded a number of climate-related partnerships and programs with other agencies and national governments, including the United Nations Framework Convention on Climate Change (UNFCCC), the Global Environment Facility (GEF), Carbon Finance, the Energy Sector Management Assistance Program (ESMAP), the Asia Alternative Energy Program (ASTAE), the Global Facility for Disaster Reduction and Recovery (GFDRR), the Vulnerability and Adaptation Resources Group (VARG), and the Global Gas Flaring Reduction partnership (GGFR).

The UNFCCC is an international treaty through which countries consider ways to reduce global warming and cope with inevitable temperature increases. The World Bank is an observer to the UNFCCC and also takes part in a number of technical discussions conducted by the UNFCCC Secretariat, such as by the Subsidiary Body for Implementation and the Subsidiary Body for Scientific and Technological Advice.

The GEF is the financing mechanism for the UNFCCC, as well as other key international environmental agreements. As a GEF-implementing agency, the World Bank helps identify, prepare, and implement projects that reduce poverty and benefit the local and global environment. Climate change was the second most active focal area of the GEF active portfolio at the end of fiscal year 2006.

The World Bank's Carbon Finance Unit offers a means of leveraging new private and public investment into projects that serve to mitigate climate change by reducing greenhouse gas emissions, while promoting sustainable development. Projects relate to rural electrification, renewable energy, energy efficiency, urban infrastructure, waste management, pollution abatement, forestry, and water resource management.

The atrium of the World Bank in Washington, D.C. It has funded a number of climate-related partnerships and programs.

ESMAP, cosponsored by the World Bank and the United Nations Development Programme, is a global technical assistance program that provides policy advice on sustainable development issues to the governments of developing countries and economies in transition. ESMAP also contributes to technology and knowledge transfer in energy sector management and, since its creation in 1983, has operated in 100 different countries through some 450 different activities. Recently, a new window in ESMAP has opened to support the goals of the Clean Energy Investment Framework.

The ASTAE program, established in 1992, aims at mainstreaming renewable energy and energy efficiency in the World Bank's lending operations in the power sector in Asia. The World Bank is cooperating actively with the GFDRR, which aims at integrating hazard risk reduction strategies in development processes at local and national levels. The potential exacerbation of extreme climatic events as climate changes suggests significant overlap between the areas of adaptation to climate change and disaster risk reduction.

The VARG is an informal network of multi- and bilateral development institutions that aim to facilitate the integration of adaptation to climate change in the development process. The World Bank is one of 19 organizations that are partnering in this open-knowledge network.

The GGFR partnership, a World Bank–led initiative, facilitates and supports national efforts to use currently flared gas by promoting effective regulatory frameworks and tackling the constraints on gas utilization, such as insufficient infrastructure and poor access to local and international energy markets, particularly in developing countries. Launched at the World Summit on Sustainable Development in August 2002, GGFR brings around the table representatives of governments of oil-producing countries, state-owned companies, and major international oil companies, so that together they can overcome the barriers to reducing gas flaring by sharing global best practices and implementing country-specific programs.

THE SUSTAINABLE DEVELOPMENT NETWORK

Another World Bank initiative is the Sustainable Development Network, a part of which includes a Climate Change Team within the Environment Department of the Bank. The Climate Change Team provides resources and expertise for the World Bank's participation in international climate change negotiations under the UNFCCC and provides technical advice to the World Bank's GEF Program on the preparation of GEF climate change mitigation projects in energy efficiency and renewable energy and on the development of strategic initiatives with the GEF. The team also is leading the Bank's efforts related to climate change vulnerability and adaptation issues for its client countries.

The Bank recognizes that achieving objectives related to climate change is a long-term process requiring the integration of the greenhouse gas (GHG) mitigation and the vulnerability and adaptation agendas into mainstream operational work. These instruments include planning, policy dialogue, generation and dissemination of knowledge, and investment lending, all of which are primarily aimed at promoting national development priorities. Bank support to clients for better managing climate change occurs in three key areas: mitigation of GHG emissions, reduction of vulnerability and adaptation to climate change, and capacity building. In the area of GHG mitigation, the bank promotes policy and regulations, as these tend to have large and sustainable effects on improving the efficiency of resource use and, consequently, reduction of GHG emissions. In the context of these reforms, the Bank mobilizes resources from the GEF and the Prototype Carbon Fund to support GHG abatement measures that simultaneously address poverty reduction and sustainable development goals.

In areas of vulnerability and adaptation, where the decision on UNFCCC support is pending, the Bank will mobilize donor financing for a Vulnerability and Adaptation Facility to better prepare for climate change. Over the medium term, the Bank will focus on improving the understanding of the potential effects of climate change and on identifying and implementing no-regrets measures to reduce vulnerability to current climate and to climate change. Finally, the Bank will assist clients in building the capacity needed to deal with GHG abatement and with vulnerability and adaptation.

As part of its work in climate change, the World Bank has developed a variety of resource and training materials addressing the fundamental issues underlying climate change, examples of successful mainstreaming of climate change concerns into project work or underlying analysis, and basic tools for accurately identifying the climate change effects of projects, baselines, and alternatives. The climate risk screening toolkit is referred to as ADAPT (Assessment and Design for Adaptation to Climate Change), which is a prototype tool that will screen proposed development projects for potential risks posed by climate change and variability. The Bank has also developed a variety of tools and examples to help its staff and clients more readily address the methodological, technical, and economic issues underlying the incorporation of GHG issues in project development and economic analysis. In the area of renewable energy, the Renewable Energy Toolkit comprises a range of tools to help Bank staff and country counterparts improve the design and implementation of renewable energy projects. It aims at pro-

viding practical implementation needs at each stage of the project cycle and also helps project staff determine sustainable business models, financing mechanisms, and regulatory approaches.

SEE ALSO: Climate Change, Effects.

BIBLIOGRAPHY. World Bank, *An Adaptation Mosaic: A Sample of the Emerging World Bank Work in Climate Change Adaptation (Parts 1 and 2)* (World Bank, 2004); World Bank, *Come Hell or High Water—Integrating Climate Change Vulnerability and Adaptation into Bank Work* (World Bank, 1999); World Bank, *Contributions from the National Strategy Studies Program to COP6 Negotiations Regarding CDM and JI* (World Bank, 2001); World Bank, *Look Before You Leap: A Risk Management Approach for Incorporating Climate Change Adaptation in World Bank Operations* (World Bank, 2004); World Bank, *Making Sustainable Commitments: Environment Strategy for the World Bank (Annex F—Climate Change)* (World Bank, 2001); World Bank, *Poverty and Climate Change: Reducing the Vulnerability of the Poor through Adaptation (Report, Parts 1 and 2)* (World Bank, 2003); World Bank, *Sustainable Development and Global Environment—The Experience of the World Bank Group-Global Environment Facility Program* (World Bank, October 2002).

MICHAEL J. SIMSIK
U.S. PEACE CORPS

World Business Council for Sustainable Development

THE WORLD BUSINESS Council for Sustainable Development (WBCSD), based in Geneva, Switzerland, is a global association of businesses focused on market-oriented solutions to sustainable development and is appropriately structured along the lines of a chief executive officer (CEO)–led business. The WBCSD's mission statement asserts its purpose: "Our mission is to provide business leadership as a catalyst for change toward sustainable development, and to support the business license to operate, innovate and grow in a world increasingly shaped by sustainable development issues." The WBCSD arose from ideas forwarded at the United Nations Conference on Environment and Development (UNCED), also known as the Rio de Janeiro Earth Summit of 1992, where Swiss businessman Stephan Schmidheiny was appointed chief adviser for business and industry to the secretary general of the UNCED. Schmidheiny created a forum called Business Council for Sustainable Development that eventually published the book *Changing Course*, in which the phrase eco-efficiency—doing more with less—was first used. The WBCSD was constituted in its current form when it merged (1995) with the World Industry Council for the Environment.

The WBCSD has approximately 60 national and regional business councils and has its North America office in Washington, D.C. Only companies committed to sustainable development are invited, by the executive council, to join the association.

General Motors, DuPont, 3M, Deutsche Bank, Coca-Cola, Sony, Oracle, BP, and Royal Dutch/Shell are among approximately 200 member companies from 35 countries representing 20 industrial sectors. Members make their knowledge, experience, and some limited human resources available to the WBCSD and are asked to base their business development within the parameters of economic, social, and environmental sustainability. Member company CEOs act as executive council members, cochair working groups, and promote the objectives of the WBCSD within their companies.

OBJECTIVES OF THE COUNCIL

The council's objectives are to be a leading business advocate on sustainable development, participate in policy development to create the right framework conditions for business to make an effective contribution to sustainable human progress, develop and promote the business case for sustainable development, demonstrate the business contribution to sustainable development solutions and share leading-edge practices among members, and contribute to a sustainable future for developing nations and nations in transition. To achieve these objectives the WBCSD has four focus areas (FAs): energy and climate, development, the business role, and ecosystems.

The WBCSD initiated its Energy and Climate FA in 1999 by creating and promoting ways for business to prosper while being socially and environmentally conscious within a sustainable development framework that prepares business for the carbon-con-

strained future necessary for the planet to counter anthropogenic (human-induced) global warming and climate change. This required innovation, education, and dialogue seeks to create an efficient approach to energy sustainability and climate change through changing and adapting policy frameworks, business plans and markets, advancing technologies, and the interaction between business, government, and economies regionally, nationally, and globally. One focus of this initiative was/is the large greenhouse gas (GHG) emitters (United States, China, India, and Russia). The WBCSD interacts with the UN Environment Programme (UNEP), the UNEPs Intergovernmental Panel on Climate Change (IPCC), and the UN Framework Convention on Climate Change, and also focuses on the development of a post-2012 flexible Kyoto mechanism. Energy and climate became the primary WBCSD FA in 2005.

WBSCDs *Facts and Trends to 2050* (2004) and *Pathways to 2050* (2005) proposed potential pathways to GHG stabilization and to creating a low GHG economy. Its *Policy Directions to 2050: A Business Contribution to the Dialogues on Cooperative Action* (2007) promotes these goals and the averting of a climate change catastrophe through the coordinated and sustained actions of governments, businesses, and consumers. The WBCSD's Greenhouse Gas Protocol Initiative joint initiative with the World Resources Institute attempts to standardize all GHG accounting.

The WBCSD contends that stable and sustainable societies cannot and must not tolerate poverty among their citizens and that businesses, economies, governments, and societies must work together to ensure open and fair access to all markets and opportunities. The WBCSD supports this focus through its Sustainable Livelihoods project, which advocates the changes in societal, policy, and business frameworks necessary to achieve this focus while promoting business-led development targeting the long-term alleviation of poverty.

The WBCSD Business Role FA began in 1992 with the publication of the book *Changing Course*, the basic premise of which was that the only solution to the challenges of environmental deterioration and climate change is the active participation of business. The WBCSD tried to ignite a discussion on the role that business should or might play in any future economic, social, and environmental sustainable society with its

2006 publication of *From Challenge to Opportunity: The Role of Business in Tomorrow's Society.* The WBCSD secretariat and its FA Core Team (FACT), composed of 11 member companies led by two cochairs, creates and implements the WBCSD Action Plan seeking to provide platforms for discussion and debate among business leaders, facilitate future thinking on how business might support sustainable development, use the available WBCSD resources more appropriately, and shape the message of the WBCSD to the business world.

Ecosystems became a WBCSD FA in March 2007 and, as a FA, is built on the WBCSD's Sustaining Ecosystems Initiative created in November 2005 to rally and direct business to address the ecosystem opportunities and challenges outlined in the July 2005 *Millennium Ecosystem Assessment* that the WBCSD partially authored. The goal is to encourage business to address the risks inherent in the accelerating degradation of the ecosystem and the loss of ecosystem services and to adopt mitigation and market-based strategies for enhancing sustainable management and ecosystems use. The WBCSD began addressing these issues by demonstrating how business could integrate these strategies into management systems when the WBCSD and the World Conservation Union (IUCN) jointly produced (1997) two reports on business and biodiversity: *A Guide for the Private Sector* and *A Handbook for Corporate Action.* The WBCSD and the IUCN continued to promote the benefits of integrating biodiversity into business after they partnered with the World Conservation Union to host a workshop on the subject in Bangkok, Thailand, in November 2004.

SEE ALSO: Climate; Climate Change, Effects; Developing Countries; Ecosystems; Greenhouse Gases; Intergovernmental Panel on Climate Change (IPCC); Sustainability; United Nations Environmental Programme (UNEP); World Resources Institute (WRI).

BIBLIOGRAPHY. Livio D. Desimore, Frank Popoff, and the World Business Council for Sustainable Development, *Eco-Efficiency: The Business Link to Sustainable Development* (MIT Press, 1997); David L. Rainey, *Sustainable Business Development: Inventing the Future Through Strategy, Innovation, and Leadership* (Cambridge University Press, 2006); Anne-Marie Sacquet, *World Atlas of Sustainable Development: Economic, Social and Environmental Data* (Anthem Press, 2005); Stephan Schmidheiny, Federico J. L. Zorraquin, and the World Business Council for Sustainable Development, *Financing Change* (MIT Press, 1996); "World Business Council for Sustainable Development," www.wbcsd.org (cited November 2007).

RICHARD MILTON EDWARDS
UNIVERSITY OF WISCONSIN COLLEGES

World Climate Research Program

THE WORLD CLIMATE Research Program (WCRP) is sponsored by the International Council for Science (ICSU), the World Meteorological Organization (WMO), and the Intergovernmental Oceanographic Commission (IOC) of the United Nations Educational, Scientific and Cultural Organization (UNESCO). The program brings together the intellectual and structural potentialities related to climate and climate change of more than 185 countries. The program thus aims to work as an international forum to share scientific discoveries and facilities to advance the understanding of the phenomena that influence climate. The two underlying objectives of the WCRP are to determine the predictability of climate and to assess the effect of human activities on it.

These two objectives stem from the needs identified by the UN Framework Convention on Climate Change. To achieve its objectives, the WCRP adopts a multidisciplinary approach, organizes extensive observational and modeling projects, and encourages researches on aspects of climate too large and complex to be addressed by any one nation or single scientific discipline. The WCRP is not open exclusively to scientists. On the contrary, it aims to involve different subjects such as policymakers, information end-users, and sponsors in a scientifically accurate debate on climate change and variability.

The WCRP was established in 1980. It was initially joint-sponsored by the ICSU and the WMO. Since 1993, the IOC of UNESCO has also become a sponsor of the program. Since its establishment, the WCRP has contributed to the advancement of climate science. Thanks to WCRP researchers, climate scientists can monitor, simulate, and project global climate with improved accuracy so that climate information can be used for governance, in decision making, and in support of a wide range of practical applications. In 2005, after 25 years of serving science and society, the WCRP launched its Strategy Framework 2005–2015, which expresses the program's commitment to working efficiently and effectively toward strengthening knowledge and increasing capabilities with regard to climate variability and change. Titled the Coordinated Observation and Prediction of the Earth System, the framework aims "to facilitate analysis and prediction of Earth system variability and change for use in an increasing range of practical applications of direct relevance, benefit and value to society." The WCRP is thus devoted to provide a larger series of products and services to an ever-increasing group of users. The WCRP intends to reach this goal through the integration of observations and models to generate new understanding and improve climate predictions.

Today, the WCRP covers studies of the different parts of the Earth's climate system: global atmosphere, oceans, sea and land ice, the biosphere, and the land surface. The major core projects, diverse working groups, various cross-cutting activities, and many cosponsored activities of the WCRP all aim to improve scientific understanding of processes that can enable better forecasts.

The Global Energy and Water Cycle Experiment (GEWEX) project studies the dynamics and thermodynamics of the atmosphere, the atmosphere's interactions with the Earth's surface (especially over land), and the global water cycle. GEWEX uses suitable models to represent and forecast the variations of the global hydrological regime and its effect on atmospheric and surface dynamics. GEWEX also focuses on variations in regional hydrological processes and water resources and their response to changes in the environment, such as the increase in greenhouse gases. GEWEX projects are divided into three focus areas corresponding to the key elements in the global energy and water cycle: radiation, hydrometeorology, and modeling and prediction.

The Climate Variability and Predictability (CLIVAR) project, set up in 1995, specifically targets climate variability. Its mission is to observe, simulate, and predict the Earth's climate system, with a focus on ocean-atmosphere interactions. CLIVAR seeks to develop predictions of climate variations

on seasonal to centennial timescales and to refine the estimates of anthropogenic climate change. CLIVAR also includes a Working Group on Seasonal to Interannual Prediction, which oversees development of improved models, assimilation systems, and observing system requirements for seasonal prediction.

The section of WCRP dealing with the Stratospheric Processes and Their Role in Climate (SPARC), founded in 1993, carries out research on the chemistry of the climate system. In particular, it focuses on the interaction of dynamic, radiative, and chemical processes. SPARC's projects include the construction of stratospheric reference climatology and the improvement of understanding of trends in temperature, ozone, and water vapor in the stratosphere. SPARC also studies gravity wave processes, their role in stratospheric dynamics, and how these may be represented in models.

The Climate and Cryosphere (CliC) project, founded in 2000, measures the effects of climatic variability and change on components of the cryosphere and their consequences for the climate system. CliC is also charged with the task of improving the management of data and information relating to the cryosphere and climate, and with making data more readily available for use by the broader scientific community. To this end, CliC has established a Web-based Data and Information Service for CliC.

The Surface Ocean–Lower Atmosphere Study aims to quantify the key biogeochemical-physical interactions and feedbacks between the ocean and the atmosphere. WCRP cosponsors the project jointly with the Commission on Atmospheric Chemistry and Global Pollution, the International Geosphere-Biosphere Programme and the Scientific Committee on Oceanic Research. The project investigates biogeochemical interactions and feedbacks between ocean and atmosphere, exchange processes at the air-sea interface and the role of transport and transformation in the atmospheric and oceanic boundary layers, and air-sea fluxes of carbon dioxide and other long-lived radiatively active gases.

The Working Group on Surface Fluxes was established in 2007 to review the requirements of the different WCRP schemes for surface sea fluxes including biogeochemical fluxes, to coordinate the various related research initiatives, and to encourage research and facilitate operational activities on surface fluxes.

SEE ALSO: Carbon Dioxide; Climate; Climate Models.

BIBLIOGRAPHY. Olav Slaymaker and Richard Kelly, *Cryosphere and Global Environmental Change* (Wiley & Sons, 2007); World Climate Research Program, http://wcrp.wmo.int/index.html.

LUCA PRONO
UNIVERSITY OF NOTTINGHAM

World Health Organization

THE WORLD HEALTH Organization (WHO), established in 1948 by the United Nations (UN), has a mission to promote international cooperation for improved health conditions of all peoples. The headquarters for WHO are in Geneva, Switzerland. Operations fall under the World Health Assembly (the policymaking body holds annual meetings), an executive board of health (specialists elected for three-year terms by the assembly) and a secretariat with regional offices and staff throughout the world. Financing for the services provided by WHO comes from member governments based on ability to pay and, since 1951, from an allocation from the technical assistance program of the UN.

The auspices of the WHO include providing a central clearinghouse for information and research as relates to health, sponsoring measures for the control of disease, and strengthening and expanding the public health administrations of member nations.

Overall health depends on potable water for drinking and washing, sufficient nutrition, and shelter and protection from weather extremes. Increased temperatures, rising sea levels, and changing weather patterns are factors affecting food, water, and shelter. Excellent health services and living conditions would have a protective or palliative effect on those populations. Substandard health services and living conditions would have a negative effect on populations in poverty or in disaster situations.

For prevention and preparedness, the WHO coordinates review of scientific evidence on the links between climate, climate change, and health. On the basis of the information available, WHO established a list of possible health effects resulting from rapid climate change, especially in vulnerable populations. The effect of warmer temperatures would increase the incidence of heat-related illnesses and lead to higher concentrations of ground-level ozone pollution, causing respiratory illnesses (diminished lung function, asthma, and respiratory inflammation). Flooding associated with storms and rising sea levels could increase the risk of contracting certain infectious diseases from water contamination or disease-carrying vectors, especially for the malnourished.

WORLD HEALTH PROGRAMS AND ACTIVITIES

WHO supports actions to reduce human influence on the global climate while still recognizing the effect past emissions and human action have on the likelihood of warming and more variable climate for at least several decades. To alleviate this issue as well as reduce health vulnerability to future climate change, WHO supports programs for combating infectious diseases, improving water and sanitation, and ensuring response to natural disasters. In addition, research and effort are being put into building the capacity of health services and information to help adapt to climate changes.

Primary prevention includes actions to prevent the onset of disease from environmental disturbances in an otherwise unaffected population (mosquito nets, vector control, early weather-watch warning systems). Secondary prevention includes early-response action (disease surveillance). Tertiary prevention includes diagnosis and treatment to lessen the morbidity or mortality caused by disease.

Adaptation refers to actions taken to lessen the effects of the anticipated changes in climate. The ultimate goal of adaptation interventions is the reduction, with the least cost, of diseases, injuries, disabilities, suffering, and death from climate change. Public health programs should anticipate the health effects of climate change such as, for instance, those on infectious diseases. For example, surveillance systems could be improved in sensitive geographic areas. Such regions include those bordering areas of current distribution of vector-borne diseases that could themselves experience epidemics under certain climatic conditions. Vaccination programs could be intensified, and pesticides for vector control and drugs for prophylaxis and treatment could be stockpiled.

To prepare for climate change, the WHO encourages and fosters international cooperation through collaborating with partners (UN agencies and other international organizations, donors, civil society, and the private sector). WHO provides support to countries to implement programs using best technical guidelines and practices and helping to establish health priorities and strategy.

To prepare appropriately for climate change, climate and environmental monitoring are essential. To be most effective, health professionals must interact with the environmental sectors to evaluate the changing climate and assess risks. Researchers and monitoring organizations interested in health impact assessment have requested assistance from health researchers to ensure an accurate risk assessment. Health researchers must also be encouraged to access environmental data or be given the tools necessary to find the information on air and water pollution levels. Additional monitoring should include food contamination and identification of emission sources or contamination sites. The WHO/ECEH (European Centre for Environment and Health) is developing a Health and Environment Geographic Information System (HEGIS) to identify areas and issues of priority for the environment and health. With an initial focus on demographics and air quality, the information system has the potential for expansion to include climate change data and health effects.

The agenda of WHO includes promoting development of health-promoting activities, fostering health security, strengthening health systems, and collecting and disseminating research information.

Health development is directed by the ethical principle of equity: access to lifesaving or health-promoting interventions should not be denied for unfair (economic or social) reasons. Despite how health has become a

Children in Uganda: The World Health Organization coordinates review of scientific evidence on the links between climate change and health. Increased temperatures, rising sea levels, and changing weather patterns would affect food, water, and shelter.

standard factor in perception of social and economic progress, as well as receiving additional resources and funding, poverty and poor health are still widespread around the world. WHO activities are aimed at health development and at making health outcomes in poor, disadvantaged, or vulnerable groups a priority. The organization's agenda is to focus on the prevention and treatment of chronic disease in all populations and to research and address tropical diseases as part of their Millennium Development goals.

WHO intends to foster health security even for vulnerable populations by demanding action from the international community in addressing preventable diseases. Forty percent of the global population is very poor, many suffer from water shortages, more than a billion lack safe drinking water, and water-related diseases cause between 2 and 5 million deaths each year. Public health and preventive measures are available, and action is justified, including vaccination or control measures such as improved sanitation and temperature control of foods. In addition there is the risk of outbreaks from emerging and epidemic-prone disease.

Such outbreaks are occurring in increasing numbers, fuelled by such factors as rapid urbanization, environmental mismanagement, the way food is produced and traded, and the way antibiotics are used and misused. The threat of climate change also demands action for building capacity and adaptabilities to climate change. WHO provides workshops in vulnerable countries and works with policymakers to ensure adequate healthcare is available for treatment. These activities have a twofold benefit: action taken now both improves current health conditions and provides a foundation for additional adaptation measures in the future to address climate changes.

As an ultimate measure for ensuring adequate health care not only in the future but in the present as well, health systems around the world must be strengthened. To function as a method for reducing poverty, health services must be available to all populations, including those that are now underserved because of economic reasons. WHO works with countries to ensure an adequate number of trained professional and additional staff to meet the healthcare needs, funding to provide services, and access to medication and supplies as well as the technology to provide appropriate care.

For collecting and disseminating research information, the WHO maintains a searchable database on the Web on a broad range of health subjects including links to journal articles, research undertaken, and recommendations. As mentioned before, the WHO/ECEH also is developing a HEGIS to identify areas and issues of priority for environment and health. With an initial focus on demographics and air quality, the information system has the potential for expansion to include climate change data and health effects.

SEE ALSO: Food Production; Population; United Nations.

BIBLIOGRAPHY. World Health Organization, *Early Human Health Effects of Climate Change and Stratospheric Ozone Depletion in Europe* (WHO, April 9, 1993); World Health Organization, "The WHO Agenda," www.who.int; Intergovernmental Panel on Climate Change, "Potential Impacts of Climate Change," www.ciesin.columbia.edu/docs/001-011/001-011.html; Worldwatch Institute, *Vital Signs 2006–2007* (Norton, 2006).

LYN MICHAUD
INDEPENDENT SCHOLAR

World Meteorological Organization

THE WORLD Meteorological Organization (WMO) is a specialized agency of the United Nations (UN). It is the UN voice on the state and behavior of the Earth's atmosphere, its interaction with the oceans, the climate it produces, and the resulting distribution of water resources. The WMO has a membership of 188 member states and territories. It originated from the International Meteorological Organization, which was founded in 1873. Established in 1950, WMO became the specialized agency of the United Nations in 1951 for meteorology (weather and climate), operational hydrology, and related geophysical sciences. It is based in Geneva, Switzerland.

The WMO seeks to provide the framework for an international cooperation regarding climate matters. The organization points out that weather, climate, and the water cycle know no national boundaries, so international cooperation at a global scale is essential for the development of meteorology and operational hydrology. Since its establishment, the WMO has devised programs and services attempting to contribute to the preservation of the environment and the welfare of humanity. The National Meteorological and Hydrological Services have sought to protect life and property against natural disasters, to preserve the environment, and to encourage the economic and social well-being of all sectors of society in areas such as food security, water resources, and transport.

The WMO supports cooperation to establish networks for meteorological, climatological, hydrological, and geophysical observations, as well as for the exchange, processing, and standardization of related data. It also provides technology transfer, training, and research and fosters collaboration between the National Meteorological and Hydrological Services and its members. The organization sponsors the application of meteorology to public weather services, agriculture, aviation, shipping, the environment, water issues, and the mitigation of the effects of natural disasters.

WMO aids the free exchange of data and information, products, and services in real- or near-real-time on topics relating to safety and security of society, economic welfare, and the protection of the environment. It contributes to policymaking in these areas at national and international levels. In the specific case of disasters related to weather, climate, and water, which represent nearly 90 percent of all natural catastrophes, WMO's programs try to provide advance warnings that save lives and reduce damage to property and the environment. The organization is also committed to reducing the effects of human-induced disasters, such as those associated with chemical and nuclear accidents, forest fires, and volcanic ash. Thus, the WMO is central in international efforts to monitor and protect the environment. In collaboration with other UN agencies and the National Meteorological and Hydrological Services, the WMO supports the implementation of a number of environmental conventions and is instrumental in providing advice and assessments to governments on related issues. The organization claims that its activities contribute toward ensuring the sustainable development and well-being of nations.

Global warming is a major WMO concern. The organization supports intergovernmental legal agreements on major global environmental concerns such as ozone-layer depletion, climate change, desertification, and biodiversity. WMO also coordinates the observing systems that provide the necessary data to assess atmospheric-ocean processes and interactions, such as El Niño/La Niña, and water-resources availability. Most significantly for global warming, the WMO lists, among its programs, the Global Atmosphere Watch (GAW). The WMO's interest in a program of atmospheric chemistry and the meteorological aspects of air pollution dates back to the 1950s. This included assuming responsibility for standard procedures for uniform ozone observations and establishing the Global Ozone Observing System during the 1957 International Geophysical Year.

In the late 1960s, the Background Air Pollution Monitoring Network was set up and was subsequently consolidated with the Global Ozone Observing System into the current GAW in 1989. The GAW monitoring scheme includes a coordinated global network of observing stations along with supporting facilities. GAW provides data for scientific measurements of changes in the chemical composition and related physical characteristics of the atmosphere that may negatively affect our environment. The priorities of the scheme have been identified in greenhouse gases for possible climate change, ozone and ultraviolet radiation for both climate and biological concerns, and certain reactive gases and the chemistry of precipitation.

GAW is intended to provide accessible, high-quality atmospheric data to the scientific community. These components include measurement stations, calibration and data-quality centers, data centers, and external scientific groups for program guidance. Support for these components is obtained, largely, by individual WMO member countries that directly participate in the program. Additional resources come from international funding and the WMO secretariat's internal budget.

SEE ALSO: Climate; Global Warming; Weather.

BIBLIOGRAPHY. Global Atmosphere Watch, http://www.wmo.ch/pages/prog/arep/gaw/gaw_home_en.html; World Meteorological Organization, http://www.wmo.ch/pages/index_en.html.

LUCA PRONO
UNIVERSITY OF NOTTINGHAM

World Resources Institute

THE WORLD RESOURCES Institute (WRI) is a nonpartisan and nonprofit environmental organization that tries to find practical applications for theoretical research on the protection of the Earth and the improvement of people's lives. WRI supplies information and proposals for policies that promote a sustainable development both in environmental and social terms. WRI is based in Washington, D.C., and has a staff of more than 100 scientists, economists, policy experts, business analysts, statistical analysts, mapmakers, and communicators.

The institute takes as its point of departure for action the fact that the shift to low-carbon technology will only occur by persuading owners and shareholders of its profitability. To WRI, business investors are instrumental in solving the climate crisis. WRI produces a highly respected biennial publication, the *World Resources* report, which supplies data and in-depth analysis on current environmental issues, including, for example, the importance of efficient ecosystem management for rural poverty relief. The report is a collaborative product of WRI with the World Bank, United Nations Environment Programme, and UN Development Programme. *World Resources* was launched in 1986 to bridge the gap in information about the conditions of the world's natural resources.

GOALS AND ACTIVITIES

The activities of the institute are structured around four key areas and goals. In the field of people and ecosystems, the institute aims to reverse the process of rapid degradation of ecosystems and guarantee their ability to provide humans with needed goods and services. Regarding access to information, WRI attempts to improve public knowledge about decisions on natural resources and the environment. In the area of climate protection, energy, and transport, WRI is active in the protection of the global climate system from further harm caused by emissions of greenhouse gases. The institute is also committed to help humanity and the natural world adapt to the climate change that is already taking place. As for markets and enterprise, WRI seeks to promote an economic development that will increase social opportunities and, at the same time, protect the environment. WRI has worked with the private sector to find profitable

solutions that have both economic and environmental benefits. WRI also created the accounting system which companies all over the world use to account for their greenhouse gas emissions.

WRI was founded in June 1982 by American lawyer and environmental activist James Gustave Speth. Speth, a former chairman of the U.S. Council on Environmental Quality and later professor of law at Georgetown University, also acted as WRI's first director. He held this position until January 1993, when he became director of the UN Development Programme and was then succeeded by Jonathan Lash, senior staff attorney at the Natural Resources Defense Council from 1978 to 1985. The institute was created thanks to the John D. and Catherine T. MacArthur Foundation of Chicago, which provided $15 million to help finance the first five years of WRI. The institute was organized as a nonprofit Delaware corporation that could receive tax-deductible contributions under the U.S. Internal Revenue Code. WRI was founded not as an activist environmental membership organization but as an independent institution that should carry out scientifically sound research and suggest viable policies.

In 1985, WRI was one of the first research centers to organize an international meeting on the rising emissions of carbon dioxide and other greenhouse gases into the atmosphere. During its over two decades of activity, WRI has attracted other centers that have decided to merge with the institute, including the North American office of the International Office for Environment and Development and the Management Institute for Environment and Business. In 1990, the UN Development Programme commissioned WRI with a study that eventually resulted in the creation of the Global Environment Facility. Throughout the 1990s, WRI played a key role in initiatives aimed to contain the phenomenon of global warming. In 1992, the institute made important contributions to the development of the Convention on Biological Diversity, which was then signed at the Earth Summit in Rio de Janeiro.

In 1999, WRI committed to stopping its own emissions of carbon dioxide. The institute has also created several important networks such as the Global Forest Watch, which monitors the conditions of forests; the Access Initiative, a global forum of civil society organizations committed to improving citizen access to information and favor their participation in decisions that affect the environment;

and the Green Power Market Development Group, a partnership of Fortune 500 companies devoted to establishing corporate markets for renewable energy. WRI has also been responsible, in partnership with Mexico City, for the creation of the Bus Rapid Transit Corridor, a system of transport designed to reduce environmental damages. The institute is collaborating with metropolises such as Shanghai, Hanoi, and Istanbul for the creation of similar systems.

SEE ALSO: Carbon Dioxide; Carbon Emissions; Climate Change, Effects; Global Warming; Population.

BIBLIOGRAPHY. EarthTrends—Environmental Information, World Resources Institute, http://earthtrends.wri.org; World Resources Institute, http://www.wri.org.

Luca Prono
University of Nottingham

World Systems Theory

AN UNDEREXAMINED ASPECT of global warming is the geographic separation between the people who generate greenhouse gases and the people who are most likely to be hurt by global warming. This article explains how World Systems Theory (WST) can help us understand how climate change is both a cause and a consequence of existing social and regional inequalities. This essay begins by explaining the basic concepts of WST. It then uses WST to explain how geopolitical structures in the global economy ensure that the benefits and costs of burning fossil fuels are not shared equally.

WST evolved to counter free-market economists such as Walter Rostow, who argued that countries were poor because of deficiencies within those countries. Less developed countries could advance by emulating wealthy countries, which had moved through several stages of development.

DEPENDENCY THEORY

A chorus of critics challenged this view because, from their perspective, adherents to the stages of development view blamed the poor countries for their poverty. In response, critics argued that the lack of development in places such as South America was a result of the economic and political structures imposed by wealthy countries through colonial and lingering postcolonial relationships. Foremost among these scholars was Andre Gunder Frank.

His dependency theory argued that as capitalism diffused outward from the economic core of Western Europe over the past 500 years, it transformed the places at the periphery that were drawn into the growing capitalist world system. These new territories in Africa and Latin America were thrust into a subordinate role, and their economies were restructured to suit the needs of the colonial powers in Europe. Frank summarized this subordination as the "development of underdevelopment." Capitalist penetration into new regions underdevelops or undermines the economic potential of these places. Frank wanted to shift the blame for poverty away from the processes within poor countries and to place responsibility for poverty on the structural relationships imposed by colonial, and later, neocolonial, core powers.

Building on dependency theory, scholars such as Immanuel Wallerstein proposed WST to provide more historical context into the processes leading to the development of inequality. WST sorts the countries of the world into three broad categories according to the type of economic processes that predominate in those countries. Countries in the periphery are largely dependent on agriculture and other forms of natural resource extraction. Social inequality is high because most jobs are typically low skill and low wage. Core countries have diverse economies based on manufacturing, services, and information technology. An intermediate category called the semiperiphery includes countries that are more economically diverse than peripheral countries but that have invested less in manufacturing and other tertiary sectors. Hence the level of development is less than that of core countries.

WST improves on the rigid core-periphery dichotomy of dependency theory by adding the semiperipheral category. This modification helps to explain the historical reality of peripheral countries such as Korea ascending to core status, or how the global economy can relegate former colonial powers such as Portugal to semiperipheral status. WST tells us that the international division of labor creates a system of unequal exchange and income inequality between core, semiperipheral and peripheral countries. Peripheral countries extract raw materials and ship them largely unprocessed to the semiperipheral and core countries for processing. The finished goods are then shipped back to the peripheral countries at much higher prices. This system of unequal exchange is perpetuated by protectionist international trade agreements; copyright and patent laws, which limit the diffusion of processing technology to the periphery; the role of multinational corporations, which repatriate profits to the core that were earned by extracting resources in the periphery; and the role of wealthy local elites in poor countries, who benefit from the status quo relationships between the core and periphery.

The creation and perpetuation of global economic and political inequality has direct implications for understanding the effects of global climate change. As noted above, countries in the periphery are far more reliant on agriculture than are countries in the core. Countries in the core are much more invested in manufacturing. As a consequence, they produce disproportionately large amounts of greenhouse gases. That means that there is a geographic separation between the largest producers of greenhouse gases and the regions that will suffer the most harm from global warming. In short, climate changes will have a disproportionately negative impact on the economies and communities in the periphery. The fact that countries at the periphery have the fewest resources to adapt to climate change makes matters even worse. WST is a useful counterbalance to traditional neoclassical discussions of market-based solutions to greenhouse gas reductions because it brings a geopolitical perspective to our understanding of climate change.

SEE ALSO: Developing Countries; Economics, Cost of Affecting Climate Change.

BIBLIOGRAPHY. Andre Gunder Frank, *Capitalism and Underdevelopment in Latin America* (Monthly Review Press, 1967); Karen O'Brien and Robin M. Leichenko, "Winners and Losers in the Context of Global Change," *Annals of the Association of American Geographers* (v.93/1); James Rice, "Ecological Unequal Exchange: Consumption, Equity, and Unsustainable Structural Relationships Within the Global Economy," *International Journal of Comparative Sociology* (v48/1, 2007); Timmons Roberts, "Global Inequality and Climate Change," *Society and Natural Resources* (v.14/6, 2001); Walter Rostow, *The Stages of Economic Growth: A Non-Communist Manifesto* (Cambridge University Press, 1960); Immanuel Wallerstein, *The Capitalist World-Economy* (Cambridge University Press, 1979).

CHRISTOPHER D. MERRETT
WESTERN ILLINOIS UNIVERSITY

Worldwatch Institute

THE WORLDWATCH INSTITUTE is "dedicated to fostering the evolution of an environmentally sustainable and socially just society, where human needs are met in ways that do not threaten the health of the natural environment or the prospects of future generations." It describes itself as "an independent, globally focused environmental and social policy research organization" with a "unique blend of interdisciplinary research and accessible writing." Worldwatch is essentially a think tank, with its closest environmental movement analogues being Resources for the Future, the World Resources Institute, and the Earth Policy Institute. The latter is headed by Lester Brown, who founded Worldwatch in 1974 and served as its president through 2000. Its current president is Christopher Flavin.

Worldwatch prides itself on its accessible writing style and its fact-based analysis of critical global issues. It focuses on the underlying causes of these issues and seeks, through education and dissemination of information, to inspire people to act in positive ways. A search of its website produces large numbers of publications regarding climate change, which it has addressed in its publications since at least 1984. *Worldwatch Papers*, one of its signature publications, has sought to educate the public regarding "pressing economic, environmental, and social issues" since 1975. Worldwatch has published *State of the World*, a widely read and widely influential annual report, beginning in 1984. Although Worldwatch does not lobby Congress directly, this comprehensive report is read by legislators as well as world leaders, students, and ordinary individuals and has been translated into

25 languages. In 1992, *Vital Signs: The Trends That Are Shaping Our Future* came into being—an annual series designed to be even more accessible, with its "brief, digestible glimpses into more than 50 issues affecting the world each year." The group publishes a bimonthly magazine, *World Watch,* and has moved to Web-based education recently with *Vital Signs Online*. It also produces an occasional series of books on specialized issues.

Worldwatch is not a one-issue organization, having written about a very wide range of environmental issues including energy, water pollution and availability, soil erosion and other agricultural concerns, population, biodiversity, materials recycling and conservation, forests, toxic materials, and so on. However, it seeks to foster recognition that these issues are inextricably tied to issues of social justice and peace. It began paying consistent attention to the relationship between social and environmental issues, particularly in international settings, much earlier than most environmental organizations. It began calling attention to the need for a sustainable society in at least 1982, five years before "sustainability" began to gain widespread attention with the publication of the Brundtland Commission report, *Our Common Future*. One of the features of its website is an item entitled "Natural Disasters and Peacemaking." A premise is that natural disasters can serve as a means of breaking down social and political barriers, leading to opportunities for peace.

A desire to inspire change in societal attitudes and actions from a grassroots perspective is a hallmark of this organization. It seeks to effect change not by force from the top but by educating the public and thereby inspiring them to demand change.

Worldwatch spends 78.4 percent of its overall budget to pay for the programs and services it exists to provide. Its administrative expenses use 6.7 percent of its budget, and fundraising accounts for another 14.7 percent. Worldwatch spends $0.19 to raise each dollar it earns. Christopher Flavin's compensation for fiscal year 2005 was $95,000, which amounted to 3.38 percent of the group's total budget—a larger percentage (though a smaller amount) than the leaders of many other environmental organizations such as World Resources Institute, Environmental Defense, or the Natural Resources Defense Council.

SEE ALSO: Developing Countries; Resources for the Future (RFF); Sustainability; World Resources Institute.

BIBLIOGRAPHY. Charity Navigator, http://www.charitynavigator.org (cited August 1, 2007); Worldwatch Institute, Linda Starke ed., *State of the World 2007: An Urban Planet* (W.W. Norton, 2007); Worldwatch Institute, http://www.worldwatch.org (cited October 31, 2007).

PAMELA RANDS
GORDON RANDS
WESTERN ILLINOIS UNIVERSITY

World Weather Watch

THE WORLD WEATHER Watch is the central program of the World Meteorological Organization (WMO), the United Nations' agency for cooperation among national weather bureaus, founded in 1950. The Fourth World Meteorological Congress approved the idea of the program in 1963, and the WMO, which has 188 member countries and territories, subsequently established the World Weather Watch to make available an integrated worldwide weather-forecasting system.

The World Weather Watch includes the Tropical Cyclone Program, the Antarctic Activities Program, an Emergency Response Activities Program for environmental emergencies, and the Instruments and Methods of Observation Program to ensure the quality of the observations that are vital for weather forecasting and climate monitoring.

Through the World Weather Watch, a system is in place for countries around the world to obtain daily weather forecasts. The core components of the World Weather Watch—the Global Observatory System (GOS), the Global Telecommunications System (GTS), and the Global Data-Processing and Forecasting System (GDPFS)—enable the World Weather Watch to provide basic meteorological data to the WMO and other related international organizations.

The GOS allows for observing, documenting, and communicating data about the weather and climate for the creation of forecast and warning services. Monitoring the climate and the environment is a priority of the WMO, and the GOS is critical to the effective and efficient operations of the WMO. Long-term objectives of the GOS include the standardization of observation practices and the optimization of global observation systems.

The GTS consists of land and satellite telecommunication links that connect meteorological telecommunication centers. The GTS provides efficient and reliable communication service among the three World Meteorological Centers in Melbourne, Moscow and Washington, and the 15 Regional Telecommunication Hubs that make up the Main Telecommunication Network. The six Regional Meteorological Telecommunication Networks, covering Africa, Asia, South America, North America, Central America and the Caribbean, South–West Pacific, and Europe, ensure the collection and distribution of data to members of the WMO. The National Meteorological Telecommunication Networks make it possible for the National Meteorological Centers to collect data and to receive and disseminate weather information on a national level.

The primary aim of the GDPFS is to prepare and provide meteorological analyses to members in the most cost-effective manner possible. The GDPFS is organized to implement functions at international, regional, and national levels through the World Meteorological Centers, the Regional Specialized Meteorological Centers, and the National Meteorological Centers. Real-time functions include preprocessing and postprocessing of data and the preparation of forecast products. Non-real-time functions include long-term storage of data and the preparation of products for climate-related analysis.

Increasingly, the World Weather Watch provides support for developing international programs related to global climate and other environmental issues and to sustainable development. The entire continent of Africa has only 1,150 World Weather Watch stations—one per 26,000 sq. km., (10,038 sq. mi.)—even though the continent's land mass is as large as North America, Europe, Australia, and Japan put together. This represents coverage eight times lower than the WMO's recommended minimum level. The changing climate of Africa necessitates greater capacity building on the part of institutions prepared to address the likely crises that lie ahead. The World Weather Watch is vital in developing that capacity.

The World Weather Watch and its parent organization, the WMO, through the development of a permanent global weather data network, have proven critical to defining global warming as a given. As a consequence, the political and policymaking debates about climate change and its very real consequences, such as those facing Africa, have moved to a new arena. Although the World Weather Watch cannot compel individual governments to act on its findings, it can and has framed the issue of climate change on a truly global scale.

SEE ALSO: Climate; United Nations.

BIBLIOGRAPHY. Paul N. Edwards, "Meteorology as Infrastructural Globalism," *OSIRIS* (v.21, 2006); *The Environment in the News*, The United Nations Environment Programme, _www.unep.org/cpi/briefs/2006Nov20.doc; "The View from Space," *Weatherwise* (v.48/3, 1995); WMO in Brief, www.wmo.ch/pages/about/index_en.html.

ROBIN K. DILLOW
ROTARY INTERNATIONAL ARCHIVES

World Wildlife Fund

ALTHOUGH THE WORLD Wildlife Fund around the world has changed its name to the World Wide Fund for Nature, the original name remains the official one in the United States and Canada. As an international nongovernmental organization, it was founded in 1961 in Switzerland to help with the conservation, research, and restoration of the natural environment, changing its name in 1986, although still keeping its initials (WWF) around the world.

Although over many years the WWF became famous for its protection of endangered fauna—its symbol remains a panda bear—it has also been keen to preserve natural environments, seeing its role as helping endangered flora as much as fauna, with the change in its name reflecting this. Indeed the WWF now recognizes that the single biggest threat to the environment today comes from global warming, and as a result it has campaigned for companies and individuals to reduce their greenhouse gas emissions.

In December 2007, the World Wildlife Fund issued a report titled "Antarctic Penguins and Climate Change."

The major area where the WWF initially concentrated its energies was in reducing deforestation, especially in Brazil, Central Africa, and the Russian Far East. This has seen U.S. experts from the WWF-U.S. taking part in projects in these regions, and also in other parts of the world. They have been involved in recording the level of deforestation, and in many cases illegal logging, and notifying the relevant governments as well as bringing extreme levels of deforestation to world attention.

Traditionally, the WWF has organized throughout the United States at a city, town, and village level, with the education of young people being at the forefront of its approach. This means that the WWF has devoted much of its time and energy to encouraging students to gain a greater interest in the environment and the threat of global warming through the provision of resource kits, booklets, and lectures. Many of these items have been available free of charge, or heavily subsidized, with many schoolchildren becoming interested in the world of the WWF through television documentaries and other media sources such as the internet.

This has seen the developing of educational problems to allow more students to plan ways of reducing carbon dioxide emissions. Among the children who have been involved in WWF projects have been some of those displaced by Hurricane Katrina, who have been better able to understand the problems leading to the hurricane. To that end, the WWF hosts the Southeast Climate Witness Program, which allows students to attend a Climate Camp in June 2008 and to take part in the Youth Summit in Washington, D.C., in the following month. Many schools around the United States also raise money for the WWF that is used for the campaign against climate change.

Although it has long been a community movement, the WWF has also started working heavily with businesses. This change has seen the WWF and some of its partners collaborating with 12 prominent companies including the Collins Companies, IBM, Johnson & Johnson, Nike, Polaroid, and Sony. These large corporations have agreed to work toward reducing their carbon dioxide emissions by over 10 million tons each year, which, as a result, has led to many smaller companies becoming aware of their effect on the generating of greenhouse gases and working to reduce their emissions. The WWF has also tried to get, with less success, the energy utility companies to reduce the emissions of their operations.

SEE ALSO: Animals; Climate Change, Effects; Deforestation; Economics, Cost of Affecting Climate Change.

BIBLIOGRAPHY. J. Brooks Flippen, *Conservative Conservationist: Russell E. Train and the Emergence of American Environmentalism* (Louisiana State University Press, 2006); World Wildlife Fund, www.worldwidlife.org (cited November 2007).

JUSTIN CORFIELD
GEELONG GRAMMAR SCHOOL, AUSTRALIA

Wyoming

WYOMING IS 97,819 sq. mi. (253,350 sq. km.) with inland water making up 714 sq. mi. (1,849 sq. km.). The Continental Divide (the separation mark of the Pacific and Atlantic watersheds) passes through Wyoming, making Wyoming a source for the Missouri-Mississippi, the Great Basin, the Columbia, and the Colorado drainage systems.

The climate in Wyoming is relatively cool and depends on the elevation; for example, the summers are moderately warm at lower elevations. Early freezes and a late spring provide long winters and a short growing season (from approximately 80 days in the northwest to 120 days in the plains). Precipitation also varies with elevation. Snow falls from November to May—the yearly snowfall can be as little as 10 in. or less in the basins and 15 to 20 in. in the Plains to over 60 to 70 in. at the higher elevations.

Wyoming has 22 state parks, Grand Teton and Yellowstone national parks, and national forests set aside for preservation. The largest aquifer in the world, Ogallala Aquifer lies underground beneath eight states including Wyoming. Major industries include agriculture (most of the agricultural land is used for grazing, though dryland wheat and some irrigated crops are grown) and mining for coal, natural gas, and petroleum. Industries that are important to Wyoming's economy include petroleum refining, chemical industries, food processing, industrial machinery, and wood products.

Although climate models vary on the amount of temperature increase possible, potential risks include having decreased water supplies; increased risk for wildfires; changes in food production, with agriculture improving in cooler climates and decreasing in warmer climates; change in rain pattern to downpours, with the potential for flash flooding and health risks of certain infectious diseases from water contamination or disease-carrying vectors such as mosquitoes, ticks, and rodents, and heat-related illnesses.

Wyoming may benefit from changing climate. Shorter, milder winters could mean longer growing seasons and increasing crop yields, though higher temperatures may mean changing crops produced to those that are better adapted to a warmer climate and that are more drought resistant. The milder climate could attract more tourists. Taking advantage of sun and wind to produce electricity could provide economic benefits. The effect of climate change on agriculture will be mixed, and some crops such as potatoes and wine grapes could be negatively affected by rising temperatures, decreasing yields. By comparison, the orchard crops will mature more quickly at height temperatures, with increased quality and market-share value. Some areas may need to change crops for those with higher drought resistance and adaptability to a warmer climate.

Wyoming's glaciers are melting at a rapid pace because of milder temperatures brought on by global warming. Warmer temperatures also mean less snowpack in the mountains and earlier snowmelt, leading to more winter runoff and reduced summer flows in many Wyoming streams. Snowpack also stores much of Wyoming's clean water supply for drinking, agriculture, and wildlife. Any reduction in snow would increase pressures on this valuable and scarce resource.

On the basis of energy consumption data from Energy Information Administration's State Energy Consumption, Price, and Expenditure Estimates, released June 1, 2007, Wyoming's total CO_2 emissions from fossil fuel combustion for 2004 was 63.54 million metric tons, made up of contributions from commercial (0.86 million metric tons), industrial (10.12 million metric tons), residential (0.86 million metric tons), transportation (8.07 million metric tons), and electric power (43.62 million metric tons).

Wyoming's current incentive programs and tax breaks targeted at reducing carbon emissions are designed to encourage energy efficiency and the use of renewable energy sources such as wind and solar power. New laws enacted in 2007 include authorizing a clean-coal task force and providing funding for the clean-coal research, extending a tax benefit for an additional four years (until 2012) for renewable power generation facilities such as wind farms, and improved funding for the Wyoming Game and Fish Department—the state's agency for managing both game and nongame species of wildlife. In addition, the legislature approved project money drawn on the interest in the Wyoming Wildlife and National Resources Trust Fund, established in 2005, to enhance aspen and sage on the Bates Creek watershed and aspen, sage, bitterbrush, and sumac near Lander, both for wildlife foraging.

SEE ALSO: Climate Models; Greenhouse Gases.

BIBLIOGRAPHY. Mark Bowen, *Thin Ice: Unlocking the Secrets of Climate in the World's Highest Mountains* (Henry Holt, 2005); Alfred N. Garwood, *Weather America Latest Detailed Climatological Data for Over 4,000 Places and Rankings* (Toucan Valley Publications, 1996); National Wildlife Federation, "Global Warming and Wyoming" (June 25, 2007); University of Wyoming, "Continued Global Warming Could Destroy Existing Climates and Create New Ones," www.uwyo.edu/news (cited March 26, 2007).

LYN MICHAUD
INDEPENDENT SCHOLAR

Yemen

LOCATED IN THE southern region of the Arabian Peninsula, the Republic of Yemen has a land area of 203,849 sq. mi. (527,968 sq. km.), with a population of 22,389,000 (2006 est.) and a population density of 104 people per sq. mi. (40 people per sq. km.). Only 3 percent of the land in Yemen is arable, with 13 percent of that area being under permanent cultivation. In addition, 34 percent of the country is used as meadows or pasture, and 8 percent of the country is forested.

Yemen has a very low rate of per capita carbon dioxide emissions, being 0.8 metric tons per person in 1991, rising gradually to 1.03 metric tons in 2004. The entire electricity production in the country comes from fossil fuels, with liquid fuels being responsible for 96 percent of Yemen's carbon dioxide emissions. In terms of the sector producing the carbon dioxide in the country, 48 percent comes from transportation, with 20 percent from electricity and heat production, 20 percent from residential use, and 7 percent from manufacturing and construction.

The effect of global warming and climate change has already been dramatic in Yemen, with the alienation of some marginal arable land. The rise in temperature has made it harder to grow crops in arid parts of the country, and there has been extensive bleaching of coral reefs along the Red Sea coastline and the Socotra Archipelago.

The Yemen government of Ali Abdullah Saleh took part in the United Nations Framework Convention on Climate Change, signed in Rio de Janeiro in May 1992, and accepted the Kyoto Protocol to the UN Framework Convention on Climate Change on September 15, 2004, with it entering into force on February 16, 2005.

SEE ALSO: Climate Change, Effects; Drought.

BIBLIOGRAPHY. Madawi al-Rasheed and Robert Vitalis, eds., *Counter-Narratives: History, Contemporary Society, and Politics in Saudi Arabia and Yemen* (Palgrave Macmillan, 2004); Andrew Cockburn, "Yemen," *National Geographic* (v.197/4, April 2000); Nicholas Pilcher and Abdullah Alsuhaibany, "Regional Status of Coral Reefs in the Red Sea and the Gulf of Aden," in Clive Wilkinson, ed., *Status of Coral Reefs of the World* (Global Coral Reef Monitoring Network, Australian Institute of Marine Science, 2000); World Resources Institute, "Yemen—Climate and Atmosphere," http://earthtrends.wri.org (cited October 2007).

ROBIN S. CORFIELD
INDEPENDENT SCHOLAR

Younger Dryas

MARKING THE BOUNDARY between the Holocene and Pleistocene epochs, the Younger Dryas, a period of glacial conditions between 12,900 and 11,500 years ago, is named for *Dryas octopetala*, a flower that is adapted to the cold. *Dryas'* pollen is found in abundance in strata of this age. *Dryas'* pollen is also found in older strata, necessitating the term Younger Dryas to distinguish this time from older periods in which *Dryas'* pollen is abundant. Locked in an ice age, earth had finally warmed and the glaciers had begun to retreat 15,000 years ago. Counteracting this warming trend, the Younger Dryas reduced temperatures 50 degrees F in only a decade. Glaciers once more advanced in North America and Europe. Rainfall diminished, and frigid winds carried dust from central Asia throughout Europe.

Climatologists have identified three causes of the Younger Dryas, though it is uncertain whether all three operated at the same time. The fact that the Southern Hemisphere cooled before the Northern Hemisphere suggests that some mechanism cooled the south, whereas no mechanism was then operating in the north. The rapid change in climate that was the Younger Dryas may have caused the extinction of large mammals in North America and the collapse of the first Native American culture. In western Asia, the Younger Dryas may have prompted humans to invent agriculture. The end of the Younger Dryas ushered in the modern climate.

The Younger Dryas was part of the Cenozoic Ice Age, which locked the world in glaciers 100,000 years ago. The climate was particularly cold as recently as 18,000 years ago. From these frigid conditions, the climate gradually warmed until 15,000 years ago, the glaciers began to retreat. The Younger Dryas interrupted this warming trend, restoring glacial conditions to earth.

Climatologists have advanced three causes of the Younger Dryas. The leading explanation focuses on ocean currents. The Gulf Stream brings warm water from the tropics to the North Atlantic Ocean, warming the Atlantic coasts of North America and Europe. The Gulf Stream remained undisturbed as the North American glacier began to retreat north 15,000 years ago. In the initial centuries of retreat, the ice sheet emptied its water down the Mississippi River and into the Gulf of Mexico. By 12,900 years ago, however, the North American glacier had retreated to the Great Lakes. Melted water no longer flowed south down the Mississippi River but now went east along the St. Lawrence River to the Atlantic Ocean. This cold water shut down the Gulf Stream, robbing North America and Europe of its warmth and returning the climate to glacial conditions.

Climatologists have identified a second cause in the impact of an asteroid near the Great Lakes 12,900 years ago. Upon impact, the asteroid ejected enormous amounts of debris, dust, and ash into the atmosphere, blocking out sunlight and cooling the Earth. A third cause might have been the sudden, and unexplained, cessation of El Niño. Every two to seven years, warm water from the western Pacific Ocean and Indian Ocean flows east, warming the west coasts of South and North America. Without El Niño, these continents cooled, returning them to glacial conditions. Possibly more than one cause initiated the Younger Dryas.

The Younger Dryas ended as abruptly as it had begun, when temperatures rose 50 degrees F (28 degrees C) in just 10 years. Glaciers retreated to Antarctica, Greenland, and the North Pole, and rainfall again became abundant. The Cenozoic Ice Age, having cooled the Earth for 100,000 years, finally ended with the close of the Younger Dryas. Forests returned to Scandinavia, Germany, and North America. The return of warmth and rainfall, along with the invention of agriculture, allowed humans to settle in communities. With some exceptions, humans were no longer nomads. The end of the Younger Dryas initiated the modern climate. Although temperatures fluctuated in modernity, the retreat of glaciers has so far been permanent. Perhaps glaciers will return one day, though there is no evidence that they will come soon.

SEE ALSO: Climate; Cretaceous Era; Greenhouse Effect; Greenhouse Gases.

BIBLIOGRAPHY. John D. Cox, *Climate Crash: Abrupt Climate Change and What It Means for Our Future* (Joseph Henry Press, 2005); Lynn J. Rothschild and Adrian M. Lister, eds., *Evolution on Planet Earth: The Impact of the Physical Environment* (Academic, 2003).

CHRISTOPHER CUMO
INDEPENDENT SCHOLAR

Zambia

THE REPUBLIC OF Zambia lies in the interior of southern Africa and shares its borders with the Democratic Republic of Congo, Angola, Zimbabwe, and five other countries. From 1924, the country was known as Northern Rhodesia and was administered as a British protectorate by the United Kingdom. It achieved independence in 1964 and was renamed the Republic of Zambia for the Zambezi River.

Zambia is a landlocked country with a tropical climate. Most of the country consists of flat plateau with altitudes of 3,281–4,921 ft. (1,000–1,500 m.), which contributes to a milder climate. Average maximum temperatures during the hot, rainy season (November to March) range from 79–95 degrees F (26–35 degrees C); the cooler, dry season (April to August) brings high temperatures of 77–82 degrees F (25–28 degrees C). Annual rainfall ranges from 750 mm. in the south to more than 1,300 mm. in the north.

Rising temperatures and erratic rainfall are the primary symptoms of climate change. The most significant risks of climate change include water scarcity, reduced agricultural productivity, spread of vector-borne diseases such as malaria, risk of forest fires, reduced fish and wildlife stocks, and increased flooding and droughts. Farmers, rural households, and communities that depend on natural resources for their livelihoods are the most vulnerable to the effects of climate change. The government has embarked on a National Adaptation Programme for Action, with support from the United Nations Development Programme.

With declining per capita food production, food security is already a grave concern. Food-insecure people requiring humanitarian assistance number roughly one million. Food security may be further threatened by the effects of climate change, especially declining rainfall, and is exacerbated by the existing challenges of chronic poverty, HIV/AIDS, and an ineffective food distribution system. Northern Zambia is prone to flooding, whereas the south is increasingly dry.

Maize is an extremely important crop—the cornmeal-based *nshima* is served with nearly every meal—but maize is highly vulnerable to drought. Shorter rainy seasons, later start of rains, and declining rainfall are associated with reduced productivity of maize and other crops. Some farmers have switched to earlier-maturing, drought-resistant crops such as sweet potatoes as an adaptation strategy, but many are reluctant to pass up government maize subsidies and the steady demand for cornmeal.

Deforestation is a continuing problem, and the clearing of forests for agriculture, construction, and industrialization is likely to increase the risks of climate change. Forests protect watersheds, provide erosion control, and absorb carbon that would otherwise contribute to global warming. Deforestation rates are estimated between 200,000 and 300,000 hectares per year. Extensive forest exploitation is related to the production of charcoal, which contributes to carbon monoxide emissions. Eighty-three percent of urban households use charcoal for cooking fuel.

Miombo woodlands provide an important source of timber and nontimber forest products. Although Miombo woodlands require regular burning, rising temperatures and declining rainfall may lead to increases in wildfires. Wildfires and deforestation threaten traditional bark-hive beekeeping and the reproductive capacity of Mophane caterpillar products, an important source of nutrition and income for rural households.

SEE ALSO: Climate Change, Effects; Current; Deforestation; Drought.

BIBLIOGRAPHY. R.T. Mano, J. Arntzen, S. Drimie, P. Dube, J.S.I. Ingram, C. Mataya, M.T. Muchero, E. Vhurumuku, and G. Ziervogel, "Global Environmental Change and the Dynamic Challenges Facing Food Security Policy in Southern Africa" GECAFS Working Paper 5 (2007); Joseph Schatz, "Farmers Scorched by Climate Change," *Zambia Daily Mail* (July 11, 2007); Newton Sibanda, "Charcoal Burners Feeding Urban Poor, Deforestation," *Zambia Daily Mail* (April 2, 2007).

ROBERT B. RICHARDSON
MICHIGAN STATE UNIVERSITY

Zimbabwe

A LANDLOCKED COUNTRY located in southern Africa, Zimbabwe (formerly Rhodesia) has a land area of 150,871 sq. mi. (390,757 sq. km.), with a population of 13,349,000 (2006 est.) and a population density of 85 people per sq. mi. (33 people per sq. km.). Its economy has been heavily reliant on agriculture, with 7 percent of the country being arable, 13 percent being used as meadows or pasture, and 62 percent being forested.

In Zimbabwe, some 53.3 percent of electricity generation comes from fossil fuels, with 46.7 percent from hydropower. This extensive use of hydropower, much of it from the Kariba Dam, as well as the declining economy in the country, has led to Zimbabwe having a low rate of per capita carbon dioxide emissions—1.6 metric tons per person in 1990, falling to 0.81 metric tons in 2004. In 1999, electricity production contributed 55 percent of Zimbabwe's carbon dioxide emissions, with manufacturing and construction making up 23 percent and transportation a further 20 percent. In terms of the source of carbon dioxide emissions, 86 percent were from solid fuels, with 10 percent from liquid fuels, and most of the remainder from gaseous fuels. Problems with electricity supplies in the country have led to many people reverting to the use of private generators. The economy has also been badly damaged by a decline in the demand for tobacco—the most important single export in the country.

The effects of global warming on the country have been a steady alienation of marginal arable land, which, together with a general decline in the economy, has seen widespread impoverishment. The rise in temperature has also affected the productivity of fish farms. The Zimbabwean government of Robert Mugabe took part in the United Nations Framework Convention on Climate Change, signed in Rio de Janeiro in May 1992, but the government has so far not expressed an opinion on the Kyoto Protocol to the UN Framework Convention on Climate Change. However C. Madova, co-vice president of the African Region at the World Bank addressed the Fifth World Bank Conference on Environmentally and Socially Sustainable Development in Washington, D.C., in December 1997, supporting measures to reduce world emissions.

SEE ALSO: Carbon Dioxide; Climate Change, Effects.

BIBLIOGRAPHY. Peter Godwin, "Zimbabwe's Bitter Harvest" *National Geographic* (v.204/2, August 2003); Ian Scoones, "The Dynamics of Soil Fertility Change: Historical Perspectives on Environmental Transformation From Zimbabwe" *Geographical Journal* (v.163/2, July 1997); World Resources Institute, "Zimbabwe—Climate and Atmosphere," www.earthtrends.wri.org (cited October 2007).

ROBIN S. CORFIELD
INDEPENDENT SCHOLAR

Resource
Guide

BOOKS

Abrahamson, D.E. (ed.). *The Challenge of Global Warming* (Island Press, 1989)

Adger, N., et al. *Climate Change 2007: Impacts, Adaptation and Vulnerability Working Group II* (Intergovernmental Panel on Climate Change, 2007)

Aguado, E., and Burt, James E. *Understanding Weather and Climate* (Prentice Hall, 2006)

Ahrens, C. Donald. *Meteorology Today* (Thomson Brooks/Cole, 2007)

Archer, David. *Global Warming: Understanding the Forecast* (Blackwell Publishing, 2007)

Attfield, Robin. *Environmental Ethics: An Overview for the Twenty-First Century* (Polity, 2003)

Bailey, R.A., Clark, H.M., Ferris, J.P., Krause, S., and Strong, R.L. *Chemistry of the Environment* (Academic Press, 2002)

Baumert, Kevin, Pershing, Jonathan, Herzog, Timothy, and Markoff, Matthew. *Climate Data: Insight and Observations* (Pew Center on Global Climate Change, 2004)

Beatley, Timothy. *Green Urbanism: Learning from European Cities* (Island Press, 1999)

Brennan, Scott, and Withgott, Jay H. *Environment: The Science Behind the Stories* (Pearson Education, Inc., 2004)

Burton, I., Diringer, N., and Smith, J. *Adaptation to Climate Change: International Policy Options* (Pew Center on Global Climate Change, 2006)

Ciambrone, David F. *Environmental Life Cycle Analysis* (CRC, 1997)

Consultative Group on International Agricultural Research. *Global Climate Change: Can Agriculture Cope?* (CGIAR, 2007)

Coward, Harold, and Hurka, Thomas. (eds.). *Ethics and Climate Change: The Greenhouse Effect* (Laurier Press, 1993)

Cudworth, Erica. *Environment and Society* (Routledge, 2003)

Curran, Mary Ann. *Environmental Life-Cycle Assessment* (McGraw-Hill, 1996)

Dardo, Mario. *Nobel Laureates and Twentieth-Century Physics* (Cambridge University Press, 2004)

Dryzek, John. *The Politics of the Earth: Environmental Discourses* (Oxford University Press, 1997)

Dunne, Thomas, and Leopold, Luna B. *Water in Environmental Planning* (W.H. Freeman and Company, 1978)

Energy Information Administration, *Impact of the Kyoto Protocol on U.S. Energy Markets and Economic Activity* (U.S. DOE, 1998)

Feenstra, J., Burton, I., Smith, J.B., and Tol, R. (eds.). *Handbook on Methods for Climate Change Impact*

Assessment and Adaptation Strategies (United Nations Environment Programme and Institute for Environmental Studies, 1998)

Frakes, Lawrence. *Climate Throughout Geologic Time* (Elsevier/North-Holland, 1979)

Frederick, Kenneth D., and Gleick, Peter H. *Water and Global Climate Change: Potential Impacts on U.S. Water Resources* (Pew Center on Global Change, 1999)

Gasch, Robert, and Twele, Jochen (eds.). *Wind Power Plants: Fundamentals, Design, Construction and Operation* (Earthscan Publications Ltd., 2004)

Gipe, Paul. *Wind Power, Revised Edition: Renewable Energy for Home, Farm, and Business* (Chelsea Green Publishing Company, 2004)

Gordon, David. *Green Cities: Ecologically Sound Approaches to Urban Space* (Black Rose Books Ltd., 1990)

Gottlieb, Roger. *Forcing the Spring: The Transformation of the American Environmental Movement* (Island Press, 1993)

Graedel, Thomas. *Atmosphere, Climate and Change* (W. H. Freeman, 1995)

Hart, David M. *The Forged Consensus: Science, Technology, and Economic Policy in the United States* (Princeton University Press, 1998)

Hartmann, Dennis L. *Global Physical Climatology* (Academic Press, 1994)

Harvey, L. D. Danny. *Climate and Global Environmental Change* (Prentice-Hall, 2000)

Hay, Peter. *Main Currents in Western Environmental Thought* (Indiana University Press, 2002)

Held, D., McGrew, A., Goldblatt, D., and Peraton, J. *Global Transformations: Politics, Economics, Culture* (Cambridge Polity Press, 1999)

Hughes, Donald J. *An Environmental History of the World: Humanity's Changing Role in the Community of Life* (Routledge, 2001)

Intergovernmental Panel on Climate Change (IPCC). *Climate Change, Contribution of Working Group I to the Third Assessment Report* (Cambridge University Press, 2001)

IPCC. *Climate Change 1994, Radiative Forcing of Climate Change* (Cambridge University Press, 1995)

IPCC. *Climate Change 1995, The Science of Climate Change* (Cambridge University Press, 1996)

IPCC. *Climate Change 2001: The Scientific Basis* (Cambridge University Press, 2001)

IPCC. *Climate Change 2007: The Physical Basis. Contribution of Working Group I to the Fourth Assessment Report of the Intergovernmental Panel on Climate Change* (Cambridge University Press, 2007)

Jain, R.L., Urban, L.V., Stacey, G.S., and Balbach, H. *Environmental Assessment*, 2nd ed. (McGraw-Hill, 2002)

Jenks, Mike, and Dempsey, Nicola. *Future Forms and Design for Sustainable Cities* (Architectural Press, 2005)

Kendrew, Wilfrid. *The Climates of the Continents* (Clarendon Press, 1961)

Kiely, Gerard. *Environmental Engineering* (McGraw-Hill, 1996)

Krech, Shepard, McNeill, John Robert, and Merchant, Carolyn. *Encyclopedia of World Environmental History* (Routledge, 2004)

Leroy, Francis (ed). *A Century of Nobel Prize Recipients: Chemistry, Physics, and Medicine (Neurological Disease and Therapy)* (CRC, 2003)

Linacre, Edward, and Geerts, Bart. *Climates and Weather Explained* (Routledge, 1997)

Maroto, M., and Valer, M. (eds). *Environmental Challenges and Greenhouse Gas Control for Fossil Fuel Utilization in the 21st Century* (Kluwer Academic/Plenum Publishers, 2002)

Masters, G.M. *Introduction to Environmental Engineering and Science*, 2nd ed. (Prentice-Hall, New Jersey, 1990)

McNeill, John Robert. *Something New Under the Sun. An Environmental History of the Twentieth Century World* (Norton, 2000)

Merchant, Carolyn (ed). *Major Problems in American Environmental History*, 2nd ed. (Houghton Mifflin, 2005)

Moran, Emilio F. *People and Nature. An Introduction to Human Ecological Relations* (Blackwell Publishing, 2006)

Oxfam. *Adapting to Climate Change, What's Needed in Poor Countries and Who Should Pay,* (Oxfam Briefing Paper, v.104, 2007)

Petty, G.W. *A First Course in Atmospheric Radiation* (Sundog Publishing, 2004)

Phelps, Edmund. *Private Wants and Public Needs.* (W.W. Norton & Company Inc., 1962)

Ponting, Clive. *A Green History of the World. The Environment and the Collapse of Great Civilizations* (Penguin, 1991)

Rampino, Michael R. *Climate: History, Periodicity, and Predictability* (Van Nostrand Reinhold, 1987)

Raupach, M.R., Marland, G., Ciais, P., Quéré, J. C. Le, Canadell, G., Klepper, G., and Field, C. B. *Global and Regional Drivers of Accelerating CO_2 Emissions* (PNAS, 2007)

Register, Richard. *Ecocities: Rebuilding Cities in Balance With Nature* (New Society, 2006)

Richards, John F. *The Unending Frontier: An Environmental History of the Early Modern World* (University of California Press, 2003)

Robinson, Peter, and Henderson-Sellers, Ann. *Contemporary Climatology* (Prentice-Hall, 1999)

Sawyer, C.N., McCarty, P.L., and Parkin, G.F. *Chemistry for Environmental Engineering and Science,* 5th ed. (McGraw-Hill, 2003)

Shanley, Robert A. *Presidential Influence and Climate Change* (Greenwood Press, 1992)

Siedler, Gerold, Church, John, and Gould, John (eds). *Ocean Circulation and Climate: Observing and Modelling the Global Ocean* (Academic Press, 2001)

Singer, S. Fred, and Avery, Dennis T. *Unstoppable Global Warming: Every 1,500 Years* (Rowman and Littlefield, 2007)

Singh, Ram Babu. *Urban Sustainability in the Context of Global Change: Towards Promoting Healthy and Green Cities* (Science Publishers, 2001)

Solomon, S., Qin, D., Manning, M., Chen, Z., Marquis, M., Avery, K.B., Tignor M., and Miller, H.L. (eds). *Climate Change 2007: The Physical Science Basis. Contribution of Working Group I to the Fourth Assessment Report of the Intergovernmental Panel on Climate Change* (Cambridge University Press, 2007)

Spray, Sharon. *Global Climate Change* (Rowman & Littlefield, 2002)

Stern, Nicholas. *The Economics of Climate Change: The Stern Review* (Cambridge University Press, 2007)

Stow, Dorrik. *Oceans: An Illustrated Reference* (University of Chicago Press, 2005)

Szasz, Andrew. *EcoPopulism: Toxic Waste and the Movement for Environmental Justice* (University of Minnesota Press, 1994)

Udall, Stewart L. *The Quiet Crisis* (Holt, Rinehart and Winston, 1963)

United Nations Framework Convention on Climate Change. *An Introduction to the Kyoto Protocol Compliance Mechanism* (UNFCC, 2006)

Viessman, Warren Jr., and Lewis, Gary L. *Introduction to Hydrology* (Prentice Hall, 2003)

Weart, Spencer. *The Discovery of Global Warming* (Harvard University Press, 2004)

Weir, Gary E. *Ocean in Common: American Naval Officers, Scientists, and the Ocean Environment (Texas A & M University Military History Series)* (Texas A & M University Press, 2001)

Williams, Mary E. *Global Warming: An Opposing Viewpoints Guide* (Greenhaven Press, 2006)

Wise/TranTolo. *Bioremediation of Contaminated Soils (Environmental Science and Pollution Control)* (CRC, 2000)

Worster, Donald. *Nature's Economy: A History of Ecological Ideas* (Cambridge University Press, 1994)

Yamin, Farhana, and Depledge, Joanna. *The International Climate Change Regime: A Guide to Rules, Institutions and Procedures* (Cambridge University Press, 2004)

Yoshino, Masatoshi. *Climates and Societies: A Climatological Perspective* (Kluwer Academic Publishers, 1997)

ARTICLES

Bates, Diane. "Environmental Refugees? Classifyng Human Migrations Caused by Environmental Change." *Population and Environment* (v. 23/5, May 2002)

Bindoff, N., Willebrand, V., Artale, A., Cazenave, J., Gregory, S., Gulev, K., Hanawa, C., Quéré, S. Le, Levitus, Y., Nojiri, C., Shum, L., Talley, A., and Unnikrishnan, A. "Observations: Oceanic Climate Change and Sea Level", in *Climate Change 2007: The Physical Science Basis. Contribution of Working Group I to the Fourth Assessment Report of the Intergovernmental Panel on Climate Change* [Solomon, S., Qin, D., Manning, M., Chen, Z., Marquis, M., Avery, K., Tignor, M., and Miller, H.L. (eds.)], (Cambridge University Press, 2007)

Brown, Donald. "The ethical dimensions of global environmental issues." *Daedalus* (2001, pp. 59-77)

Bryden, Harry L., Longworth, Hannah R., and Cunningham, Stuart A. "Slowing of the Atlantic Meridional Overturning Circulation at 25° N," *Nature* (v.438, Dec. 2005)

Caldeira, Ken, and Wickett, Michael E. "Anthropogenic carbon and ocean pH." *Nature* (v.425/6956, September 2003)

Coughlin, Steven S. "Educational intervention approaches to ameliorate adverse public health and environmental effects from global warming," *Ethics in Science and Environmental Politics* (2006, pp. 13-14)

Crowley, T.J. "Causes of climate change over the past 1000 years." *Science* (v.289, pp. 270-277)

Dybas, Cheryl. "Increase in Rainfall Variability Related to Global Climate Change." *Earth Observatory, NASA* (December 12, 2002)

Ganopolski, A. and Rahmstorf, S. "Rapid changes of glacial climate simulated in a coupled climate model." *Nature* (v.409, pp. 153-158, 2001)

Gardiner, Stephen. "Ethics and global climate change." *Ethics* (2004, pp. 555-600)

Gordon, Arnold L. "Inter-ocean Exchange of Thermocline Water." *Journal of Geophysical Research* (v.91, 1986)

Guha, Ramachandra. "Radical American Environmentalism and Wilderness Preservation: A Third World Critique." *Environmental Ethics*, Vol. 11, No.1 (Spring 1989, pp. 71-83)

Haque, C., and Burton, I. "Adaptation Options Strategies for Hazards and Vulnerability Mitigation An international Perspective", *(Mitigation and Adaptation Strategies for Global Change.* (v.10, 2005)

Hay, J., and Mimura, N. "Sea Level Rise : Implications for water Resources Management." *Mitigation and Adaptation Strategies for Global Change.* (v.10, 2005)

Keeling, Charles David. "Is Carbon Dioxide from Fossil Fuels Changing Man's Environment?" *Proceedings of the American Philosophical Society* (v.114/1: 10-17, 1970)

McCarthy, Michael. "Charles David Keeling – Climate Scientist Who First Charted the Rise of Greenhouse Gases," *The Independent.* (June 27, p. 32, 2005)

Meier, Mark F., Dyurgerov, Mark, Ursula, Rick, O'Neel, Shad, Pfeffer, W. Tad, Anderson, Robert, Anderson, Suzanne, and Glazovsky, Andrey. "Glaciers Dominate Eustatic Sea-Level Rise in the 21st Century." *Science (*v.298/5602, December, 2002)

Mirza, M. M. Q., Warrick, R. A. and Ericksen, N. J. "The Implications of Climate Change on Floods of the Ganges, Brahmaputra and Meghna Rivers in Bangladesh." *Climate Change.* (v.57(3), pp. 287-318, 2003)

Myhre, M., Highwood, E.J., Shine, K.P. and Stordal, F. "New estimates of radiative forcing due to well mixed greenhouse gases," *Geophysical Research Letters* (v.25(14), pp. 2715-18, 1998)

Neumann, James, Yohe, Gary, Nicholls, Robert and Manion, Michelle. *Sea-Level Rise & Global Climate Change: A Review of Impacts to U.S. Coasts* (Pew Center on Global Climate Change, 2000)

Ormerod, W.G., Ferund, P., and Smith, A. "Ocean Storage of CO_2." (IEA Greenhouse Gas R&D Program, Cheltenham, UK, 2002)

Poulsen, C.J., Seidov, D., Barron, E.J., and Peterson, W.H. "The Impact of Paleogeographic Evolution on the Surface Oceanic Circulation and the Marine Environment Within the Mid-Cretaceous Tethys." (*Paleoceanog*, 13, 546-559)

Randerson, James. "Should governments play politics with science?" *New Scientist*, 184.2468 (10/9/2004) 12-14

Schubert, C. "Global Warming Debate Gets Hotter," *Science News*, 159.24 (06/16/2001), 372

Simmonds, Mark P., and Isaac, Steven J. "The Impacts of Climate Change on Marine Mammals: Early Signs of Significant Problems", *Oryx* (v.41, 2007, pp.19-26)

Supreme Court of the United States, *Opinion of the Court: Commonwealth of Massachusetts, et al., v. U.S. Environmental Protection Agency, et al.* (549 U.S. No. 05.1120, 2007)

UNESCO. "Changes in Climate." *Arid Zone Research* (No. 20, 1963)

PERIODICALS

Energy and Environment – Multi-Science Publishing

Environmental Ethics – Center for Environmental Philosophy

Environmental Justice: Issues, Policies, and Solutions

Environmental Law – Oxford University Press

Environmental Management – Academic Press

Environmental Politics – Frank Cass

Environmental Science and Technology – Center for Environment and Energy Research and Studies

EPA Journal – Environmental Planning Agency

Global Environmental Change – Royal Society of Chemistry

Global Environment Politics – MIT Press

Hazardous Waste – BPI News

Journal of Environmental Economics and Management – Academic Press

Journal of Environmental Education – Heldref Education

Journal of Environmental Management – Academic Press

Journal of Environment and Development – SAGE Publications

Journal of Forestry – Oxford University Press

Journal of Geochemical Exploration – Elsevier Science

Journal of Geophysical Research – American Geophysical Union

National Geographic – National Geographic Society

Nature – Palgrave Macmillan

Oryx – Fauna and Flora International

Paleoceanography – American Geophysical Union

Proceedings of the American Philosophical Society – The American Philosophical Society

Science Trends in Ecology and Evolution – Oxford University Press

INTERNET RESOURCES

http://eia.doe.gov/cneaf/solar.renewables/page/trends/table1.html

http://paoc.mit.edu

http://query.nytimes.com/gst/fullpage.html?res=9502EFD6103CF935A25755C0A9649C8B63

http://unfccc.int/kyoto_protocol/mechanisms/items/1673.php

http://unfccc.int/resource/process/guideprocess-p.pdf

http://weather.nmsu.edu

http://web.mit.edu

http://web.mit.edu/globalchange

http://www.epa.gov/climatechange/emissions/usinventoryreport.html

www.agu.org

www.alaskacoast.state.ak.us/ACMPGrants/EGS_05/pdfs/

www.bbc.co.uk

www.climatecrisis.net

www.climatescience.gov

www.cnn.com

www.epa.gov

www.epa.gov/air/oaq_caa.html/

www.fauna-flora.org

www.gcrio.org

www.irinnews.org/webspecials/DR/Definitions.asp

www.isa-research.co.uk/docs/ISA-UK_Report_07-01_carbon_footprint.pdf

www.life-cycle.org

www.marshal.org

www.naesco.org

www.nationalacademies.org

www.nationalgeographic.com/

www.ncar.ucar.edu

www.newurbanism.org/

www.nsf.gov

www.ucsusa.org

www.worldbank.org/

Glossary

A

Acid Deposition

Acidic aerosols in the atmosphere are removed from the atmosphere by wet deposition (rain, snow, fog) or dry deposition (particles sticking to vegetation). Acidic aerosols are present in the atmosphere primarily due to discharges of gaseous sulfur oxides (sulfur dioxide) and nitrogen oxides.

Aerosol

A collection of airborne solid or liquid particles, with a typical size between 0.01 and 10 micrometers (µm) and residing in the atmosphere for at least several hours. Aerosols may be of either natural or anthropogenic origin. Aerosols may influence climate in two ways: directly through scattering and absorbing radiation, and indirectly through acting as condensation nuclei for cloud formation or modifying the optical properties and lifetime of clouds.

Afforestation

The planting of new forests on land which historically had been covered by forest.

Albedo

The fraction of solar radiation reflected by a surface or object, often expressed as a percentage. Most snow-covered surfaces have a high albedo; the albedo of soils ranges from high to low; vegetation covered surfaces and oceans have a low albedo. The Earth's albedo varies mainly through varying cloudiness, snow, ice, leaf area, and land cover changes.

Alleroed

A village in Denmark whose name is used for a warm period at the end of the last glacial.

Alliance of Small Island States (AOSIS)

The group of Pacific and Caribbean nations who call for relatively fast action by developed nations to reduce greenhouse gas emissions. The AOSIS countries fear the effects of rising sea levels and increased storm activity predicted to accompany global warming. Its plan is to hold Annex I Parties to a 20 percent reduction in carbon dioxide emissions by 2005.

Allometric Equation

An equation that uses known growth measurements to estimate related unknown growth measurements.

Alternative Energy

Energy derived from nontraditional sources (e.g., compressed natural gas, solar, hydroelectric, wind, and others).

Annex I Parties

Industrialized countries that, as parties to the Framework Convention on Climate Change, have pledged to reduce their greenhouse gas emissions by the year 2000 to 1990 levels. Annex I Parties consist of countries belonging to the Organisation for Economic Cooperation and Development (OECD) and countries designated as Economies-in-Transition.

Anthropogenic

Made by people or resulting from human activities. Usually used in the context of emissions that are produced as a result of human activities.

Atmosphere

The gaseous envelope surrounding the Earth. The dry atmosphere consists almost entirely of nitrogen (78.1 percent volume mixing ratio) and oxygen (20.9 percent volume mixing ratio), together with a number of trace gases, such as argon (0.93 percent volume mixing ratio), helium, radiatively active greenhouse gases such as carbon dioxide (0.035 percent volume mixing ratio), and ozone. In addition the atmosphere contains water vapor, whose amount is highly variable but typically 1 percent volume mixing ratio. The atmosphere also contains clouds and aerosols. The atmosphere can be divided into a number of layers according to its mixing or chemical characteristics, generally determined by its thermal properties (temperature). The layer nearest the Earth is the troposphere, which reaches up to an altitude of about about 5 mi. (8 km) in the polar regions and up to nearly 11 mi. (17 km) above the equator. The stratosphere, which reaches to an altitude of about 31 mi. (50 km.) lies atop the troposphere. The mesosphere which extends up to 50–56 mi. (80–90 km.) is atop the stratosphere, and finally, the thermosphere, or ionosphere, gradually diminishes and forms a fuzzy border with outer space.

Atmospheric Lifetime

The lifetime of a greenhouse gas refers to the approximate amount of time it would take for the anthropogenic increment to an atmospheric pollutant concentration to return to its natural level (assuming emissions cease) as a result of either being converted to another chemical compound or being taken out of the atmosphere via a sink. This time depends on the pollutant's sources and sinks as well as its reactivity. The lifetime of a pollutant is often considered in conjunction with the mixing of pollutants in the atmosphere; a long lifetime will allow the pollutant to mix throughout the atmosphere. Average lifetimes can vary from about a week (sulfate aerosols) to more than a century (chlorofluorocarbons [CFCs], carbon dioxide).

B

Baseline Emissions

The emissions that would occur without policy intervention (in a business-as-usual scenario). Baseline estimates are needed to determine the effectiveness of emissions reduction programs.

Berlin Mandate

A ruling negotiated at the first Conference of the Parties (COP 1), which took place in March, 1995, concluding that the present commitments under the Framework Convention on Climate Change are not adequate. Under the Framework Convention, developed countries pledged to take measures aimed at returning their greenhouse gas emissions to 1990 levels by the year 2000.

Biogeochemical Cycle

The chemical interactions that take place among of key chemical constituents essential to life, such as carbon, nitrogen, oxygen, and phosphorus.

Biomass

Organic nonfossil materials that are biological in origin, including organic material (both living and dead) from above and below ground, for example, trees, plants, crops, roots, and animals and animal waste.

Biomass Energy

Energy produced by combusting renewable biomass materials such as wood. The carbon dioxide emitted from burning biomass will not increase total atmospheric carbon dioxide if this consumption is done on a sustainable basis (i.e., if in a given period of time, regrowth of biomass takes up as much carbon dioxide as is released from biomass combustion). Biomass energy is often suggested as a replacement for fossil fuel combustion, which has large greenhouse gas emissions.

Biome
A naturally occurring community of flora and fauna (or the region occupied by such a community) adapted to the particular conditions in which they occur (e.g., tundra).

Biosphere
The region of land, oceans, and atmosphere inhabited by living organisms.

Black Carbon
Operationally defined species based on measurement of light absorption and chemical reactivity and/or thermal stability; consists of soot, charcoal, and/or possible light-absorbing refractory organic matter.

Borehole
Any exploratory hole drilled into the Earth or ice to gather geophysical data. Climate researchers often take ice core samples, a type of borehole, to predict atmospheric composition in earlier years.

Bubble
A system that lets several countries meet a reduction target together while having different individual targets.

C

Capital Stocks
The accumulation of machines and structures that are available to an economy at any point in time to prune goods or render services. These activities usually require a quantity of energy that is determined largely by the rate at which that machine or structure is used.

Carbon Cycle
The global scale exchange of carbon among its reservoirs, namely the atmosphere, oceans, vegetation, soils, and geologic deposits and minerals. This involves components in food chains, in the atmosphere as carbon dioxide, in the hydrosphere, and in the geosphere.

Carbon Dioxide (CO_2)
The greenhouse gas whose concentration is being most affected directly by human activities. CO_2 also serves as the reference to compare all other greenhouse gases.

The major source of CO_2 emissions is fossil fuel combustion. CO_2 emissions are also a product of forest clearing, biomass burning, and non-energy production processes such as cement production. Atmospheric concentrations of CO_2 have been increasing at a rate of about 0.5 percent per year and are now about 30 percent above preindustrial levels.

Carbon Equivalent (CE)
A metric measure used to compare the emissions of the different greenhouse gases based upon their global warming potential (GWP). Greenhouse gas emissions in the United States are most commonly expressed as "million metric tons of carbon equivalents" (MMTCE). Global warming potentials are used to convert greenhouse gases to carbon dioxide equivalents.

Carbon Sequestration
The uptake and storage of carbon. Trees and plants, for example, absorb carbon dioxide, release the oxygen and store the carbon. Fossil fuels were at one time biomass and continue to store the carbon until burned.

Carbon Sinks
Carbon reservoirs and conditions that take in and store more carbon (carbon sequestration) than they release. Carbon sinks can serve to partially offset greenhouse gas emissions. Forests and oceans are common carbon sinks.

Chlorofluorocarbons and Related Compounds
This family of anthropogenic compounds includes chlorofluorocarbons (CFCs), bromofluorocarbons (halons), methyl chloroform, carbon tetrachloride, methyl bromide, and hydrochlorofluorocarbons (HCFCs). These compounds have been shown to deplete stratospheric ozone, and therefore are typically referred to as ozone-depleting substances. The most ozone-depleting of these compounds are being phased out under the Montreal Protocol.

Clean Development Mechanisms (CDM)
Article 12 of the Kyoto Protocol provides for the CDM whereby developed countries are able to invest in emissions-reducing projects in developing countries to obtain credit to assist in meeting their assigned

amounts. The details of the CDM have yet to be negotiated at the international level.

Climate

The average weather for a particular region and time period. Climate is not the same as weather, but rather, it is the average pattern of weather for a particular region. Climatic elements include precipitation, temperature, humidity, sunshine, wind velocity, phenomena such as fog, frost, and hail storms, and other measures of the weather.

Climate Change

The term *climate change* refers to all forms of climatic inconsistency. Climate change has been used synonymously with the term *global warming*.

Climate Change Action Plan

Unveiled in October 1993 by President Clinton, the CCAP is the U.S. plan for meeting its pledge to reduce greenhouse gas emissions under the terms of the Framework Convention on Climate Change (FCCC). The goal of the plan was to reduce U.S. emissions of greenhouse gases to 1990 levels by 2000.

Climate Feedback

An atmospheric, oceanic, terrestrial, or other process that is activated by the direct climate change induced by changes in radiative forcing. Climate feedbacks may increase (positive feedback) or diminish (negative feedback) the magnitude of the climate change.

Climate Lag

The delay that occurs in climate change as a result of some factor that changes only very slowly.

Climate Model

A quantitative way of representing the interactions of the atmosphere, oceans, land surface, and ice.

Climate Modeling

The simulation of the climate using computer-based models.

Climate Sensitivity

The equilibrium response of the climate to a change in radiative forcing; for example, a doubling of the carbon dioxide concentration.

Climate System (or Earth System)

The five physical components (atmosphere, hydrosphere, cryosphere, lithosphere, and biosphere) that are responsible for the climate and its variations.

Cloud Condensation Nuclei

Airborne particles that serve as an initial site for the condensation of liquid water and which can lead to the formation of cloud droplets.

CO_2 Fertilization

The enhancement of plant growth as a result of elevated atmospheric CO_2 concentrations.

Coalbed Methane

Coalbed methane is methane contained in coal seams, and is often referred to as virgin coalbed methane, or coal seam gas. For more information, visit the Coalbed Methane Outreach program site.

Coal Mine Methane

Coal mine methane is the subset of CBM that is released from the coal seams during the process of coal mining. For more information, visit the Coalbed Methane Outreach program site.

Cogeneration

The process by which two different and useful forms of energy are produced at the same time. For example, while boiling water to generate electricity, the leftover steam can be sold for industrial processes or space heating.

Compost

Decayed organic matter that can be used as a fertilizer or soil additive.

Conference of the Parties (COP)

The supreme body of the United Nations Framework Convention on Climate Change (UNFCCC). It comprises more than 180 nations that have ratified the Convention. Its first session was held in Berlin, Germany, in 1995 and it is expected to continue meeting on a yearly basis. The COP's role is to promote and review the implementation of the Convention.

Cryosphere

One of the interrelated components of the Earth's system, the cryosphere is frozen water in the form of snow, permanently frozen ground (permafrost), floating ice, and glaciers. Fluctuations in the volume of the cryosphere cause changes in ocean sea level, which directly impact the biosphere.

D

Deforestation

Those practices or processes that result in the change of forested lands to nonforest uses. This is often cited as one of the major causes of the enhanced greenhouse effect for two reasons: (1) the burning or decomposition of the wood releases carbon dioxide; and (2) trees that once removed carbon dioxide from the atmosphere in the process of photosynthesis are no longer present and contributing to carbon storage.

Desertification

The progressive destruction or degradation of existing vegetative cover to form desert. This can occur due to overgrazing, deforestation, drought, and the burning of extensive areas.

Diurnal Temperature Range

The difference between maximum and minimum temperature over a period of 24 hours.

E

Economic Potential

The portion of the technical potential for GHG emissions reductions or energy-efficiency improvements that could be achieved cost-effectively in the absence of market barriers. The achievement of the economic potential requires additional policies and measures to break down market barriers.

Eddy Mixing

Mixing due to small scale turbulence processes (eddies). Such processes cannot be explicitly resolved by even the finest-resolution atmosphere-ocean general ciculation models currently in uses and so their effects must be related to the larger-scale conditions.

El Niño

A climatic phenomenon occurring irregularly, but generally every three to five years. El Niños often first become evident during the Christmas season (El Niño means Christ-child) in the surface oceans of the eastern tropical Pacific Ocean. The phenomenon involves seasonal changes in the direction of the tropical winds over the Pacific and abnormally warm surface ocean temperatures. The changes in the tropics are most intense in the Pacific region; these changes can disrupt weather patterns throughout the tropics and can extend to higher latitudes.

Emission Permit

A nontransferable or tradeable allocation of entitlements by a government to an individual firm to emit a specific amount of a substance.

Emission Quota

The portion or share of total allowable emissions assigned to a country or group of countries within a framework of maximum total emissions and mandatory allocations of resources or assessments.

Emissions

The release of a substance (usually a gas when referring to climate change) into the atmosphere.

Emission Standard

A level of emission that under law may not be exceeded.

Energy Intensity

Ration of energy consumption and economic or physical output. At the national level, energy intensity is the ratio of total domestic primary energy consumption or final energy consumption to gross domestic product or physical output.

Enhanced Greenhouse Effect

The natural greenhouse effect has been enhanced by anthropogenic emissions of greenhouse gases. Increased concentrations of carbon dioxide, methane, and nitrous oxide, CFCs, HFCs, PFCs, SF_6, NF_3, and other photochemically important gases caused by human activities such as fossil fuel consumption and adding waste to landfills trap more infrared radiation, thereby exerting a warming influence.

Equilibrium Response

The steady state response of the climate system (or a climate model) to an imposed radiative forcing.

Evapotranspiration

The sum of evaporation and plant transpiration. Potential evapotranspiration is the amount of water that could be evaporated or transpired at a given temperature and humidity, if there was water available.

F

Fluorocarbons

Carbon-fluorine compounds that often contain other elements such as hydrogen, chlorine, or bromine. Common fluorocarbons include chlorofluorocarbons (CFCs), hydrochlorofluorocarbons (HCFCs), hydrofluorocarbons (HFCs), and perfluorocarbons (PFCs).

Forcing Mechanism

A process that alters the energy balance of the climate system, i.e., changes the relative balance between incoming solar radiation and outgoing infrared radiation from Earth. Such mechanisms include changes in solar irradiance, volcanic eruptions, and enhancement of the natural greenhouse effect by emissions of greenhouse gases.

G

Geosphere

The soils, sediments, and rock layers of the Earth's crust, both continental and beneath the ocean floors.

Glacier

A multiyear surplus accumulation of snowfall in excess of snowmelt on land and resulting in a mass of ice at least 0.1 km2 in area that shows some evidence of movement in response to gravity. A glacier may terminate on land or in water. Glaciers are found on every continent except Australia.

Global Warming

Global warming is an average increase in the temperature of the atmosphere near the Earth's surface and in the troposphere, which can contribute to changes in global climate patterns. Global warming can occur from many causes, both natural and human induced.

Global Warming Potential (GWP)

Defined as the cumulative radiative forcing effects of a gas over a specified time horizon resulting from the emission of a unit mass of gas relative to a reference gas. The GWP-weighted emissions of direct greenhouse gases in the U.S. inventory are presented in terms of equivalent emissions of carbon dioxide, using units of teragrams of carbon dioxide equivalents.

Greenhouse Effect

Trapping and buildup of heat in the atmosphere (troposphere) near the Earth's surface. Some of the heat flowing back toward space from the Earth's surface is absorbed by water vapor, carbon dioxide, ozone, and several other gases in the atmosphere and then reradiated back toward the Earth's surface. If the atmospheric concentrations of these greenhouse gases rise, the average temperature of the lower atmosphere will gradually increase.

Greenhouse Gas (GHG)

Any gas that absorbs infrared radiation in the atmosphere. Greenhouse gases include, but are not limited to, water vapor, carbon dioxide (CO_2), methane (CH_4), nitrous oxide (N_2O), chlorofluorocarbons (CFCs), hydrochlorofluorocarbons (HCFCs), ozone (O_3), hydrofluorocarbons (HFCs), perfluorocarbons (PFCs), and sulfur hexafluoride (SF_6).

H

Halocarbons

Compounds containing either chlorine, bromine, or fluorine and carbon. Such compounds can act as powerful greenhouse gases in the atmosphere. The chlorine and bromine containing halocarbons are also involved in the depletion of the ozone layer.

Hydrocarbons

Substances containing only hydrogen and carbon. Fossil fuels are made up of hydrocarbons.

Hydrochlorofluorocarbons (HCFCs)

Compounds containing hydrogen, fluorine, chlorine, and carbon atoms. Although ozone-depleting substances, they are less potent at destroying stratospheric ozone than chlorofluorocarbons (CFCs). They have been introduced as temporary replacements for CFCs and are also greenhouse gases.

Hydrologic Cycle

The process of evaporation, vertical and horizontal transport of vapor, condensation, precipitation, and the flow of water from continents to oceans. It is a major factor in determining climate through its influence on surface vegetation, the clouds, snow and ice, and soil moisture. The hydrologic cycle is responsible for 25 to 30 percent of the midlatitudes' heat transport from the equatorial to polar regions.

Hydrosphere

The component of the climate system comprising liquid surface and subterranean water, such as oceans, seas, rivers, freshwater lakes, underground water.

I

Ice Core

A cylindrical section of ice removed from a glacier or an ice sheet in order to study climate patterns of the past. By performing chemical analyses on the air trapped in the ice, scientists can estimate the percentage of carbon dioxide and other trace gases in the atmosphere at a given time.

Infrared Radiation

Radiation emitted by the Earth's surface, the atmosphere and the clouds. It is also known as terrestrial or longwave radiation. Infrared radiation has a distinctive range of wavelengths longer than the wavelength of the red color in the visible part of the spectrum.

Intergovernmental Panel on Climate Change (IPCC)

The IPCC was established jointly by the United Nations Environment Program and the World Meteorological Organization in 1988. The purpose of the IPCC is to assess information in the scientific and technical literature related to all significant components of the issue of climate change. With its capacity for reporting on climate change, its consequences, and the viability of adaptation and mitigation measures, the IPCC is also looked to as the official advisory body to the world's governments on the state of the science of the climate change issue.

L

Landfill

Land waste disposal site in which waste is generally spread in thin layers, compacted, and covered with a fresh layer of soil each day.

Longwave Radiation

The radiation emitted in the spectral wavelength greater than 4 micrometers corresponding to the radiation emitted from the Earth and atmosphere.

M

Methane (CH_4)

A hydrocarbon that is a greenhouse gas with a global warming potential most recently estimated at 23 times that of carbon dioxide (CO_2). Methane is produced through anaerobic (without oxygen) decomposition of waste in landfills, animal digestion, decomposition of animal wastes, production and distribution of natural gas and petroleum, coal production, and incomplete fossil fuel combustion.

Metric Ton

Common international measurement for the quantity of greenhouse gas emissions. A metric ton is equal to 2,205 lbs. or 1.1 short tons.

Mount Pinatubo

An active volcano located in the Philippine Islands that erupted in 1991. The eruption of Mount Pinatubo ejected enough particulate and sulfate aerosol matter into the upper atmosphere to block some of the incoming solar radiation from reaching Earth's atmosphere.

N

Natural Gas

Underground deposits of gases consisting of 50 to 90 percent methane (CH_4) and small amounts of heavier gaseous hydrocarbon compounds such as propane (C_3H_8) and butane (C_4H_{10}).

Nitrogen Oxides (NOx)

Gases consisting of one molecule of nitrogen and varying numbers of oxygen molecules. Nitrogen oxides are produced in the emissions of vehicle exhausts and from power stations. In the atmosphere, nitrogen oxides can contribute to formation of smog, can impair visibility, and have health consequences.

Nitrous Oxide (N_2O)

A powerful greenhouse gas with a global warming potential of 296 times that of carbon dioxide (CO_2). Major sources of nitrous oxide include soil cultivation practices, especially the use of commercial and organic fertilizers, fossil fuel combustion, nitric acid production, and biomass burning.

O

Oxidize
To chemically transform a substance by combining it with oxygen.

Ozone (O_3)
Ozone, the triatomic form of oxygen (O_3), is a gaseous atmospheric constituent. In the troposphere, it is created both naturally and by photochemical reactions involving gases resulting from human activities (photochemical smog). In high concentrations, tropospheric ozone can be harmful to a wide range of living organisms. Tropospheric ozone acts as a greenhouse gas. In the stratosphere, ozone is created by the interaction between solar ultraviolet radiation and molecular oxygen (O_2). Stratospheric ozone plays a decisive role in the stratospheric radiative balance. Depletion of stratospheric ozone, due to chemical reactions that may be enhanced by climate change, results in an increased ground-level flux of ultraviolet (UV-) B radiation.

Ozone-Depleting Substance (ODS)
A family of man-made compounds that includes chlorofluorocarbons (CFCs), bromofluorocarbons (halons), methyl chloroform, carbon tetrachloride, methyl bromide, and hydrochlorofluorocarbons (HCFCs). These compounds have been shown to deplete stratospheric ozone, and therefore are typically referred to as ODSs.

Ozone Layer
The layer of ozone that begins approximately 9 mi. (15 km.) above Earth and thins to an almost negligible amount at about 31 mi. (50 km.), shields the Earth from harmful ultraviolet radiation from the Sun.

Ozone Precursors
Chemical compounds, such as carbon monoxide, methane, nonmethane hydrocarbons, and nitrogen oxides, which in the presence of solar radiation react with other chemical compounds to form ozone, mainly in the troposphere.

P

Particulate Matter (PM)
Very small pieces of solid or liquid matter such as particles of soot, dust, fumes, mists, or aerosols.

Parts per Billion (ppb)
Number of parts of a chemical found in one billion parts of a particular gas, liquid, or solid mixture.

Parts per Million (ppm)
Number of parts of a chemical found in one million parts of a particular gas, liquid, or solid.

Perfluorocarbons (PFCs)
A group of human-made chemicals composed of carbon and fluorine only. These chemicals were introduced as alternatives, along with hydrofluorocarbons, to the ozone-depleting substances.

Photosynthesis
The process by which plants take CO_2 from the air (or bicarbonate in water) to build carbohydrates, releasing O_2 in the process. There are several pathways of photosynthesis with different responses to atmospheric CO_2 concentrations.

Precession
The comparatively slow torquing of the orbital planes of all satellites with respect to the Earth's axis, due to the bulge of the Earth at the equator which distorts the Earth's gravitational field.

R

Radiation
Energy transfer in the form of electromagnetic waves or particles that release energy when absorbed by an object.

Radiative Forcing
Radiative forcing is the change in the net vertical irradiance (expressed in Watts per square meter: Wm-2) at the tropopause due to an internal change or a change in the external forcing of the climate system, such as, for example, a change in the concentration of carbon dioxide or the output of the Sun.

Recycling
Collecting and reprocessing a resource so it can

be used again. An example is collecting aluminum cans, melting them down, and using the aluminum to make new cans or other aluminum products.

Reforestation

Planting of forests on lands that have previously contained forests but that have been converted to some other use.

Residence Time

The average time spent in a reservoir by an individual atom or molecule. With respect to greenhouse gases, residence time usually refers to how long a particular molecule remains in the atmosphere.

Respiration

The biological process whereby living organisms convert organic matter to CO_2, releasing energy and consuming O_2.

S

Short Ton

Common measurement for a ton in the United States. A short ton is equal to 2,000 lbs. or 0.907 metric tons.

Sink

Any process, activity, or mechanism which removes a greenhouse gas, an aerosol, or a precursor of a greenhouse gas or aerosol from the atmosphere.

Soil Carbon

A major component of the terrestrial biosphere pool in the carbon cycle. The amount of carbon in the soil is a function of the historical vegetative cover and productivity, which in turn is dependent in part upon climatic variables.

Solar Radiation

Radiation emitted by the Sun. It is also referred to as shortwave radiation. Solar radiation has a distinctive range of wavelengths (spectrum) determined by the temperature of the Sun.

Stratosphere

Region of the atmosphere between the troposphere and mesosphere, having a lower boundary of approximately 5 mi. (8 km.) at the poles to 9 mi. (15 km.) at the equator and an upper boundary of approximately 31 mi. (50 km.).

Depending upon latitude and season, the temperature in the lower stratosphere can increase, be isothermal, or even decrease with altitude, but the temperature in the upper stratosphere generally increases with height due to absorption of solar radiation by ozone.

Streamflow

The volume of water that moves over a designated point over a fixed period of time. It is often expressed as cubic feet per second (ft3/sec).

Sulfate Aerosols

Particulate matter that consists of compounds of sulfur formed by the interaction of sulfur dioxide and sulfur trioxide with other compounds in the atmosphere. Sulfate aerosols are injected into the atmosphere from the combustion of fossil fuels and the eruption of volcanoes.

Sulfur Hexafluoride (SF_6)

A colorless gas soluble in alcohol and ether, slightly soluble in water. A very powerful greenhouse gas used primarily in electrical transmission and distribution systems and as a dielectric in electronics.

T

Thermohaline Circulation

Large-scale density-driven circulation in the ocean, caused by differences in temperature and salinity. In the North Atlantic the thermohaline circulation consists of warm surface water flowing northward and cold deep water flowing southward, resulting in a net poleward transport of heat.

Trace Gas

Any one of the less common gases found in the Earth's atmosphere. Nitrogen, oxygen, and argon make up more than 99 percent of the Earth's atmosphere. Other gases, such as carbon dioxide, water vapor, methane, oxides of nitrogen, ozone, and ammonia, are considered trace gases.

Troposphere

The lowest part of the atmosphere from the surface to about 6 mi./10 km. in altitude in mid-latitudes (ranging from 5.5 mi. [9 km.] in high latitudes to 10 mi. [16 km.] in the tropics on average) where clouds and "weather" phenomena occur.

U

Ultraviolet Radiation (UV)

The energy range just beyond the violet end of the visible spectrum. Although ultraviolet radiation constitutes only about 5 percent of the total energy emitted from the sun, it is the major energy source for the stratosphere and mesosphere, playing a dominant role in both energy balance and chemical composition.

United Nations Framework Convention on Climate Change (UNFCCC)

The Convention on Climate Change sets an overall framework for intergovernmental efforts to tackle the challenge posed by climate change. It recognizes that the climate system is a shared resource whose stability can be affected by industrial and other emissions of carbon dioxide and other greenhouse gases.

W

Wastewater

Water that has been used and contains dissolved or suspended waste materials.

Water Vapor

The most abundant greenhouse gas, it is the water present in the atmosphere in gaseous form. Water vapor is a part of the natural greenhouse effect.

Weather

Atmospheric condition at any given time or place. It is measured in terms of such things as wind, temperature, humidity, atmospheric pressure, cloudiness, and precipitation.

FERNANDO HERRERA
UNIVERSITY OF CALIFORNIA, SAN DIEGO

Appendix

GRAPHIC PLOTS AND TEXT PREPARED BY
ROBERT A. ROHDE
UNIVERSITY OF CALIFORNIA, BERKELEY

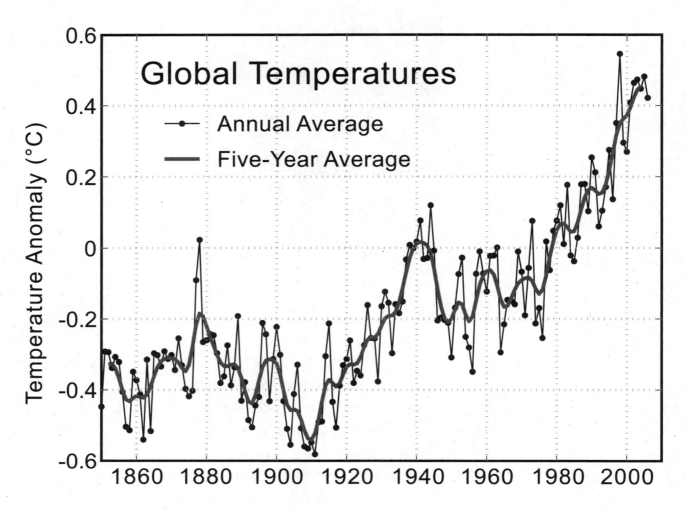

This image shows the instrumental record of global average temperatures as compiled by the Climatic Research Unit of the University of East Anglia and the Hadley Centre of the UK Meteorological Office. Data set Had-CRUT3 was used, which follows the methodology outlined by Brohan et al. (2006). Following the common practice of the IPCC, the zero on this figure is the mean temperature from 1961 to1990.

The uncertainty in the analysis techniques leading to these measurements is discussed in Foland et al. (2001) and Brohan et al. (2006). They estimate that global averages since ~1950 are within ~0.05 degrees C of their reported value with 95 percent confidence. In the recent period, these uncertainties are driven primarily by considering the potential impact of regions where no temperature record is available. For averages prior to ~1890, the uncertainty reaches ~0.15 degrees C driven primarily by limited sampling and the effects of changes in sea surface measurement techniques. Uncertainties between 1880 and 1890 are intermediate between these values.

Incorporating these uncertainties, Foland et al. (2001) estimated the global temperature change from 1901 to 2000 as 0.57 ± 0.17 degrees C, which contributed to the 0.6 ± 0.2 degrees C estimate reported by the Intergovernmental Panel on Climate Change (IPCC 2001a, [1]). Both estimates are 95 percent confidence intervals.

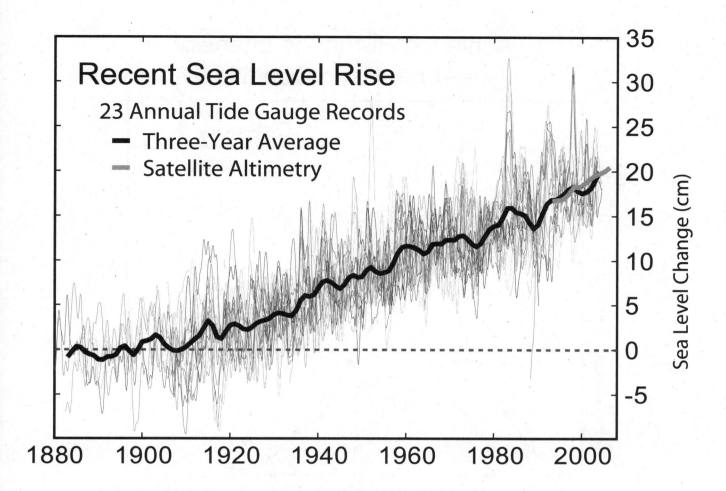

This figure shows the change in annually averaged sea level at 23 geologically stable tide gauge sites with long-term records as selected by Douglas (1997). The thick dark line is a three-year moving average of the instrumental records. This data indicates a sea level rise of ~18.5 cm. from 1900–2000. Because of the limited geographic coverage of these records, it is not obvious whether the apparent decadal fluctuations represent true variations in global sea level or merely variations across regions that are not resolved.

For comparison, the recent annually averaged satellite altimetry data from TOPEX/Poseidon are shown in the thick gray line. These data indicate a somewhat higher rate of increase than tide gauge data, however the source of this discrepancy is not obvious. It may represent systematic error in the satellite record and/or incomplete geographic sampling in the tide gauge record. The month-to-month scatter on the satellite measurements is roughly the thickness of the plotted gray curve.

Much of recent sea level rise has been attributed to global warming.

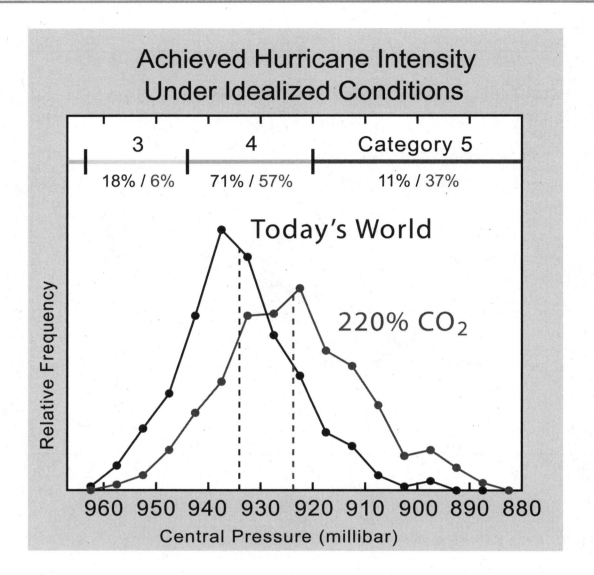

This figure, which reproduces one of the key conclusions of Knutson and Tuleya (2004), shows a prediction for how hurricanes and other tropical cyclones may intensify as a result of global warming. Specifically, Knutson and Tuleya performed an experiment using climate models to estimate the strength achieved by cyclones allowed to intensify over either a modern summer ocean or over an ocean warmed by carbon dioxide concentrations 220 percent higher than present day. A number of different climate models were considered as well as conditions over all the major cyclone-forming ocean basins. Depending on site and model, the ocean warming involved ranged from 0.8 to 2.4 degrees C. Results, which were found to be robust across different models, showed that storms intensified by about one-half category (on the Saffir-Simpson Hurricane Scale) as a result of the warmer oceans. This is accomplished with a ~6 percent increase in wind speed or equivalently a ~20 percent increase in energy (for a storm of fixed size). Most significantly these result suggest that global warming may lead to a gradual increase in the probability of highly destructive category 5 hurricanes. This work does not provide any information about future frequency of tropical storms. Also, since it considers only the development of storms under nearly ideal conditions for promoting their formation, this work is primarily a prediction for how the maximum achievable storm intensity will change. Hence, this does not directly bear on the growth or development of storms under otherwise weak or marginal conditions for storm development (such as high upper-level wind shear). However, it is plausible that warmer oceans will somewhat extend the regions and seasons under which hurricanes may develop.

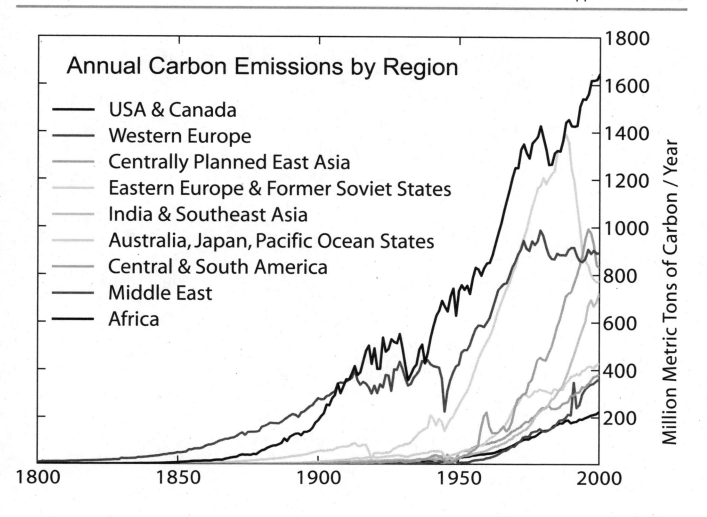

Annual Carbon Emissions by Region

— USA & Canada
— Western Europe
— Centrally Planned East Asia
— Eastern Europe & Former Soviet States
— India & Southeast Asia
— Australia, Japan, Pacific Ocean States
— Central & South America
— Middle East
— Africa

This figure shows the annual fossil fuel carbon dioxide emissions, in million metric tons of carbon, for a variety of non-overlapping regions covering the Earth. Data source: Carbon Dioxide Information Analysis Center. Regions are sorted from largest emitter (as of 2000) to the smallest:

United States and Canada
Western Europe (plus Germany)
Communist East Asia (China, North Korea, Mongolia, etc.)
Eastern Europe, Russia, and Former Soviet States
India and Southeast Asia (plus South Korea)
Australia, Japan and other Pacific Island States
Central and South America (includes Mexico and the Caribbean)
Middle East
Africa

Ice Age Temperature Changes

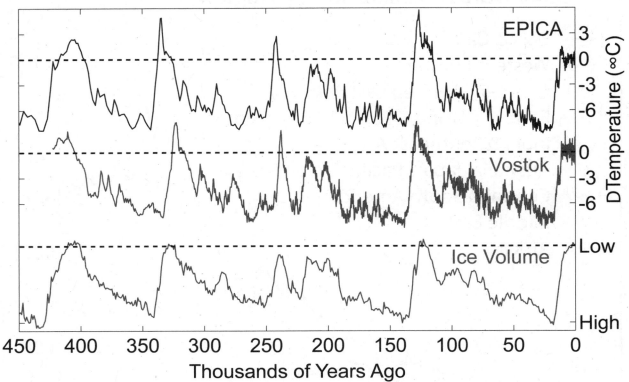

This figure shows the Antarctic temperature changes during the last several glacial/interglacial cycles of the present ice age and a comparison to changes in global ice volume. The present day is on the right.

The first two curves show local changes in temperature at two sites in Antarctica as derived from deuterium isotopic measurements (δD) on ice cores (EPICA Community Members 2004, Petit et al. 1999). The final plot shows a reconstruction of global ice volume based on $\delta 18O$ measurements on benthic foraminifera from a composite of globally distributed sediment cores and is scaled to match the scale of fluctuations in Antarctic temperature (Lisiecki and Raymo 2005). Note that changes in global ice volume and changes in Antarctic temperature are highly correlated, so one is a good estimate of the other, but differences in the sediment record do not necessarily reflect differences in paleotemperature. Horizontal lines indicate modern temperatures and ice volume. Differences in the alignment of various features reflect dating uncertainty and do not indicate different timing at different sites.

The Antarctic temperature records indicate that the present interglacial is relatively cool compared to previous interglacials, at least at these sites. It is believed that the interglacials themselves are triggered by changes in Earth's orbit known as Milankovitch cycles and that the variations in individual interglacials can be partially explained by differences within this process. For example, Overpeck et al. (2006) argues that the previous interglacial was warmer because of increased solar radiation at high latitudes. The Liesecki and Raymo (2005) sediment reconstruction does not indicate significant differences between modern ice volume and previous interglacials, though some other studies do report slightly lower ice volumes/higher sea levels during the 120 ka and 400 ka interglacials (Karner et al. 2001, Hearty and Kaufman 2000). It should be noted that temperature changes at the typical equatorial site are believed to have been significantly less than the changes observed at high latitude.

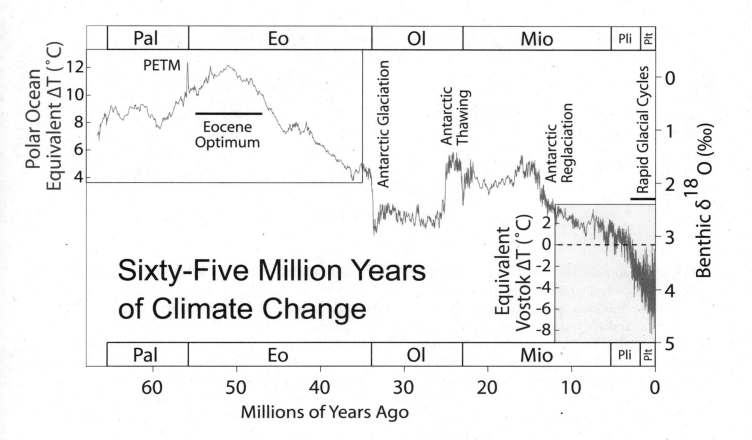

This figure shows climate change over the last 65 million years. The data is based on a compilation of oxygen isotope measurements ($\delta 18O$) on benthic foraminifera by Zachos et al. (2001), which reflect a combination of local temperature changes in their environment and changes in the isotopic composition of seawater associated with the growth and retreat of continental ice sheets.

Because it is related to both factors, it is not possible to uniquely tie these measurements to temperature without additional constraints. For the most recent data, an approximate relationship to temperature can be made by observing that the oxygen isotope measurements of Lisiecki and Raymo (2005) are tightly correlated to temperature changes at Vostok, Antarctica as established by Petit et al. (1999). Present day is indicated as 0. For the oldest part of the record, when temperatures were much warmer than today, it is possible to estimate temperature changes in the polar oceans (where these measurements were made) based on the observation that no significant ice sheets existed and hence all fluctuation in ($\delta 18O$) must result from local temperature changes (as reported by Zachos et al.).

The intermediate portion of the record is dominated by large fluctuations in the mass of the Antarctic ice sheet, which first nucleates approximately 34 million years ago, then partially dissipates around 25 million years ago, before re-expanding toward its present state 13 million years ago. These fluctuations make it impossible to constrain temperature changes without additional controls. Significant growth of ice sheets did not begin in Greenland and North America until approximately 3 million years ago, following the formation of the Isthmus of Panama by continental drift. This ushered in an era of rapidly cycling glacials and interglacials (upper right). Also appearing on this graph are the Eocene Climatic Optimum, an extended period of very warm temperatures, and the Paleocene-Eocene Thermal Maximum (labeled PETM). Due to the coarse sampling and averaging involved in this record, it is likely that the full magnitude of the PETM is underestimated by a factor of 2 to 4 times its apparent height.

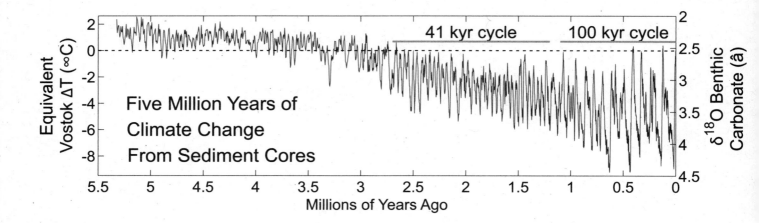

This figure shows the climate record of Lisiecki and Raymo (2005) constructed by combining measurements from 57 globally distributed deep-sea sediment cores. The measured quantity is oxygen isotope fractionation in benthic foraminifera, which serves as a proxy for the total global mass of glacial ice sheets.

Lisiecki and Raymo constructed this record by first applying a computer-aided process of adjusting individual "wiggles" in each sediment core to have the same alignment (i.e.. wiggle matching). Then the resulting stacked record is orbitally tuned by adjusting the positions of peaks and valleys to fall at times consistent with an orbitally driven ice model (see Milankovitch Cycles). Both sets of these adjustments are constrained to be within known uncertainties on sedimentation rates and consistent with independently dated tie points (if any). Constructions of this kind are common, however, they presume that ice sheets are orbitally driven, and hence data such as this can not be used in establishing the existence of such a relationship.

The observed isotope variations are very similar in shape to the temperature variations recorded at Vostok, Antarctica, during the 420 kyr for which that record exists. Hence the right-hand scale of the figure was established by fitting the reported temperature variations at Vostok (Petit et al. 1999) to the observed isotope variations. As a result, this temperature scale should be regarded as approximate and its magnitude is only representative of Vostok changes. In particular, temperature changes at polar sites, such as Vostok, frequently exceed the changes observed in the tropics or in the global average. A horizontal line at 0 degrees C indicates modern temperatures (circa 1950).

Labels are added to indicate regions where 100 kyr and 41 kyr cyclicity is observed. These periodicities match periodic changes in Earth's orbital eccentricity and obliquity, respectively, and have been previously established by other studies (not relying on orbital tuning).

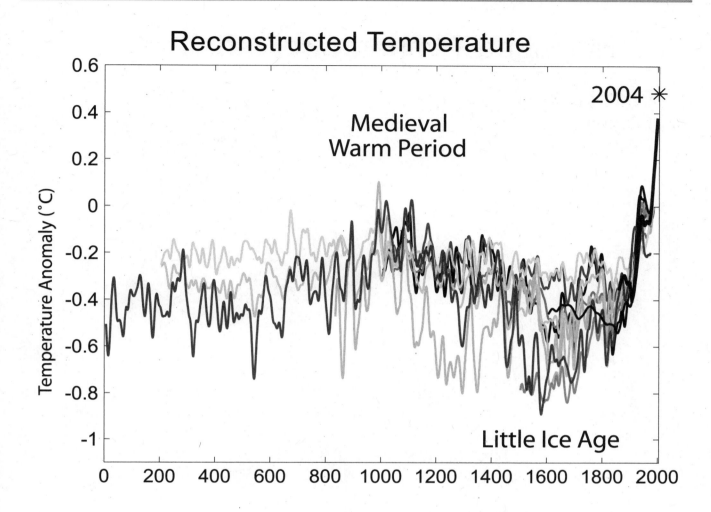

This image is a comparison of 10 different published reconstructions of mean temperature changes during the last 2,000 years. More recent reconstructions are plotted toward the front and in redder colors, older reconstructions appear towards the back and in bluer colors. An instrumental history of temperature is also shown in black. The medieval warm period and Little Ice Age are labeled at roughly the times when they are historically believed to occur, though it is still disputed whether these were truly global or only regional events. The single, unsmoothed annual value for 2004 is also shown for comparison.

It is unknown which, if any, of these reconstructions is an accurate representation of climate history; however, these curves are a fair representation of the range of results appearing in the published scientific literature. Hence, it is likely that such reconstructions, accurate or not, will play a significant role in the ongoing discussions of global climate change and global warming.

For each reconstruction, the raw data has been decadally smoothed with a $\sigma = 5$ yr Gaussian weighted moving average. Also, each reconstruction was adjusted so that its mean matched the mean of the instrumental record during the period of overlap.

Holocene Temperature Variations

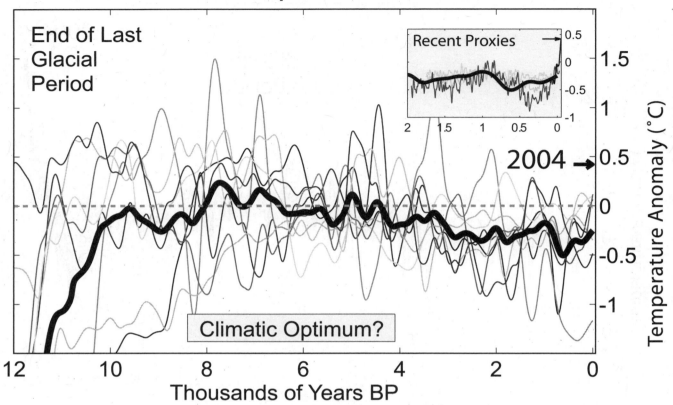

The main figure shows eight records of local temperature variability on multi-centennial scales throughout the course of the Holocene, and an average of these (thick dark line). The records are plotted with respect to the mid-20th-century average temperatures, and the global average temperature in 2004 is indicated. The inset plot compares the most recent two millennia of the average to other high resolution reconstructions of this period. At the far left of the main plot climate emerges from the last glacial period of the current ice age into the relative stability of the current interglacial. There is general scientific agreement that during the Holocene itself temperatures have been quite stable compared to the fluctuations during the preceding glacial period. The average curve above supports this belief. However, there is a slightly warmer period in the middle which might be identified with the proposed Holocene climatic optimum. The magnitude and nature of this warm event is disputed, and it may have been largely limited to summer months and/or high northern latitudes.

Because of the limitations of data sampling, each curve in the main plot was smoothed, and consequently, this figure can not resolve temperature fluctuations faster than approximately 300 years. Further, while 2004 appears warmer than any other time in the long-term average, an observation that might be a sign of global warming, it should also be noted that the 2004 measurement is from a single year. It is impossible to know whether similarly large short-term temperature fluctuations may have occurred at other times but are unresolved by the resolution available in this figure. The next 150 years will determine whether the long-term average centered on the present appears anomalous with respect to this plot. Since there is no scientific consensus on how to reconstruct global temperature variations during the Holocene, the average shown here should be understood as only a rough, quasi-global approximation to the temperature history of the Holocene. In particular, higher resolution data and better spatial coverage could significantly alter the apparent long-term behavior.

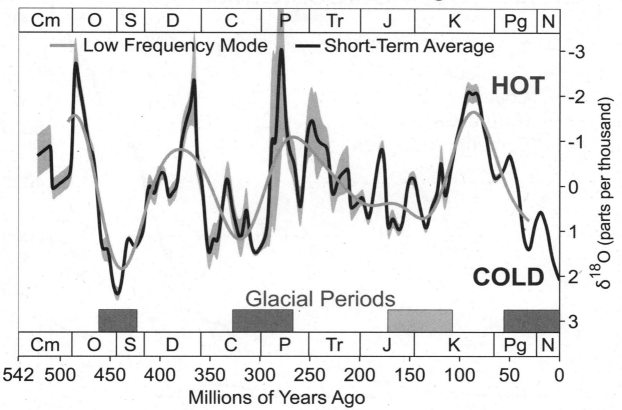

Phanerozoic Climate Change

This figure shows the long-term evolution of oxygen isotope ratios during the Phanerozoic eon as measured in fossils, reported by Veizer et al. (1999), and updated online in 2004 [1]. Such ratios reflect both the local temperature at the site of deposition and global changes associated with the extent of continental glaciation. As such, relative changes in oxygen isotope ratios can be interpreted as rough changes in climate. Quantitative conversion between this data and direct temperature changes is a complicated process subject to many systematic uncertainties, however, it is estimated that each 1 part per thousand change in $\delta 18O$ represents roughly a 1.5–2 degrees C change in tropical sea surface temperatures (Veizer et al. 2000). Also shown on this figure are blue bars showing periods when geological criteria (Frakes et al. 1992) indicate cold temperatures and glaciation as reported by Veizer et al. (2000). All data presented here have been adjusted to the 2004 ICS geologic timescale. The "short-term average" was constructed by applying a $\sigma = 3$ Myr Gaussian weighted moving average to the original 16,692 reported measurements. The gray bar is the associated 95 percent statistical uncertainty in the moving average. The "low frequency mode" is determined by applying a band-pass filter to the short-term averages in order to select fluctuations on timescales of 60 Myr or greater.

On geologic time scales, the largest shift in oxygen isotope ratios is due to the slow radiogenic evolution of the mantle. It is not possible to draw any conclusion about very long-term (>200 Myr) changes in temperatures from this data alone. However, it is usually believed that temperatures during the present cold period and during the Cretaceous thermal maximum are not greatly different from cold and hot periods during most of the rest the Phanerozoic. Some recent work has disputed this (Royer et al. 2004) suggesting instead that the highs and lows in the early part of the Phanerozoic were both significantly warmer than their recent counterparts. Common symbols for geologic periods are plotted at the top and bottom of the figure for reference.

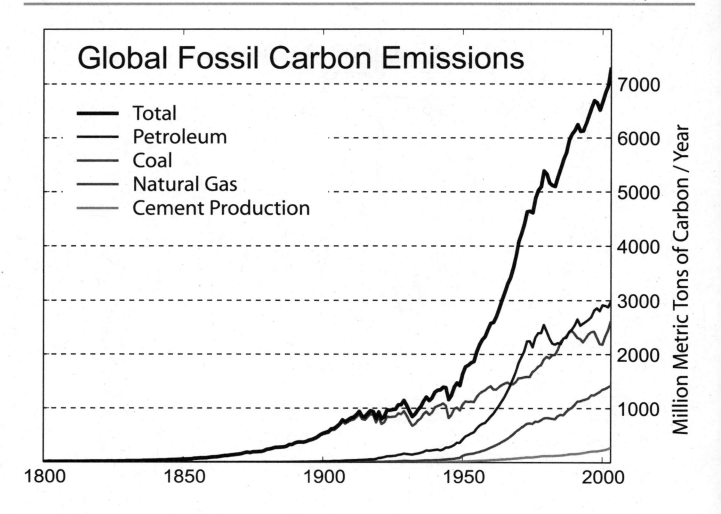

Global annual fossil fuel carbon dioxide emissions, in million metric tons of carbon, as reported by the Carbon Dioxide Information Analysis Center.

Original data: [full text] Marland, G., T.A. Boden, and R. J. Andres (2003). "Global, Regional, and National CO2 Emissions" in *Trends: A Compendium of Data on Global Change.* Oak Ridge, Tenn., U.S.A.: Carbon Dioxide Information Analysis Center, Oak Ridge National Laboratory, U.S. Department of Energy.

The data is originally presented in terms of solid (e.g., coal), liquid (e.g., petroleum), gas (i.e., natural gas) fuels, and separate terms for cement production and gas flaring (i.e., natural gas lost during oil and gas mining). In the plotted figure, the gas flaring (the smallest of all categories) was added to the total for natural gas. Note that the carbon dioxide releases from cement production result from the thermal decomposition of limestone into lime, and so technically are not a fossil fuel source.

Global Warming Predictions

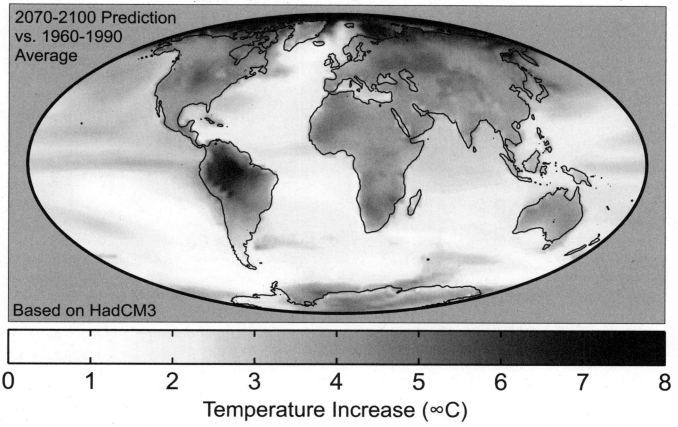

2070-2100 Prediction
vs. 1960-1990
Average

Based on HadCM3

0   1   2   3   4   5   6   7   8

Temperature Increase (∞C)

This figure shows the predicted distribution of temperature change due to global warming from the Hadley Centre HadCM3 climate model. These changes are based on the IS92a ("business as usual") projections of carbon dioxide and other greenhouse gas emissions during the next century, and essentially assume normal levels of economic growth and no significant steps are taken to combat global greenhouse gas emissions.

The plotted gray tints show predicted surface temperature changes expressed as the average prediction for 2070–2100 relative to the model's baseline temperatures in 1960–90. The average change is 3.0 degrees C, placing this model on the lower half of the Intergovernmental Panel on Climate Change's 1.4-5.8 degrees C predicted climate change from 1990 to 2100. As can be expected from their lower specific heat, continents are expected to warm more rapidly than oceans with an average of 4.2 degrees C and 2.5 degrees C in this model, respectively. The lowest predicted warming is 0.55 degrees C south of South America and the highest is 9.2 degrees C in the Arctic Ocean (points exceeding 8 degrees C are plotted as black).

This model is fairly homogeneous except for strong warming around the Arctic Ocean related to melting sea ice and strong warming in South America related to predicted changes in the El Niño cycle and Brazillian rainforest. This pattern is not a universal feature of models, as other models can produce large variations in other regions (e.g., Africa and India) and less extreme changes in places like South America.

Global Warming Projections

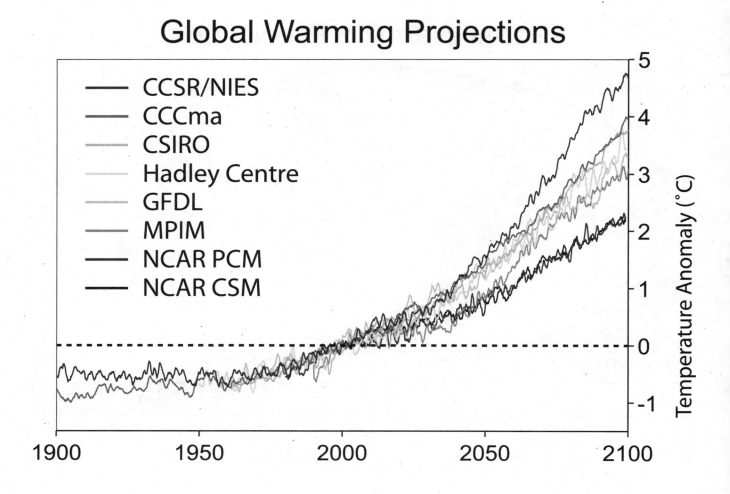

This figure shows climate model predictions for global warming under the SRES A2 emissions scenario relative to global average temperatures in 2000. The A2 scenario is characterized by a politically and socially diverse world that exhibits sustained economic growth but does not address the inequities between rich and poor nations, and takes no special actions to combat global warming or environmental change issues. This world in 2100 is characterized by large population (15 billion), high total energy use, and moderate levels of fossil fuel dependency (mostly coal). At the time of the IPCC Third Assessment Report, the A2 scenario was the most well-studied of the SRES scenarios.

The IPCC predicts global temperature change of 1.4-5.8 degrees C due to global warming from 1990 to 2100 (IPCC 2001a). As evidenced above (a range of 2.5 degrees C in 2100), much of this uncertainty results from disagreement among climate models, though additional uncertainty comes from different emissions scenarios.

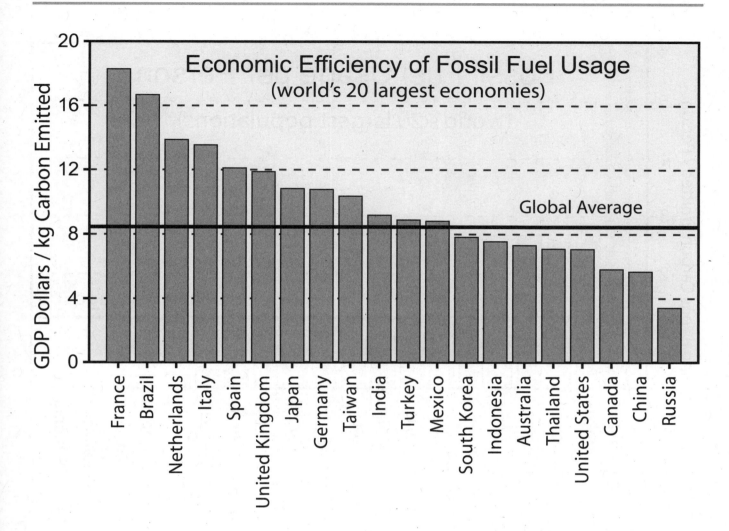

This figure shows an estimate of how efficiently the world's 20 largest economies convert fossil fuel usage into wealth as expressed by the ratio of their gross domestic product (calculated by the method of purchasing power parity in U.S. dollars) over the number of kilograms of fossil fuel carbon released into the atmosphere each year. The relatively narrow range of variation between most countries in this figure suggests that the pursuit of wealth in the present world is strongly tied to the availability of fossil fuel energy sources.

As countries may be reluctant to combat fossil fuel emissions in ways that cause economic decline, this figure serves to suggest the degree to which different large economies can decrease emissions through short-term improvements in efficiency and alternative fuel programs.

The two countries that produce the highest GDP per kilogram carbon, Brazil and France, are heavily reliant on alternative energy sources, hydroelectric and nuclear power, respectively.

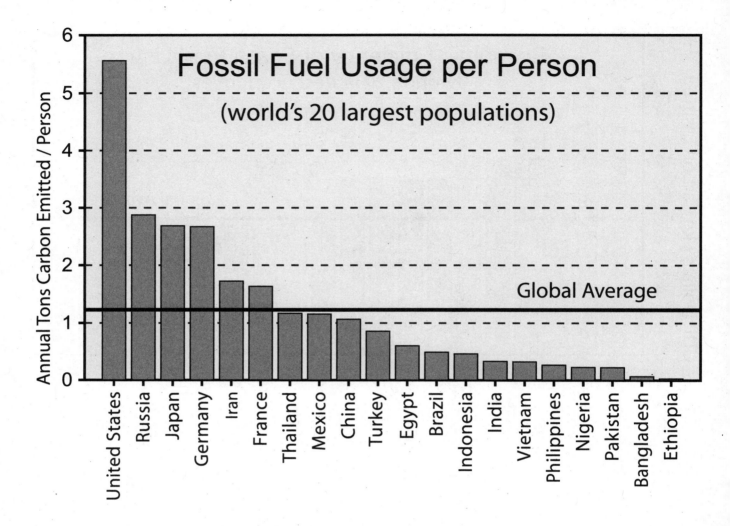

This figure shows the disparity in fossil fuel consumption per capita for the countries with the twenty largest populations. The large range of variation is indicative of the separation between the rich, industrialized nations and the poor/developing nations. The global average is also shown.

As most countries desire wealth and aim to develop that wealth through the development of industry, this figure suggests the degree to which poor nations may strive to increase their emissions in the course of trying to match the industrial capacity of the developed world. Managing such increases and dealing with the apparent social inequality of the present system will be one of the challenges involved in confronting global warming.

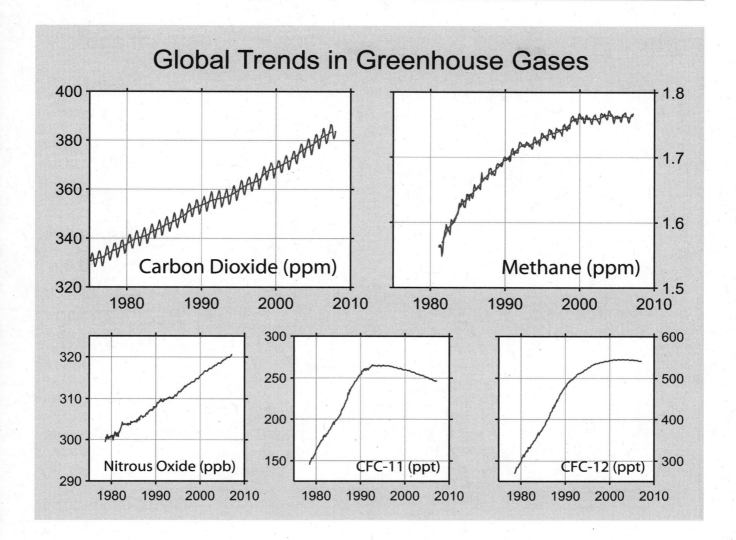

Global trends in major greenhouse gas concentrations. The rise of greenhouse gases, and their resulting impact on the greenhouse effect, are believed to be responsible for most of the increase in global average temperatures during the last 50 years. This change, known as global warming, has provoked calls to limit the emissions of these greenhouse gases (e.g., Kyoto Protocol). Notably, the chlorofluorocarbons CFC-11 and CFC-12 shown above have undergone substantial improvement since the Montreal Protocol severely limited their release due to the damage they were causing to the ozone layer.

At present, approximately 99 percent of the 100-year global warming potential for all new emissions can be ascribed to just the three gases: carbon dioxide, methane, and nitrous oxide.

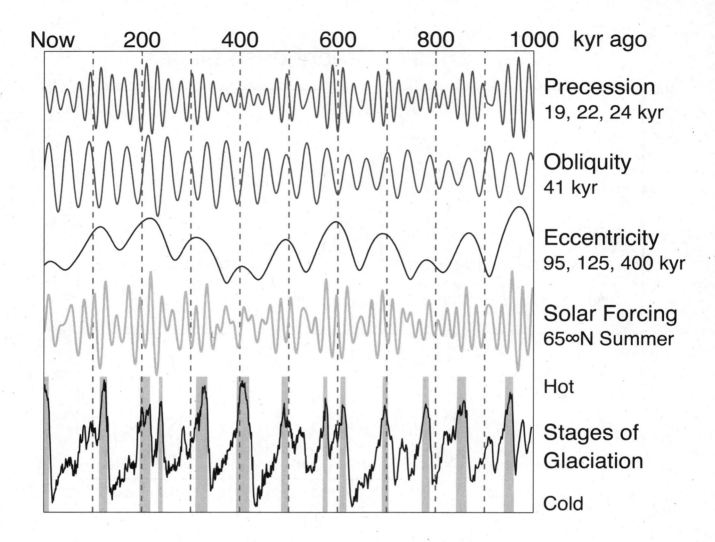

Milankovitch Cycles

The Earth's orbit around the sun is slightly elliptical. Over time the gravitational pull of the moon and other planets causes the Earth's orbit to change following a predictable pattern of natural rhythms, known as Milankovitch cycles. Over a ~100,000 year cycle the Earth migrates from an orbit with near-zero eccentricity (a perfect circle) to one with approximately 6 percent eccentricity (a slight ellipse). In addition, the tilt of the Earth axis, known as its obliquity, varies from 21.5 to 24.5 degrees with a 41,000 year rhythm. And lastly, the orientation of the Earth's axis rotates with a ~20,000-year cycle relative to the orientation of the Earth's orbit. This cycle, known as "precession", affects the intensity of the seasons.

The figure shows the pattern of changes in each of the three modes of orbital variability: eccentricity, obliquity, and precession. These changes in the Earth's orbit lead to a complex series of changes in the amount of sunlight that a given location on Earth can expect to receive during a given season. An example is shown for summer sunlight near the Arctic circle. Sunlight at this location is believed to influence the growth and decay of ice sheets during ice ages. The last line shows measured changes in climate during the last million years with warm interglacials highlighted in gray bands. As can be seen, such interglacials appear to preferentially occur near maxima in eccentricity and slightly following times of maximum summer sunlight.

Index

Note: Page numbers in **boldface** refer to volume numbers and major topics. Article titles are in **boldface**.

population, **3:**1062
tourism, **3:**1063
UNFCCC, **3:**1063
Villach Conference, 3:1063–64
Brundtland Report, **3:**1064
GHG features, **3:**1064
participants, **3:**1064
Virginia, 3:1064–66
agriculture, **3:**1066
CO$_2$ emissions, **3:**1065
geographical regions, **3:**1065
GHG, **3:**1065
VLS. *See* **Vermont Law School (VLS)**
Vo Chi Cong, **3:**1063
VOCs. *See* volatile organic compounds (VOCs)
volatile organic compounds (VOCs), **2:**467
volcanism, 3:1066–68
climate change and, **3:**1066
discharges, **3:**1067
hydrogen fluoride gas, **3:**1067
ozone depletion, **3:**1067
sulphur compounds, **3:**1066
volcanic dust, **3:**1066
Von Neumann, John (1903-1957), 3:1068–69
Fat Man atomic bomb, **3:**1069
game theory, **3:**1068
nuclear strategy, **3:**1069
publications, **3:**1068
stored-program digital computer and, **3:**1068
Vostok core, 3:1069–70
glacial cycles, **3:**1070
polar snowfall, **3:**1070
research, **3:**1070

W
Wackernagel, Mathis, **1:**342
Wadhams, Peter, **2:**484
Walker, Gilbert Thomas (1868-1958), 3:1071–72
honors, **3:**1072
monsoons and, **3:**1072
Southern Oscillation and, **3:**1071–72
Walker Circulation, 3:918, 3:1072–73
atmospheric circulation, **3:**1073
transient warming, **3:**1073
weakening, **3:**1073
Walker, Gilbert Thomas (1868-1958), 3:1071–72
honors, **3:**1072
monsoons and, **3:**1072
Southern Oscillation and, **3:**1071–72
Wallack, Lawrence, **2:**639
Wallerstein, Immanuel, **3:**1110

walruses, **2:**620
wars
Bosnian War, **1:**130
Gulf War, **1:**29
Sierra Leone civil war, **3:**896
study of, **3:**865
Washington, 3:1073–76
CO$_2$ emissions, **3:**1075
Mount St. Helens, **3:**1074
rising sea levels, **3:**1075
rising temperatures, **3:**1075
transportation, **3:**1075
Western Regional Climate Action Initiative, **3:**1075
See also **University of Washington**
Washington, Warren (1936-), 3:1076–77
climate prediction computer models, **3:**1076–77
greenhouse effect, **3:**1076–77
human-induced global warming, **3:**1077
publications, **3:**1077
water vapor (WV), **1:**64
water vapor feedback, **1:**217
waves
internal, **3:**1079–80
"long waves" influence, **1:**188
microwaves, **1:**237, **3:**840–41
radiation, **2:**501
sunlight, **1:**163
waves, **3:**984
waves, gravity, 3:1077–78
climate models, **3:**1078
cloud patterns, **3:**1078
generation, **3:**1077
mesoscale processes, **3:**1078
spatial/temporal extent, **3:**1078
wave-ducting processes, **3:**1078
waves, internal, 3:1078–80
causes, **3:**1079
defined, **3:**1078
energy, **3:**1079
heat transfer, **3:**1079
highs, **3:**1079
studies, **3:**1079–80
waves, Kelvin, 3:1080–81
Coriolis Effect, **3:**1080
equatorial, **3:**1080
propagation, **3:**1080
tidal cycles, **3:**1080
weather prediction, **3:**1080–81
waves, planetary, 3:1081–82
features, **3:**1081
formation, **3:**1081

Photo Credits